ANGIOPLASTY

EDITED BY

G. DAVID JANG, M.D.

Director, Cardiovascular Radiology
Loma Linda University
School of Medicine
Loma Linda, California

McGraw-Hill Book Company

New York   St. Louis   San Francisco   Auckland   Bogotá   Guatemala   Hamburg   Johannesburg
Lisbon   London   Madrid   Mexico   Montreal   New Delhi   Panama   Paris
San Juan   São Paulo   Singapore   Sydney   Tokyo   Toronto

ANGIOPLASTY

1234567890HALHAL89876

ISBN 0-07-032286-4

This book was set in Times Roman by General Graphic Services, Inc.; the editors were Beth Kaufman Barry and Stuart D. Boynton; the production supervisor was Avé McCracken; the design was done by Irving Perkins Associates.

Halliday Lithograph Corporation was printer and binder.

Library of Congress Cataloging in Publication Data
Main entry under title:

Angioplasty.

   Includes bibliographies and index.
   1. Transluminal angioplasty. 2. Cardiovascular
system—Diseases—Treatment. I. Jang, G. David.
[DNLM: 1. Angioplasty, Transluminal. 2. Coronary
Disease—therapy. 3. Vascular Diseases—therapy.
WG 500 A5885]
RD598.5.A555 1986   617'.413   85-13239
ISBN 0-07-032286-4

To all the pioneers
who have contributed to
the development of
transluminal angioplasty.

Contents

PART I   GENERAL VASCULAR ANGIOPLASTY

PART II   CORONARY ANGIOPLASTY

List of Contributors

Kurt Amplatz, M.D.
Professor of Radiology
Director of Cardiovascular Radiology
University of Minnesota
Minneapolis, Minnesota

Carlos A. Ayers, M.D.
Professor of Medicine, Division of Cardiology
University of Virginia School of Medicine
Charlottesville, Virginia

C. Roger Bird, M.D.
Assistant Professor of Radiology
USC School of Medicine
Director of Neuroradiology
Childrens Hospital of Los Angeles
Los Angeles, California

Peter C. Block, M.D., F.A.C.C.
Associate Professor of Medicine
Harvard Medical School
Director of Cardiac Catheterization Laboratory
Massachusetts General Hospital
Boston, Massachusetts

William J. Casarella, M.D.
Professor and Chairman
Department of Radiology
Emory University School of Medicine
Atlanta, Georgia

Michael J. Cowley, M.D.
Associate Professor of Medicine
Cardiology Division
Medical College of Virginia
Richmond, Virginia

Andrew Cragg, M.D.
Department of Radiology
University of Minnesota
Minneapolis, Minnesota

Joseph M. Craver, M.D.
Associate Professor of Cardiothoracic Surgery
Emory University School of Medicine
Atlanta, Georgia

David P. Faxon, M.D.
Associate Professor of Medicine
Boston University School of Medicine
Director of Cardiovascular Laboratories
University Hospital
Boston, Massachusetts

James C. Finn, M.D.
Department of Medicine
Stanford University Medical Center
Stanford, California

Garrett A. FitzGerald, M.D.
Associate Professor of Medicine and Pharmacology
Vanderbilt University School of Medicine
Nashville, Tennessee

Thomas J. Fogarty, M.D.
Chief, Cardiovascular Surgery
Sequoia Hospital
Redwood City, California

Henry Gewirtz, M.D.
Associate Professor of Medicine
Rhode Island Hospital
Providence, Rhode Island

Herman K. Gold, M.D.
Associate Professor of Medicine
Harvard Medical School
Co-Director of Gray IIIA Cardiac Catheterization
  Laboratory
Massachusetts General Hospital
Boston, Massachusetts

William H. Hartz, M.D.
Assistant Professor of Radiology
University of Pennsylvania School of Medicine
Philadelphia, Pennsylvania

Geoffrey O. Hartzler, M.D.
Consulting Cardiologist
Mid-America Heart Institute
St. Luke's Hospital, Kansas City
Clinical Associate Professor of Medicine
University of Missouri
Kansas City, MO

Jay Hollman, M.D.
Director of Interventional Cardiology
Cleveland Clinic Foundation
Cleveland, Ohio

Anton N. Hasso, M.D.
Professor of Radiology
Director, Neuroradiology
Loma Linda University Medical Center
Loma Linda, California

G. David Jang, M.D.
Associate Professor of Radiology
Director, Cardiovascular Radiology
Associate Director, Cardiovascular Laboratories
Loma Linda University Medical Center
Loma Linda, California

Kenneth R. Jutzy, M.D.
Assistant Professor of Medicine
Loma Linda University Medical Center
Loma Linda, California

Saadoon Kadir, M.D.
Associate Professor of Radiology
John Hopkins University School of Medicine
Baltimore, Maryland

Frederick S. Keller, M.D.
Associate Professor of Radiology
Chief of Angiography and Interventional Radiology
University of Alabama
Birmingham, Alabama

Howard Knapp, M.D., Ph.D.
Assistant Professor of Medicine
Vanderbilt University School of Medicine
Nashville, Tennessee

Eric C. Martin, M.A. (Oxon)
Professor of Clinical Radiology
Director of Cardiovascular Laboratories and
    Interventional Radiology
College of Physicians and Surgeons
Columbia University
New York, NY

Gordon K. McLean, M.D.
Associate Professor of Radiology
Section Chief of Angiographic/Interventional
    Radiology
University of Pennsylvania Hospital
Philadelphia, Pennsylvania

Noel L. Mills, M.D.
Head, Section of Cardiovascular Surgery
Ochsner Clinic
Director of Extracorporel Technology Program
Ochsner Foundation Hospital
Clinical Associate Professor of Surgery
Tulane University Medical Center
New Orleans, Louisiana

Albert S. Most, M.D.
Professor of Medicine
Brown University
Physician in Charge, Division of Cardiology
Rhode Island Hospital
Providence, Rhode Island

Douglas A. Murphy, M.D.
Assistant Professor of Cardiothoracic Surgery
Emory University School of Medicine
Atlanta, Georgia

Richard K. Myler, M.D.
Medical Director, San Francisco Heart Institute
Seton Medical Center
Clinical Professor of Medicine
University of California, San Francisco
San Francisco, California

Stephen N. Oesterle, M.D.
Clinical Assistant Professor of Medicine/Cardiology
Director of Unit for Coronary Intervention
Stanford University Medical Center
Stanford, California

Louis Roy, M.D.
Assistant Professor of Medicine
Vanderbilt University School of Medicine
Nashville, Tennessee

John B. Simpson, M.D.
Interventional Cardiology
Sequoia Hospital
Redwood City, California

S. Matts E. Sjolander, Ph.D.
Physicist
Loma Linda University Medical Center
Loma Linda, California

Simon H. Stertzer, M.D.
Director of Medical Research
San Francisco Heart Institute
Chief, Hemodynamics Laboratory
Seton Medical Center
Associate Clinical Professor of Medicine
University of California, San Francisco
San Francisco, California

Charles J. Tegtmeyer, M.D.
Professor of Radiology
Associate Professor of Anatomy
Director of Special Procedures
University of Virginia School of Medicine
Charlottesville, Virginia

George W. Vetrovec, M.D.
Associate Professor of Medicine
Director of Adult Cardiac Catheterization Laboratory
Medical College of Virginia
Richmond, Virginia

David O. Williams, M.D.
Associate Professor of Medicine
Brown University
Director of Cardiovascular Laboratories
Rhode Island Hospital
Providence, Rhode Island

William H. Willis, Jr., M.D.
Professor of Medicine
Director, Cardiovascular Laboratories
Loma Linda University Medical Center
Loma Linda, California

Preface

Transluminal angioplasty has an important part to play in modern medicine, not because it employs a new or high technology but because it deals with cardiovascular disease, which is the number one cause of death in the United States. However, transluminal angioplasty has only a limited application among the numerous medical and surgical modalities for treatment of cardiovascular disease, and should be employed in the context of broader approaches.

Despite this qualification, there is unquestionably a need for a book like this one, which offers a comprehensive assessment of the topic for a broad spectrum of readers: beginners and advanced operators, whether radiologists, cardiologists, or physicians in general medical and surgical practice. The book provides a systematic and detailed documentation of the numerous technical developments and advances that have marked the evolution of this field over the last few years. In essence each chapter reflects the state of the art in a specific aspect or topic of transluminal angioplasty at the time it was written.

The text has two parts: general vascular angioplasty and coronary angioplasty. Part I has 10 chapters on general vascular angioplasty, and Part II has 14 chapters on coronary angioplasty. Some of the chapters in each part may contain material that overlaps that of another, but we felt that this was preferable to extensive cross-referencing between the parts. Each chapter is therefore reasonably complete and accessible in itself, while the entire volume provides unprecedented coverage of both branches of transluminal angioplasty in a single source.

The medical sciences are constantly changing and improving as their practitioners strive to make them current and useful. The microcosm of transluminal angioplasty is no exception: its workers too are constantly introducing improvements in technique. As a consequence, by the time a book such as this is published there are new developments that arrived too late to be included. Within those practical limits, however, this book is complete and up to date.

Angioplasty is a testimony to cooperation and goodwill among the contributing authors. It would not have been possible to bring together the wealth of information found here were it not for the dedication and hard work of each individual author.

As we introduce the book, it is only proper to acknowledge some key people who helped make it possible. The concept could not have been made into reality without the insight and confidence of Mr. Robert P. McGraw at the inception. The combined efforts of Mr. Joseph J. Brehm and Ms. Beth Kaufman Barry, editors in the Health Professions Division at McGraw-Hill Book Company, played a key role in bringing this text to fruition. Credit for design quality and timely publication schedule belongs to Mr. Stuart Boynton, development editor, and the staff at McGraw-Hill Book Company. I am greatly indebted to Ms. Roxie L. Poser, who made the final preparations of the manuscripts, and to Mrs. Vickie L. Thornell, who assisted me in early stages of preparation. My sincerest gratitude and appreciation go to James M. Slater, M.D., professor and chairman, Department of Radiation Sciences, Loma Linda University School of Medicine, for his personal support and encouragement for my part in this project. I also would like to acknowledge all those unnamed secretaries and typists who so patiently and willingly helped the individual authors in preparing their manuscripts.

As this text is being published, I would like to give special recognition, on behalf of all the authors of this book, to our surgical colleagues in the cardiovascular specialties; without their support and participation transluminal angioplasty could not have evolved to the state in which we know it today.

G. DAVID JANG, M.D.

1

Historical Background and Development Stages of Transluminal Angioplasty

G. DAVID JANG, M.D.

INTRODUCTION

Vascular intervention, either for diagnostic or for therapeutic purpose, has come a long way through numerous stages of development over a period stretching nearly a century. Transluminal angioplasty, as it stands today, is at the peak of a pyramid of these numerous technical and medical achievements made over this period.

It would be appropriate to look through a historical reading glass to see how the individual contributions fit into the larger perspective of these developments. What happened in the past that brought us transluminal angioplasty as we know it today and what is happening today with regard to vascular intervention will ultimately influence and determine future generations of new developments and breakthroughs. As we look into the past with the knowledge of our current health care technology and capability, it will also help us foresee how future generations will regard our current "state of the art" as still evolving or even rudimentary achievements.

Transluminal angioplasty, for both general vascular and coronary systems, depends on two basic technological groundworks that were made in the past: (1) medical imaging based on ionizing energy that we call x-ray and (2) access into the vascular lumen with a tool that we call catheter. A long period of countless individual contributions eventually produced our modern-day angioplasty techniques and other interventional vascular approaches. In chronological sequence, diagnostic procedures generally led to its therapeutic applications.

MEDICAL IMAGING

Wilhelm Konrad Roentgen reported in 1895 his discovery of the x-ray beam, without which we cannot imagine the state of diagnostic or therapeutic advances known and practiced today. For his discovery, Roentgen received the Nobel Prize in 1901 and opened a new era of in vivo imaging of the human internal organs. This x-ray energy can be manufactured with a special vacuum tube with two opposing electrodes and high-voltage electricity. With this beam, the internal silhouette of a living specimen can be photographed into a still picture. However, the x-ray beam does not distinguish one water density from another water density; in other words, vascular structures could not be imaged onto a plain x-ray film.

It was not until 1929 that dos Santos et al. (1) published the first report of x-ray visualization of the aorta by using contrast material injected directly into the lumbar aorta. With this episode, diagnostic roentgenology of the vascular system slowly took off with the use of contrast media.

Along with the concept of contrast media in vascular imaging, access to the vascular chamber by means of a catheter was being conceptualized at this period. In the same year that dos Santos used contrast media for aortic visualization with the x-ray beam, Forssmann (2) passed 65 cm of a catheter into his own heart via one of his antecubital veins, guiding it with an x-ray fluoroscopic image behind him reflected on a mirror. Although Forssmann ultimately became a urologist, he inspired a new era of cardiac catheterization.

In 1938, Robb and Steinberg (3) demonstrated that the heart and central circulation, as well as the aorta, can be visualized satisfactorily on x-ray films by peripheral venous injection of the contrast media. A year later, Castellanos and Pereiras (4) described an aortogram technique using retrograde brachial artery injection of contrast media. In 1941, Farinas (5) was the first to describe an aortogram technique using a catheter introduced into the femoral artery by means of surgical exposure to the artery and puncture with a trocar through which a catheter was threaded into the abdominal aorta. In 1948, Radner (6) introduced a radial artery catheter technique for retrograde thoracic aortography. However, the level of imaging technology at this stage was far short of the quality and sophistication required for transluminal angioplasty.

During the period between the late 1950s and mid-1970s, fluoroscopic imaging quality underwent revolutionary improvement through the introduction of a generation of image-intensifying tubes, especially the cesium iodide tubes of the mid-1970s. During the same period, x-ray fluoroscopy adapted the TV imaging chain; a combination of high-resolution intensifier tubes and TV monitoring systems vastly improved the quality of fluoroscopic images and allowed rapid development of general and cardiac angiographies and interventional techniques. As far as imaging technology is concerned, modern techniques of transluminal angioplasty could not have emerged as aggressively and as rapidly as they did were the immediate image-replay systems not available. These devices include disc-based recorders, video tape recorders, and the new laser disc recorder. Instant replay of higher-quality images improved the safety and speed of interventional vascular procedures, as well as the diagnostic quality.

VASCULAR CATHETERIZATION

Interventional vascular procedures require a direct invasion of the vascular system, both the right and left sides of the heart, with a catheter or its modification. A diagnostic catheter has two purposes: (1) collecting hemodynamic data from the vascular system and (2) delivering contrast media for diagnostic imaging under the x-ray beam. On the other hand, an interventional catheter expands these two fundamental functions by adding a balloon, by delivering a foreign object, fluids, or drugs into the vascular system.

Developments of the catheter technique are the heart of interventional vascular procedures. Most interventional vascular techniques have their predecessor in the diagnostic catheter technique.

Putting the catheter into a vascular system has more than one approach: surgical exposure of the vessel or percutaneous entry into the vessel. Among these major approaches and their derivatives, the percutaneous technique first introduced by Seldinger in 1953 (7) is the most significant and most widely used today. Another major improvement in the diagnostic catheter technique is the development of preshaped catheters, first popularized by radiologists in general angiography in the 1960s and later incorporated into coronary and cardiac catheters. Likewise, as the diagnostic catheter techniques continued to improve and gained sophistication, a new generation of interventional and therapeutic catheter techniques began to emerge.

From that point, modern vascular techniques developed into two distinctive branches of angiographic and interventional specialties. One specialty was led by radiologists, who focused on the general vascular system and were slightly ahead of another group, cardiologists, who concentrated on applications to the heart and central circulatory system. Developments relating to the general vascular system progressed at a very rapid pace in the 1960s and 1970s, whereas cardiac interventions developed in the 1970s and 1980s. There has been some overlap of the individual advances made by radiologists and cardiologists in their respective fields. We will look into some of the key developments in both general vascular intervention and cardiac intervention.

VASCULAR RECANALIZATION

Along with the rapid development of general diagnostic angiography, therapeutic concepts were introduced into vascular intervention. Rastelli et al. (8) in 1959 demonstrated contrast extravasation in an experimental model of acute gastrointestinal bleeding. Artificial embolization of an arteriovenous malformation in the brain was reported in 1960 by Luessenhop and Spence (9). The concept of transluminal angioplasty was first introduced by Dotter, who gained the nickname "Crazy Charlie" in some academic circles. The first series of transluminal angioplasties was reported by Dotter and Judkins in 1964 (10). This radical new concept did not catch on in the United States but received lukewarm acceptance in Europe. Vasopressin infusion for variceal bleeding in patients with portal hypertension was reported in 1967 by Nusbaum and his colleagues (11). This was followed by intra-arterial vasopressin infusion for acute arterial bleeding in the gastrointestinal tract reported by Nusbaum et al. in 1969 (12). Rosch and his colleagues (13) in 1972 reported selective arterial embolization for arterial bleeding in the gastrointestinal tract. A new balloon dilatation catheter was developed by Gruentzig and Hopff (14) in 1974 for recanalization of arterioscleortic obstruction in the general vascular system, including the renal, iliofemoral, and popliteal arteries. With the balloon catheter, transluminal angioplasty will enter a new era of acceptance of the concept and rapid growth in popularity and demand.

On the cardiac side, diagnostic catheterization developed in parallel with general vascular angiography. Cournand (15,16) and Richards (17) reported a series on right-heart physiology in the early 1940s. In 1956, the Nobel prize for medicine was presented to Forssmann, Cournand, and Richards (18). In 1950, Zimmerman (19) and Lason (20) reported separately on retrograde left-heart catheterization. Transseptal technique was also reported separately in 1959 by Ross (21,22) and by Cope (23). Selective coronary arteriography was developed by Sones (24,25) in 1959 and cine angiogram became an important diagnostic imaging technique in cardiac angiography in contrast to the still-cut film imaging widely used in general vascular angiography. The selective coronary arteriogram technique was modified for percutaneous approach using a preshaped catheter by many contributors, including Ricketts and Abrams (26) in 1962 and Judkins in 1967 (27). In 1970, Swan and Ganz (28) introduced a flow-guided catheter with an air-filled catheter tip ballon for use in- and outside the cardiac catheterization laboratory. Miniaturization of Gruentzig's balloon catheter in 1976 (29) opened for the first time the possibility of transluminal dilatation of coronary arteries, initially in canine models and later in human cadavers (29–32).

CORONARY ANGIOPLASTY

In September 1977, Gruentzig performed the first in vivo human percutaneous transluminal coronary angioplasty in Zurich, Switzerland (33,34), using his new miniaturized balloon catheter. Earlier in the year and prior to this event, the first intraoperative human coronary angioplasty was performed in San Francisco by Gruentzig, Myler, Hanna, and Turina (35). Shortly after the first percutaneous coronary angioplasty, Gruentzig performed another successful coronary angioplasty with Kaltenbach, and Kober (36). In March 1978, Myler and Stertzer introduced percutaneous coronary angioplasty to the United States (36). The over-the-wire catheter technique used since the 1960s by radiologists in general vascular angiography was incorporated into the coronary balloon catheter by Simpson and Robert (37). This over-the-wire catheter concept led to eventual development of the steerable guidewire–balloon system for coronary angioplasty.

With demonstration of the feasibility, safety, and potential of transluminal coronary angioplasty as a therapeutic procedure, the vascular recanalization concept, first introduced by Dotter but ignored by American physicians, suddenly became a hot topic in the late 1970s. Both radiologists and cardiologists, in their respective vascular fields, were drawn into the "new procedure" of technology.

Although great advances in angioplasty procedures in the general vascular system resulted from the eventual availability of the new balloon catheter, it is in the coronary system that balloon angioplasty has made spectacular advances. Coronary angioplasty has captured the public's imagination to a greater extent than has general vascular angioplasty. The coronary artery system has presented much tougher technical demands for transluminal angioplasty than the general vascular system has, largely as a result of a number of built-in factors.

Anatomically, the coronary artery system has smaller luminal caliber, and frequent branching that is more tortuous and more remote from the point of catheter entry, especially with the transfemoral approach, than is true of the general vascular system. Physiologically, coronary arteries affect cardiac function directly; thus, cardiac arrest, myocardial infarction, and even death become immediate possibilities should something go wrong with coronary artery intervention. Furthermore, until the time of the first coronary angioplasty, physiological effects of sudden coronary artery occlusion in living human patients were largely unknown. Unlike general vascular angioplasty, coronary angioplasty could not have evolved without the standby availability of coronary bypass surgery, which was developed more than a decade earlier. Strategically, coronary artery intervention had greater epidemiological and economic significance because coronary heart disease is the number one cause of death (38) and the most costly disease category in the United States. Overcoming all these categorical demands with regard to transluminal angioplasty of the coronary artery system indeed required an accumulation of all the medical and technological developments preceding coronary intervention.

As an incidental footnote, the role that the National Heart, Lung, and Blood Institute Registry (39) played in coronary angioplasty is noteworthy for its registration and dissemination of angioplasty data (40) in its early stages of development. However, it is quite a political irony that even by 1985 there was no sign that a prospective study design would be sponsored by the National Institute of Health to compare the clinical results of coronary angioplasty and coronary bypass surgery despite the massive annual budget of NIH.

FUTURE OF VASCULAR INTERVENTION

In the future, vascular intervention, both general and coronary, will continue to be in high demand simply because atherosclerosis and its side effects will continue to be major factors in health care in this country as its population continues to grow older. As for transluminal angioplasty, technical tools should improve in the future for balloon angioplasty, as will new techniques that are not yet clinically popular.

One such technology may involve laser application, currently a vague hope. There is nothing mysterious about the application of laser energy to recanalization of atherosclerotic blood vessels: laser energy travels a straight-line course from its point of departure and is able to vaporize tissue, whether the normal tissue of a vessel wall or atheromatous material in a diseased artery. It can be delivered to a vessel through a fiber-optic channel built into a catheter, and the depth of tissue vaporization can be controlled. Applying such energy safely in a tortuous and irregularly formed lumen of a diseased coronary artery presents an enormous challenge. It may be that applying laser energy to recanalization of coronary artery disease lies in developing a laser beam guidance system to follow the anatomical contour of a coronary artery lumen, rather than taming the laser energy itself.

We trust that the future will continue to bring new pioneers and new advances. Today's pioneers, including many of the contributors to this book, will bring tomorrow's breakthroughs.

CONCLUSION

Reviewing the development stages of interventional vascular procedures that led to today's transluminal angioplasty is an inspiring and meaningful experience. In summarizing this review of historical events, three major milestones stand out: (1) Dotter's concept of recanalizing the atherosclerotic artery with catheter, (2) Seldinger's technique of percutaneous catheter insertion, and (3) Gruentzig's percutaneous dilatation of coronary artery.

This book is dedicated to those who made the breakthroughs in the past and to those who are pioneering new advances today.

REFERENCES

1. Dos Santos R, Lamas AC, Perieira-Caldas J: Arteriografia da aorta e dose vasos abdominais. Med Contemp, 1929, p 47.
2. Forssmann, W: Die Sondierung des rechten Herzens. Klin Wochenschr 8:2085, 1929.
3. Robb RG, Steinberg I: A practical method of visualization of chambers of the heart, the pulmonary circulation and the great blood vessels in man. J Clin Invest 17:507, 1938.
4. Castellanos A, Pereiras R: Counter-current aortography. Rev Cubana Cardiol 2:187, 1939.
5. Farinas PL: A new technique for arteriographic examination of the abdominal aorta and its branches. Am J Roentgenol 46:641–645, 1941.
6. Radner S: Thoracal aortography by catheterization from the radial artery. Acta Radiol [Diagn] (Stockh) 29:178–180, 1948.
7. Seldinger SI: Catheter replacement of the needle in percutaneous arteriography. Acta Radiol [Diagn] (Stockh) 39:368–376, 1953.
8. Rastelli GC, Magnani L, Bocchialini C: L'impiego dell'arteriografia nella diagnostic delle emorragie del tubo tigerente. Minerva Chir 14:1188, 1959.
9. Luessernhop A, Spence W: Artificial embolization of cerebral arteries: Report of use in a case of arteriovenous malformation. JAMA 172:1153, 1960.
10. Dotter CT, Judkins MP: Transluminal treatment of arteriosclerotic obstruction: Description of a new technic and a preliminary report of its application. Circulation 30:654, 1964.
11. Nusbaum M, Baum S, Sakiyalak P, Blakemore WS: Pharmacologic control of portal hypertension. Surgery 62:299, 1967.
12. Nusbaum M, Baum S, Blakemore WS: Clinical experience with the diagnosis and management of gastrointestinal hemorrhage by selective mesenteric catheterization. Ann Surg 170:506, 1969.
13. Rosch J, Dotter CT, Brown MN: Selective arterial embolization: New method for control of acute gastrointestinal bleeding. Radiology 102:303, 1972.

14. Gruentzig AR, Hopff H: Perkutane Rekanalisation chronischer arterieller Verschlusse mit einem neuen Dilatations-Katheter. Dtsch Med Wochenschr 99:2502, 1974.
15. Cournand AF, Ranges HS: Catheterization of the right auricle in man. Proc Soc Exp Biol Med 46:462, 1941.
16. Cournand AF, Riley RL, Breed ES, Baldwin EF, Richards DW: Measurement of cardiac output in man using the technique of catheterization of the right auricle. J Clin Invest 24:106, 1945.
17. Richards DW: Cardiac output in the catheterization technique in various clinical conditions. Fed Proc 4:215, 1945.
18. Cournand AF: Nobel Lecture, December 11, 1956, in *Nobel Lectures, Physiology and Medicine 1942-1962*. Amsterdam-London-New York, Elsevier, 1964, p. 529.
19. Zimmerman HA, Scot RW, Becker ND: Catheterization of the left side of the heart in man. Circulation 1:357, 1950.
20. Limon Lason R, Bouchard A: El cateterismo intracardico: catheterization de las cavidades izquierdas en el hombre. Registro simultaneo de presion y electrocardiograma intracavetarios. Arch Inst Cardiol Mex 21:271, 1950.
21. Ross J Jr: Transseptal left heart catheterization: a new method of left atrial puncture. Ann Surg 149:395, 1959.
22. Ross J Jr, Braunwald E, Morrow AG: Transseptal left atrial puncture: A new method for the measurement of left atrial pressure in man. Am J Cardiol 3:653, 1959.
23. Cope C: Technique for transseptal catheterization of the left atrium: preliminary report. J Thorac Surg 37:482, 1959.
24. Sones FM Jr, Shirey EK, Proudfit WL, Westcott RN: Cine coronary arteriography. Circulation 20:773, 1959.
25. Sones FM Jr, Shirey EK: Cine coronary arteriography. Mod Concepts Cardiovasc Dis 31:735, 1962.
26. Rickets JH, Abrams HL: Percutaneous selective coronary cine arteriography. JAMA 181:620, 1962.
27. Judkins MP: Selective coronary arteriography: a percutaneous transfemoral technique. Radiology 89: 815, 1967.
28. Swan HJC, Ganz W, Forrester J, Marcus H, Diamond G, Chonette D: Catheterization of the heart in man with use of a flow directed balloon-tipped catheter. N Engl J Med 283:447, 1970.
29. Gruentzig A: Perkutane Dilatation von Koronarstenosen: Beschreibung eines neuen Kathetersystems. Klin Wochenschr 54:543, 1976.
30. Gruentzig A, Schneider HJ: Die perkutane dilatation chronischer Koronarstenosen-Experiment und Morphologic. Schweiz Med Wochenschr 107:1588, 1977.
31. Gruentzig A, Riedhammer HH, Turina M: Eine neue Methode zur Perkutanen. Verh Dtsch Ges Kreislaufforsch 42:282, 1976.
32. Gruentzig AR, Turina MI, Schneider JA: Experimental percutaneous dilatation of coronary artery stenosis. Circulation 54:81, 1976.
33. Gruentzig AR: Translumination dilatation of coronary artery stenosis. Lancet 1:263, 1978.
34. Gruentzig AR, Senning A, Siegenthaier WE: Non-operative dilatation of coronary artery stenosis. Percutaneous transluminal coronary angioplasty. N Engl J Med 301:61, 1979.
35. Gruentzig AR, Myler RK, Hanna EH, Turina MI: Coronary transluminal angioplasty, abstracted. Circulation 55–56 (suppl III): III–84, 1977.
36. Myler RK, Gruentzig AR, Stertzer SH: Coronary angioplasty, in Rapaport (ed): *Cardiology Update 1983*. New York, Elsevier, chap 1, p 5.
37. Simpson JB, Baim DS, Robert EW, Harrison DC: A new catheter system for coronary angioplasty. Am J Cardiol 49:216–222, 1982.
38. American Heart Association: *Heart Facts-1983*. Dallas, 1983.
39. Mullin SM, Passamani ER, Mock MB: Historical background of the National Heart, Lung, and Blood Institute Registry for percutaneous transluminal coronary angioplasty. Am J Cardiol 53:3C–6C, June 15, 1984.
40. Proceedings of the National Heart, Lung, and Blood Institute Workshop on the outcome of percutaneous transluminal coronary angioplasty. Am J Cardiol 53:1C–145C, June 15, 1984.

PART

I

GENERAL VASCULAR ANGIOPLASTY

2

Guidelines and Quality Standards for General Vascular Angioplasty

WILLIAM J. CASARELLA, M.D.

INTRODUCTION

Twenty years of clinical experience have been accumulated since the introduction of percutaneous transluminal angioplasty (PTA) in 1964 by Dotter and Judkins (1). During that time, practical approaches to patient selection have evolved for percutaneous transluminal angioplasty of femoral, iliac, and renal arteries. Angiographers working in different centers have developed successful protocols for the clinical care of patients before, during, and after the angioplasty procedures, and general guidelines have been established regarding the optimal facility for angioplasty, as well as the types of paraprofessional personnel required to perform this beneficial, albeit invasive, procedure. Qualifications for the angiographer doing percutaneous angioplasty, as well as a functioning team approach to the referral pattern and care of patients undergoing the procedure, have evolved on a pragmatic basis at many centers. This chapter will attempt to summarize these issues and to propose working solutions to the problems that these issues may present.

PATIENT SELECTION

Femoral Artery

The major selection criterion for performing angioplasty in a femoral artery depends on the presence of a significant arterial lesion documented by patient symptomatology and by noninvasive flow studies performed by the vascular laboratory. Significant decreases in pulse volume recordings, doppler blood pressure measurements and brachial-to-femoral indices are important not only to document the severity of the lesion, but also to verify the efficacy of the procedure immediately afterwards and to follow the course of the patient's illness in the intermediate to long term after the procedure (2). Following clinical evaluation and review of data from the noninvasive vascular laboratory, patients with significant peripheral vascular disease should undergo femoral angiography.

The diagnostic angiography is performed usually from the femoral approach, but also is appropriately accomplished by the translumbar or axillary route. It is important to film the peripheral vascular tree from the lumbar aorta all the way to the distal vascular arches of the foot in order to clearly diagnose the location and severity of any lesions in the peripheral vasculature. As in coronary angiography, this frequently will require more than one view and perhaps more than one set of films timed at various intervals in order to completely capture the circulation with its collateral blood supply and reconstitution of arteries, as well as to accurately assess both the runoff from the inflow to the primary affected area.

Following the clinical and laboratory evaluation,

the patient's entire situation can be discussed by his primary physician, vascular surgeon, and angiographer. The appropriate decision regarding bypass surgery, angioplasty, or conservative medical therapy can then be made. Factors that influence the decision include the length of the lesion, the state of the distal vessels, the presence or absence of diabetes mellitus, the general appearance of the artery to be bypassed, and the availability of the patient's native saphenous veins for bypass surgery. Strong factors that favor angioplasty over bypass surgery are: short lesions less than 5 cm in length, localization of the lesion to one or two foci, and the lack of suitable saphenous veins for bypass surgery. Modifiers of the favorable prognosis for angioplasty include the presence of diabetes mellitus, poor distal runoff, and threatened limb loss. These factors are equally detrimental to the prognosis of surgery as they are to percutaneous angioplasty.

Iliac Artery

The approach to patient selection for percutaneous angioplasty of iliac artery lesions is somewhat different than that for femoral lesions. Iliac angioplasty carries a considerably better primary and long term success rate than does femoral angioplasty (3,4). In addition, most lesions are accessible to angioplasty in that they are relatively short segmental stenoses or occlusions that readily can be traversed using selective catheterization techniques. The iliac lesions are accessible by the standard retrograde femoral approach that is used in most cases of peripheral angiography, and frequently it is possible to perform a bilateral iliac angioplasty from a single femoral artery puncture. Because the retrograde femoral puncture is used for both diagnostic angiography and iliac angioplasty, it is frequently convenient to perform iliac angioplasty at the time diagnostic angiography is performed. Accomplishing this coordinated task necessitates prior consultation on the part of the angiographer and the referring surgeon, as well as obtaining the patient consent. Usually, the presence of iliac disease can be detected clinically by the patient's history, as well as a decrease in femoral pulses. The appropriate angiographic approach can then be planned in order to facilitate iliac angioplasty immediately after the diagnostic arteriogram.

We have found that the measurement of pressure gradients across iliac lesions is helpful, both in determining the hemodynamic significance and the severity of stenosis and in assessing the adequacy of transluminal dilatation. It is important to eliminate the resting gradient in order to achieve a good func-

tional result in this group of patients. Bifurcation lesions are best managed by bilateral retrograde femoral catheterization and simultaneous balloon inflation at the bifurcation of the aorta. With most iliac lesions, at least an 8–10 mm (in diameter) balloon is essential to eliminate the pressure gradient and to provide adequate dilatation. This type of dilatation frequently results in intimal injury and the appearance of intimal tears appears on the postangioplasty films.

Iliac stenoses or short segmental occlusions that result in resting pressure gradients are good candidates for iliac angioplasty (5–7). This is especially true in the context of accompanying distal disease that will necessitate either a bypass graft or femoral angioplasty and requires excellent inflow from the iliac artery in order to maintain its patency.

Although the vast majority of iliac angioplasty has been performed in the external and common iliac arteries, moderate success has been achieved in the internal iliac for the management of buttock claudication and of impotence (8). The combination of selective arterial infusions of streptokinase (or urokinase) in long total occlusions of the iliac artery followed by balloon angioplasty has been a successful treatment in several reported series (9,10).

Renal Artery

Selection of patients for renal angioplasty obviously depends primarily on the discovery of patients with renovascular hypertension. It is our feeling that patients with severe hypertension, whether or not it is controlled by heavy doses of potent antihypertensive drugs, should undergo evaluation to determine the cause of their hypertension. In our practice, this consists of peripheral renin determinations followed by selective renal vein renin sampling and intravenous or intra-arterial digital aortography. The presence of lateralizing renin production and a concommitant stenosis on the affected side is, in virtually all cases, an indication for renal angioplasty. The majority of lesions are local segmental stenoses in the proximal portion of the main renal artery with features that are highly favorable for renal angioplasty in most cases. The most favorable candidates for renal angioplasty are those patients with fibromuscular disease and focal atherosclerosis of the main renal artery with lateralizing renin values from the ipsilateral kidney (11–13). Patients with bilateral disease and unilateral disease without lateralization of renins have a considerably less favorable outlook, whether the lesion is repaired by angioplasty or surgery. Patients with stenosis of branch vessels in the renal parenchyma are also candidates for angioplasty, and considerable success has been reported

the border between the diagonal branches and the obtuse marginal branches is located very close to the silhouette crest of left ventricle in the RAO-30° projection.

These approaches to artery mapping can be used for the right coronary, circumflex, and left anterior descending arteries. However, coronary artery mapping of this kind requires a conceptual understanding of coronary anatomy in three dimensions as it relates to the surface geometry of left ventricle.

CONCEPT OF QUANTIFICATION

In this discussion one should assume that CDF has been defined before as

$$CDF = \frac{A_r}{A_t} \qquad (1)$$

where CDF = coronary distribution fraction
A_r = regional LV coronary artery distribution area
A_t = global LV coronary artery distribution area

Although coronary artery distribution pertains to an epicardial surface geometry of the left ventricle, the coronary distribution fraction CDF, as defined above, is also a measurement of the segmental myocardial mass distribution.

If we define the myocardial mass fraction by the equation:

$$MMF = \frac{M_r}{M_t} \qquad (2)$$

where MMF = myocardial mass fraction
M_r = regional LV myocardial mass
M_t = total LV myocardial mass

we can show that $CDF = MMF$ in the following way: Assuming that the LV muscle has a constant thickness in end-diastole and that the LV has a symmetrical shape we can state that muscle volume is proportional to epicardial surface area, that is:

$$V = k \cdot A \qquad (3)$$

where V = myocardial volume
k = proportionality constant
A = epicardial surface area

Introducing myocardial density ρ and using Eq. (3) we find the myocardial mass M by

$$M = \rho \cdot V = \rho \cdot k \cdot A \qquad (4)$$

Combining Eqs. (1), (2), and (4) we have

$$MMF = \frac{M_r}{M_t} = \frac{\rho \cdot k \cdot A_r}{\rho \cdot k \cdot A_t} = \frac{A_r}{A_t} = CDF$$

When coronary artery distribution is expressed in a fraction or a percentage, it implies a fraction of left ventricular muscle mass perfused by a coronary artery, as illustrated in Figures 5 and 6.

Since coronary distribution fraction is a numerical unit, a given distribution fraction of the right coronary artery may be directly compared with the same distribution fraction of the left coronary artery. Likewise, a distribution fraction of one species may be compared with that of another species.

From the case illustrated in Figure 2, the following data were obtained by the computer algorithm described in this paper:

$$A_r = 91 \text{ cm}^2$$
$$A_t = 215 \text{ cm}^2$$
$$CDF = 0.42$$
$$\left(\text{or } 42\% = \frac{91}{215} \right)$$

where A_r is the regional distribution area, A_t is the total distribution area, and CDF is the coronary distribution fraction.

LAD SURFACE DISTRIBUTION

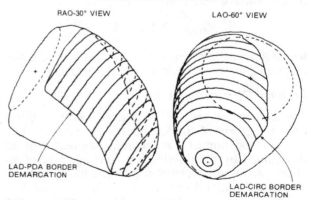

Figure 5. These drawings illustrate the epicardial surface distribution of the anterior descending artery of Figure 2. The two perpendicular views, RAO-30° and LAO-60°, give an added sense of dimension that a single view cannot. The LAD-CIRC border is demonstrated better in the LAO-60° view, whereas the LAD-PDA border is clearer in the RAO-30° view. These graphics are computer-generated.

CUSTOM-FIT FRUSTUMS   **FRUSTUM OF RIGHT CONE**

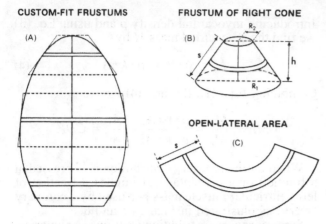

OPEN-LATERAL AREA

Figure 8. These drawings illustrate how a frustum of a right cone approximates the lateral surface of a thin (i.e., 1.0 = mm width) cross-sectional slice of the left ventricle. **A.** The custom-fit frustums of right cones for the lateral surface of individual cross-sectional slices. **B.** Dimensional designations for the equation of a frustum of right cone. **C.** Open lateral area of a frustum of right cone. A narrow frustum has its own respective dimensions, like a larger frustum.

where A_i = lateral surface of frustum
  R_1, R_2 = radii of bottom and top cut surfaces of frustum, respectively
  S = slant chord of frustum
  h = height of frustum

The sum of the lateral surface areas of all the individual slices will represent the total left ventricular surface:

$$A_t = (\Sigma A_i) \div M^2$$

where A_t is the total left ventricular surface, M is the magnification factor, and A_i is as defined above.

Regional Distribution Area

The distribution area of a coronary artery occupies only a segment of the left ventricular surface. The size and location of a myocardial segment will be determined by the distribution zone and size of an artery. A regional distribution area has a vertical spread and a lateral spread.

The *vertical spread* pertains to an extent from the proximal (from base) to the distal (to apex) ends along the major axis of left ventricle. The vertical spread is accounted for by the number of ventricular muscle slices included in the target distribution zone.

The *lateral spread* is a circumferential extent of the epicardial surface of a cross-sectioned left ventricular disc, which is occupied by the distribution territory of a target coronary artery. This lateral spread occupies an arc of the cross-sectional circumference, defined by the border demarcation already determined by the mapping process (Fig. 4). Therefore, a lateral spread is the circumferential extent of the arc that is defined by the angle of the sector. The angle of a sector is determined by the vectors that connect the two lateral borders to the center of the cross-sectional circle.

The angles of the two vectors OB and OC can be defined by the following equations (Fig. 4):

$$\alpha_1 = \arcsin (OB')$$
$$\alpha_2 = \arcsin (OC')$$

The lateral coronary spread and the lateral surface area of each slice are expressed in the following equations:

$$LCS = \frac{\alpha_2 - \alpha_1}{2\pi}$$
$$L_i = (\alpha_2 - \alpha_1) \times \frac{A_i}{2\pi}$$

where LCS = lateral coronary spread
  A_i = total lateral surface area of each muscle slice
  $\alpha_2 - \alpha_1$ = angle of sector
  L_i = lateral surface area distributed by target artery in each muscle slice

Summation of coronary distribution areas of all the individual muscle slices will give a regional distribution area:

$$A_r = (\Sigma L_i) \div M^2$$

where A_r is the regional distribution area, M is the magnification factor, and L_i is as defined above.

Coronary Distribution Fraction

If the regional distribution area is divided by the total left ventricular surface area, it gives the coro-

the border between the diagonal branches and the obtuse marginal branches is located very close to the silhouette crest of left ventricle in the RAO-30° projection.

These approaches to artery mapping can be used for the right coronary, circumflex, and left anterior descending arteries. However, coronary artery mapping of this kind requires a conceptual understanding of coronary anatomy in three dimensions as it relates to the surface geometry of left ventricle.

CONCEPT OF QUANTIFICATION

In this discussion one should assume that CDF has been defined before as

$$CDF = \frac{A_r}{A_t} \qquad (1)$$

where CDF = coronary distribution fraction
A_r = regional LV coronary artery distribution area
A_t = global LV coronary artery distribution area

Although coronary artery distribution pertains to an epicardial surface geometry of the left ventricle, the coronary distribution fraction CDF, as defined above, is also a measurement of the segmental myocardial mass distribution.

If we define the myocardial mass fraction by the equation:

$$MMF = \frac{M_r}{M_t} \qquad (2)$$

where MMF = myocardial mass fraction
M_r = regional LV myocardial mass
M_t = total LV myocardial mass

we can show that $CDF = MMF$ in the following way: Assuming that the LV muscle has a constant thickness in end-diastole and that the LV has a symmetrical shape we can state that muscle volume is proportional to epicardial surface area, that is:

$$V = k \cdot A \qquad (3)$$

where V = myocardial volume
k = proportionality constant
A = epicardial surface area

Introducing myocardial density ρ and using Eq. (3) we find the myocardial mass M by

$$M = \rho \cdot V = \rho \cdot k \cdot A \qquad (4)$$

Combining Eqs. (1), (2), and (4) we have

$$MMF = \frac{M_r}{M_t} = \frac{\rho \cdot k \cdot A_r}{\rho \cdot k \cdot A_t} = \frac{A_r}{A_t} = CDF$$

When coronary artery distribution is expressed in a fraction or a percentage, it implies a fraction of left ventricular muscle mass perfused by a coronary artery, as illustrated in Figures 5 and 6.

Since coronary distribution fraction is a numerical unit, a given distribution fraction of the right coronary artery may be directly compared with the same distribution fraction of the left coronary artery. Likewise, a distribution fraction of one species may be compared with that of another species.

From the case illustrated in Figure 2, the following data were obtained by the computer algorithm described in this paper:

$$A_r = 91 \text{ cm}^2$$
$$A_t = 215 \text{ cm}^2$$
$$CDF = 0.42$$
$$\left(\text{or } 42\% = \frac{91}{215} \right)$$

where A_r is the regional distribution area, A_t is the total distribution area, and CDF is the coronary distribution fraction.

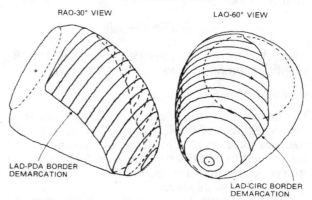

LAD SURFACE DISTRIBUTION

RAO-30° VIEW   LAO-60° VIEW

LAD-PDA BORDER DEMARCATION

LAD-CIRC BORDER DEMARCATION

Figure 5. These drawings illustrate the epicardial surface distribution of the anterior descending artery of Figure 2. The two perpendicular views, RAO-30° and LAO-60°, give an added sense of dimension that a single view cannot. The LAD-CIRC border is demonstrated better in the LAO-60° view, whereas the LAD-PDA border is clearer in the RAO-30° view. These graphics are computer-generated.

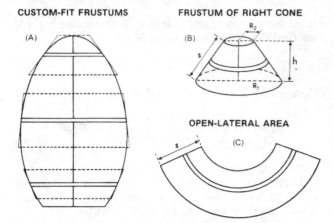

CUSTOM-FIT FRUSTUMS   FRUSTUM OF RIGHT CONE

(A)   (B)

OPEN-LATERAL AREA

(C)

Figure 8. These drawings illustrate how a frustum of a right cone approximates the lateral surface of a thin (i.e., 1.0 = mm width) cross-sectional slice of the left ventricle. **A.** The custom-fit frustums of right cones for the lateral surface of individual cross-sectional slices. **B.** Dimensional designations for the equation of a frustum of right cone. **C.** Open lateral area of a frustum of right cone. A narrow frustum has its own respective dimensions, like a larger frustum.

where A_i = lateral surface of frustum
R_1, R_2 = radii of bottom and top cut surfaces of frustum, respectively
S = slant chord of frustum
h = height of frustum

The sum of the lateral surface areas of all the individual slices will represent the total left ventricular surface:

$$A_t = (\Sigma A_i) \div M^2$$

where A_t is the total left ventricular surface, M is the magnification factor, and A_i is as defined above.

Regional Distribution Area

The distribution area of a coronary artery occupies only a segment of the left ventricular surface. The size and location of a myocardial segment will be determined by the distribution zone and size of an artery. A regional distribution area has a vertical spread and a lateral spread.

The *vertical spread* pertains to an extent from the proximal (from base) to the distal (to apex) ends along the major axis of left ventricle. The vertical spread is accounted for by the number of ventricular muscle slices included in the target distribution zone.

The *lateral spread* is a circumferential extent of the epicardial surface of a cross-sectioned left ventricular disc, which is occupied by the distribution territory of a target coronary artery. This lateral spread occupies an arc of the cross-sectional circumference, defined by the border demarcation already determined by the mapping process (Fig. 4). Therefore, a lateral spread is the circumferential extent of the arc that is defined by the angle of the sector. The angle of a sector is determined by the vectors that connect the two lateral borders to the center of the cross-sectional circle.

The angles of the two vectors OB and OC can be defined by the following equations (Fig. 4):

$$\alpha_1 = \text{arcsin (OB')}$$
$$\alpha_2 = \text{arcsin (OC')}$$

The lateral coronary spread and the lateral surface area of each slice are expressed in the following equations:

$$\text{LCS} = \frac{\alpha_2 - \alpha_1}{2\pi}$$
$$L_i = (\alpha_2 - \alpha_1) \times \frac{A_i}{2\pi}$$

where LCS = lateral coronary spread
A_i = total lateral surface area of each muscle slice
$\alpha_2 - \alpha_1$ = angle of sector
L_i = lateral surface area distributed by target artery in each muscle slice

Summation of coronary distribution areas of all the individual muscle slices will give a regional distribution area:

$$A_r = (\Sigma L_i) \div M^2$$

where A_r is the regional distribution area, M is the magnification factor, and L_i is as defined above.

Coronary Distribution Fraction

If the regional distribution area is divided by the total left ventricular surface area, it gives the coro-

with stenosis in patients with renal transplants. The potential for reversal of renal failure may be expected in approximately 25 percent of patients with renovascular disease (14).

Subclavian Artery

Patients with stenosis at the origin of the subclavian artery due either to atherosclerosis or to Takayasu's disease are candidates for percutaneous angioplasty (15). Although there are theoretical possibilities of embolization to the vertebral artery, and therefore to the circulation of the central nervous system, no complications of this sort have yet been reported. There is insufficient immediate and follow-up data to precisely evaluate the efficacy of subclavian angioplasty.

Miscellaneous Categories

Successful results of transluminal dilatation techniques have been reported in patients with pulmonary valvular stenosis (16), peripheral pulmonary artery stenosis (17), postoperative stenosis in patients with coarctation of the aorta (18), stenoses of various shunts such as AV fistulas for angio access in renal failure (19) and portal systemic shunts for portal hypertension (20), as well as for stenoses in various vascular bypass grafts. Stenotic lesions of the biliary tree, gastrointestinal tract, and genitourinary tract have all been successfully dilated by angioplasty techniques. The experience with these miscellaneous lesions outside of the vascular system indicate that the balloon catheter based on the original Gruentzig design is superior to many of the alternative devices that have been developed over the years for dilating nonvascular strictures.

PATIENT CARE AND PREPARATION BEFORE PROCEDURE

It is vital that the patient be informed regarding the risks and advantages of percutaneous angioplasty. We find it helpful for one of our Interventional Radiology nurses to visit the patient in order to explain the procedure and give the patient a previously prepared booklet describing the techniques and some of its complications. This visit can then be followed up by a subsequent interview with the responsible angiographer who will perform the procedure. The preliminary visit tends to answer many of the patient's questions and more efficiently utilizes the physician–patient time during the preangioplasty visit. The only preoperative medications that have been

widely used are aspirin, 325 mg to begin 12–24 hours prior to the procedure, and a mild sedative given parenterally about 1 hour before the procedure. Usual preoperative orders are identical to those used for diagnostic angiography. In all angioplasties, and especially renal angioplasty, we make every effort to hydrate the patient by liberal use of clear liquids orally or intravenous fluids prior to the procedure. This appears to play a beneficial role in preventing renal complications of contrast media.

During the procedure, an interested, knowledgeable and caring staff is probably the most important ingredient in patient acceptance and reassurance. We have found that a mixture of nurses and technologists who perform the procedure on a regular basis is the ideal staff for angioplasty. We choose to closely monitor every patient, including EKG, systemic blood pressure, and catheter tip pressure during the procedure. The peripheral pulses are assessed with a doppler stethoscope before, during, and after the procedure in order to evaluate the hemodynamic changes made during the angioplasty procedure. For most angioplasty cases, a heparinized saline drip is used through the catheter, and a low-dose parenteral bolus of heparin (usually 3000 units) is given during the procedure. Vasodilators such as parenteral nitroglycerin, oral nifedipine, and parenteral xylocaine are readily available for use as needed. Streptokinase and urokinase are also kept in the laboratory in the event that an acute thrombosis occurs which would require fibrinolytic therapy.

Compression of the punctured artery and hemostasis are obtained in a holding area immediately outside the laboratory, and the patients are observed for approximately 30 minutes prior to their transport back to their rooms. The patient is put on strict bed rest for 8 hours, and the puncture site is closely monitored. It is important to monitor systemic blood pressure in patients undergoing renal angioplasty. There is a tendency for an acute fall in blood pressure to occur in the early hours following successful renal angioplasty. This transient hypotension responds usually to administration of intravenous fluids.

Peripheral and renal angioplasty are unusual therapeutic procedures in that they are often performed by physicians other than the patient's primary doctor. Most angioplasty procedures are performed by radiologists or other angiographers who are not primary physicians and may not have clinical practices. Therefore, the preoperative evaluation and postoperative follow-up are usually performed by the patient's referring physician, whether he be an internist, nephrologist, cardiologist, or vascular surgeon. In the case of peripheral angioplasty, most patients should be evaluated by a vascular surgeon, who will make the primary decision as to whether the patient indeed has vascular disease and whether an intervention, either surgery or angioplasty, should occur. Close rapport between the vascular surgeon and the angiographer is, therefore, critical to the proper management of the patient undergoing angioplasty of the peripheral vascular system. Close cooperation obviously is a necessity on each individual case that is being considered for an angioplasty. Regular vascular conferences, where both surgery and angioplasty cases are presented, are a very helpful adjunct in fostering the development of a true vascular group and in galvanizing the type of interpersonal relationships that lead to the development of an excellent vascular service.

In the case of renal angioplasty, the primary physician is more likely to be a nephrologist or an internist who directs the initial workup of the patient in the hopes of detecting cases of curable renovascular hypertension. The radiologist usually becomes directly involved with the renal vein sampling and angiographic studies of the renal arteries. In most cases, percutaneous angioplasty will be the treatment of choice in renovascular hypertension, since its success rate and complication rate compare favorably with those of surgery at a much lower cost and morbidity. Surgery remains the primary treatment for patients with total occlusion of the renal arteries which cannot be traversed by the usual angiographic techniques or in those patients who will be undergoing major abdominal vascular surgery, such as repair of abdominal aortic aneurysm, and can, therefore, be bypassed as a part of the other major vascular procedure. The vascular surgeon should clearly consult on all difficult cases and be willing to assist in the management of any complications. The patient's primary internist plays a critical role in the follow-up of the patient. We have found that there continues to be considerable fluctuation in blood pressure for approximately 3 months after the angioplasty and that the determination of selective renal vein renin samples 3 months after the procedure is a better indicator of the success or failure of the procedure than any sampling or clinical data which are obtained immediately following the angioplasty procedure. In a significant number of cases, early restenosis or incomplete dilatation can be corrected by a second dilatation, which may result in a more permanent cure.

ANGIOGRAPHIC FACILITY

A prerequisite for the successful performance of percutaneous transluminal angioplasty is a well-equipped angiography laboratory. Ideally, the room would be equipped with a high quality image intensifier that can be angulated in a cranial-caudal, as well as in a left-to-right fashion. The flexible image intensifier and x-ray tube allows multiple projections to be obtained without having to disturb the patient from his resting, supine position. The image intensifier should be of the highest possible quality, with a high-resolution television chain, which would include a high-resolution television tube and a 1024 line television monitor system. The generator should be of the three-phase constant potential type, and the x-ray tube should have a dual focal spot with the smaller one no greater than 0.3 mm in diameter with at least a 15-kilowatt (kW) capacity. The second focal spot on the tube can be as large as 1 mm with an 80–100 kW capacity. The table top should be freely floating and easily controllable from the operator's position and be programmable so that it may automatically change positions, as well as the appropriate x-ray exposures, during peripheral run-off studies following a single central injection. Ideally, the modern angiographic-angioplasty suite should have digital capacity in order to allow for the intravenous evaluation of the renal arteries at the time of renin sampling and for the acquisition of relatively painless, instantaneous radiographs following low-volume injections of contrast material into the arterial system. The addition of a digital imaging system to the angioplasty room is a significant advance in facilitating the overall evaluation of the vascular system.

It is also critical that adequate patient monitoring apparatus be available. For noncardiac procedures, a three-channel patient monitor to include an EKG and two arterial pressure channels would be sufficient. In addition, the laboratory should contain a well-maintained defibrillator, an emergency drug cart, and dedicated lines for oxygen, suction, and nitrous oxide.

Most importantly, the angioplasty laboratory should be equipped with properly trained and experienced personnel. A nurse or technologist thoroughly familiar with the administration of vasoactive drugs, cardiopulmonary resuscitation and pressure monitoring is essential. The technologist should be extremely facile and confident regarding the radiologic equipment so that appropriate films may be obtained in an expeditious fashion. All members of the angioplasty team work closely together and should communicate freely with the patient in order to create an affable, friendly and efficient atmosphere.

HOSPITAL FACILITIES

Most hospitals are able to provide adequate facilities, assuming that routine angiography is being adequately performed and supported. For peripheral and renal angioplasty, it has not been found necessary to have surgical teams on standby in order to perform emergency operations in case of life-threatening complications. Clearly, vascular surgical facilities and personnel need to be available in the medical center. However, the need to perform immediate, emergency vascular surgery is so rare that it would be wasteful to have this capability on standby for every case of angioplasty. This concept varies from the requirements of coronary angioplasty, where bypass surgery must be performed within 30–60 minutes following an angioplasty complication in order to prevent irreversible myocardial infarction. With peripheral or renal angioplasty, arterial occlusion can be frequently managed by streptokinase infusion or by surgery within several hours of the procedure without the patient suffering permanent tissue damage. The only life-threatening emergency which occurs during peripheral or renal angioplasty would result from rupture of either an iliac or renal artery. When this very rare complication occurs, hemostasis should be maintained by inflating the balloon catheter at the site of arterial damage in order to gain temporary hemostasis. This maneuver will allow for the patient's orderly transferral to the operating room for vascular repair. Given the existence of smoothly functioning, cooperative vascular teams consisting of vascular surgeons and radiologists, complications can usually be managed in an efficient and smooth fashion. Virtually all patients with peripheral vascular disease would have a surgical consult in the preangioplasty period. Therefore, the surgeon is always informed about the cases and presumably willing to manage any emergency situation. In renal angioplasty, a general agreement should be made between the angiographers and surgeons regarding the management of complications, since the surgeon frequently will not be involved in the postprocedure follow-up care of patients undergoing renal angioplasty. Urgent surgery would be required in less than 1 percent of the patients with renal angioplasty. Elective surgery may be required in about 5 percent of the cases, due to primary failure of percutaneous angioplasty.

In most situations, the physician who performs the diagnostic angiography in the involved vascular area should be the individual to perform the angioplasty when indicated. Clearly, if a cardiologist performs the coronary angiography in a given institution, he would be the ideal person to undertake coronary angioplasty. When the radiologist is responsible for performing angiography in the peripheral and renal circulations, his expertise in diagnostic angiography would make him the ideal person to perform angioplasty in those areas. Other individuals should demonstrate some element of specific training in angioplasty in order to be properly credentialled to perform the procedure. Fortunately, there are several excellent postgraduate courses and visiting fellowships available to experienced cardiologists to obtain specific knowledge and updates on the technique of coronary angioplasty. Similar opportunities are available for experienced angiographers in the peripheral and renal circulations.

Maintenance of an excellent data base for angioplasty is a very important component in the development of the procedure at any given laboratory. The ability to calculate the success of the procedure and, also, the relative success of individual operators in performing the angioplasty procedure is an important asset in the drive to improve the overall performance of a given laboratory.

PRACTICAL AND ECONOMIC CONSIDERATIONS FOR PTA

It is evident from the detailed studies of Abrams and Doubilet (21) that angioplasty of the iliac and femoral arteries is efficacious both medically and economically. Their recommended approach of angioplasty where possible, followed by surgery if needed, theoretically results in a significant smaller number of complications and saves considerable cost. The advent of prospective payment schemes imposed on hospitals by third-party payers should also be an economic stimulus to increase the performance of transluminal angioplasty. This type of economic incentive should stimulate the performance of angioplasty, which clearly is less costly than the alternative surgical therapy. Of course, the continuance of the economic advantage of angioplasty is dependent on the bureaucratic manipulations that hospital reimbursement is subject to. There seems to be no doubt, however, that in the modern era of cost containment, angioplasty will continue to be a favored procedure when clinically applicable. There are other factors which argue strongly in favor of the performance of transluminal angioplasty. The procedure can be repeated if restenosis or occlusion occurs. It is very unusual for a secondarily failed angioplasty to obviate the possibility of either repeating the angioplasty or performing subsequent surgery. As Abrams has shown, angioplasty is less costly than surgery and, in the aggregate, remains less costly even in those patients where subsequent surgery is necessary. Given the need to preserve saphenous veins in those patients with diffuse atherosclerosis who might require coronary bypass grafting, peripheral angioplasty is an alternative to bypass surgery with synthetic graft material. Angioplasty is clearly less costly, less debilitating, less traumatic, less risky, and less final than the alternative procedure.

It seems apparent that angioplasty should be performed in those cases where indicated, and that the more invasive surgical procedures be reserved for those patients in whom angioplasty is not possible or where the procedure has not been successful. The emergence of percutaneous transluminal angioplasty as a viable therapeutic alternative allows many patients two chances, rather than one, for successful interventional treatment of a vascular disorder.

REFERENCES

1. Dotter CT, Judkins MP: Transluminal treatment of arteriosclerotic obstruction. Circulation 30:654–670, 1964.
2. Bernstein EF: *Noninvasive Diagnostic Techniques in Vascular Disease,* 2nd ed. St. Louis, C.V. Mosby, Co, 1982.
3. Van Andel GJ: Transluminal iliac angioplasty: Long term results. Radiology 135:607–611, 1980.
4. Bachman DM, Casarella WJ, Sos TA: Percutaneous iliofemoral angioplasty via the contralateral femoral artery. Radiology 130:617–621, 1979.
5. Colapinto RF, Harris-Jones EP, Johnson KW: Percutaneous transluminal dilatation of complete iliac artery occlusion. Arch Surg 116:277–281, 1981.
6. Zeitler E: Percutaneous dilatation and recanalization of iliac and femoral arteries. Cardiovasc Intervent Radiol 3:207–212, 1980.
7. Gruentzig AR, Zeitler E: Cooperative study of results of PTA in twelve different clinics, in E Zeitler, AR Gruentzig, W Schoop (eds.): *Percutaneous Vascular Recanalization.* Berlin, Springer-Verlag, pp 118–119, 1978.
8. Castanada-Zuniga WR, Amplatz KA: Transluminal angioplasty of pelvic arteries in the management of vasculogenic erectileimpotence, in WR Castanada-Zuniga (ed.): *Transluminal Angioplasty,* New York, Thieme-Stratton, Inc, pp 192–195, 1983.
9. Dotter DT, Rosch J, Seaman AJ: Selective clot lysis with low dose streptokinase. Radiology 111:31–37, 1974.
10. Katzen BI, Van Breda A: Low dose streptokinase in the treatment of arterial occlusions. AJR 136:1171–1178, 1981.
11. Schwarten DE: Transluminal angioplasty of renal artery stenoses: 70 experiences. AJR 135:967–974, 1980.
12. Tegtmeyer CJ, Elson J, Glass TA, et al: Percutaneous transluminal angioplasty: The treatment of choice for renovascular hypertension due to fibromuscular dysplasia. Radiology 143:631–637, 1982.
13. Sos TA, Saddekni S, Sniderman KW, et al: Renal artery angioplasty: Techniques and early results. Urol Radiol 3:223–231, 1982.
14. Sones PJ, Price R, Casarella WJ: Long term follow-up results of patients undergoing renal angioplasty. Presented at the Annual Meeting of the Radiological Society of North America, November 1983.
15. Bachman DM, Kim RM: Transluminal dilatation for subclavian steal syndrome. AJR 135:995–996, 1980.
16. Kan JS, White RI Jr, Mitchell SE, Anderson SH, Gardner TJ: Percutaneous transluminal balloon valvuloplasty or pulmonary valve stenosis. Circulation 69(3):554–560, 1984.
17. Lock JE, Castaneda-Zuniga WR, Fuhrman BP, Bass JL: Balloon dilatation of hypoplastic and stenotic pulmonary arteries. Circulation 67(5):962–967, 1983.
18. Kan JS, White RI Jr, Mitchel SE, Farmlett EJ, Donahoo JS, Gardner TJ: Treatment of restenosis of coarctation by percutaneous transluminal angioplasty. Circulation 68(5):1087–1094, 1983.
19. Martin EC, Diamond NG, Casarella WJ: Percutaneous transluminal angioplasty in non-atherosclerotic disease. Radiology 135:27–33, 1980.
20. Cope C: Balloon dilatation of closed mesocaval shunts. AJR 135:989–994, 1980.
21. Doubilet P, Abrams HL: The cost of underutilization: Percutaneous transluminal angioplasty for peripheral vascular disease. NEJM 310(2):95–102, 1984.

3

Percutaneous Transluminal Angioplasty of the Renal Artery

CHARLES J. TEGTMEYER, M.D.
CARLOS A. AYERS, M.D.

INTRODUCTION

Renovascular hypertension, the type seen in approximately 4 percent of all cases of hypertension in this country (1), poses the double threat of hypertensive complications and progressive renal insufficiency. Pharmacotherapy and surgical revascularization, the traditional modes of therapy, have significant shortcomings. Drugs often only partially control blood pressure; when combinations of drugs are used, side effects and poor patient compliance can become problems (2–4). Surgery requires general anesthesia, and many patients are poor surgical risks because of severe systemic cardiovascular disease. Moreover, the results of surgery vary. There is considerable morbidity, and the mortality rate can be as high as 5.9 percent (5). Despite these shortcomings, therapy can have significant benefits (6,7), although alternative treatment modalities are desirable.

Percutaneous transluminal angioplasty (PTA) has attracted a great deal of attention in recent years. Transluminal angioplasty, however, is not a new procedure. It was originally described by Dotter and

Judkins (8) some 20 years ago for the treatment of atherosclerotic peripheral vascular disease. Then Andreas Gruentzig (9) revolutionized the technique in 1974 when he developed the balloon catheter. Gruentzig (10) performed the first balloon dilation of a renal artery stenosis in 1978. In the relatively short time that has elapsed since, several large series detailing the results of percutaneous transluminal renal angioplasty (PTRA) have been reported (11–19). The preliminary data suggest that renal angioplasty is a highly successful technique for correcting renal artery stenoses. Although only a few reported series have described the long-term results of renal angioplasty (20–23), more definitive assessments of the long-term results of renal artery angioplasty are now emerging.

ETIOLOGY OF RENAL ARTERY STENOSIS

There are many causes of the renal artery stenosis. However, the majority of renal artery lesions are atherosclerotic, and most of the remainder are fibromuscular in origin. Of the 2442 hypertensive patients who were studied in the Cooperative Study on Renovascular Hypertension (5), 884 had renal artery lesions. Atherosclerosis was the etiology in 557 (63 percent), fibromuscular hyperplasia in 286 (32.4 percent), and miscellaneous disease entities in 41 (4.6 percent) of the patients. In the University of Virginia series (24), atherosclerosis was the etiology in 75 patients who had 93 lesions dilated. Fibromuscular dysplasia was the cause of 30 renal artery stenoses in 27 patients. Seven patients had stenoses in the arteries to their transplanted kidney. Three patients had their saphenous bypass grafts dilated. One patient had his native artery dilated after his saphenous bypass graft occluded, and one patient had three stenoses due to previous radiation therapy.

Atherosclerosis

Atherosclerosis is the most common cause of stenoses in the renal arteries. It is usually seen in patients over 40 years of age and is more common in males than in females. Atheromatous disease characteristically involves the proximal third, or the *orifice*, of the renal artery. Atherosclerotic involvement of the renal artery is frequently accompanied by atherosclerosis of the aorta. Aortic plaques may engulf the origin of the renal artery, causing stenosis or occlusion. Because of the propensity of atherosclerotic disease to involve the origin of the renal artery, oblique films are necessary to visualize adequately the origin of many renal arteries in order to exclude renal artery lesions. Atherosclerotic lesions are usually localized within a single renal artery; however, 30–50 percent of patients have bilateral lesions. One side is usually more severely affected than the other. A poststenotic dilatation may be observed just distal to the stenotic segment; however, its presence does not necessarily signify that the lesion is hemodynamically flow-restricting.

Fibromuscular Dysplasia

Leadbetter and Burkland (25) first described the association of hypertension with fibromuscular dysplasia of the renal artery in 1938. Fibromuscular dysplasia is seen most frequently in the renal arteries, but it can occur in other vascular systems. It may be present in the other visceral arteries, the carotid arteries, the iliac arteries, and even the peripheral vessels. Fibromuscular dysplasia occurs more frequently in women than in men, with the ratio of incidence approximately 4 or 5 to 1. The

patients are frequently 30–50 years of age; however, fibromuscular dysplasia is also seen in the pediatric age group. The disease typically involves the middle and distal thirds of the renal arteries and may extend into the proximal branches. It is bilateral in 40–70 percent of cases. When unilateral, the disease involves the right renal artery more frequently than the left.

McCormack et al. classified the lesions of fibromuscular dysplasia according to the primary site of involvement in the arterial wall and correlated the histological findings with the angiographic appearance (26,27). Their classification includes intimal fibroplasia, medial fibroplasia, fibromuscular hyperplasia, and subadventitial fibroplasia. The classification can be simplified when based solely on the angiographic appearance of the lesion. Intimal fibroplasia consists of symmetrical narrowing of the artery with poststenotic dilatation or an irregular dilated segment. Because of the considerable difficulty in separating intimal fibroplasia from fibromuscular hyperplasia angiographically, these lesions can be grouped together as *intimal fibroplasia* (28). Medial fibroplasia consists of the classic "string of beads" appearance, with the diameter of the beads exceeding the expected diameter of the renal artery in the affected region. When the lesions appear as areas of severe uneven stenoses, they are classified as *subadventitial fibroplasia*. Beading of the artery may be produced in subadvential fibroplasia, but the width of the beads does not exceed the original diameter of the vessel. Medial fibroplasia is the most common form of fibromuscular dysplasia. When the lesions are localized and caused by intimal fibroplasia and involve the proximal third of the renal artery, they may be difficult to distinguish from atherosclerotic disease. However, when the renal arteries are affected by atherosclerotic disease, the aorta is usually involved also.

Neurofibromatosis

Neurofibromatosis is a generalized disease characterized by abnormalities of the nervous system, muscles, bones, and skin. The skin contains multiple soft tissue tumors and is usually pigmented (*café au lait spots*). The patients have an unusually high incidence of sarcomas, pheochromocytomas, and brain tumors. This condition is a rare, but important, cause of renal artery stenosis. The stenoses in the renal artery usually result directly from fibrous proliferation of the intima or media. In some cases, however, neurofibrous tissue has been demonstrated within the adventitia of the artery, producing periarterial fibrosis. Angiographically there is a smooth constriction of the vessel, frequently associated with a tubular segment of dilatation beyond the stenosis. The renal artery stenoses are bilateral in 40 percent of cases. They may be accompanied by aneurysms to the renal artery and occasionally by narrowing of the abdominal aorta.

Hypertension in patients with neurofibromatosis may be associated with the presence of pheochromocytoma. In patients below the age of 18, renal artery stenosis is usually the etiology of the hypertension, although in those over 18 years, pheochromocytoma is likely to be the cause of the elevated blood pressure.

Arteritis

Takayasu's disease is a rare cause of segmental arteritis involving the thoracic aorta, brachiocephalic vessels, and occasionally the abdominal aorta and renal arteries. It occurs primarily in women and usually appears during the second and third decades of life. Clinical symptoms are manifested as hypertension or vascular insufficiency of the upper extremities. The early stage of the disease is nonspecific and commonly associated with fever, myalgia, arthralgia, nausea, sweating, weight loss, pain, and skin rash. Histologically, Takayasu's disease is characterized by chronic arteritis, causing marked thickening of the intima, media, and adventitia. The changes in the intima are more pronounced than those in the media and adventitia. Premature atherosclerosis may be associated with Takayasu's involvement of the arteries. The arteriographic picture is usually that of a narrowing of the segment of the abdominal aorta adjacent to the renal arteries with concomitant involvement of the proximal third of the renal arteries.

Periarteritis nodosa may also involve the renal arteries. The renal vessels are involved in about 80 percent of cases. The histological picture is that of a necrotizing inflammatory process that involves primarily the arterioles in the smaller intrarenal branches of the renal artery. The inflammatory process results in formation of granulation tissue, fibrosis, and vessel wall destruction in some areas. Hemorrhage, thrombosis, and aneurysm formation may result. The demonstration of multiple aneurysms in the small- and medium-sized arteries on the angiogram is highly suggestive of periarteritis nodosa. The medium and smaller arteries may be stretched, attenuated, and occluded with secondary recanalization. Small aneurysms may also be present in the splenic, hepatic, and other visceral vessels.

The renovascular bed may be involved by other types of vasculitis, such as drug abuse, ergot, Buerger's disease, serum sickness, and Wegener's granulomatosis. These entities may all produce similar abnormalities of the blood vessels within the kidney.

Transplant Renal Artery Stenosis

Hypertension is a common complication of renal transplantation, occurring in 24–60 percent of patients (29,30). Hypertension in renal transplant patients is mediated by many factors. The incidence of hypertension after renal transplantation is higher in patients receiving cadaver kidneys than in those receiving transplants from related living donors (30). Exogenous steroids may cause persistent post transplantation hypertension. Native kidneys remaining after transplantation may also cause persistent hypertension. Many patients undergoing rejection also have elevated renin levels and hypertension (30). Finally, stenoses in the transplant arteries may lead to hypertension.

The stenoses may be proximal to, at, or distal to the anastomosis. Proximal stenoses in the hypogastric arteries are usually atherosclerotic in origin. Anastomotic stenoses are largely due to surgical technique, perfusion injury, or local reaction to the suture material. Distal stenosis is common and may result from altered hemodynamics, vessel trauma, immunological factors, and arterial kinking or extrinsic compression. The intimal hyperplasia often seen in this area may represent localized rejection (30).

Radioisotope renal scanning can be a sensitive means of detecting decreased flow to the transplant kidney. False-positive results may be seen, however, if there is chronic rejection or if stenosis and rejection coexist. Renin sampling may be helpful, but elevated renin secretion is not always seen, even when a significant stenosis is present. Conversely, hyper-reninemia may be present in rejection without an accompanying renal artery stenosis. Therefore, transplant renal artery stenosis can be reliably identified only by arteriography.

Radiation-Induced Renovascular Hypertension

Radiation is known to produce irreversible changes in the small vessels and interstitium of the kidneys resulting in hypertension and chronic renal failure. Irradiation, however, can also produce stenoses of the major renal arteries with resultant renal artery stenoses (31). The damage may be produced by accelerated atherosclerosis, increased deposition of collagen in the intima, or fibrosis. One or a combination of these mechanisms may result in correctable renal artery stenosis. Therefore, arteriography should be considered in patients who become hypertensive after abdominal radiation therapy.

Renal Bypass Grafts

Reconstructive vascular surgery is a well-established procedure in the treatment of renovascular hypertension. Angiography is often performed in the postsurgical patient when there is persistent hypertension. Ekelund et al. (32) reviewed the postoperative angiographic appearance in 128 patients. He found that after saphenous vein bypass grafts, there were 30 patients with dilated grafts, 29 with stenoses, 9 with occlusions, 2 with aneurysms, and 2 with renal infarctions. Dilatation of the grafts should not be interpreted as a complication because it did not lead to hypertension in the patients studied.

The stenoses tend to occur either at the proximal or distal anastomoses, and most are not hemodynamically significant. However, significant stenoses and occlusions may follow saphenous bypass graft insertions. Small outpouchings seen in the wall of the saphenous bypass graft are normal, being caused by areas where the branches of the graft were ligated. Synthetic grafts have much higher stenosis and occlusion rates than saphenous vein grafts. When studying patients who have undergone bypass grafts, it is important to obtain multiple oblique projections to exclude with certainty the stenosis at the anastomosis junction.

Miscellaneous Renal Artery Lesions

Numerous other lesions may result in renal mediated hypertension. *Renal infarction* can result from embolism or thrombosis. Infarction of the kidney may not lead to hypertension. If the collateral vessels supply an ischemic area, however, renal hypertension will result. *Renal artery aneurysms* may cause hypertension. Aneurysms are usually unilateral but can be bilateral. They may be congenital, atherosclerotic, traumatic, or dissecting, or they may be due to fibromuscular dysplasia or polyarteritis. Most renal artery aneurysms are small and asymptomatic. However, larger aneurysms may be associated with pain, hematuria, and hypertension. Approximately 30–60 percent of renal artery aneurysms calcify; noncalcified aneurysms are more prone to rupture spontaneously, and incidence of rupture increases during pregnancy. Hypertension is associated with aneurysms in approximately 15 percent of cases (33).

Arteriovenous fistula of the kidney may be associated with renovascular hypertension. It is thought to produce an area of localized renal ischemia resulting in renovascular hypertension. Fistulas may be either congenital or acquired; they can be associated with cardiomegaly, high-output failure, and abdominal bruit and hematuria.

Renal mediated hypertension may also be associated with *other conditions*, such as chronic pyelonephritis, arteriolar nephrosclerosis, chronic glomerulonephritis, renal trauma, renal cysts, polycystic disease, and fibrous bands compressing the renal artery. Infrequent causes of renal hypertension include hydronephrosis, renal tuberculosis, renal tumors, xanthogranulomatosis, pyelonephritis, and renal vein thrombosis.

RENIN PHYSIOLOGY

Renal hypertension can be defined as hypertension caused by obstruction of the main renal artery or one of its branches. The incidence rates have been stated to be from 0.2 to 9.5 percent in several clinics around the world, but most of these statistics are from patients referred to specialty clinics (34,35). The true incidence of renovascular hypertension is very difficult to determine for many reasons, not the least being the invasive methods needed to diagnose an anatomical obstruction of the renal artery. The difficulty of making a diagnosis of true renovascular hypertension begins only after identification of an anatomical obstruction. The incidence of significant obstruction of the renal artery in the normotensive population is known to be high from autopsy material. Homer Smith in 1956 pointed out that only 26 percent of patients having a nephrectomy for apparent unilateral renovascular hypertension were normotensive at the end of 1 year (36). The reason for the disparity between finding significant obstruction and curing hypertension by alleviating the obstruction probably involves multiple factors, but a great deal of the answer can be found by reviewing the information gained from the study of renovascular hypertension in animals.

There have been many studies on renovascular hypertension in animals since Dr. Harry Goldblatt's demonstration that hypertension could be produced by partial obstruction of the renal artery. It was subsequently shown that at least 70 percent reduction of the cross-sectional area of an artery is necessary to decrease flow (37). The sequence of events that occurs after partial obstruction of the renal artery in the dog probably holds the answer as to why significant obstruction of the renal artery can occur without producing hypertension.

When the renal artery of a one-kidney "in" conscious dog is suddenly obstructed to produce a mean gradient of 25 mmHg, there is an immediate decrease in renal resistance to flow, and the renal blood flow is maintained at normal (37). Within 5 minutes renin secretion and mean arterial pressure increase. Within 10 minutes, the resistance to flow in the renal microvasculature increases, and renal blood flow begins to decrease. At 24–48 hours, the peripheral plasma renin activity has returned to normal, and the mean arterial pressure remains elevated at about 25 mmHg above base line. The renal blood flow is reduced to about two-thirds normal, and the mean renal artery pressure returns to preconstriction level (38).

The renal microvascular (arteriolar) constriction is mediated initially by the intrarenal renin-angiotensin system. If an angiotensin-converting enzyme inhibitor is introduced into the renal artery at a dose level that gives no peripheral reaction, reduced renal resistance, producing a marked decrease in renal artery pressure distal to the stenosis, results. The reduction in renal artery pressure may be sufficient to produce anuria. After approximately 3–7 days, at which time the peripheral plasma renin activity is normal, blockade of the intrarenal renin-angiotensin system markedly decreases intrarenal pressure and releases large amounts of renin from the kidney. When the one-kidney, one clip dog is maintained on a 75-mmol sodium intake for 7–10 days, renal artery infusion of angiotensin-converting enzyme inhibitor no longer results in a drop in renal artery pressure, and the renin release from the kidney is normal. The renal vasculature can be reverted to an angiotensin-dependent state by restricting sodium intake to 10 mmol per day for 5 days. If angiotensin-converting enzyme inhibitor is administered via the renal artery, the initial finding, a decrease in intrarenal artery pressure, results. This renal microvascular constriction can be converted back and forth from angiotensin dependency to nondependency an indefinite number of times. The mechanism maintaining the renal arteriolar constriction during the period not dependent on angiotensin II is unknown at present (39). This mechanism occurs in the two-kidney, one-clip dog, with the mean arterial pressure returning to normal at 48 hours when the plasma renin activity returns to normal.

The mechanism of renal arteriolar constriction is probably of great importance in renovascular hypertension in humans. The ability of intrarenal angiotensin II to initiate renal vasoconstriction that suppresses renin release to normal and reduces the kidney's ability to release renin to normal probably explains why significant renal artery obstruction occurs in humans without producing hypertension. In those patients whose obstructed kidney's ability to release renin to normal when stimulated is not reduced, excess renin is released, and hypertension results. It has been demonstrated that when converting enzyme inhibitors are given to patients with renal artery stenosis, the renin response is a good predictor of the clinical importance of the renal artery obstruction (40,41), probably because the renal arteriolar constriction is still dependent on intrarenal angiotensin II. That arteriolar constriction occurs in humans and is important in maintaining glomerular filtration pressure and suppression of renin release is demonstrated by the development of renal failure in persons with bilateral renal artery stenosis and in renal transplant patients.

It has been stated that peripheral plasma renin activity is not a reliable test for screening for renal artery stenosis because it is often normal and has been reported to be normal in up to 40 percent of the cases (42). There may be many reasons for a normal peripheral plasma renin activity, but the most common is probably inactivity in a hospital bed. Drugs that suppress the renin secretion created by the kidney may be another cause. It has been shown by many studies that the juxtaglomerular cells of the obstructed kidney are hyperreactive to stimulation. If this characteristic is taken advantage of and the patient is studied without any medication in the ambulatory state, the peripheral plasma renin is elevated for the state of salt balance in about 85 percent of cases. An elevated peripheral plasma renin activity is strong supportive evidence in a patient selected with these criteria. Most medical centers go directly to renal vein renin sampling or renal arteriography. Many centers have shown that rapid sequence intravenous pyelography and renal scanning are associated with too many false-positive and false-negative results to be useful, in fact, if relied upon, the false-positive results would lead to a large number of studies that would potentially be harmful. Which sequence should be used for performing renal vein renin testing probably depends on the center's facilities and the distance of travel by the patient to the center. The renal vein renin studies should probably be done first because they are of least risk to the individual, who can be treated as an outpatient.

INDICATIONS FOR RENOVASCULAR WORK-UP

It has been generally accepted that studying every patient with hypertension for renal artery stenosis is neither cost-effective nor safe. The individuals selected for study vary from one medical center to another, but certain patients are at greater risk for renovascular hypertension and should be seriously considered for diagnostic work-up. They are (1) patients with documented sudden onset of hypertension; (2) individuals with no family history of hypertension and other identifiable secondary causes of hypertension, especially young women, who may have fibromuscular dysplasia; (3) patients with malignant hypertension, especially among the white population; (4) persons who are refractory to drug therapy or become refractory to drug therapy (excluding response blockers of the renin-angiotensin system); and (5) patients with hypertension and a bruit highly suggestive of renal artery obstruction.

DIAGNOSTIC RENAL ARTERIOGRAPHY

High-quality renal angiography requires a high-milliampere three-phase generator or a constant potential generator with the ability to reproduce the desired exposure factors consistently. The tube should be a 100 kW tube with a highspeed rotating anode capable of delivering 125 kV. The smaller the focal spot, the better the resolution. Standard renal arteriography can be obtained with a 0.6-mm focal spot; however, magnification requires a 0.3-mm focal spot. Renal angiography requires tubes with high heat storage capacity. Because of these requirements, the ideal angiographic laboratory often has two tubes, one tube having a 0.1- to 0.3-mm focal spot, and the other tube having a 0.6- to 1.0-mm focal spot.

The use of carbon fiber tops and rare earth film screen combinations reduces the kilowatt requirements of the tube and improves the image quality. Now, only large patients require tubes with focal spots greater than 0.6 mm.

The film changers should be capable of at least 4 films per second. For renal angiography, a 14 by 14 inch film size is desirable. The changer should hold at least 20 films. Cut film may be easier to handle than roll film, depending on personal preference.

The Seldinger technique is utilized for renal angiography, and the renal artery is usually approached from the femoral artery. Occasionally, the axillary approach is necessary when there is severe disease or a bypass graft in the distal aorta or the iliac arteries. In the axillary approach, a high brachial puncture should be utilized, staying below the inferior axillary crease. This puncture is easier to control and decreases the possibility of brachial plexus injury.

The midstream renal arteriogram is performed with a number 5 French high-flow pigtail catheter. The tip of the catheter should be placed just above the renal arteries and below the origin of the superior mesenteric artery. Filling of the superior mesenteric artery branches often obscures the renal arteries. The contrast material should be injected at 20 mL per second for a total of 60 mL. After the first run, which is performed in the anteroposterior (AP) projection, the amount of contrast can be adjusted. In small aortas, the subsequent injections can often be made at 20 mL per second for a total of 40 mL. The films are obtained at 2 films per second for 3 seconds, then 1 film per second for 4 seconds, followed by films every 2 seconds, for a total of 12 films. If oblique films are necessary to visualize the origins of the renal arteries, they are obtained at 2 films per second for a total of 12 films.

The position of the kidneys is variable, but the upper pole is usually at the level of the twelfth thoracic (T12) vertebra, and the lower pole is at the third lumbar (L3) or fourth lumbar (L4). The renal arteries typically originate between the midportion of the L1 vertebra and the upper third of the L2 vertebra. Variations of approximately one vertebral body in either direction are well known. However, on rare occasions, the renal arteries may arise from a level as high as T11 or as low as the iliac vessels. There is usually a single renal artery supplying each kidney, but in 20–40 percent of cases, multiple renal arteries are present.

The renal arteriogram, performed to rule out renal artery stenoses, must clearly depict the origin of each renal artery. The renal arteries usually originate from the lateral aspect of the abdominal aorta, however, they may originate anteriorly or posteriorly off the aorta. The right renal arteries often originate somewhat anteriorly from the aorta.

It is very important to visualize the origin of the renal artery because atherosclerosis often involves the origin of the vessel. The renal arteries are slightly wider at their origin; therefore, if the arteries do not appear larger in their proximal portion, the origin has not been visualized, and oblique films are necessary. It is also very important to visualize the origin of the renal arteries prior to selecting the approach to be used for the renal angioplasty procedure.

RENAL VEIN RENIN SAMPLING

The temptation to dilate a renal artery stenosis, once it has been discovered, is great, particularly since absence of lateralization of renins does not preclude a successful result after correction of the stenosis in 21 percent of patients (43). However, Eyler et al. (44) studied the arteriograms of normotensive and hypertensive adults and found major renal artery stenoses in both groups. Holley and associates (45) demonstrated that renal artery stenosis was present, at autopsy, in a significant percentage of patients who had been normotensive during life. Therefore, once a renal artery stenosis is identified in a hypertensive patient, the physiological importance of the lesion should be assessed before correcting it to ensure optimal results.

Because many drugs stimulate or suppress renin activity, the patient should be taken off such medications for at least 2 weeks and preferably for 1 month prior to renin sampling. Beta blockers are very potent suppressors of renin activity, and it is usually not productive to sample patients using these drugs. However, caution should be exercised to keep the diastolic pressure below 110 mmHg; if it rises above this level, the patient should be placed on antihypertensives, such as prazosin hydrochloride (Minipress [Pfizer]), which have minimal effect on renin activity. Renin secretion can be stimulated by upright posture, sodium depletion, or captopril, which can be given prior to renal vein renin sampling to enhance renin secretion. A positive renal vein renin test is present when the ratio of the renal vein renin of the obstructed kidney is 1.48:1.0, or greater than the unobstructed contralateral side, and the contralateral kidney's secretion of renin is totally suppressed. Suppression of the contralateral kidney is important because the ratio may be increased as a result of hemoconcentration rather than increased renin secretion by the kidney (46). Patients with bilateral renal artery stenosis usually exhibit the same renal vein renin findings produced by unilateral renal artery stenosis. If the stenosis is due to atherosclerosis, the kidney first obstructed is small and secretes all the renin. If only the obstruction to the kidney secreting the renin is alleviated, the contralateral kidney will then usually take over the renin secretion.

Renin sampling from the renal vein is usually performed through the right femoral vein by utilizing the *Seldinger technique*. A number 6.5 French cobra catheter with two side holes near the tip is used for renin sampling. The venous samples are obtained from the main right renal vein and from the left renal vein distal to the entry of the gonadal vein. If more than one renal vein is present, all the veins draining the kidney should be sampled. In the presence of multiple renal arteries or a stenosis in the branch of a renal artery, it may be necessary to obtain samples from the segmental veins draining the area supplied by the renal artery that is stenosed. At the University of Virginia, a sample is also obtained from the aorta. However, at many medical centers, the sample is taken from the inferior vena cava below the level of the renal veins. Contrast material containing iodine may interfere with the determination of the plasma renin activity. Therefore, the catheter is flushed with normal saline, then a separate syringe is used to withdraw 5 mL of blood, which is discarded before the renin sample is obtained with a new syringe. The correct position of the catheter is ascertained after the renin sample has been drawn by injecting a small amount of contrast material.

Samples for renal vein renin are usually obtained after angiographic demonstration of the renal artery stenosis. Harrington et al. (47) found no significant change in the renal vein renin activity 10 minutes after aortography; therefore, samples may be obtained immediately after an arteriogram has been performed. They, however, should be obtained within 15–20 minutes following the arteriogram.

An alternative approach is to screen individuals with hypertension on an outpatient basis, by obtaining a digital subtraction angiogram and following this procedure with selective renal vein renin sampling.

The renal vein renin sample should be obtained as rapidly as possible, immediately placed on ice, and sent promptly to the laboratory for analysis. The plasma renin activity is determined by the radioimmunoassay of angiotensin I (48).

INDICATIONS FOR PTRA

The indications for renal angioplasty include the correction of proven renovascular hypertension and the correction of the angiotensinogenic component in patients with essential hypertension and superimposed renovascular hypertension. Renal artery angioplasty may also be helpful in patients with critical renal artery stenoses and renal insufficiency; renal function may be preserved or improved by correcting the renal artery stenoses present in these individuals.

RENAL DILATION TECHNIQUES

A high-quality preliminary midstream arteriogram is necessary to evaluate patients prior to renal balloon dilation. If an abdominal arteriogram has not been obtained within the previous month, a repeat arteriogram should always be obtained before selecting the renal arteries. This is important because profound changes may occur in a short period of time in the presence of a tight renal artery stenosis. The preliminary midstream arteriogram determines the approach to be used.

Four different percutaneous angiographic approaches can be used to treat stenoses in the renal arteries: (1) Gruentzig guided coaxial balloon catheter system (2) Gruentzig femoral balloon catheter system: femoral approach (3) Gruentzig femoral balloon catheter system: axillary approach and (4) Gruentzig femoral balloon catheter system: sidewinder approach.

Gruentzig Guided Coaxial Balloon Catheter System

The Gruentzig guided coaxial balloon catheter system (Fig. 1) utilizes a number 8 or 9 French renal guiding catheter and a number 4.5 French coaxial balloon catheter. The guiding catheter is inserted through the femoral approach, and the orifice of the renal artery is carefully selected. The small coaxial balloon catheter is then passed through the guiding catheter and across the stenosis. The balloon catheter will accept an 0.018 inch guidewire, which can be passed across the stenosis prior to advancing the balloon catheter in very tight stenoses. Often, the coaxial catheter can be advanced across the lesion while a small amount of contrast material, is injected, and the guidewire may not be needed.

The advantage of this technique is that the number 4.5 French coaxial catheter passes through tight stenoses more readily than does the number 7 French femoral balloon catheter. Therefore, this catheter system is usually used in the renal arteries only in the presence of a very tight stenosis that cannot be crossed by the other catheter techniques. There are several disadvantages to this technique. The catheters are expensive, and it is necessary to make a number 8 or 9 French puncture wound in the femoral artery. The guiding catheter is stiff and may damage the aortic wall. The guiding catheter is inserted over a 0.063 inch guide wire, which is also quite stiff and may traumatize the aortic wall. The balloon sizes are limited to 3.7 mm or 5 mm.

The technique is also more complex than the standard femoral balloon catheter technique. The common femoral artery is punctured, and a two-part sheath is introduced over a 0.038 inch guidewire and advanced into the distal abdominal aorta. Once the

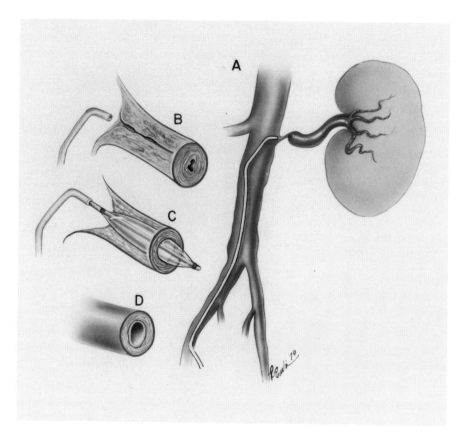

Figure 1. Technique of renal dilation with the coaxial balloon catheter system. **A.** The stenosis is shown in the left renal artery. **B.** The orifice of the renal artery is selected with the guiding catheter. **C.** The number 4.5 French dilation catheter is directed through the stenosis by the guiding catheter and then inflated, dilating the lesion. **D.** The intima is split, and a portion of the media is stretched or split, relieving the obstruction in the renal artery. (Reprinted from Ref. 17.)

sheath is in place, the inner cannula is removed, and the 0.063 inch guidewire is advanced into the abdominal aorta to the level of the diaphragm. The guiding catheter is then advanced over the guidewire through the sheath and into the abdominal aorta. The guidewire is removed, and the guiding catheter is used to select the renal artery orifice. The balloon catheter is then advanced through the guiding catheter and across the stenosis. The small diameter of the balloon catheter makes obtaining pressures across the stenosis possible. The balloon is then inflated with a mixture of one-half contrast material and one-half Renografin-60 (Squibb). The balloon is inflated, and the progress of the angioplasty can be monitored by injecting contrast either through the balloon catheter or through the guiding catheter. This is a definite advantage. However, because of the injury to the intima caused by the dilation procedure, caution should be exercised when injecting contrast material into the area of the dilation.

Gruentzig Femoral Balloon Catheter System: Femoral Approach

The Gruentzig femoral balloon catheter femoral approach (Fig. 2) uses a modification of the double lumen balloon catheter originally designed by Gruentzig for superficial femoral artery angioplasty. Only catheters with the balloon located close to the tip should be utilized in the renal arteries. The number 7 French Gruentzig balloon catheters are usually employed, however, in pediatric cases or in patients

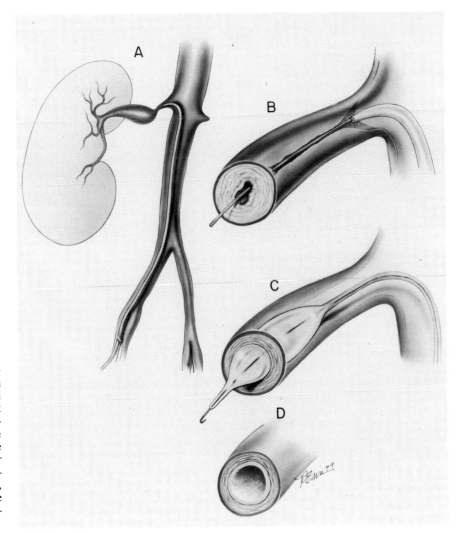

Figure 2. Technique of renal dilation with the femoral balloon catheter system, using the femoral approach. **A.** The stenotic right renal artery is catheterized selectively with a cobra catheter. **B.** The guidewire is passed through the stenosis. **C.** The selective catheter is exchanged for the dilation catheter, and the balloon is inflated, compressing the stenosis. **D.** The obstruction is now relieved. (Reprinted from Ref. 17.)

with small arteries, a number 5 French catheter system can be used.

The femoral artery is punctured, and the appropriate number 5 French diagnostic catheter is advanced into the abdominal aorta by the Seldinger technique. The orifice of the renal artery is carefully selected, and contrast material is injected to localize the lesion. Under fluoroscopic guidance, a 0.035 tight J or a 0.035 LLT (Newton) guidewire is advanced beyond the stenosis into the renal artery. If these guidewires will not pass, the stenosis can often be passed with a 15 mm J guide wire. It is very important to avoid a subintimal passage of the guidewire. Once the stenosis has been crossed, a number 5 French selective catheter is advanced across the stenosis into the renal artery. A small amount of contrast material is injected to ascertain the intraluminal position of the catheter. Once the catheter is across the lesion, 2000–3000 international units (IU) of heparin is injected through the catheter into the renal artery. A movable core tight J or a Rosen wire is then inserted through the catheter into the renal artery beyond the stenosis. The diagnostic catheter is then exchanged for the renal balloon catheter of the appropriate size. The balloon size is chosen by measuring the renal artery proximal and distal to the stenosis. The original size of the renal artery in the stenotic area is estimated; if the renal artery is estimated to have measured 5 mm in diameter, that balloon size should be used. This calculation does not take into account the magnification factor, therefore, the renal arteries are being slightly overdilated (by approximately 1 mm).

It is very important to avoid moving the guidewire back and forth in the branches of the renal artery when exchanging catheters. This movement may induce spasm or occlusion of the segmental branches of the renal artery. The catheter is positioned across the stenosis under fluoroscopic control, and the balloon is inflated either by a Schneider pump or a syringe. If a syringe is employed, a 10-mL syringe is probably ideal because it is capable of generating approximately 9.4 atmospheres of pressure when inflating the syringe; when deflating the syringe, a sufficient vacuum is created to aspirate the balloon rapidly. A pressure gauge should always be employed to ascertain the exact pressure being applied to the balloon. The balloon is inflated first with 2 atmospheres of pressure to determine the position of the balloon in relation to the stenosis. After the balloon has been properly positioned, it is inflated with 4–6 atmospheres of pressure for 30–40 seconds. It may be necessary to dilate the lesion several times. The progress of the dilation can be monitored by configuration of the balloon as it is inflated and by test injections into the artery distal to the stenosis. Immediately after angioplasty, a repeat arteriogram is performed to assess the results of the procedure. When removing the balloon catheter, it is important to deflate the balloon completely by applying a negative pressure to the balloon as it is being removed from the femoral artery.

The primary advantage of this technique is that only a number 5 or 7 French puncture wound is necessary in the femoral artery. If the severity of the stenosis permits easy passage of the balloon, this is the simplest approach. However, if the stenosis is tight or the renal artery branches from the aorta at an acute angle, it is often difficult to make the balloon catheter follow the guidewire across the lesion. The balloon catheter has a tendency to buckle in the aorta when pressure has to be applied to traverse the stenosis from this approach. This difficulty may often be overcome by advancing, first, the number 5 French diagnostic catheter, and then a stiffer number 7 French tapered catheter across the stenosis and then reinserting the balloon catheter.

Gruentzig Femoral Balloon Catheter: Axillary Approach

The axillary approach (Fig. 3) may also be used to dilate the renal arteries (49). When the renal arteries originate from the aorta at a sharp downward angle, the axillary approach greatly simplifies the procedure. The stenotic renal artery is easily selected from the axillary approach. Once the guidewire is in place across the lesion, the dilation catheter has a natural tendency to follow the gentle downward curve of the guidewire across the stenosis. Passage of the guidewire through the stenosis and of the catheter across the stenosis is often facilitated if the patient takes a deep breath. The axillary approach is also useful when severe atherosclerotic disease or bypass grafts are present in the pelvic or abdominal vessels.

The technique also uses the double lumen balloon catheter, designed by Gruentzig for superficial femoral artery angioplasty. The balloon catheter is available in 4, 5, 6, 7, and 8 mm for renal angioplasty. The balloon length is available in several sizes, but the 2 cm balloon catheter is usually used in renal angioplasty. A left axillary approach is usually utilized because this is the straightest course to the descending aorta. The balloon catheter has a tendency to buckle in the ascending aorta when the right axillary approach is attempted. The technique is similar to that utilized in the femoral approach.

Theoretically, there is an increased risk of damage to the smaller axillary artery, however, this risk can

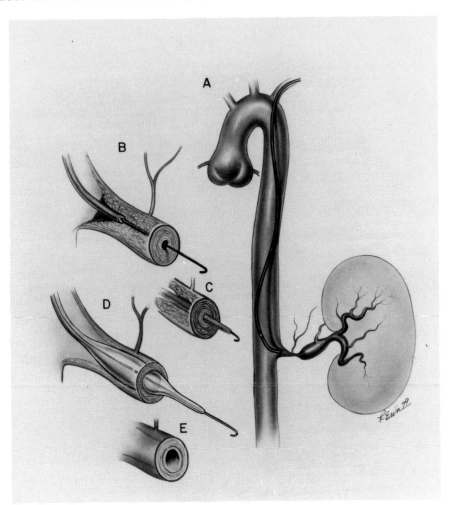

Figure 3. Technique of renal dilation through the axilla. **A.** The stenotic left renal artery is selected with a diagnostic cobra catheter. **B.** A guidewire is passed through the stenosis. **C.** The cobra catheter is advanced through the stenosis. **D.** The selective catheter is exchanged for the dilation catheter, which is inflated to dilate the lesion. **E.** The stenosis is now relieved. (Reprinted from Ref. 17.)

be minimized by deflating the balloon carefully and rotating it as it is being inserted and removed. There is also the possibility of a brachial plexus injury, however, this possibility can be decreased by using the high brachial approach. The brachial artery is entered just distal to the axillary crease. The artery is easier to control in this area because it can be compressed against the humerus.

Gruentzig Femoral Balloon Catheter System: Sidewinder Approach

The sidewinder approach (Fig. 4) combines the advantages of the femoral and axillary approaches. The same number 7 French Gruentzig double lumen balloon catheter employed in the femoral approach is also used in the sidewinder approach. The larger

of the two femoral arteries is punctured, but the renal artery is approached from above in order to take advantage of the natural curve of the renal artery as it branches from the aorta. The renal artery is selected with a number 5 French "shepherd's crook" or "sidewinder," catheter. The catheter is carefully advanced across the lesion under fluoroscopic control as contrast material is injected to keep the tip within the lumen of the artery; or a flexible tip guidewire is advanced across the stenosis, followed by the catheter. Once the catheter is across the lesion, 2000–3000 IU of heparin is injected through the catheter into the renal artery. The guidewire is exchanged for a movable core tight J guidewire or a Rosen wire. The diagnostic catheter is then exchanged for the balloon catheter of appropriate size. After the sidewinder catheter has crossed the lesion, the balloon catheter will usually cross the stenosis with ease.

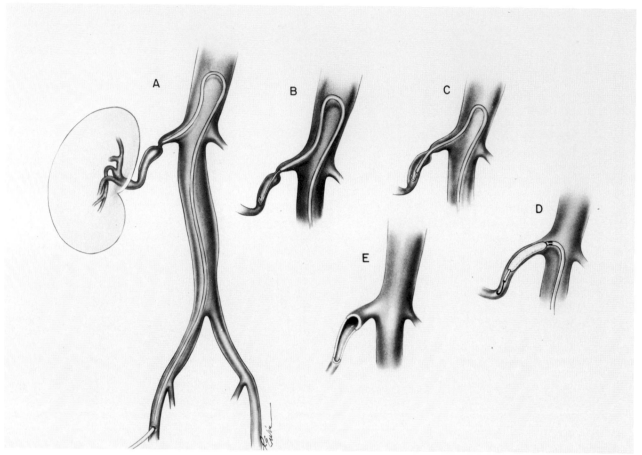

Figure 4. Technique of renal angioplasty utilizing a shepherd's crook or sidewinder, catheter. **A.** The stenotic renal artery is selected with the diagnostic catheter. **B.** The guidewire is passed across the stenosis. **C.** The guidewire is advanced across the stenosis by withdrawing the catheter. **D.** The diagnostic catheter is exchanged for the dilation catheter, which is inflated to dilate the lesion. **E.** The stenosis is ablated.

The major advantage of this technique is that withdrawing the sidewinder catheter advances the diagnostic catheter across very tight stenoses because the configuration of the Simmons' catheter exerts considerable downward force as the catheter is withdrawn over the guidewire. This technique is very helpful in traversing tight stenoses. Following balloon dilation, a midstream arteriogram is obtained to document the results of the procedure.

PATIENT MANAGEMENT

The blood pressure must be monitored very carefully during the first 24–48 hours following angioplasty because profound changes may occur. It is important to discontinue the antihypertensive medications prior to renal angioplasty, thereby helping

to prevent a precipitous drop in the blood pressure. If the diastolic pressure rises above 110 mmHg, the blood pressure should be controlled by short-acting antihypertensive drugs. A drop in blood pressure can usually be controlled by the rapid intravenous infusion of normal saline. Therefore, all patients undergoing renal angioplasty should have an intravenous line in place prior to the procedure.

If not contraindicated, the patient receives 2000 IU of heparin subcutaneously every 6 hours. The heparin is begun 8 hours after the procedure and continued for 2 days. The patients also receive 75 mg Persantine (Boehringer Ingelheim) orally twice a day and 325 mg of aspirin once a day, beginning the day before angioplasty and continuing for at least 6 months. The patients are encouraged to stop smoking. They should be followed by a hypertensionologist, either a cardiologist or a nephrologist, who specializes in the control of high blood pressure. The specialist should participate in the selection of the patients for renal angioplasty and follow them after they are dilated. The patient's blood pressure medication must be adjusted after the procedure. The blood pressure must be followed closely; if it increases in the ensuing months, a repeat arteriogram or DSA study should be performed.

COMPLICATIONS

Since the introduction of renal angioplasty by Gruentzig and associates in 1978, the technique has been refined and simplified. Renal balloon angioplasty, however, remains more complex than peripheral angioplasty. The procedure is far from innocuous and should be performed only by experienced angiographers. Inflating the balloon is simple, but successfully crossing a tight stenosis with a catheter requires great skill, and selecting the proper approach and proper balloon size requires experience. Considerable experience in performing peripheral angioplasty should be acquired before renal angioplasty is attempted. Because the potential complications of renal angioplasty are serious, the procedure should be performed only in hospitals where a skilled vascular surgeon is immediately available.

The complication rate following renal angioplasty varies between 5 and 10 percent. Most of the complications are minor, but major complications may be encountered. The most frequent major complication is transient renal insufficiency, which is clearly related to the contrast material. The frequency of this complication can be decreased by performing the diagnostic arteriogram several days prior to the therapeutic procedure in patients with renal insufficiency. Subintimal dissection with the guidewire or the diagnostic catheter may result when attempting to cross the stenosis. A small intimal flap will usually heal, however, in the presence of such an intimal dissection, angioplasty should be postponed for approximately 4–6 weeks to allow time for healing. Thrombosis of the renal artery, which may occur immediately after balloon dilatation, is an infrequent complication that may sometimes be treated by the infusion of streptokinase into the renal artery after the procedure. It is very important that sufficient collateral vessels be present so that prolonged renal ischemia does not develop during the streptokinase infusion. The renal artery may be ruptured after balloon dilatation. This rare complication may result from weakening of the arterial wall or from a subintimal placement of the balloon catheter. If rupture of the renal artery is noted immediately after

deflation of the balloon, the balloon should be reinflated, occluding the proximal renal artery, and the patient should be taken immediately to the operating room. Distal embolization is an infrequent complication, but a recently occluded renal artery should be approached with caution. Thrombi may be dislodged by the catheter, occluding several of the segmental branches.

The most frequent minor complication encountered during angioplasty is spasm, or occlusion of an arterial branch. This is usually caused when the tip of the guidewire moves back and forth within the branches of the renal artery during the procedure (Fig. 5). If possible, the guidewire should not be placed within the segmental branches of the renal artery. Focal spasm may also be produced in the area immediately adjacent to the angioplasty site. Calcium channel blockers are effective in preventing and reversing spasm in the renal arteries. Nifedipine is the most potent vasodilator of the calcium antagonist drugs and can be given in a dose of 20 mg sublingually at the onset of the procedure if spasm is encountered, and verapamil hydrochloride may be given parentally through the arterial catheter in a dose of 2.5–5 mg. Nitroglycerin is also effective in reversing vasospasm. It is either injected directly into the affected renal artery in a dose of 50–200 μg or given sublingually in a dose of 0.4–0.6 mg. Calcium channel blockers may induce hypotension, and they should be used with caution in patients with known cardiac conduction defects.

In addition to those unique to renal angioplasty, other complications may occur at the puncture site. Because the patient is anticoagulated and the axillary approach involves frequent catheter exchanges, caution should be exercised to puncture high in the brachial artery and not in the axilla itself to avoid the devastating effects of a large hematoma in the axilla and ensuing brachioplexus injury.

RESULTS

In experienced hands, renal angioplasty is a highly effective method for correcting renal artery lesions. A primary success rate greater than 90 percent should be achieved when dilating renal artery stenoses (Table 1). Technical failures usually result from an inability to cross the lesion or to dilate it adequately by means of the balloon.

The long-term results of renal angioplasty can be assessed in terms of the effect of the procedure on vessel patency, blood pressure, and renal function. The effect on vessel patency is related to the etiology and characteristics of the lesion. The patients can be divided into five distinct groups: (1) patients

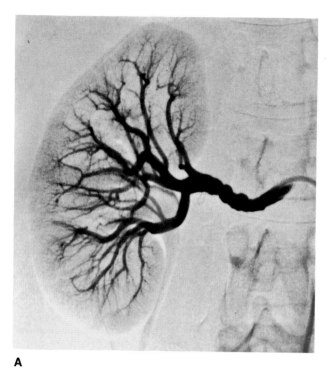

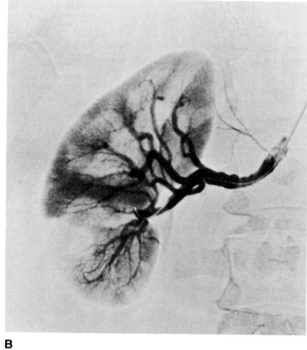

A

B

C

3 MOS

Figure 5.   Renal parenchymal defect caused by a guide-wire during the angioplasty procedure. **A.** Selective renal arteriogram reveals changes consistent with fibromuscular dysplasia. **B.** Immediately after dilation, the angiogram demonstrates a defect in the lower pole of the kidney, apparently caused by the guidewire, which is still in place. **C.** Arteriogram obtained 3 months later shows resolution of the defect. The renal arteries are widely patent. (Reprinted from Ref. 18.)

TABLE 1   Results of Percutaneous Transluminal Renal Angioplasty

	Number of Patients	Initial Success Rate (Patients), Percentage	Patency Rate S/P Initial Success (Patients), Percentage	Length Follow-Up
Colapinto (Ref. 21)	68	85	81	3 years
Katzen (Ref. 12)	17	94 (lesions)	75 (lesions)	1 year
Puijlaert (Ref. 14)	54	96	70	
Schwarten (Ref. 15)	70	93 (lesions)	71	6 months
Sos (Ref. 16)	101	79		
Tegtmeyer (Ref. 24)	108	95	93	5 years
Total	417			
Range		79–95	70–93	0.5–5 years

*Includes successful redilations. S/P = status post.

with atherosclerotic renal artery stenoses or occlusions (2) those with fibromuscular dysplasia of the renal artery (3) those with transplants (4) patients with saphenous bypass grafts and (5) those dilated primarily for renal insufficiency.

Atherosclerotic Lesions

Atherosclerotic disease is the most frequent etiology of the stenosis subjected to renal dilation. In the University of Virginia series, 94 percent of the 65 hypertensive patients with atherosclerotic lesions were helped by the procedure. Twenty-three percent were cured, and 71 percent improved after renal balloon angioplasty. Analysis of the results in the patients with atherosclerotic lesions revealed several other factors that are important when selecting patients for angioplasty. Better results are achieved in individuals with unilateral renal artery stenoses than in those with bilateral renal artery stenoses. The restenosis rate is higher in patients with severe bilateral renal artery disease than in those with unilateral renal artery stenosis (24). Sos et al. (23) also showed that success was more frequent in patients

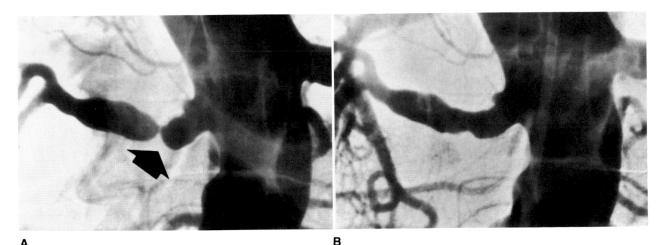

A   **B**

Figure 6.   Short isolated atherosclerotic lesions respond well to dilation as in this 67-year-old man. **A.** Predilation arteriogram reveals an 85 percent stenosis of the right renal artery (arrow). **B.** Immediately after PTRA with a 6-mm balloon, only a small plaque remains. (Reprinted from Tegtmeyer CJ, Kofler TJ, Ayers CA: Renal angioplasty: Current status. AJR 142:17–21, 1984).

with unilateral lesions than in those with bilateral disease. Furthermore, it is becoming increasingly clear that certain lesions are more amenable to balloon dilation than others. A good result can be anticipated in short isolated atherosclerotic lesions (Fig. 6). When the renal artery stenosis is caused by a large plaque in the abdominal aorta that engulfs the origin of the renal artery (Fig. 7), chances of success are diminished. The schematic drawings (Fig. 8) illustrate the type of lesion in which the diminishing results are often encountered. Cicuto et al. (50), Sos et al. (23), and Schwarten (51) all reported similar results. In the University of Virginia series, one-third of the redilated lesions were due to lesions caused by aortic plaques that engulfed the origin of the renal artery. Complete occlusions are more difficult to dilate than is the stenotic artery, and it is clear that complete blockage in a vessel that is not perfectly straight should not be attempted.

Fibromuscular Dysplasia

The best results in renal angioplasty are achieved in patients with flow-limiting lesions of fibromuscular dysplasia (Fig. 9). Fibromuscular lesions usually respond well to balloon dilation at pressures of 4 atmospheres or less. If the lesion can be crossed, a good result can be anticipated. There were 27 hypertensive patients with 30 lesions due to fibromuscular dysplasia in our series (24). All benefited from renal angioplasty, and only one patient with fibromuscular dysplasia required repeat dilation. Similar results have been achieved in several other series. Geyskes et al. (22) dilated 21 patients for fibromuscular dysplasia, and 95 percent of them were either cured or improved. There was only one failure. Sos et al. (23) achieved a technical success when dilating lesions due to fibromuscular dysplasia in 27 of their 31 patients.

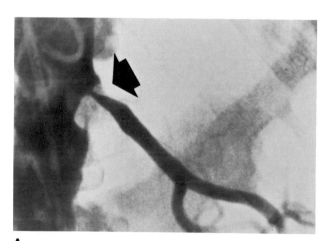

A

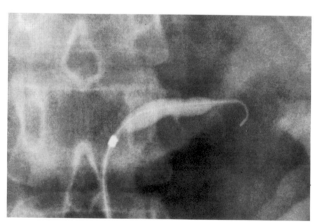

B

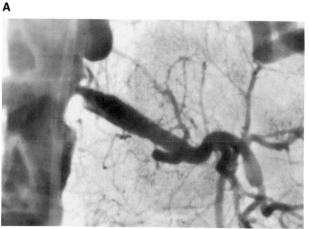

C

Figure 7. Diminished response in renal artery orifice lesions caused by plaques in the abdominal aorta in a 68-year-old woman with longstanding hypertension. She has a very tight stenosis at the origin of the left renal artery and a 60 percent stenosis in the right renal artery. **A.** Predilation arteriogram: tight stenosis at the origin of the left renal artery (arrow). **B.** The 100-mm spot radiograph shows 6 mm balloon in place (note that balloon is not completely expanded). **C.** Immediately after PTRA: artery is improved, but residual stenosis remains. (Reprinted from Tegtmeyer CJ, Kofler TJ, Ayers CA: Renal angioplasty: Current status. AJR 142:17–21, 1984).

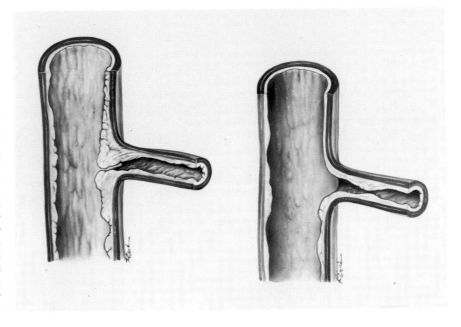

Figure 8. The type of atherosclerotic lesion in which a diminished response to balloon dilation can be anticipated. This lesion (left) is caused by an atherosclerotic plaque in the abdominal aorta that engulfs the orifice of the renal artery. Occasionally, a good result is obtained; usually results are poor when compared with those for other lesions. A good response can be anticipated in short stenoses within the renal artery (right). (Reprinted from Tegtmeyer CJ, Kofler TJ, Ayers CA: Renal angioplasty: Current status. AJR 142:17–21, 1984).

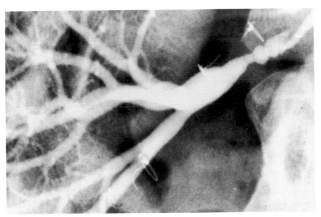

A

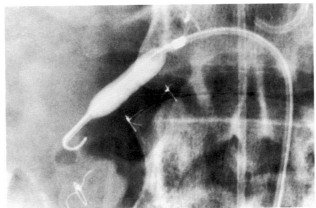

B

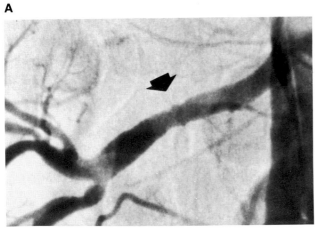

C

Figure 9. Successful renal angioplasty in a fibromuscular lesion in a 24-year-old hypertensive woman. **A.** The selective right renal arteriogram reveals a stenosis in the right renal artery. (Reprinted from Ref. 20.) **B.** Fibromuscular dysplasia is usually easy to dilate, and the 5 mm balloon expands easily, dilating the lesion. **C.** Immediately after dilation, the lesion has disappeared (arrow). (Reprinted from Ref. 20.)

Lesions in Transplanted Kidneys

There were seven patients with stenoses in the arteries to the transplanted kidney (Fig. 10) who underwent dilation in our experience. Five patients were improved, and there were two failures. One of these failures was due to the inability to dilate the lesion initially, probably because of the use of a polyvinyl chloride balloon. Now that polyethylene balloons are available, a better result would probably be achieved in similar patients. In another patient, a tight stenosis at the anastomosis of the renal artery to the hypogastric artery failed after 7 months despite three attempts at dilation. Pressures as high as 14 atmospheres were utilized. Sniderman et al. (52) attempted to dilate 15 patients with stenoses in the arteries of the transplanted kidneys. Three patients underwent repeat dilations. The procedure was technically successful in 15 of the 18 attempts in 13 of the 15 patients. Gerlock et al. (53) successfully dilated all seven patients with renal transplant stenoses in their series. Caution should be exercised when dilating lesions in patients with renal transplants because there are no collateral vessels to the kidney. If the vessel is occluded, surgery must be undertaken immediately. Therefore, in cases of renal transplant dilation, a surgeon should be available during the procedure, as in coronary angioplasty.

Stenoses in Vein Bypass Grafts

In the University of Virginia series (24) three patients had balloon dilatation of stenoses in their renal saphenous vein bypass grafts of the kidney. The lesions were successfully dilated in all three cases.

Renal Insufficiency

Under certain circumstances, alleviation of renal artery stenosis is important to preserve renal function. It is clear that renal function can be preserved if the renal artery stenosis is corrected; this procedure is probably worthwhile in unilateral renal disease if the kidney size indicates potentially significant preservation of renal mass or if bilateral renal artery stenosis is present. In the latter case, an attempt to alleviate flow obstruction to the larger kidney should be given priority even when the smaller kidney is secreting all the renin.

The results of angioplasty are less dramatic in patients treated primarily for renal insufficiency than

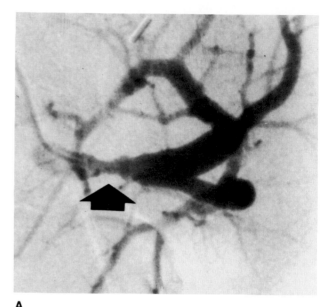

A

B

Figure 10. Percutaneous transluminal angioplasty in a 23-year-old woman with a renal transplant. **A.** Selective renal arteriogram demonstrates a tight stenosis (arrow) at the anastomosis between the renal and hypogastric arteries. **B.** The lumen of the vessel is much improved after balloon angioplasty through the axillary approach. (Reprinted from Ref. 19.)

in those treated primarily for hypertension, although they are encouraging. In our series (24), 10 patients were treated primarily for renal insufficiency. They all had atherosclerotic lesions. The mean serum creatinine was 5.2 mg per 100 mL before dilation; after angioplasty this decreased to 2.3 mg per 100 mL for the group. Five of the patients had a positive response to dilation, and five were not helped. Twenty-nine of the patients treated primarily for hypertension also had accompanying renal insufficiency. Nine of these have improved renal function following the procedure, and four have normal blood urea nitrogen (BUN) and creatinine values. In some of the patients, the improvement following angioplasty was gradual, and the full benefit of the procedure was not apparent for several months.

Summary of Results

Renal angioplasty is a clinically effective means of treating renovascular hypertension. The blood pressure response has been analyzed in the 92 hypertensive patients whose initial dilation was successful (24). The patients have been followed from 1 to 60 months (mean, 23.7 months; median 23.0 months). The mean systolic pressure was 199.74 mmHg before renal angioplasty and 140.34 mmHg afterward, a difference of 59.40; the mean diastolic pressure was 117.05 mmHg before renal angioplasty and 83.74 mmHg afterward, a difference of 33.31. Analysis of the long-term clinical results in the 98 hypertensive patients who underwent renal angioplasty to control their blood pressure reveals that 26 percent were classified as cured (defined by the Cooperative Study of Renovascular Hypertension [54] as having an average diastolic pressure less than or equal to 90 mmHg with at least a 10 mmHg decrease from the predilatation level). Sixty-seven percent were classified as improved, and their blood pressure was easier to control by medication as a result of the angioplasty procedure. Seven percent of the patients were non-responders, although one was helped for several months. A review of a large series in the world literature (Table 1) shows that if the initial dilation is successful, the vessels can be expected to remain patent in 70–90 percent of the patients. If the vessel remains patent for at least 8 months, it is likely to remain patent for at least 5 years. The recurrence rate following renal angioplasty has been variously reported as being between 12.9 percent (24) and 22.5 percent (51); it is clearly higher in atherosclerotic stenoses than in other types of stenoses. A major factor in determining the recurrence rate is the success of the initial dilation. Lesions that exhibit a 30 percent or greater residual stenosis on the immediate postdilatation films are more likely to recur than lesions in which a better result has been obtained (18,24). Therefore, it is essential that a good

initial result be obtained. In restenosis cases, a re-dilation is usually successful, and the procedure is often easier than the initial procedure.

CONCLUSION

The results of the University of Virginia series and a review of the literature indicate that percutaneous transluminal angioplasty of the renal artery is a versatile and reliable procedure. Excellent results can be obtained if the patients are carefully selected.

It is encouraging to note that the procedure has the potential for improving renal insufficiency. In the University of Virginia series (24), of the 39 patients who had high BUN and creatinine levels, 18 patients currently have improved renal function; 4 of these have normal BUN and creatinine values. In experienced hands, the results of renal angioplasty compare favorably with the surgical results for the treatment of renovascular hypertension. The success of renal angioplasty has changed the original skepticism to tempered enthusiasm. If the renal artery remains patent for at least 8 months following the dilation, a good long-term result can be anticipated. Renal angioplasty should be considered the treatment of choice in patients with hypertension and renal artery stenoses due to fibromuscular dysplasia or to short isolated atherosclerotic lesions. Good results can also be obtained in individuals with stenoses in the arteries of the transplanted kidneys and in those with stenoses in bypass grafts.

The procedure is enticing because it offers many advantages. Percutaneous transluminal angioplasty is a relatively simple procedure when compared with surgery, and it preserves renal tissue. It avoids general anesthesia and major surgery, and the patient experiences less discomfort. The procedure is relatively inexpensive and decreases the hospital stay. For the more than a million in the United States who suffer from renovascular hypertension, renal angioplasty, indeed, provides a viable alternative to surgery.

REFERENCES

1. Gifford RW Jr: Evaluation of the hypertensive patient with emphasis on detecting curable causes. Milbank Mem Fund Q 47:170–186, July 1969.
2. Genest J, Boucher R, Rojo-Ortega JM, et al: Renovascular hypertension, in Genest J, Koiw E, Kuchel O, et al (eds): *Hypertension. Physiopathology and Treatment.* New York, McGraw-Hill, 1977, chap 24.2, pp 815–840.

3. Youngerg SP, Sheps SG, Strong CG: Fibromuscular disease of the renal arteries. Med Clin North Am 61:623–641, May 1977.

4. Dollery CT, Bulpitt CJ: Management of hypertension, in Genest J, Koiw E, Kuchel O, et al (eds): *Hypertension. Physiopathology and Treatment.* New York, McGraw-Hill, 1977, chap 31.3, pp 1038–1068.

5. Foster JH, Maxwell MH, Franklin SS, et al: Renovascular occlusive disease. Results of operative treatment. JAMA 231:1043–1048, March 10, 1975.

6. Veterans Administration Cooperative Study Group on Antihypertensive Agents: Effects of treatment on morbidity in hypertension: I. Results in patients with diastolic blood pressure averaging 115 through 129 mmHg. JAMA 202:1028–1034, Dec 11, 1967.

7. Veterans Administration Cooperative Study Group on Antihypertensive Agents: Effects of treatment in morbidity in hypertension. II. Results in patients with diastolic blood pressure averaging 90 through 114 mmHg. JAMA 213:1143–1152, Aug 17, 1970.

8. Dotter CT, Judkins MP: Transluminal treatment of arteriosclerotic obstruction. Description of a new technique and a preliminary report of its application. Circulation 30:654–670, November 1964.

9. Gruentzig A, Hopff H: Perkutane Rekanalisation chronischer arterieller Verschlusse mit einem neuen Dilatationskatheter. Modifikation der Dotter-Technik. Tsch Med Wochenschr 99:2502–2505, Dec 6, 1974.

10. Gruentzig A, Kuhlmann U, Vetter W, et al: Treatment of renovascular hypertension with percutaneous transluminal dilatation of a renal-artery stenosis. Lancet 1:801–802, 1978.

11. Boomsma H: *Percutaneous Transluminal Dilatation of Stenotic Renal Arteries in Hypertension.* Groningen, 1982.

12. Katzen BT, Chang J, Knox WG: Percutaneous transluminal angioplasty with the Gruentzig balloon catheter. A review of 70 cases. Arch Surg 114:1389–399, 1979.

13. Martin EC, Mattern RF, Baer L, Fankuchen EI, Casarella WJ: Renal angioplasty for hypertension: Predictive factors for long-term success. AJR 137:921–924, 1981.

14. Puijlaert CBAJ, Boomsma JHB, Ruijs JHJ, Geyskes GG, Franken AH, Hoekstra A, Oei HY: Transluminal renal artery dilatation in hypertension: Technique, results and complications in 60 cases. Urol Radiol 2:201–210, 1981.

15. Schwarten DE: Percutaneous transluminal renal angioplasty. Urol Radiol 2:193–200, 1981.

16. Sos TA, Saddeknis, Sniderman KW, et al: Renal artery angioplasty: Techniques and early results. Urol Radiol 3:223–231, 1982.

17. Tegtmeyer CJ, Dyer R, Teates CD, et al. Percutaneous transluminal dilatation of the renal arteries. Techniques and results. Radiology 135:589–599, 1980.

18. Tegtmeyer CJ, Teates CD, Crigler N: Percutaneous transluminal angioplasty in patients with renal artery stenosis. Follow-up studies. Radiology 140:323–330, 1981.

19. Tegtmeyer CJ, Brown J, Ayers CA, Wellons HA, Stanton LW: Percutaneous transluminal angioplasty for the treatment of renovascular hypertension. JAMA 245:2068–2070, 1981.

20. Tegtmeyer CJ, Elson J, Glass TA, et al: Percutaneous transluminal angioplasty: The treatment of choice for renovascular hypertension due to fibromuscular dysplasia. Radiology 143:631–637, 1982.

21. Colapinto RJ, Stronell RD, Harries–Jones EP, et al: Percutaneous transluminal dilatation of the renal artery: Follow-up studies on renovascular hypertension. AJR 139:727–732, 1982.

22. Geyskes GG, Puijlaert CBAJ, Oei HY, Mees EJD: Follow-up study of 70 patients with renal artery stenosis treated by percutaneous transluminal dilatation. Br Med J 287:333–336, 1983.

23. Sos TA, Pickering TG, Sniderman K, et al: Percutaneous transluminal renal angioplasty in renovascular hypertension due to atheroma or fibromuscular dysplasia. N Engl J Med 309:274–279, 1983.

24. Tegtmeyer CJ, Kellum CD, Ayers C: Percutaneous Transluminal Angioplasty of the renal artery: Results and long-term follow-up. Radiology 153:77–84, 1984.

25. Leadbetter WF, Burkland CD: Hypertension in unilateral renal disease. J Urol 39:611–626, 1938.

26. McCormack LJ, Poutasse EF, Meaney TF, Noto TJ Jr, Dustan HP: A pathologic-arteriographic correlation of renal arterial disease. Am Heart J 72:188–198, 1966.

27. McCormack LJ, Dustan HP, Meaney TF: Selected pathology of the renal artery. Semin Roentgenol 2:126–138, 1967.

28. Meaney TF, Dustan HP, McCormack LJ: Natural history of renal arterial disease. Radiology 91:881–887, 1968.

29. Curtis JJ, Galla JH, Kotchen TA, et al: Prevalence of hypertension in a renal transplant population on alternate-day steroid therapy. Clin Nephrol 5:123–127, 1976.

30. Bachy CH, van Ypersele de Strihou CH, Alexandre GPJ et al: Hypertension after renal transplantation. Proc Eur Dial Transplant Assoc 12:461–470, 1976.

31. Staab GE, Tegtmeyer CJ, Constable WC: Radiation-induced renovascular hypertension. Am J Roentgenol, Radiat Ther Nucl Med 126:634–637, 1976.

32. Ekelund L, Gerlock J Jr, Goncharenko V, Foster J: Angiographic findings following surgical treatment for renovascular hypertension. Radiology 126:345–359, 1978.

33. Boijsen E, Kohler R: Renal Artery Aneurysms. Acta Radiol [Diagn] (Stockh.) 1:1077, 1963.

34. Vetter W, Vetter H, Kuhlmann U, et al: Clinical findings, diagnosis and therapy of renovascular hypertension. Schweiz Med Wochenschr 109:384–394, 1979.

35. Tucker RM: Renal artery hypertension: Diagnosis and management, in Onesti G, Brest AN (eds): *Hypertension: Mechanisms, Diagnosis and Treatment.* Philadelphia, Davis, 1978, pp 165–181.

36. Smith HW: Unilateral nephrectomy in hypertensive diseases: J Urol 76:685–701, 1956.

37. Gupta TC, Wiggers CJ: Basic hemodynamic changes produced by aortic coarctation of different degrees. Circulation 3:17–31, 1951.

38. Ayers CR, Harris RH Jr, Lefer LG. Control of renin

release in experimental hypertension. Circ Res 24–25(suppl I): I-103–I-112, 1969.

39. Ayers CR, Katholi RE, Vaughan ED Jr, Carey RM, Kimbrough HM, Yancey MR, Morton CL: Intrarenal renin-angiotension sodium interdependent mechanism controlling post-clamp renal artery pressure and renin release in the conscious dog with chronic one-kidney Goldblatt hypertension. Circ Res 3(40):238–242, 1977.

40. Case B, Laragh JH: Reactive hyperreninemia in renovascular hypertension after angiotensin blockade with saralasin or converting enzyme inhibitor. Ann Intern Med 91:153–160, 1979.

41. Re R, Novelline R, Escourrou MT, et al: Inhibition of angiotensin converting enzyme for diagnosis of renal artery stenosis. N Engl J Med 298:582–586, 1978.

42. Genest J: The renin-angiotensin-aldosterone system: Physiopathology, in Brest An, Moyer (eds): *Cardiovascular Diseases*. Philadelphia, Davis, 1968, pp 144–160.

43. Bourgoignie J, Jrz S, Catanzaro FJ, Serirat P, Perry HM: Renal venous renin in hypertension. Am J Med 48:332–342, 1970.

44. Eyler WR, Clark MD, Garman JE, et al: Angiography of the renal areas including a comparative study of renal arterial stenoses in patients with and without hypertension. Radiology 78:879–892, 1962.

45. Holley KE, Hunt JC, Brown AL Jr, et al: Renal artery stenosis: A clinical-pathologic study in normotensive and hypertensive patients. Am J Med 37:14–22, 1964.

46. Vaughan ED Jr, Buhler FR, Laragh HG, et al: Re-novascular hypertension: Renin measurements to indicate hypersecretion and contralateral suppression, estimate renal plasma flow and score for surgical curability. Am J Med 55:402–414, 1973.

47. Harrington DP, White RI, Kaufman SL, Whelton PK, Rusel RP, Walker WG. Determination of optimum methods of renal venous renin sampling in suspected renovascular hypertension. Invest Radiol 10:45, 1975.

48. Sealey JE, Gerten-Banes J, Laragh JH: The renin system: Variations in man measured by radioimmunoassay or bioassay. Kidney Int 1:240–253, 1972.

49. Tegtmeyer CJ, Ayers CA, Wellons HA: The axillary approach to percutaneous renal dilatation. Radiology 135:775–776, 1980.

50. Cicuto KP, McLean GK, Oleaga JA, et al: Renal artery stenosis. Anatomic classification for percutaneous transluminal angioplasty. AJR 137:559–601, 1981.

51. Schwarten DE: Percutaneous transluminal angioplasty of the renal arteries: Intravenous digital subtraction angiography for follow-up. Radiology 150:369–373, 1984.

52. Sniderman KW, Sos TA, Sprayregen S: Postrenal transplantation, in Castaneda-Zuniga W (ed): *Transluminal Angioplasty*. New York, Thieme-Stratton, 1983, pp 80–89.

53. Gerlock AJ Jr, MacDonnell RC Jr, Smith CW, et al: Renal transplant arterial stenosis: Percutaneous transluminal angioplasty. AJR 140:325–331, 1983.

54. Simon N, Franklin SS, Bleiffer KH, et al: Clinical characteristics of renovascular hypertension. JAMA 220:1209–1218, 1972.

4

Percutaneous Transluminal Angioplasty of the Iliac and Common Femoral Arteries and Their Accessory Vessels

SAADOON KADIR, M.D.

INTRODUCTION

Percutaneous transluminal angioplasty of the iliac arteries was first described by Dotter and Judkins in 1964 (1). Presently, iliac artery stenosis is the most frequent indication for general vascular angioplasty. Long-term patency rates comparable to those achieved by aortobifemoral bypass grafting have now been reported, making PTA an acceptable alternative to bypass surgery (2–7). In the past, the indications for iliac and common femoral artery angioplasty were limited to patients with stenoses or femoral occlusion because angioplasty of iliac and femoral artery occlusions was associated with a high incidence of thromboembolic complications (8). However, with the use of adjunctive intra-arterial infusion of thrombolytic agents, the indications for iliofemoral artery angioplasty can be extended to include recent occlusions.

The internal iliac arteries are also the most common donor vessels for renal transplantation. In the management of stenoses of the transplanted renal

arteries (other than those due to rejection) angioplasty plays an increasingly important role.

PATHOLOGY OF THE LESIONS

Atherosclerosis is by far the most common cause of occlusive disease of the iliofemoral arteries. The atherosclerotic plaque causing arterial stenosis may be soft, permitting easy dilation, or hard, leading to intimal and medial tears. Unchecked, atherosclerotic stenosis progresses to complete occlusion, by superimposition of a thrombus. However, organization of the thrombus in a diseased vessel is slow and incomplete (9). It often remains soft, permitting intrathrombus placement of catheters for thrombolytic therapy.

Other causes of stenosis are fibromuscular disease which involves the iliac arteries in 1–5 percent of patients and occurs in association with fibromuscular disease of the renal arteries, and postoperative stenoses (10,11). The latter are usually due to aortofemoral or aortoiliac bypass grafts and infrequently due to other operative procedures. Postoperative anastomotic stenoses are observed in approximately 4 percent of patients undergoing aortoiliac or aortofemoral reconstruction (12).

Transplant renal artery stenosis is the cause of renovascular hypertension in approximately 5 percent of such patients (13). Stenosis in transplant renal arteries may be due to technical factors (i.e., tight anastomosis, clamp injury), rejection, perivascular adhesions, or intimal fibrous and cellular proliferation.

CLINICAL CONSIDERATIONS

The most common symptom associated with iliac and femoral artery occlusive disease is claudication. Stenosis or occlusion of the common iliac artery manifests as unilateral buttock and thigh claudication. Occlusion of the common femoral artery manifests as thigh and occasionally as calf claudication. In the absence of occlusive disease of the superficial femoral artery, rest pain is absent. Bilateral claudication is observed in patients with donor iliac artery stenosis after a femorofemoral bypass graft. Unilateral iliac artery occlusion may also lead to pudendal steal and impotence. Stenosis or occlusion of both internal iliac arteries is responsible for impotence.

Objective criteria for clinical evaluation of peripheral vascular disease of the iliofemoral arteries include femoral pulses (by palpation) and ankle and thigh pressures (by Doppler measurement). In the presence of superficial femoral artery disease, the ankle pressure may not accurately reflect the severity of the iliac and common femoral artery disease.

In patients with renal transplants, the most common indications for arteriography and subsequent angioplasty are deterioration of renal function and increasing or poorly controlled hypertension. Preangioplasty renin determination is used to localize the source of renin in patients with recurrent hypertension.

PATIENT SELECTION, INDICATIONS, AND CONTRAINDICATIONS FOR ILIOFEMORAL PTA

Angioplasty is recommended only for patients who are either symptomatic or have other evidence of hemodynamically significant arterial stenoses. The practice of prophylactic angioplasty is to be dis-

couraged. Thus, demonstration of an arterial stenosis on the diagnostic arteriogram is not an indication for angioplasty, since arteriographic appearance does not permit assessment of hemodynamic significance.

Hemodynamic significance is assessed by evaluation of symptoms, ankle/arm index (AAI), systolic thigh pressures, and measurement of an intra-arterial pressure gradient. A resting gradient > 15 mm Hg or a gradient increase induced by priscoline or contrast medium of greater than 10–15 mm is considered hemodynamically significant.

Patients who are ideally suited for angioplasty are those with focal stenoses of the iliac and common femoral arteries. Patients with diffuse disease are more likely to benefit from surgical reconstruction. More recent experience shows that intra-arterial low-dose streptokinase can be used in patients with recent iliac and femoral artery occlusions (less than about 8 months duration) to provide thrombolysis, followed by angioplasty.

In patients with unilateral iliofemoral artery occlusion and stenosis of the contralateral iliac artery who are not candidates for either vascular reconstructive surgery or thrombolytic therapy, angioplasty can be used in conjunction with a less invasive surgical procedure. In such patients, angioplasty of the stenotic iliac artery with subsequent femoral-femoral crossover bypass graft may provide an alternate vascular reconstructive procedure. In addition, postoperative stenoses after aortoiliac and aortobifemoral bypass graft can also be managed successfully with angioplasty (14).

Selection of patients for angioplasty should be made together with the vascular surgeon. This combined approach both insures proper patient selection and prepares the surgeon for treatment of eventual failures and complications. After successful angioplasty, the patients should be followed by both the vascular surgeon and vascular radiologist.

CRITERIA FOR SELECTION OF PATIENTS FOR ILIOFEMORAL PTA

The criteria used to determine feasibility of iliac and femoral artery angioplasty are as follows:

1. *Location of lesion.* This not only determines the applicability of angioplasty but also dictates the approach to be used (ipsilateral or contralateral femoral or axillary artery approach). For example, in patients with an acute angled aortic bifurcation, the contralateral approach may be difficult.

2. *Stenoses*. Short segmental stenoses are ideally suited for angioplasty. Some patients with diffuse atherosclerotic stenoses may also be candidates for angioplasty if surgical intervention is deemed to be a high risk. Stenoses associated with ulcerated plaques or aneurysmal disease are not suited for angioplasty.

3. *Occlusions*. Although some short occlusions of the pelvic vessels have been recanalized without morbidity, the complications associated with attempted angioplasty of iliac or femoral artery occlusion may be high (8,15). Thus, in the past, such patients were considered for bypass surgery. In selected patients, intra-arterial thrombolysis followed by angioplasty may provide an excellent method of treating relatively recent occlusions (<8 months duration).

4. *Type of lesion*. Both atherosclerotic and postoperative stenoses can be treated by angioplasty. In the presence of ulcerated plaques or a history of cholesterol embolization, angioplasty should not be performed. Dilation of a stenosis distal to ulcerated plaques and traumatization of such segments may precipitate peripheral embolization of cholesterol particles. In such patients the treatment of choice is a bypass graft with exclusion of the diseased vascular segment (16,17).

5. *Associated disease*. The presence and extent of other arterial lesions, medical, or surgical problems determine whether angioplasty is performed as the primary procedure or in conjuction with surgery.

6. Heavy *vascular calcification* is not a contraindication for angioplasty.

Angioplasty is not indicated for (1) patients with long-standing iliac or femoral artery occlusions, (2) patients with ulcerated plaques in the iliac arteries or distal aorta who manifest "blue toe syndrome," and (3) patients with diffuse atherosclerotic stenosis of the iliac and femoral arteries or with focal stenoses in association with aneurysmal disease of the iliofemoral arteries. Such patients are best treated with aortobifemoral bypass grafts.

CRITERIA FOR SELECTION OF PATIENTS FOR TRANSPLANT RENAL ARTERY PTA

There are essentially four types of transplant renal artery stenosis demonstrable on arteriography.

1. *Anastomotic stenosis* secondary to technical factors (e.g., tight suture). Such patients should not be considered for angioplasty and are best managed by operation. Angiographically, a tight suture can be recognized as a focal stenosis with absence of the tapering usual with other stenoses.

2. *Segmental stenosis* due to intimal fibrosis and cellular buildup. This type of stenosis is ideal for angioplasty.

3. *Perivascular adhesions*. These result in vascular kinking. They can be approached by angioplasty, but likelihood of recurrence is high.

4. *Multiple stenosis* due to rejection. Such patients are not considered for angioplasty.

CHOICE OF PROCEDURE

Angioplasty can be performed as the only therapeutic procedure or in conjunction with surgery. In patients with focal disease and in some patients with generalized disease (who are poor candidates for surgery) angioplasty may be the only form of therapy required. In patients with tandem lesions or associated severe aortic disease, angioplasty may be used in conjunction with surgery.

Angioplasty may be the only therapeutic procedure for (1) focal stenoses of the iliac and common femoral arteries, (2) stenosis of the internal iliac arteries, (3) recent, unilateral iliac artery occlusion (after successful thrombolysis with low dose intra-arterial streptokinase), and occasionally for (4) diffuse atherosclerotic stenoses.

Adjunctive angioplasty is gaining widespread application. In one series reporting angioplasty on 293 patients, 31 procedures were performed in combination with surgery (18). In a more recent publication approximately 30 percent of procedures were performed in conjunction with surgery (19). Indications for adjunctive angioplasty are as follows:

1. Compromised inflow of iliac artery and outflow disease of superficial femoral artery. In such patients angioplasty of the iliac artery stenosis in combination with a distal bypass graft obviates a second operation (Figure 1).

2. Unilateral iliac artery occlusion with stenosis of the contralateral iliac artery. Surgical management of such patients who are a good operative risk includes an aortoiliac or aortobifemoral bypass graft. In some patients a femorofemoral bypass graft is performed. For the long-term patency of such grafts, severity of inflow disease is the single most important factor. In one series, donor iliac artery stenosis greater than 50 percent was the single most important factor responsible for failure of the femorofemoral crossover bypass

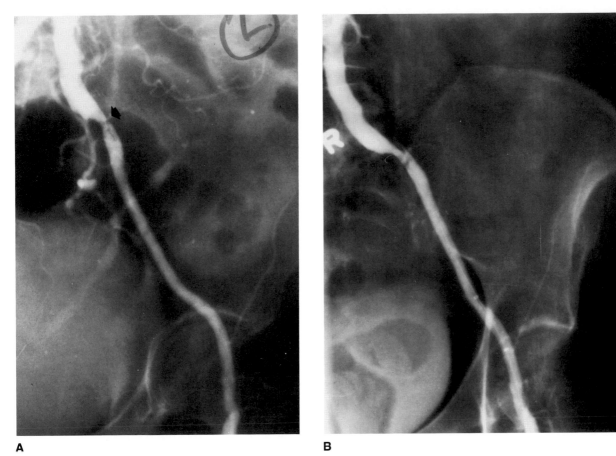

A   **B**

Figure 1. Dilatation of inflow stenosis prior to femoropopliteal bypass graft. **A.** Pelvic arteriogram shows stenosis of the proximal left external iliac artery (arrow). **B.** Arteriogram following angioplasty from the ipsilateral femoral approach. (Reproduced from Kadir S, et al: *Selected Techniques in Interventional Radiology.* Philadelphia, Saunders, 1982.)

graft (20). In such patients angioplasty of the donor vessel may be able to prevent graft failure (Figure 2).

3. Postoperative stenosis. In patients with anastomotic stenosis after iliac or femoropopliteal bypass grafts, angioplasty offers an alternative to reoperation for treatment of hemodynamically significant stenosis (Figure 3) (21).

4. Hemodynamically failed grafts. In patients with hemodynamically failed femorofemoral bypass grafts, angioplasty of the donor vessel may restore normal graft function.

5. In patients with graft occlusion secondary to inflow obstruction, angioplasty of such lesions can be followed by graft thrombectomy (Figure 4).

6. Transplant renal artery stenosis.

PREANGIOPLASTY PATIENT EVALUATION

As part of the evaluation of the patient prior to percutaneous transluminal angioplasty there is a review of the patient's symptoms and evaluation of degree of compromise caused by them. There must also be consultation and consensus with vascular surgeons regarding the indications for the procedure. A recent arteriogram (not older than 2 weeks) is reviewed, as are hemodynamic data (AAI, intra-arterial pressures or flow measurements), if available.

Informed consent is obtained at least 12 hours before (the evening before) the anticipated procedure. The consent should be obtained by the radiologist who will perform the angioplasty. The

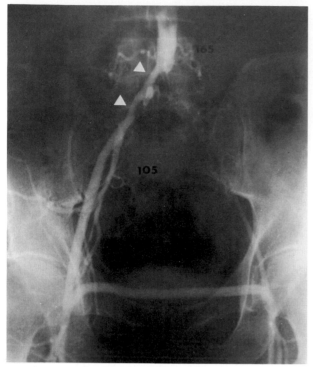

Figure 2. Angioplasty of inflow stenosis causing hemo-dynamic graft failure. Pelvic arteriogram shows left common iliac artery occlusion. There are two areas of stenosis in the right common iliac artery (arrowheads). Although the femoral bypass graft is patent, the patient had experienced a recurrence of symptoms. Before angioplasty, there was an intra-arterial resting pressure gradient of 60 mmHg. Subsequent to angioplasty from the right femoral artery approach, the pressure gradient was eliminated and normal graft functioning restored.

planned procedure, benefits, alternatives, and potential risks should be explained. Subsequently, a limited physical examination of the pertinent areas is performed. The physical examination should include:

1. Evaluation of the subcutaneous tissues at the anticipated puncture site for scar, infection, or hematoma (from previous angiogram). In the presence of a large puncture site hematoma, an alternative route may become necessary or the procedure postponed until the hematoma resolves.
2. Evaluation of the pulses (including distal pulses) and comparison of both extremities.
3. Evaluation of the distal circulation with reference to circulatory changes, e.g., skin temperature, gangrene, infection, etc.

In addition, the following laboratory data and parameters are reviewed:

4. Bleeding parameters (prothrombin time, partial prothrombin time, platelet count).
5. Hematocrit.
6. Renal function (BUN, creatinine).
7. A preangioplasty AAI is obtained.
8. Allergies.

In preparation for the angioplasty, only clear liquids are permitted the patient after midnight. If the procedure is planned for the afternoon, the patient is allowed a liquid breakfast. Patients with renal function impairment, diabetics, or others prone to contrast-induced renal function impairment are hydrated overnight by intravenous infusions. All oral hypoglycemic agents or insulin are withheld. Transient hyperglycemia is better tolerated and easier to manage than hypoglycemia.

Medications are given approximately 30 minutes before the procedure. Demerol (75-100 mg, IM) and Nembutal (100 mg, IM) are given (barring patient allergy to the drug). Nembutal should not be used for patients with obstructive lung disease and renal failure. Intramuscular atropine may or may not be useful for preventing vasovagal reactions. Side effects (dryness of the mouth, urinary retention in males) are common and its effectiveness 2–3 hours after intramuscular injection is questionable. Similarly, intramuscular injections of diazepam (Valium) should not be used because the absorption rate is unpredictable. During the procedure, a cocktail of intravenous diazepam 2.5 mg (Valium) and 12.5 mg Demerol is used as the need arises.

Premedication with platelet aggregation inhibitors, which was reported by some authors and was transiently used at our institution, is not recommended as a routine. Diminished pulses, previous femoral artery punctures for diagnostic arteriography, and multiple attempts at femoral artery puncture for angioplasty predispose to the formation of large puncture site hematomas. In addition, intramural hematomas have been observed at the site of iliac angioplasty that may become large enough to compromise the arterial lumen, and occasionally require operation.

An IV line is started on the patient's floor, approximately 2 hours before angioplasty unless overnight hydration is required.

All patients with a renal transplant or a single kidney should have an indwelling bladder catheter. The catheter is placed approximately ½ hour before angioplasty. This serves to detect complications which may lead to arterial occlusion, thereby resulting in diminished renal output.

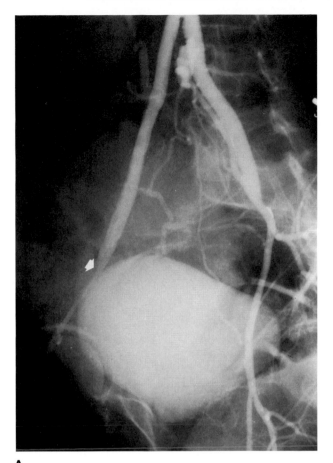
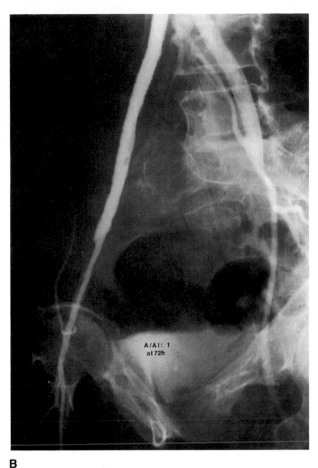

A

B

Figure 3. Angioplasty of anastomotic stenosis following aortoiliac bypass graft. **A.** Right posterior oblique pelvic arteriogram demonstrates a stenosis at the distal anastomosis of the aortoiliac bypass graft (arrow). Resting pressure gradient across this lesion was 35 mmHg. **B.** After angioplasty from the right femoral approach, the gradient was eliminated. (From Kadir S, et al: *Selected Techniques in Interventional Radiology*. Philadelphia, Saunders, 1982.)

All patients for transplant renal artery angioplasty should be observed in an intensive care facility for 24–28 hours afterward. The blood pressure, fluid intake, and output must be monitored.

POSTANGIOPLASTY CARE AND DRUG THERAPY

Bedrest

Following angioplasty, a 24-hour bedrest is prescribed. The extremity used for access is immobilized for at least 8 hours. The puncture site and peripheral pulses (if present) are monitored for hematoma formation, occlusion, or distal embolization: every 15 minutes for 1 hour, every ½ hour for 2 hours, and every hour for 4 hours.

After transaxillary angioplasty, the arm is immobilized in a sling and sensory and motor function is evaluated. The patient is encouraged to take oral fluids. Ambulation is permitted after 24 hours. If a hematoma develops after the patient has returned to the floor, renewed arterial compression is applied for 20–30 minutes and the period of bedrest is prolonged by 4–6 hours.

Drug Therapy

An intra-arterial bolus of 5000 U heparin is given immediately after the stenosis is traversed. After angioplasty, the patient receives intravenous heparin at full systemic doses (1000 U per hour) for at

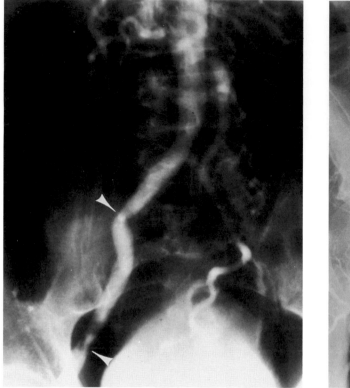

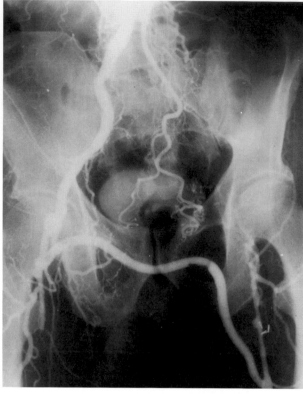

A   **B**

Figure 4. Occlusion of femoral-femoral bypass graft due to significant stenosis of donor iliac artery. **A.** Distal abdominal aortogram via left axillary approach demonstrates occlusion of the left common iliac artery. Two areas of stenosis are seen involving the right common and external iliac arteries (arrowheads). (Reproduced from Kadir S, et al: *Selected Techniques in Interventional Radiology.* Philadelphia, Saunders, 1982.) **B.** Following angioplasty of both lesions, the graft was thrombectomized leading to restoration of normal graft function.

least 24 hours via a constant infusion pump (I-MED). At the same time oral medication with platelet aggregation inhibitors is begun and is continued for at least 3 and preferably for 6 months. The dosages used are as follows: Persantine, 75 mg tid, and aspirin, 325 mg qd.

Follow-up

The patient can usually be discharged from the hospital on the second or third postangioplasty day. Following discharge, the patients are seen in the vascular clinic. The peripheral pulses and Doppler-derived ankle-arm index is obtained routinely. Recurrence of symptoms is evaluated by stress AAI and subsequently by intravenous digital subtraction angiography.

DEFINITION OF ARTERIAL ANATOMY PRIOR TO ANGIOPLASTY

The standard lower-extremity arteriogram is obtained in the AP projection. In this view, tortuous iliac arteries are often foreshortened and common iliac and common femoral artery bifurcations obscured by superimposition of blood vessels. In addition, significant arterial stenosis may be obscured (Figure 5). Complete evaluation of pelvic and common femoral arteries must include arteriograms in the oblique projection. The course of the iliac arteries and the internal iliac artery orifice is best defined by an ipsilateral posterior oblique (RPO for right iliac artery) (Figure 6). For definition of the common femoral artery bifurcation, the ipsilateral anterior oblique projection is used (Figure 7).

Exact definition of arterial anatomy prior to an-

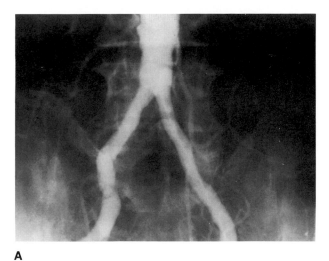
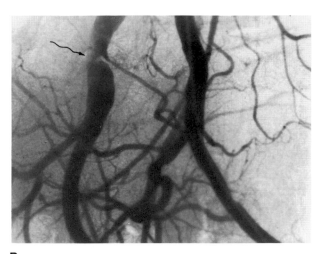

A   **B**

Figure 5.   Value of oblique arteriography for definition of arterial anatomy. **A.** AP pelvic arteriogram in patient with right buttock and thigh claudication does not demonstrate any obvious stenosis. **B.** Oblique pelvic arteriogram shows a hemodynamically significant right common iliac artery stenosis (arrow).

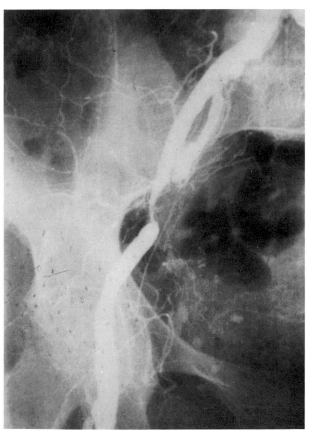
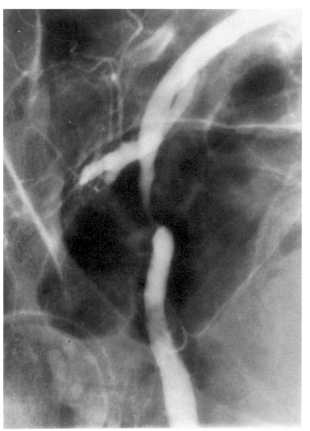

A   **B**

Figure 6.   Oblique pelvic arteriography for definition of arterial anatomy prior to angioplasty. **A.** AP pelvic arteriogram demonstrates a mid right external iliac artery stenosis. **B.** Left posterior oblique arteriogram does not define the arterial stenosis adequately.

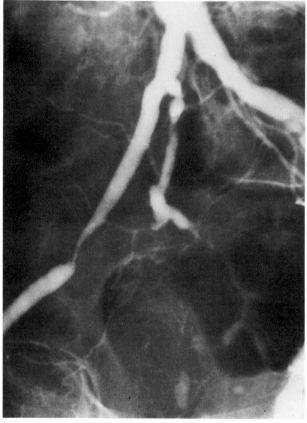

C

Figure 6 (*Continued*).   C. Right posterior oblique arteriogram lays out the external iliac artery stenosis. In addition, the orifice of the internal iliac artery is well defined. (Reproduced from Kadir S, et al: *Selected Techniques in Interventional Radiology*. Philadelphia, Saunders, 1982.)

gioplasty, i.e., the course of the vessels and location of the stenosis, is essential. This not only helps to determine the approach but is essential for determination of balloon size and length.

Once the course of the vessel has been defined, this projection (on the C arm) or placing the patient in that particular obliquity facilitates catheterization. In addition, in a patient with severe stenoses or tortuous iliac arteries a single curve or cobra catheter can be placed from the opposite side, in the iliac artery proximal to the stenosis. At the time of guidewire and catheter manipulation, small contrast injections through this catheter can be used as an aid in identifying the stenotic lumen. Alternatively, the road-mapping technique, if available, may be used.

ANGIOPLASTY TECHNIQUE (TABLE 1)

Selection of Dilation Catheters

The diameter and length of the balloon to be used for angioplasty is determined by measurement of the arterial lumen proximal and distal to the stenosis. If poststenotic dilatation is present, only the proximal measurement is used. In the presence of aneurysmal disease, measurement of a dilatated segment cannot be used for choosing a balloon. For iliac and common femoral artery angioplasty, polyethylene balloons between 6 and 8 mm in diameter are used. The preferred balloon length is 3 cm. Tapered Teflon catheters (Staple–van Andel) are occasionally used in patients with severe stenoses or diffuse disease. These are used only to predilate, prior to balloon angioplasty.

Guidewires

The guidewires used most commonly for traversing the iliofemoral stenoses are:

0.035 in. straight LLT
0.035 in. 7.5 mm J movable core
0.035 in. 1.5 mm J LLT
0.035 in. Kadir double-flex wire (straight LT taper on the one end and a 1.5 mm J on the opposite end; Cook, Inc., Bloomington, IN)
0.035 in. 1.5 mm J heavy duty exchange wire (200-260 cm)

For tortuous iliac arteries, the double-curve wire (Kadir-Sloan double curve wire, Cook, Inc., Bloomington, IN) may be used (Figure 8).

Accessory Materials

For balloon inflation and deflation, the ideal syringe is a 20-ml disposable plastic syringe. The pressure generated by this syringe is 7.1 atmospheres and the suction 0.95 atmospheres. The pressure generated by smaller syringes is significantly higher and may contribute to overinflation and rupture of angioplasty balloons. As an aid and to monitor inflation pressure, mechanical gauges can be used.

A coaxial "Y" adapter and an 0.021 inch J guidewire is always used for stenting the catheter during angioplasty (Figure 9). This serves a dual purpose: First, during angioplasty of tortuous vessels, the

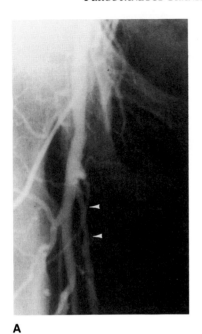

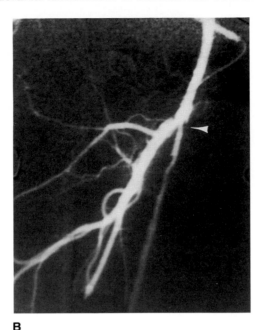

Figure 7. Definition of common femoral artery bifurcation anatomy. **A.** AP pelvic arteriogram shows minimal atherosclerotic narrowing of the proximal right superficial femoral artery (arrowheads). **B.** Intra-arterial digital subtraction angiogram in the right posterior oblique projection demonstrates severe orifice stenosis of the right superficial femoral artery (arrowhead).

A   B

stenting guidewire prevents the catheter tip from burrowing under a plaque or into the arterial wall. Second, with the wire in place, intra-arterial pressure gradients can be evaluated. The wire obviates the repeat manipulation used to get past the arterial segment subjected to angioplasty.

PTA OF UNILATERAL COMMON ILIAC ARTERY STENOSIS

This lesion is best reached from the ipsilateral femoral artery, with retrograde technique (Figure 10). After femoral artery puncture, a Teflon sheath is inserted in the distal external iliac artery and the intra-arterial pressure is measured. Subsequently, a guidewire is inserted and an attempt made to traverse the stenosis. In iliac arteries with a straight course a straight LLT wire may be tried. In vessels that are tortuous a 1.5 mm J wire or a 7.5 mm J wire is used. In severely tortuous vessels the double-curve wire is used.

Once the guidewire is past the stenosis, the balloon dilation catheter is inserted and placed in the abdominal aorta. Intra-arterial heparin is given and aortic pressure measured to derive a pressure gradient across the stenosis. A stenting guidewire (0.021 inch J) is inserted past the catheter tip for a distance of approximately 10–15 cm. A coaxial sidearm "Y"

adapter is used to enable constant pressure monitoring and to permit catheter flushing. The balloon is positioned in the area or areas of stenosis and at each location it is inflated and deflated 2 or 3 times. Subsequently, the catheter is advanced into the aorta for measurement of a pullback pressure.

Following angioplasty, the catheter is readvanced into the aorta for arteriography. The straight angioplasty catheter can usually be used for arteriography if appropriately positioned (catheter tip not against aortic wall) and a flow rate of 7 ml per second is not exceeded. A higher flow rate is not recommended because the end hole jet from the catheter may lead to complications such as subintimal contrast injection or plaque dislodgement. In aortas with severe atherosclerosis, a pigtail catheter should be used for arteriography. The use of intra-arterial DSA has provided diagnostic postprocedure angiograms without the need for exchange to a pigtail catheter.

PTA OF BILATERAL COMMON ILIAC ARTERY STENOSIS

A single and a double balloon technique has been described for angioplasty of bilateral proximal common iliac artery stenoses (Figure 11). In our experience consecutive dilations of both iliac arteries provide similar results. The atherosclerotic plaque

TABLE 1 PTA of the Iliac and Common Femoral Arteries: Approaches, Guidewires, and Catheters

Artery	Entry Catheter	Length cm	Guidewires	PTA Catheter Type	PTA Catheter Length cm	Balloon* Diam. mm	Balloon* Length cm
Common iliac							
Ipsilateral approach	6–7 Fr. straight	60	0.035 in.	7–8 Fr.	60	6–10	2–3
Contralateral approach	5–6 Fr. Cobra, Cobra in loop configuration, single curve, Sidewinder I	60	Straight LLT, 7.5 mm J movable, 1.5 mm J LLT, Double curve wire	7–8 Fr.	60	6–10	2–3
Axillary approach	5 or 6.5 Fr. Head-hunter	100	(260 cm J exchange wire)	7 Fr.	110	6–8	2–3
External iliac							
Ipsilateral approach	6–7 Fr. straight	60	0.035 in.	7 Fr.	60	6–8	2–3
Contralateral approach	5–6 Fr. Cobra, Cobra in loop configuration, single curve, Sidewinder I	60	Straight LLT, 7.5 mm J movable, 1.5 mm J LLT	7 Fr.	60	6–8	2–3
Internal iliac							
Contralateral approach	5–6 Fr. Cobra, Cobra in loop configuration, single curve, Sidewinder I	60	0.035 in. Straight LLT, 7.5 mm J movable, 1.5 mm J LLT	7 Fr.	60	6–8	1.5–2
Axillary approach	5–6.5 Fr. Headhunter	100	(250 cm J exchange wire)	5–7 Fr.	110	6–8	1.5–2
Common femoral							
Contralateral approach	5–6 Fr. Cobra, Cobra in loop configuration, single curve, Sidewinder I	60	0.035 in. J Straight LLT, 7.5 mm J movable, 1.5 mm J LLT	7 Fr.	60	6–8	2–3
Axillary approach	5–6.5 Fr. Headhunter	100	(250 cm J exchange wire)	5–7 Fr.	110	6–8	2–3
Renal transplant: Internal iliac anastomosis							
Contralateral femoral approach	5–6 Fr. Cobra, Cobra in loop configuration, single curve, Sidewinder I	60	0.035 in. Straight LLT, 7.5 mm J movable, 1.5 mm J LLT	5–7 Fr.	60	4–6	1.5–2
Axillary approach	5–6.5 Fr. Headhunter	100	(250 cm J exchange wire)	5–7 Fr.	110	4–6	1.5–2

TABLE 1 PTA of the Iliac and Common Femoral Arteries: Approaches, Guidewires, and Catheters (Continued)

Artery	Entry Catheter	Length cm	Guidewires	PTA Catheter Type	PTA Catheter Length cm	Balloon* Diam. mm	Balloon* Length cm
Renal transplant: External iliac anastomosis							
Ipsilateral femoral approach	5–6 Fr. Hockey stick shape (multipurpose catheter)	60	0.035 in. Straight LLT, 7.5 mm J movable, 1.5 mm J LLT	5–7 Fr.	60	4–6	1.5–2
Contralateral femoral approach	5–6 Fr. Cobra, Cobra in loop configuration, single curve, Sidewinder I	60	0.035 in. Straight LLT, 7.5 mm J movable	5–7 Fr.	60	4–6	1.5–2

*Selection of balloon diameter should be based on actual measurements of the arterial lumen diameter.

is usually hard and cannot be displaced toward the opposite common iliac artery during balloon angioplasty of one iliac artery. Indeed, the use of the single balloon technique may be advantageous as only half the aortic bifurcation is stressed at a given time.

Both femoral arteries are punctured, short arterial sheaths are placed, and intra-arterial pressures are recorded. First one stenosis is traversed and the balloon catheter is inserted. Again, the intra-arterial pressure is recorded. The area of stenosis is dilated after which the angioplasty catheter is readvanced into the aorta for measurement of a pullback pressure. Next the catheter is placed in the distal aorta. Subsequently, the opposite iliac artery stenosis is traversed and intra-arterial heparin is administered. After angioplasty, a pullback pressure gradient is measured and an arteriogram performed with the catheter in the distal abdominal aorta.

After angioplasty of one iliac artery, if there is difficulty in localizing the stenotic lumen in the opposite vessel, contrast medium is injected through the catheter in the aorta to define the anatomy in order to facilitate guidewire manipulation. (See earlier section on definition of arterial anatomy.)

PTA OF DISTAL COMMON ILIAC OR EXTERNAL ILIAC ARTERY STENOSIS

Such lesions can be approached by either the retrograde or antegrade (crossover) techniques (Figure 12). For lesions of the distal common iliac or proximal external iliac artery, the retrograde technique is recommended. If the lesion is located distally, the antegrade approach is more feasible.

In the antegrade technique, the opposite common femoral artery is punctured and a single curve catheter is placed in the ipsilateral common iliac artery. Intra-arterial pressure is measured and, subsequently, a guidewire (straight LLT or 7.5 mm J movable core) is inserted past the stenosis. Once the guidewire is past the stenosis, a balloon catheter is inserted. Following removal of the guidewire, intra-arterial heparin is administered and the distal pressure is measured. Again, a stenting guidewire (0.021

TABLE 2 PTA of Iliac and Common Femoral Arteries: Potential Complications
Arterial dissection
Arterial thrombosis
Arteriovenous fistula
Bleeding
Distal embolization
False aneurysm
Intramural hematoma
Puncture site hematoma
Systemic complications (ATN, contrast allergy)

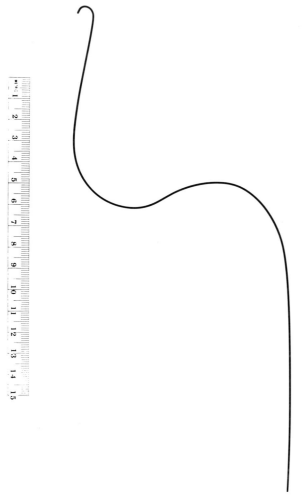

Figure 8. Kadir-Sloan double curve wire. (From S Kadir, Cath Cardiovasc Diagn 9:625–627, 1983.)

Figure 9. Coaxial "Y" adapter (A). A coaxial guidewire (G) is used for stenting the catheter during angioplasty. The sidearm is used to monitor pressure (P) through a three-way stopcock.

inch J) is placed approximately 5-10 cm distal to the catheter tip and balloon angioplasty is performed. A sidearm coaxial adapter is used for constant pressure measurement and to enable catheter flushing.

Upon completion of the study, a pullback gradient is obtained. The catheter is withdrawn to at least 5 cm proximal to the lesion and an arteriogram is obtained to evaluate success.

As an alternative to the single-curve catheter, a cobra catheter, cobra in the loop configuration, or a sidewinder catheter may be used to catheterize the common iliac artery proximal to a stenosis (22). A tip-deflecting catheter/guidewire system has been used by some authors. Because a soft tip J guidewire to lead the catheter is lacking and contrast injection at the time of catheterization is not possible, this

technique is not recommended in patients with atherosclerotic disease of the distal aorta and common iliac arteries.

The antegrade technique may not be feasible in patients with acute angled aortic bifurcation. In such patients, there is a tendency for guidewire and catheter to buckle and form a loop in the aorta rather than advance through the stenosis. This occurs more commonly with the stiffer polyethylene catheter. If this occurs, the stenosis is predilated as follows:

A curve is placed on a 6.3 or 7 F tapered Staple–van Andel Teflon catheter (Figure 13). The proximal portion of the 65 cm catheter is held between the index finger and the thumb of each hand. While applying pressure with the thumb, the hand is moved towards the catheter tip. With this maneuver, a permanent curve can be placed on the catheter which conforms more readily to the angle of the aortic bifurcation. The stiffness of the catheter together with the curvature permits advancement of the catheter through the stenosis without buckling in the aorta. The tapered tip of the catheter advances more readily through the stenosis and at the same time the lesion is predilated, which facilitates passage of the balloon catheter.

Once the Teflon catheter is past the stenosis, the guidewire is removed and heparin is injected. A bend is placed on a heavy duty exchange guidewire approximately 25-30 cm proximal to the wire tip. The wire is inserted through the catheter and is placed with the bend at the aortic bifurcation. The Teflon catheter is removed and the balloon dilation catheter is inserted for angioplasty.

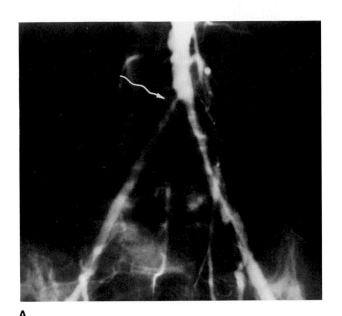

A

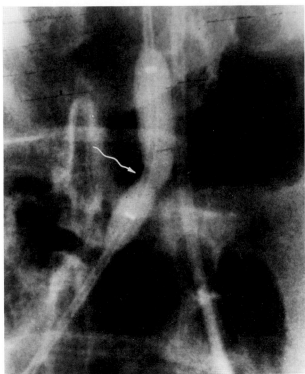

B

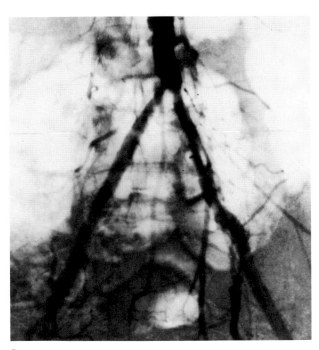

C

Figure 10. Angioplasty of the unilateral common iliac artery stenosis. **A.** Pelvic arteriogram demonstrates severe stenosis of the proximal right common iliac artery (arrow). **B.** Spot films show constriction of the balloon by the hard atherosclerotic stenosis (arrow). Note overdistention of the nonrestricted portion of the balloon in the aorta. **C.** Subtraction filming from the arteriogram following angioplasty shows relief of the stenosis.

PTA OF COMMON FEMORAL ARTERY STENOSIS

Common femoral artery stenosis is best approached from the opposite femoral artery or from the axillary

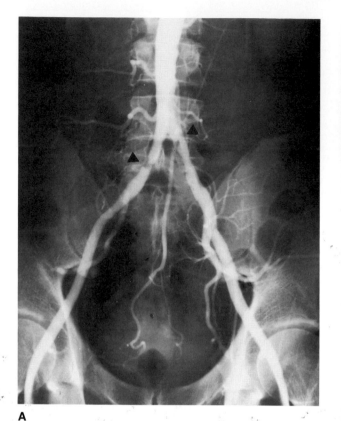

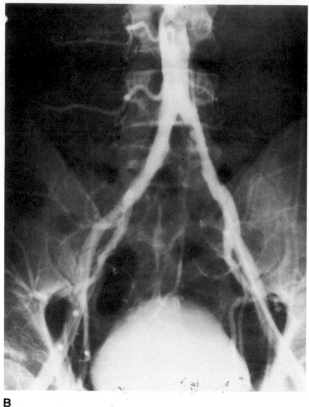

A

B

Figure 11. Angioplasty of bilateral common iliac artery stenosis. **A.** Pelvic arteriogram demonstrates bilateral common iliac artery stenoses (arrowheads). Bilateral retrograde femoral artery punctures were used to dilate both of the stenoses. **B.** Pelvic arteriogram following successful angioplasty.

artery (Figure 14). However, it must be kept in mind that traversing a severely stenotic lesion may not be easy from these approaches due to inability to advance the catheter past the stenosis. The major problem is buckling of the catheter in the aorta. As an alternative to angioplasty, focal stenoses of the common femoral artery can be treated by endarterectomy under local anesthesia.

From the contralateral femoral approach a cobra catheter is placed across the aortic bifurcation. A J guidewire is inserted and advanced to the stenosis. The catheter is then advanced to approximately 5 cm of the stenosis. The patient or the U-arm is placed in that particular oblique position that showed the stenosis best. After measurement of proximal arterial pressures, an attempt is made to traverse the stenosis. For vessels with a tortuous course a tight J (1.5 mm J) or a 7.5 mm movable core J are used. In vessels with a straight course a straight LLT or a 7.5 mm movable core wire is used. Once the lesion has been traversed, the catheter is advanced over

the wire. Following removal of the wire, intra-arterial heparin is administered and distal pressure is recorded. Subsequently, the angioplasty catheter is inserted over a heavy-duty exchange wire.

Problems may arise during guidewire manipulation (to traverse stenosis) or catheter exchange. For the former, definition of the lesion by oblique arteriography has been extremely helpful. Problems associated with catheter exchanges are managed as discussed above (see angioplasty of distal common and external iliac arteries).

PTA OF INTERNAL ILIAC ARTERY STENOSIS

The internal iliac arteries can be approached either from a contralateral femoral artery puncture or via the left axillary artery (Figure 15). From the contralateral femoral artery, the technique is similar to

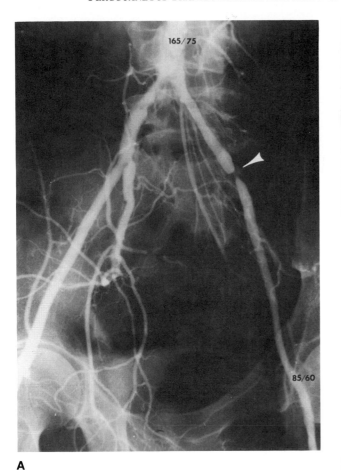

A

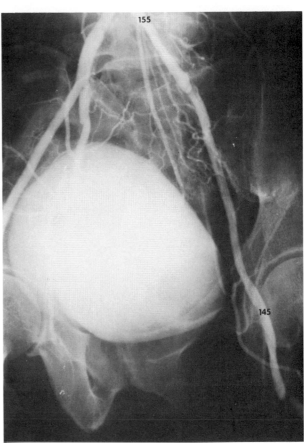

B

Figure 12.   Angioplasty of distal common iliac artery stenosis using the ipsilateral retrograde femoral artery technique. **A.** Oblique pelvic arteriogram shows severe stenosis of the distal common iliac artery (arrowhead). There was a resting pressure gradient of 80 mmHg. **B.** Following angioplasty, the resting pressure gradient was reduced to 10 mmHg.

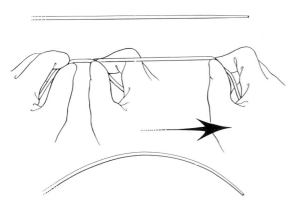

Figure 13.   Technique for placing a curve on a tapered Teflon catheter. See text for discussion. (Reproduced with permission of Georg Thieme Verlag, Stuttgart.)

that used for a common or external iliac artery angioplasty. A cobra catheter is placed across the aortic bifurcation and arterial pressure is measured. A J guidewire is inserted and the catheter is advanced into the external iliac artery. The patient is placed in the ipsilateral posterior oblique (e.g., RPO for right internal iliac artery). The internal iliac artery orifice is localized by small injections of contrast medium as the catheter is slowly withdrawn. Once the internal iliac artery orifice has been found, a straight LLT or a 7.5 mm movable core J guidewire is used to traverse the stenosis. After the guidewire is past the stenosis the catheter is advanced over it. Distal pressure is measured and intra-arterial heparin is administered. A heavy-duty J exchange wire is inserted and the cobra catheter is exchanged for the balloon dilation catheter. Again, a coaxial "Y"

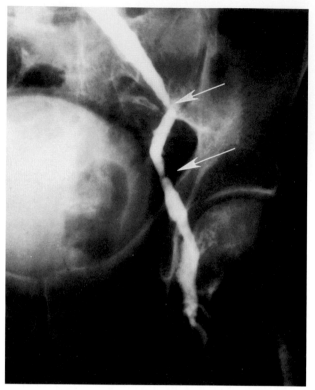

A

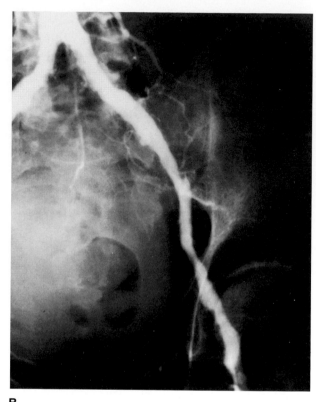

B

Figure 14. Angioplasty of stenoses of external iliac and common femoral arteries. **A.** Left iliac arteriogram demonstrates severe stenoses of the external iliac and common femoral arteries (arrows). **B.** Pelvic arteriogram following angioplasty.

adapter is used for placement of a 0.021 inch J stenting guidewire. Following successful angioplasty, a pullback pressure is recorded to evaluate for residual gradient. Subsequently, the catheter is placed in the common iliac artery and an oblique pelvic arteriogram is obtained.

From the left axillary approach a 100 cm long, 5 or 6.5 F HlH catheter is used to catheterize the internal iliac artery.

PTA OF TRANSPLANT RENAL ARTERY STENOSIS

Transplant Renal Artery–External Iliac Artery Anastomosis

The approach to an anastomotic stenosis involving a transplant renal artery and external iliac artery is determined by the angle of origin of the anastomosed vessel and its location (Figures 16, 17). For transplant renal arteries anastomosed to the distal external iliac artery with a perpendicular (to the external iliac) or inferiorly directed course, the contralateral approach is used. For transplant renal arteries showing a craniad course the ipsilateral femoral artery approach is recommended.

Ipsilateral Femoral Artery Approach

From the ipsilateral approach, a multipurpose catheter (hockey-stick shape, 45° angle of the distal limb) is inserted. A catheter is used to find the orifice of the transplant renal artery. A straight LLT or 7.5 mm J movable core guidewire is used to traverse the stenosis. Subsequently the catheter is advanced past the stenosis. Intra-arterial heparin is administered and the balloon catheter is inserted over a 1.5 mm J heavy duty exchange wire. Subsequent to balloon angioplasty, an external iliac arteriogram is performed to evaluate success. For tight stenoses, predilation with a tapered Teflon catheter is recommended. Such a catheter can be shaped like a multipurpose catheter and can also be used as the initial catheter to localize the transplant renal artery orifice.

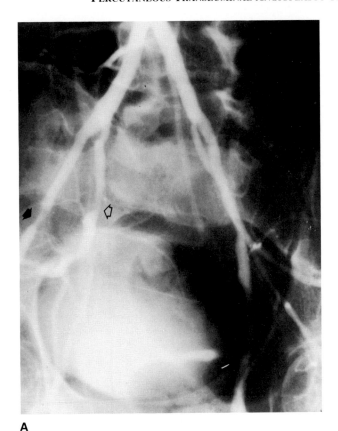

A

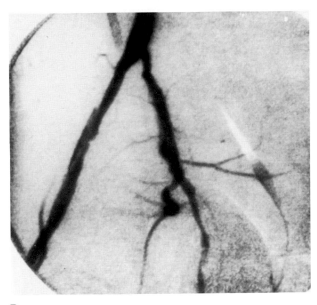

B

Figure 15. Angioplasty of internal iliac artery stenosis. **A.** Right posterior oblique pelvic arteriogram demonstrates a severe stenosis of the distal right internal iliac artery (open arrow). In addition, there is a severe stenosis of the right external iliac artery (arrow). **B.** Arterial digital subtraction angiogram following angioplasty of both external iliac and internal artery stenoses. Both lesions were dilated from the contralateral femoral approach.

Contralateral Femoral Artery Approach

From the contralateral approach, a cobra catheter is placed across the aortic bifurcation. A heavy-duty J exchange wire is inserted for placement of a headhunter catheter. With the patient placed in an ipsilateral posterior oblique position, the headhunter catheter is used to locate the orifice of the transplant renal artery. A straight LLT or 7.5 mm J movable core wire is used to negotiate the stenosis and the catheter is advanced past. Intra-arterial heparin is administered and again the heavy-duty J exchange wire is used for insertion of a balloon dilation catheter. A coaxial ''Y'' adapter is used for placement of a stenting 0.021 inch J guidewire. Following PTA, an iliac arteriogram is obtained.

PTA of Renal Transplant–Internal Iliac Artery Anastomosis

For transplant renal arteries anastomosed to the internal iliac artery either the contralateral femoral or the left axillary artery approach can be used.

Contralateral Femoral Artery Approach

This is similar to the technique used for internal iliac artery angioplasty. Following femoral artery puncture, a cobra catheter is placed across the aortic bifurcation. A J guidewire is inserted into the external iliac artery and the catheter is advanced over it. Occasionally the J guidewire enters the internal iliac artery. In this case, the catheter is advanced into the vessel. The patient is placed in the ipsilateral posterior oblique (e.g., RPO for right internal iliac artery–renal transplant artery anastomosis) or the obliquity that demonstrates the stenosis best. A tight J (1.5 mm J LLT) or a 7.5 mm movable J guidewire is used to traverse the stenosis. Straight wires are rarely necessary because the anastomosis most frequently lies in a curvature. Once the wire has traversed the stenosis, the cobra catheter is inserted over it. The wire is removed and intra-arterial heparin is administered. Subsequently a heavy-duty J exchange wire is used to exchange for a balloon angioplasty catheter and a coaxial ''Y'' adapter is used for placement of a stenting 0.021 inch J guide-

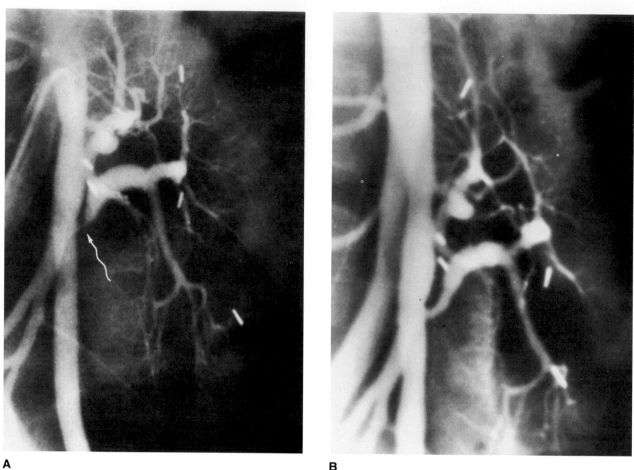

A   **B**

Figure 16. Angioplasty of the transplant renal artery stenosis. **A.** Steep left posterior oblique pelvic arteriogram demonstrates severe stenosis of the lower transplant renal artery (arrow). **B.** Arteriogram following angioplasty shows relief of the stenosis. (Reproduced from Kadir S, et al: *Selected Techniques in Interventional Radiology*. Philadelphia, Saunders, 1982.)

wire. Balloon angioplasty is performed followed by measurement of pullback pressure. The catheter is withdrawn into the proximal internal iliac artery for arteriography.

Left Axillary Artery Approach

From the left axillary artery, a 100 cm headhunter HlH catheter is used to catheterize the internal iliac artery. Again, a 1.5 mm J LLT or 7.5 mm movable core J guidewire is used to traverse the stenosis. The catheter is advanced over the wire, past the stenosis, and a heavy duty J exchange guidewire is used for placement of the balloon angioplasty catheter. Subsequent to the angioplasty, an internal iliac arteriogram is performed.

RESULTS

Iliofemoral PTA

An angioplasty is termed successful if there is elimination or significant reduction in the intra-arterial pressure gradient, symptomatic relief, and improvement in the skin perfusion or an increase in systolic thigh pressure and/or ankle arm index of greater than 0.15. In addition, some degree of cosmetic improvement, as determined by the postangioplasty arteriogram, is desirable.

At The Johns Hopkins Hospital, between 1978 and 1984, 250 patients underwent iliac and common femoral angioplasty. The initial success rate has been

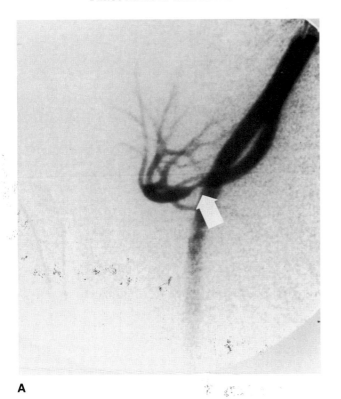

A

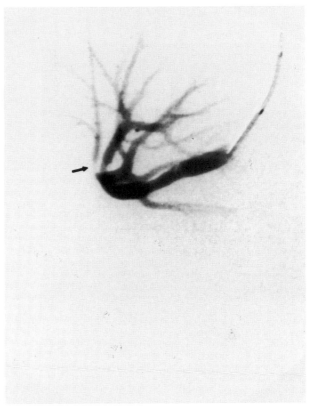

B

Figure 17. Angioplasty of transplant renal artery-internal iliac artery anastomosis stenosis. **A.** Right posterior oblique digital subtraction arteriogram shows severe, tapered stenosis at the anastomosis of the transplant renal artery with the internal iliac artery. **B.** Digital subtraction arteriogram following angioplasty. Segmental spasm of a branch renal artery is seen as a result of guidewire trauma (arrow).

maintained around 96 percent. Review of the first 141 iliac angioplasties revealed an accumulated 1-year patency rate of 91.3 percent. The 2- and 3-year accumulated patency rates were 89 percent (5). These compare with 89 percent and 90 percent patency rates for aortobifemoral bypass grafts (6,7).

Evaluation of our data suggests that a residual intra-arterial pressure gradient and the presence of outflow disease are two most significant factors adversely affecting long term patency.

Renal Transplant PTA

In angioplasty of renal transplant arteries, arterial pressure measurements are not as important as for iliofemoral artery angioplasty. Pressure recording distal to a stenosis with the catheter traversing the narrowed segment does not reflect the true gradient. However, whenever possible such pressures should be obtained, especially after the procedure, as these can aid in the evaluation of the success of angioplasty in obliteration of a hemodynamically significant gradient. In renal transplant artery angioplasty, an effort should be made to achieve a good cosmetic result.

Transplant renal artery angioplasty is considered technically successful if (1) there is no gradient across the stenosis after the procedure, or (2) the postoperative arteriogram shows an anatomically "perfect" or near-perfect result.

Long term success is evaluated by the following criteria (23):

Cured: Blood pressure returns to normal (<140/90 mmHg) off medication.
Improved: Diastolic pressure is decreased by at least 10 mmHg or the medication requirement is reduced by 50 percent.

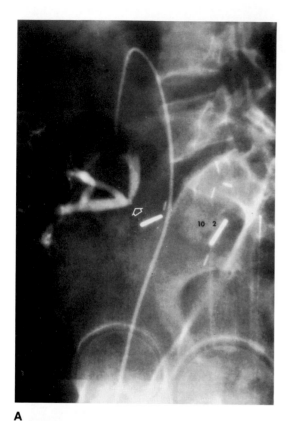

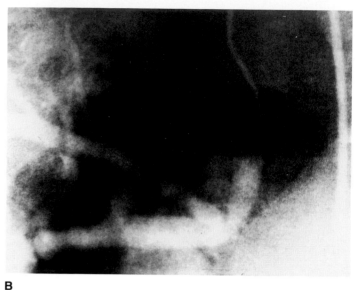

B

A

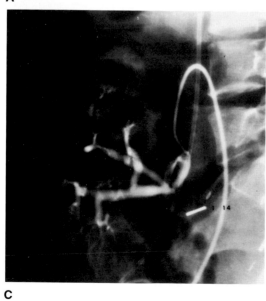

C

Figure 18. Transplant renal artery stenosis due to kinking. **A.** Steep right posterior oblique internal iliac arteriogram shows a kink at the internal iliac–transplant renal artery anastomosis (open arrow). **B.** Arteriogram following angioplasty shows some cosmetic improvement. **C.** Repeat arteriography prompted by recurrent symptoms demonstrates restenosis at the kink.

Unimproved: If the above changes were not observed.

Recurrence: Increase in medication requirement or elevation of brachial blood pressure on two consecutive clinical visits after an initial cure or improvement (Figure 18).

Favorable long term results have been reported after angioplasty of transplant renal artery stenosis (24,25).

COMPLICATIONS

Potential complications of iliofemoral and transplant renal angioplasty are listed in Table 2.

Puncture Site Hematoma and Bleeding

Most puncture site hematomas and bleeding are related to multiple attempts at arterial puncture, or may be due to an unusually high arterial puncture site (above inguinal ligament), faulty compression, or use of anticoagulants or platelet aggregation inhibitors. In our own experience, premedication with platelet aggregation inhibitors with subsequent heparinization is associated with a higher incidence of groin hematoma and postangioplasty bleeding. At the same time, the technical and short-term success is not different in patients in whom platelet aggregation inhibitory medication was not used. Thus, the use of platelet aggregation inhibitors as a premedication is no longer recommended. In addition, patients receiving such medication may be predisposed to intramural hematomas at the site of angioplasty (Figure 19). Manual compression of the femoral artery puncture site should extend over a duration of at least 20–30 minutes. The use of compression bandages or mechanical compression devices is not recommended.

Arterial Occlusion

Subintimal passage of catheter or guidewire may lead to dissection, plaque embolization, or occlu-

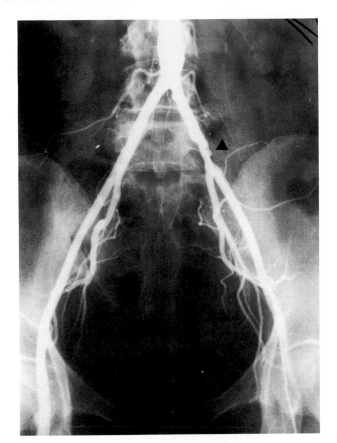

A

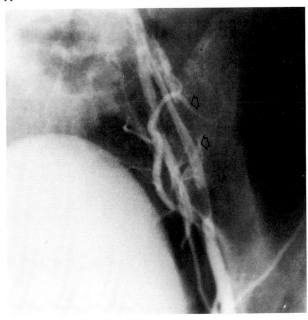

B

Figure 19. Intramural hematoma following iliac artery angioplasty. **A.** Preangioplasty pelvic arteriogram demonstrates stenosis of the left external iliac artery (arrowhead). **B.** Following angioplasty, an intramural hematoma is seen partially occluding the proximal external iliac artery lumen (open arrows). The femoral pulse remained diminished for approximately 24 hours. Subsequently the pulse returned to normal and the patient had an excellent result.

sion. Obliteration of the femoral pulse at the time
of arterial compression may also impede blood flow
significantly and result in thrombus formation and
arterial occlusion. Nonoccluding thrombi, on the
other hand, are occasionally seen at the site of angioplasty. These are most likely a consequence of
the intimal trauma sustained during balloon angioplasty. These observations underline the necessity
for systemic heparinization in the immediate postangioplasty period.

Dissection

This may be a consequence of the use of force while
attempting to traverse a stenosis or secondary to the
use of an inappropriate guidewire (e.g., straight wire
for tortuous vessels) (26). Attempts at probing a
stenosis should be gentle and force should never be
necessary. The only time when some degree of force
may become necessary is at the time of insertion of
the dilation catheter over a guidewire.

Arterial Rupture and False Aneurysm Formation

These complications occur after balloon rupture or
as a consequence of overdilatation (4,27). The most
common cause of balloon rupture is the use of smaller
syringes (smaller than 20 ml) which generate pressures that exceed the limits recommended by balloon manufacturers.

Distal Embolization

Distal embolization following angioplasty of iliac artery stenosis has been observed in 1 percent of patients (Figure 20) (15). Following angioplasty of occlusions, up to 25 percent of patients may experience
clinically significant distal embolization.

Systemic Complications

Systemic complications are very infrequent. In our
experience, the most common systemic complication of iliofemoral angioplasty is transient renal failure. The most likely explanation is increased contrast toxicity due to dehydration of the patients or
use of excessive volumes of contrast medium. The

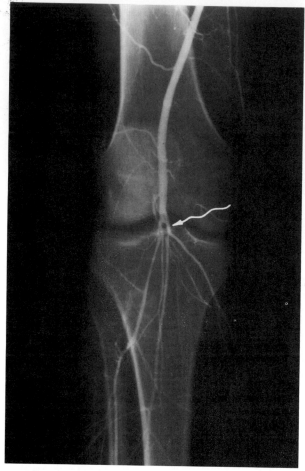

Figure 20.   Plaque embolization following iliac artery angioplasty. Femoral arteriogram demonstrates occlusion of
the popliteal artery at the knee joint space (arrow). Balloon embolectomy retrieved an atheromatous plaque. (Reproduced from Kadir S, et al: *Selected Techniques in
Interventional Radiology*. Philadelphia, Saunders, 1982.)

latter situation may arise if the diagnostic angiography and angioplasty are performed at the same
sitting.

THERAPY FOLLOWING PTA

The primary goal of angioplasty is the elimination
of hemodynamically significant obstructions. Systemic heparinization for 24–48 hours provides pro-

tection from thrombus formation at both the angioplasty and the arterial puncture site. Treatment with platelet aggregation inhibitors should also be initiated soon after angioplasty. Such treatment should be continued for at least 3 months, and preferably for 6 months, to permit the repair process at the angioplasty site to be completed. Oral anticoagulation with coumarin is not considered necessary, as there is sufficient clinical data to suggest that coumarin or its derivatives do not offer any added protection after angioplasty.

In addition to drug therapy, graduated exercise programs (walking, cycling, etc.) to increase the blood flow through the arterial segments subjected to angioplasty should be instituted. Risk factors (cigarette smoking, treatment of diabetes mellitus, hypertension, and hypercholesterolemia) should also be dealt with appropriately.

CONCLUSIONS

With appropriate patient selection and meticulous technique, iliofemoral artery angioplasty provides long-term patency rates comparable to those achieved by aortobifemoral bypass grafts. Thus, in patients with focal, hemodynamically significant obstructive lesion, angioplasty is the treatment of choice. In addition, this technique is applicable to some patients with diffuse disease or in conjunction with operation.

Angioplasty may also be indicated in some patients with transplant renal artery stenosis. Our own experience suggests that an anastomotic stenosis due to intimal fibrous build-up can be managed successfully by angioplasty. Anastomotic stenoses due to technical problems, such as tight suture line, are best managed by reoperation.

REFERENCES

1. Dotter CT, Judkins MP: Transluminal treatment of atherosclerotic obstruction. Description of a new technic and a preliminary report of its application. Circulation 30:654–679, 1964.
2. Zeitler E: Percutaneous dilatation and recanalization of iliac and femoral arteries. J Cardiovasc Intervent Radiol 3:207–212, 1980.
3. Spence RK, Freiman DB, Gatenby R, et al: Long term results of transluminal angioplasty of the iliac and femoral arteries. Arch Surg 116: 1377–1386, 1981.
4. Waltman AC: Percutaneous transluminal angioplasty of iliac and deep femoral arteries. AJR 125:921–925, 1980.
5. Kadir S, White RI Jr, Kaufman SL, et al: Long term results of aortoiliac angioplasty. Surgery 94:10–14, 1983.
6. Malone JW, Moore WS, Goldstein J: The natural history of bilateral aortofemoral bypass grafts for ischemia of the lower extremities. Arch Surg 110:1300–1306, 1975.
7. Brewster DC, Darling RC: Optimal methods of aortoiliac reconstruction. Surgery 84:739–748, 1972.
8. Ring EJ, Freiman DB, McLean GK, et al: Percutaneous recanalization of common iliac artery occlusions: An unacceptable complication rate? Am J Roentgen 139:587–589, 1982.
9. Leu JH, Gruentzig A: Histopathologic aspects of transluminal recanalization, in E Zeitler, A Gruentzig, W Schoop (eds): *Percutaneous Vascular Recanalization*, Heidelberg, Springer-Verlag, 1978, pp 39–50.
10. Wylie EJ, Brinkley FM, Palubinskas AJ: Extrarenal fibromuscular hyperplasia. Am J Surg 112:149–155, 1966.
11. Stanley JC, Gewertz BL, Bove EL, et al: Arterial fibroplasia: Histopathologic character and current etiologic concepts. Arch Surg 110:561–566, 1975.
12. Thompson WM, Johnsrude IS, Jackson DC, et al: Late complications of abdominal aortic reconstructive surgery. Ann Surg 185:326–334, 1977.
13. Doyle TJ, McGregor WR, Fox PS, et al: Homotransplant renal artery stenosis. Surgery 77:53–60, 1975.
14. Mitchell S, Kadir S, Kaufman SL, et al: Percutaneous transluminal angioplasty of aortic graft stenoses. Radiology 149:439–444, 1983.
15. Kadir S, White RI Jr, Kaufman SL, et al: Prevention of thromboembolic complications of angioplasty by critical patient selection. 82nd Annual Meeting, Am Roentgen Ray Soc, New Orleans, May 10–14, 1982.
16. Kempczinski RF: Lower extremity arterial emboli from ulcerating atherosclerotic plaques. JAMA 241:807–810, 1979.
17. Rosenberg MW, Shah DM: Bilateral blue toe syndrome. A case report. JAMA 243, 365–366, 1980.
18. Brunner U, Gruentzig A: Das Dilatationsverfahren zur perkutanen Rekannalization chronischer arterieller verschlusse in Gefasschirurgischer Sicht. VASA 4:334–337, 1975.
19. Kadir S, Smith GW, et al: Percutaneous transluminal angioplasty as an adjunct to vascular surgical reconstruction. Ann Surg 195:786–795, 1982.
20. Flanigan DP, Pratt DG, Goodreau FJ: Hemodynamic and angiographic guidelines in selection of patients for femoro-femoral bypass. Arch Surg 113:1257–1262, 1978.
21. Zajko AB, McClean GK, Freiman DB, et al: Percutaneous puncture of venous bypass grafts for transluminal angioplasty. Am J Roentgen 137:799–802. 1981.
22. Kadir S: Loop catheter technic. Med Radiogr Photogr (KODAK) 57:22–30, 1981.
23. Kadir S, Russell RP, Kaufman SL, et al: Renal artery angioplasty: Technical considerations and results Fortschr Röntgenstr 141:378–383, 1984.

24. Sniderman KW, Sos TA, Sprayregen S, et al: Percutaneous transluminal angioplasty in renal transplant arterial stenosis for relief of hypertension. Radiology 135:23–26, 1980.

25. Gerlock AJ, MacDonnell RC Jr, Smith CW, et al: Renal transplant arterial stenosis: Percutaneous transluminal angioplasty. Am J Roentgen 140:325–331, 1983.

26. Kadir S, Kaufman SL, Barth K, et al: *Selected Techniques in Interventional Radiology,* Philadelphia, Saunders, 1982, pp 142–207.

27. Simmonetti G, Rossi P, Passariello R, et al: Iliac artery rupture. A complication of transluminal angioplasty. Am J Roentgen 140:989–990. 1983.

5

Percutaneous Angioplasty of the Femoral, Popliteal, and Tibial Arteries

FREDERICK S. KELLER, M.D.

INTRODUCTION

Atherosclerotic obstructions of the lower-extremity vascular systems are a small part of a generalized systemic disease involving multiple, and frequently more critical, vascular beds. Only a small percentage of patients with intermittent claudication of the lower extremities will eventually require amputation, however, approximately 50 percent of those undergoing femoropopliteal grafting will be dead within 5 years and approximately 75 percent within 10 years, with most deaths due to other cardiovascular events (1,2). In January 1964, Dotter and Judkins introduced a new method of treating atherosclerotic obstructions of the lower extremities (3). Their first patient was an 84-year-old woman with severe rest pain and a tight, focal stenosis of the superficial femoral artery. She was considered an unacceptable candidate for femoropopliteal bypass surgery because of inadequate distal runoff. The procedure, called *percutaneous transluminal angioplasty* by Dotter, was successfully accomplished by carefully passing a guidewire through the stenosis and then advancing a number 8 French and a coaxial number 12 French catheter over the guidewire. The leg was saved; recognizing the potential of transluminal angioplasty, Dotter hailed it as a revolutionary new procedure. His enthusiasm, which met

with skepticism in the United States, was transmitted to several European radiologists, who began to employ the "Dotter" procedure (4). Wide acceptance of percutaneous transluminal angioplasty by the radiologic community came in the late 1970s after Gruentzig developed a double lumen balloon catheter on a flexible shaft (5).

INDICATIONS

The indications for transluminal angioplasty of the femoral and popliteal arteries should be nearly the same as those used to select patients for surgical revascularization. Limb-threatening ischemia can exhibit claudication, constant pain at rest, paresthesias, or ulceration and frank gangrene. Any of these symptoms mandates some form of therapy, either percutaneous transluminal angioplasty or surgical revascularization. The indications for intervention with symptoms of intermittent claudication are, however, somewhat less clear-cut. If the claudication is disabling in that it significantly interferes with the patient's everyday life, especially employment, most vascular surgeons and radiologists agree that intervention should be undertaken (6) (Fig. 1 A–D). If, on the other hand, intermittent claudica-

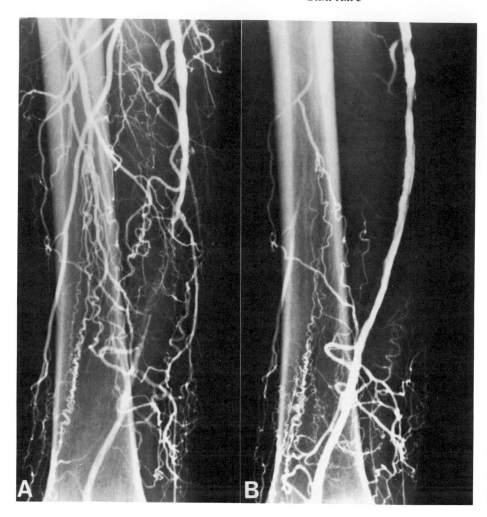

Figure 1. Radiographs of a 56-year-old mailman with bilateral intermittent claudication forcing him to consider early retirement. **A,B.** Control and follow-up right superficial femoral recanalization.

tion is not severe enough to interfere with everyday activity but merely restricts various recreational activities, there is disagreement as to whether intervention is indicated.

PREANGIOPLASTY EVALUATION

Prior to transluminal angioplasty, patients should have a complete history and physical examination, with special attention to vascular status. The degree of ischemia should be classified along with any objective signs of ischemic changes. Peripheral pulses should be palpated and recorded, as well as the presence of bruits. Routine hematologic screening including a coagulation profile and serum chemistries is obtained. Noninvasive vascular laboratory evaluation, which in our institution includes segmental systolic Doppler pressures at rest and after exercise

in both extremities compared to the systolic pressure in the brachial artery and directional Doppler flow analysis of the lower extremities, is performed (7). If history, physical examination, and noninvasive vascular laboratory evaluation demonstrate the likelihood of significant vascular disease requiring some form of intervention, diagnostic angiography is indicated.

Whether diagnostic angiography is performed by using the standard film-screen combination or by intravenous or intra-arterial digital angiography depends on the type of radiographic equipment available and the individual angiographer's preference. Each of these methods has advantages. Greatest spatial resolution and clearest images are obtained with standard film-screen angiography. Furthermore, fewer injections of contrast material and fewer exposures are needed since the field size is significantly greater. This standard film-screen angiogra-

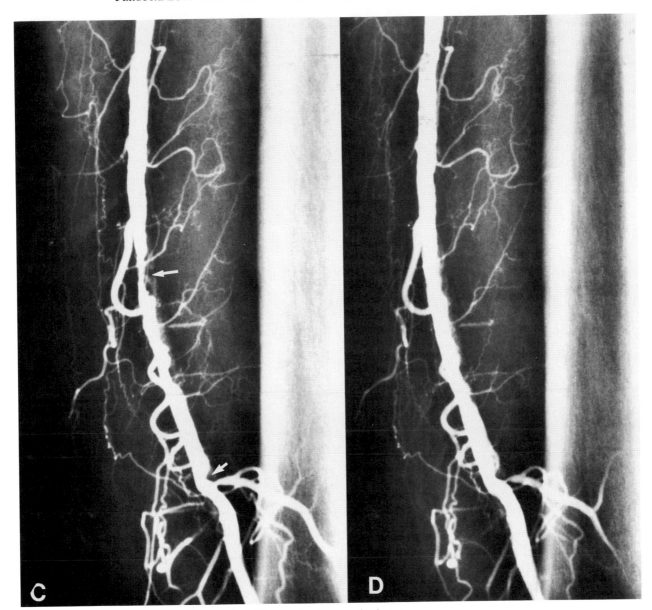

Figure 1 *(Continued).*   **C,D.** Left superficial femoral artery angioplasty.

phy, however, causes most patient discomfort since undiluted contrast is injected intra-arterially. (In our institution, 1 mL of 2% xylocaine is mixed with every 10 mL of contrast material prior to injection to decrease patient discomfort.) Intravenous digital angiography does not require arterial catheterization; therefore, the patient can usually leave the hospital 2 hours after this type of examination. Its drawbacks include the large volumes of contrast that must be used since multiple injections are required to cover the abdominal aorta and peripheral vessels because of the small field size. In addition, images generated are often suboptimal and frequently non-diagnostic. An advantage of intra-arterial digital angiography of the lower extremities is that it produces images superior to those obtained by intravenous digital angiography (but not as good as film-screen images). Furthermore, the contrast material can be diluted with saline or 5% dextrose to one-third or one-fourth its original concentration, thereby decreasing patient discomfort and the total amount (in grams of iodine) of contrast administered. Since the contrast material is diluted, smaller-size (number 4 French) catheters may be used, permitting outpatient examinations. There are considerable savings in film cost with both intravenous and intra-arterial digital angiography.

The procedure of percutaneous transluminal angioplasty, including benefits, risks, possible complications, and the rare need for emergency surgery, must be explained to the patient. When angioplasty is undertaken for limb salvage on lesions that are not well suited for it, the patient should be informed that it is a measure of last resort to avert an inevitable amputation and that chances of success are guarded.

If diagnostic angiography has not yet been done, the arterial approach is selected on the basis of the expected pathologic anatomy and the best possible route to approach the lesion for transluminal angioplasty. For obstructive disease of femoral, popliteal, and tibial arteries, the contralateral femoral approach is usually chosen for the diagnostic angiogram. This leaves the ipsilateral femoral artery available for an antegrade, ''downstream'' approach. Abdominal aortography is routinely performed on all patients with suspected lower-extremity vascular disease. After the abdominal aortogram, the catheter is withdrawn to a site just proximal to the aortic bifurcation, and a peripheral angiogram extending from the aortic bifurcation to the ankles is done. Oblique views of the pelvis are frequently required for optimal demonstration of the origins of the superficial and deep femoral arteries.

MORPHOLOGIC CRITERIA FOR ANGIOPLASTY

The decision whether the patient is a candidate for transluminal therapy (angioplasty or recanalization), surgery, both, or neither depends on the morphologic characteristics of the lesion or lesions demonstrated by the diagnostic angiogram. Percutaneous transluminal angioplasty involves decreasing the degree of arterial stenosis, whereas transluminal recanalization entails reestablishing the lumen in an occluded arterial segment.

Length of Lesion

The ideal lesion for transluminal angioplasty is a short (up to 2 cm), focal stenosis (Figs. 2A–D, 3A–B). The longer the lesion to be dilated or recanalized, the greater the rate of initial failure and complication (8). Even if initial success is obtained with long-segment occlusions of the superficial femoral

artery, overall long-term success is less frequent since recurrences are common. Occlusions 12 cm or less in length are amenable to successful recanalization (Fig. 4A–B); however, if the occlusion is longer than 12 cm and the patient is a good candidate for femoropopliteal bypass, surgery is the preferred therapy (9) (Fig. 5A–B).

Location of Lesion

Lesions located in the superficial femoral, deep femoral, and popliteal arteries are amenable to angioplasty or recanalization. More distal lesions involving the tibioperoneal trunk or the trifurcation vessels should not routinely be considered for transluminal angioplasty or recanalization because of the small caliber of these vessels and the increased likelihood of spasm and complication. Their accidental occlusion frequently causes an exacerbation of symptoms and may precipitate amputation. Therefore, for vessels more distal than the popliteal artery, we reserve transluminal angioplasty or recanalization for patients with severe ischemia threatening the limb in patients who are not candidates for surgical reconstruction and are facing amputation (10) (Figs. 6A–B, 7A–B). In some patients with occlusions of the superficial femoral artery that are not suitable for transluminal angioplasty who also have a stenosis of the deep femoral artery, angioplasty of the latter lesion probably will result in some symptomatic improvements (11) (Fig. 8A–B). Dilation or recanalization of superficial femoral stenoses or occlusions in patients with absent calf arteries allows perfusion of distal collaterals with increased pressure. In a high percentage of patients the increased blood flow to the foot is sufficient for healing of ischemic ulcers (Fig. 9A–D).

In addition to unacceptable morphology for percutaneous transluminal therapy, angioplasty and recanalization are usually contraindicated in patients with very fresh occlusions that may have soft, unorganized thrombus proximally and in patients with the ''blue toe syndrome'' (recurrent distal emboli from an ulcerated plaque in the superficial femoral artery) (9). The chances of embolic complications with either of these conditions are significantly increased.

On the day prior to angioplasty, the patient is started on antiplatelet medication. If no contraindication to aspirin exists, we usually give 325 mg of aspirin (one tablet) a day and advise the patient to continue taking it after the angioplasty. For patients who cannot tolerate aspirin, Persantine (Boehringer Ingelheim), 75 mg three times a day, is given. If the location of the lesion to be dilated is known to

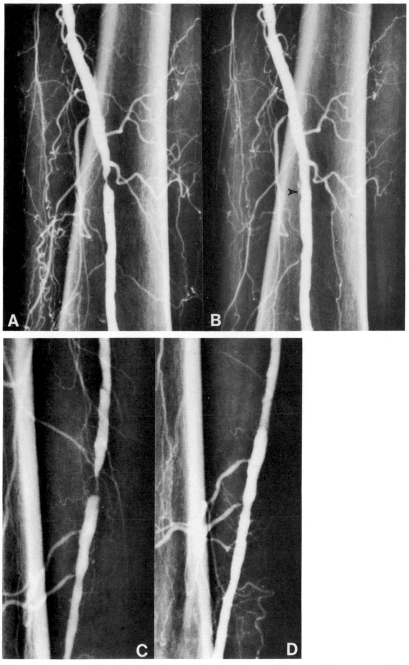

Figure 2. **A–D.** Ideal lesions for angioplasty in an 82-year-old woman with rest pain. **A.** Control right superficial femoral angiogram demonstrates a short, focal, concentric stenosis. **B.** Postangioplasty the lesion is successfully dilated, as evidenced by widened lumen and intimal disruption (arrowhead). **C.** Slightly longer, focal, concentric stenosis of the left superficial femoral artery in the same patient. **D.** The follow-up angiogram reveals successful dilatation.

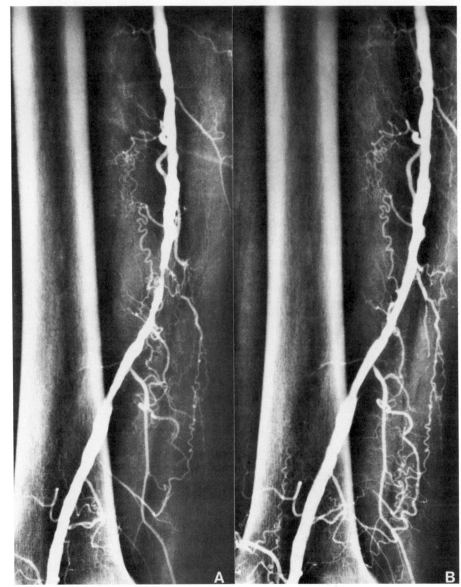

Figure 3.   A,B. Ideal for angioplasty: Short, focal, concentric stenosis of the right superficial femoral artery (**A**) before and (**B**) after successful angioplasty.

be in or distal to the popliteal arteries, nifedipine, 10 mg PO three times a day, is given on the day preceding and the morning of angioplasty as an antispasmodic measure since these vessels are very susceptible to spasm (Fig. 10A–D).

PROXIMAL STENOSES

For stenoses at or near the origin of the superficial femoral artery or in the deep femoral artery, the contralateral femoral approach is used. A V-shaped catheter is hooked over the aortic bifurcation and then exchanged over a guidewire for a catheter with a 30–45° angle approximately 1 cm from its tip. This catheter is then advanced to the origin of the artery (either the superficial or the deep femoral) to be dilated. If the origins of these two vessels are superimposed, positioning the patient in the contralateral posterior oblique will ''open up'' the femoral bifurcation and allow catheterization of the appropriate vessel.

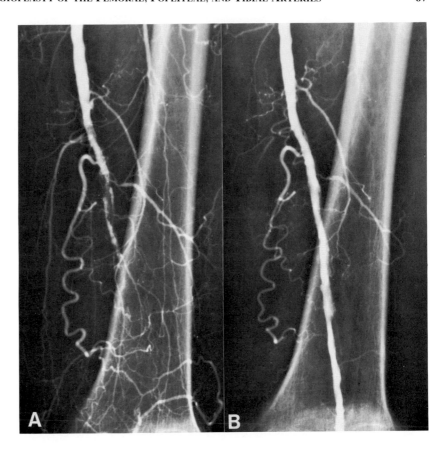

Figure 4. Good lesion for transluminal recanalization in a 59-year-old veteran; status post two separate coronary artery bypass graft surgeries. **A.** Control angiogram shows a 9-cm occlusion of the right superficial femoral artery beginning at the level of the adductor canal and extending into the popliteal artery. **B.** Postrecanalization study reveals an excellent result.

Since most stenoses and occlusions of the femoral arteries are located distal to the origin of the superficial femoral artery, an antegrade approach is the simplest and most direct. Although Bachman advocates using a contralateral approach, we find this method to be more cumbersome and difficult, since advancement of the catheter around the bifurcation and through a tight stenosis or occlusion can be difficult if not impossible (12). Furthermore, catheter control and "feel" of the guidewire are greatly diminished in the contralateral approach.

ANTEGRADE ARTERIAL PUNCTURE

In performing an antegrade puncture we try to enter the common femoral artery as proximal as possible, yet making sure the arterial entry site is caudal to the inguinal ligament (9,13–15). If an arteriogram of the involved extremity is available, the location of the superficial femoral–deep femoral bifurcation in relation to the bony landmarks of the proximal femur can be used to plan an appropriate site for skin and arterial entry. Antegrade arterial puncture of the femoral artery below the inguinal ligament yet proximal to the bifurcation of the superficial and deep femoral arteries often requires an almost vertical puncture. In obese patients, elevating the buttocks with a pillow or folded sheets facilitates femoral artery entry. In cases of extreme obesity, the antegrade approach may not be possible.

To assure the proper placement of the femoral artery puncture, the skin entry site is located under fluoroscopy, using the diagnostic angiogram as a guide. The skin and subcutaneous tissues are anesthetized with a liberal amount of local anesthetic. We try to make certain that the patient has no pain at the puncture site during the entire procedure, since pain may precipitate arterial spasm. Occasionally, in obese patients, an extra-long arterial needle is required. Once the artery has been entered and a strong pulsatile flow returns, the guidewire is inserted. We usually use a small-radius (1.5 mm) J wire initially, unless there is a very proximal stenosis in the superficial femoral artery, in which case we would start with a long, tapered straight (Newton) wire (Fig. 11A–B). Greenfield (14) has described preferential passage of the guidewire down the superficial femoral artery with the patient's foot abducted and medially rotated; however, in our ex-

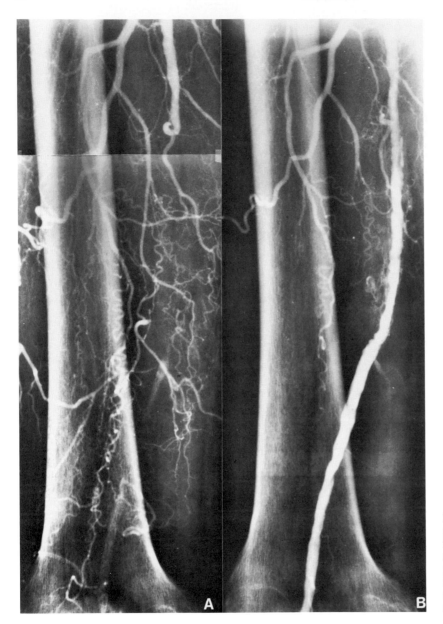

Figure 5. Long occlusion of right superficial femoral artery, a poor lesion for recanalization. **A.** Because of unavailability of saphenous vein for bypass graft, recanalization was attempted before planned prosthetic graft surgery. **B.** Follow-up angiogram demonstrates successful recanalization.

perience, this has been less useful. If the small-radius J wire goes down the superficial femoral artery, it is advanced and carefully kept proximal to the stenosis. The subcutaneous tissue is then dilated, and a number 6.5 French catheter with a 30–45° angle approximately 1 cm from the tip is inserted and positioned proximal to the stenosis. If, however, the guidewire enters the deep femoral artery, it is withdrawn and the hub of the needle is angulated laterally (placing the tip medially) in an effort to direct the guidewire into the superficial femoral ar-

tery. If this is unsuccessful, another maneuver, which sometimes aids in making the guidewire enter the superficial femoral artery, is moving the hub of the needle medially in an effort to "bounce" the J wire off the lateral wall of the common femoral artery and to direct it into the medially positioned superficial femoral artery. If, despite these maneuvers, the guidewire persistently enters the deep femoral artery, a small number 5 French catheter with a curve close to its tip is inserted. A test injection of contrast material under fluoroscopy with the patient

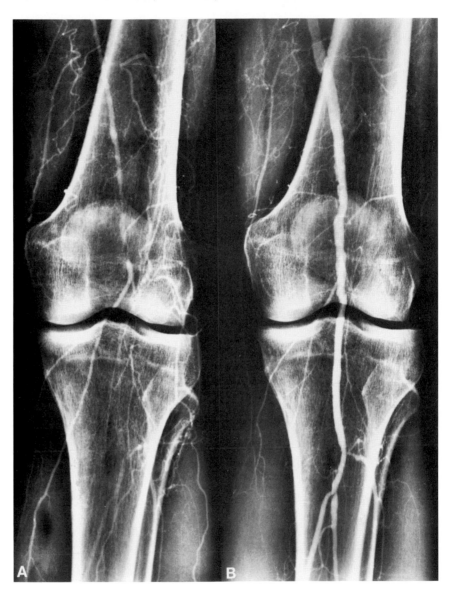

Figure 6. Angiograms of a 64-year-old woman with absent runoff vessels and pregangrenous changes who is not a candidate for surgery. **A.** Control left superficial femoral angiography demonstrates long-segment stenosis of left superficial femoral artery with occlusion of the popliteal artery extending into run-off vessels. **B.** After angioplasty and recanalization, continuity between superficial femoral artery and calf vessels is established.

in the contralateral posterior oblique position will locate the femoral artery bifurcation and help the angiographer determine whether the arterial entry site is in the common femoral artery or in the deep femoral. If the common femoral artery has been entered, the catheter is slowly withdrawn until its tip is immediately proximal to the bifurcation and then rotated medially toward the origin of the superficial femoral artery. With the catheter in this position, the guidewire almost always enters the superficial femoral artery and can be advanced distally to a position just proximal to the stenosis or occlusion. If, however, the arterial puncture has been into the deep femoral artery, it will be difficult or impossible to maneuver the catheter up to the origin

of the superficial femoral artery and then down the artery. Therefore, the catheter should be removed and another puncture made after appropriate compression. Should a large hematoma develop in attempts to achieve selective catheterization of the superficial femoral artery, the procedure should be discontinued and rescheduled at a later date to allow for resolution of the hematoma.

CROSSING THE LESION

Concentric Stenosis

The most critical part of transluminal angioplasty or recanalization of the femoral or popliteal vessels,

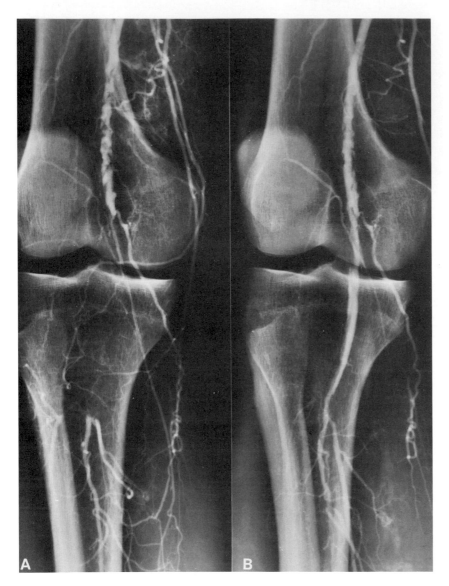

Figure 7. **A.** Angiogram showing occlusion of right popliteal artery extending into the calf vessels with single vessel runoff in a 71-year-old woman not considered an acceptable surgical candidate. **B.** After recanalization patency has been established.

or, for that matter, all other arteries, is traversing the lesion with the guidewire. If the stenosis is concentric, the catheter is placed several inches proximal to the lesion and a long, tapered straight guidewire is gently and slowly advanced through the stenosis (Fig. 12A–B). Careful monitoring of this step under fluoroscopy is essential to ascertain that the guidewire advances smoothly and does not "hang up" on the lesion. Once the guidewire is safely distal to the stenosis, the angiographic catheter is advanced over the guidewire through the lesion. Blood is aspirated through the catheter, and a small amount of dilute contrast material is injected to make certain the catheter position is still intraluminal.

Eccentric Stenosis

If the stenosis is eccentric, the catheter is placed immediately proximal to the lesion and rotated so the guidewire will find the proper intraluminal pathway (Fig. 13A–B). Frequently this requires changing direction of the catheter tip within a lesion, i.e., rotating the catheter medially to start through a tortuous eccentric stenosis and then laterally to pass the distal part of the lesion (Fig. 14A–B). The guidewire should advance smoothly; if its end stops and begins to buckle, the angiographer knows that the tip of the guidewire is not aligned with the luminal pathway but against the plaque. If continued ad-

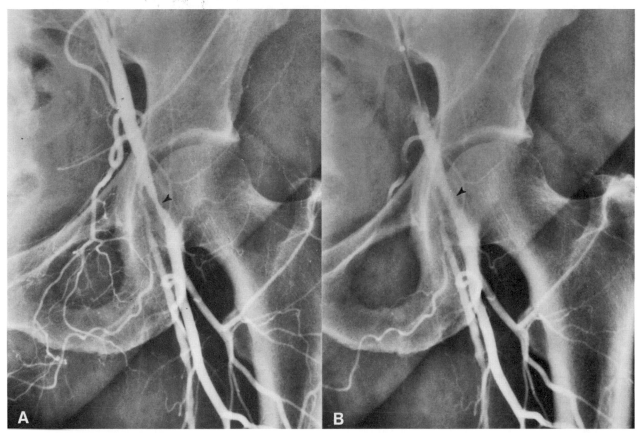

Figure 8. Profunda femoral artery angioplasty in 42-year-old obese diabetic woman with occluded superficial femoral artery. **A.** Control angiogram demonstrates a focal profunda stenosis (arrowhead). **B.** Stenosis was successfully dilated from contralateral approach, relieving symptoms of intermittent claudication.

vancement is made without redirecting the guidewire, subintimal dissection will probably result. Frequently, it is advisable to assess an eccentric lesion under fluoroscopy not only in anteroposterior (AP) but also oblique and lateral projections in order to appreciate the anatomy of the stenosis completely. As with a concentric stenosis, once the guidewire has been smoothly advanced distal to the lesion, the catheter is passed over it, and an intraluminal position is ascertained.

Occlusions

The technique we use to cross occlusions is slightly different than that used for stenoses. The angiographic catheter is advanced to the level of the occlusion and rotated away from the origin of any large collateral branch that may be near the proximal aspect of the occlusion and toward the expected lumen of the vessel (Fig. 15A–B). Again, we prefer a straight long tapered or regular straight guidewire to go through the occlusion. Other angiographers use a small-radius J wire to pass through an occlusion (8,10). Frequently there is a feeling of "grittiness" as the guidewire passes through the occlusion. Rarely (except in long-standing calcified lesions that are sometimes seen in diabetic patients) is force required to advance the guidewire through the lesion. Once the guidewire is distal to the occluded segment, it moves more smoothly. The catheter should then be advanced over the guidewire through the occlusion; after appropriate aspiration of blood, a small test injection of dilute contrast material is performed under fluoroscopy to ensure that the catheter is intraluminal.

During negotiation of the stenosis or occlusion, should the guidewire or catheter become extraluminal in a subatheromatous plane or perforate the vessel itself, the procedure is usually terminated,

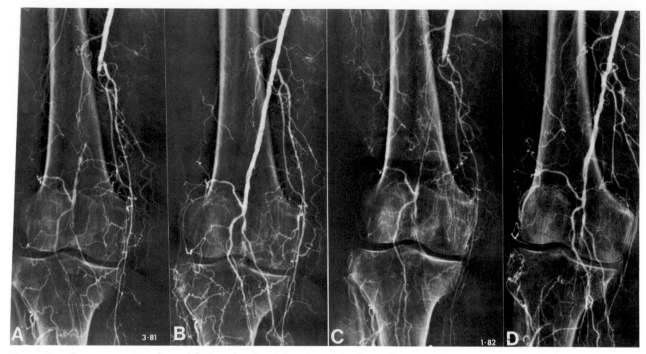

Figure 9. Constant rest pain and ischemic ulcerations on right foot of a 91-year-old woman caused her to consider amputation. **A.** Control angiogram (March 1981) reveals occlusion of superficial femoral artery over an 8-cm segment with reconstitution of the popliteal artery and absence of runoff vessels. **B.** Recanalization of occlusion allowed perfusion of distal collaterals with increased pressure, relieving rest pain and promoting healing of ulcer. **C.** Nine months later rest pain returned, and angiography demonstrates restenosis of previously recanalized area. **D.** After redilation, patient remained asymptomatic until death 2 years later.

and the vascular status of the limb will usually be no worse than it was previously before. Repeated attempts to reenter the true lumen of the vessel are most often unsuccessful and jeopardize important collateral channels near the proximal or distal ends of the lesion. Unlike extraluminal or extravascular passages of the catheter that do not worsen the patient's condition, occlusion of these collateral vessels frequently causes acute deterioration in the blood supply to the distal limb. The angiographer should carefully try to avoid injuring these collaterals. Therefore, once the guidewire or catheter is extraluminal, either a limited, gentle attempt to find the appropriate intraluminal pathway is made or the procedure is discontinued and rescheduled in 2–3 weeks.

Once the catheter has been successfully advanced through the stenosis, additional heparin (2000–5000 U) is given by either the intra-arterial or intravenous route. If the lesion is in, or distal to, the popliteal artery, nitroglycerin, 100–200 μg, can be given intra-arterially to prevent spasm (9). If, during advancement of the number 6.5 French catheter through the lesion, significant resistance results, preliminary dilation with a number 7.0 or 8.0 French teflon cath-

eter well tapered to the guidewire is indicated. This technique will facilitate advancement of the dilating balloon catheter.

BALLOON DILATION

The next step in the procedure is insertion of the dilating balloon catheter. A number of types of dilating catheters with balloons of different diameters and lengths are available (16). Choice of the proper dilating catheter depends on the diameter of the normal, uninvolved artery, both proximal and distal to the lesion, and the length of the lesion. For the superficial femoral or popliteal arteries we usually use a dilating catheter with a number 7 French shaft and a balloon with a diameter of either 4, 5, or 6 mm and a length varying from 4 to 10 cm (17). Prior to insertion of the dilating catheter the balloon is tested, evacuated, and carefully "wrapped" around the catheter shaft. A number 8 French arterial sheath with a side arm for flushing is helpful in introducing the dilating catheter through the soft tissues of the groin, particularly if any scarring or fibrosis from

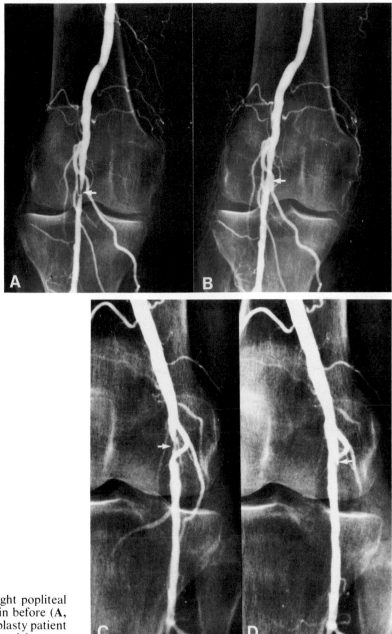

Figure 10.   A–D. Bilaterally symmetrical tight popliteal stenoses in 84-year-old woman with rest pain before (**A, C**) and after (**B, D**) angioplasty. Before angioplasty patient received oral nifedipine to prevent popliteal arterial spasm.

previous surgery is present. If preliminary dilation with a number 8 French teflon catheter was done, the sheath is required to prevent loss of blood through the arterial entry site. During insertion of the sheath and dilating catheter, it is imperative to maintain the guidewire distal to the lesion and as stationary as possible. Continuous fluoroscopy of the tip of the guidewire is required throughout these manipula-tions. Once the distal radiopaque marker on the balloon catheter has been advanced beyond the stenosis or occlusion, the guidewire can be withdrawn for flushing but should be reinserted prior to the actual dilation. When the guidewire is back in place, dilation can proceed.

We usually use a 10-mL syringe for balloon inflation and monitor the shape of the balloon under fluoroscopy. Any change from the normal "sausage" shape to an eccentric or spherical silhouette signifies impending balloon rupture. If this occurs, the balloon should be deflated immediately and the catheter exchanged for another. Pressure gauges to ensure that the balloon is not inflated with pressures that exceed the manufacturer's guidelines are recommended, although many angiographers experienced in transluminal angioplasty rely on the appearance of the balloon under fluoroscopy (16). Use of a syringe smaller than 10 mL produces very high pressures since the amount of force generated is inversely proportional to the cross-sectional area of the syringe piston (16). With the newer polyethylene balloon catheters, balloon rupture is not so traumatic to the artery as it was when the older polyvinyl chloride balloons ruptured. Nonetheless, we try to avoid rupturing the dilating balloon since there is always a chance of irreversible arterial injury, and, furthermore, a ruptured balloon may be difficult to remove through the entry site of the artery (18).

The site of the lesion can be marked with a radiopaque marker attached to the patient's skin. Once the balloon is positioned across the lesion, it is inflated with contrast material diluted with normal saline to half strength. A characteristic "sausage-link" indentation further indicates that the balloon is well positioned with respect to the lesion. Balloon inflation is monitored fluoroscopically, and a sudden release in the indented portion of the balloon indicates that the lesion has been successfully dilated. Once this occurs, no further inflations are indicated. Occasionally, despite multiple inflations, the lesion continues to indent the balloon, signifying that complete dilation with intimal cracking at the site of the plaque has not been achieved. When this occurs, the angiographer has a few choices. The balloon catheter can be exchanged for a type that will accept more pressure or for a catheter with a slightly larger diameter balloon, and the lesion can be redilated. If the catheter is exchanged, it is imperative that a guidewire remain distal to the lesion during the ex-

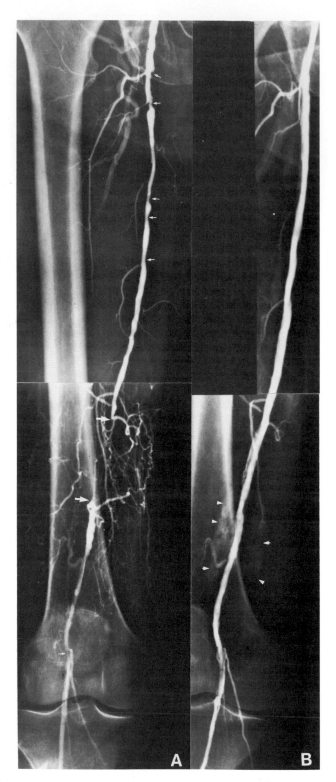

Figure 11. **A.** Angiograms of 66-year-old diabetic woman with left above-knee amputation and right leg rest pain and foot ulceration. **A.** Control angiogram reveals proximal superficial femoral artery stenosis (top small arrow) that could be negotiated only with a straight wire. Multiple tandem stenoses (small arrows) are present in superficial femoral artery as well as a 6-cm occlusion (large arrows) at adductor canal. There is also a tight popliteal stenosis. **B.** Follow-up angiogram after successful angioplasty and recanalization has small area of contrast extravasation (arrowheads) from arterial perforation by guidewire.

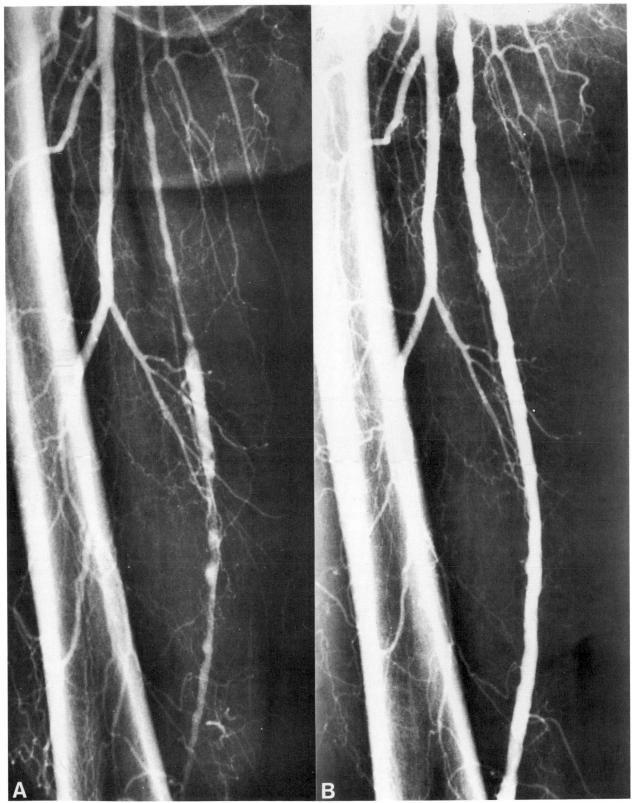

Figure 12. **A.** Multiple concentric stenoses in right superficial femoral artery of 73-year-old man. Straight guidewire easily negotiated these stenoses. **B.** Angiogram after successful angioplasty.

75

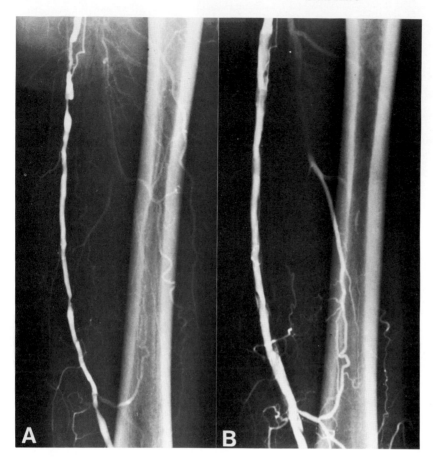

Figure 13. **A.** Multiple eccentric stenoses in left superficial femoral artery before angioplasty; each lesion negotiated by rotating a catheter with a short curve at its tip in the direction of arterial lumen and advancing long, tapered straight guidewire through lesion. **B.** Follow-up angiogram reveals a successful result.

change in order to avoid retraversing the freshly dilated lesion with the guidewire tip, a maneuver that carries a high risk of dissection.

When apparent successful dilation has taken place, the balloon is completely deflated, and the catheter is withdrawn over a stationary guidewire. Contrast material is injected through a side-arm adapter, and the result of the procedure is examined. If it appears satisfactory, the guidewire can be removed and a final follow-up angiogram performed. On the other hand, if the lesion is not completely dilated, the catheter, or one that may be more appropriate, is advanced over the guidewire which, during all the previous maneuvers, has remained distal to the lesion. We do not use pressure measurements to assess the result of femoropopliteal angioplasty or recanalization. They are not valid because the distal pressures must be determined with a catheter through the involved area, a procedure that greatly exaggerates any pressure difference by partially occluding the lumen.

Once the guidewire and catheter have been withdrawn proximal to the stenosis or occlusion, no further attempts should be made to recross it, no matter

how tempting it is to improve the result based on the findings of the postangioplasty arteriogram (Fig. 16A–B): "The enemy of good is better." With time, arterial remodeling and healing occur, and an angiogram several months after the procedure usually has a much improved appearance compared to the immediate postprocedure study.

After the procedure, the catheter is removed and appropriate compression applied at the entry site until hemostasis is obtained. Generally, resistance increases during withdrawal of the balloon through the arteriotomy site and soft tissues or through the sheath. However, with gentle traction and rotation of the catheter shaft in the direction specified by the manufacturer, the balloon can be removed easily. Usually we don't reverse the heparin, unless bleeding persists after 15 or 20 minutes of compression. With a compression bandage over the puncture site the patient remains flat in bed with the involved leg straight for at least 6 hours. Patients who are hypertensive or are using anticoagulants are kept at bed rest for 12–18 hours.

Barring complications, the patient is discharged on the day following angioplasty. Prior to discharge,

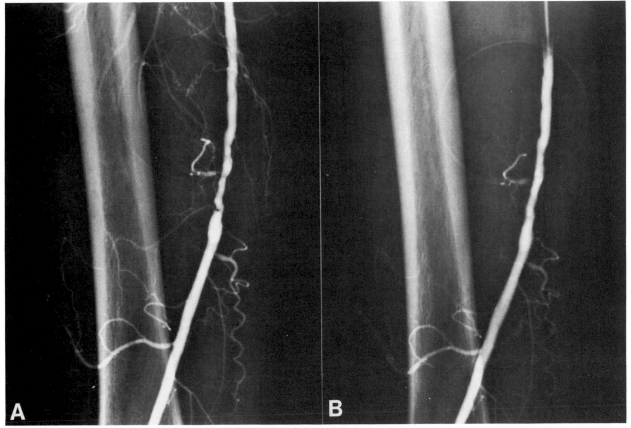

Figure 14.   A,B. Angioplasty of mildly tortuous right superficial femoral artery stenosis. To negotiate this lesion catheter tip was rotated in several directions while traversing the stenosis to avoid subintimal dissection.

the noninvasive vascular examination is repeated, and the patient receives counseling concerning smoking, exercise habits, and proper diet. Discharge medication includes aspirin, 325 mg daily, or, if a history of aspirin intolerance is present, Persantine, 75 mg tid.

COMPLICATIONS

Complications of carefully performed femoral, popliteal, and tibial angioplasty are, fortunately, few; however, they do occur. Most complications can be grouped according to their location, i.e., at the arterial entry, the lesion, or proximal or distal to these two sites (19). Those complications occurring at the arterial entry site include hematoma, occlusion, false aneurysm, and arteriovenous fistula. Aside from hematoma, which is not infrequent, the other arterial entry site complications are relatively rare. It is not surprising that these same conditions are also complications of diagnostic angiography.

Subintimal dissection, arterial rupture, and thrombosis are the complications that occur at the site of the lesion. Subintimal dissection can develop during attempts to pass the guidewire through the lesion. Once dissection has been recognized, it is crucial for the angiographer not to persist too vigorously in attempting to reestablish an intraluminal pathway. To do so risks occlusion of important collaterals near the proximal and distal ends of the lesion and subsequent thrombosis of the vessel. Arterial rupture, usually caused by inflation of a balloon that is too large for the vessel, results in occlusion with thrombosis or false aneurysm formation at the site of the rupture.

Retroperitoneal hemorrhage and distal embolization are also complications associated with lower-extremity transluminal angioplasty. If the arterial entry site is above the inguinal ligament, effective compression cannot be applied since the artery is retroperitoneal at this location. Therefore, hemorrhage from the puncture site will dissect retroperitoneally, rather than into the groin. Usually, the

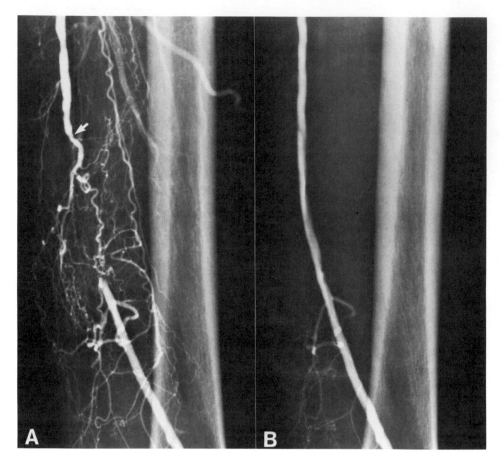

Figure 15. Transluminal recanalization of 7-cm left superficial femoral artery occlusion. **A.** Guidewire repeatedly entered large collateral (arrow) near proximal end of occlusion and had to be directed away from it by a catheter with a small curve at its tip. **B.** Angiogram following angioplasty shows good result.

bleeding will stop with conservative therapy; however, occasionally the patient will require surgical repair of the arteriotomy site. Distal embolization, occurring infrequently, is due to dislodgement of atherosclerotic debris during the procedure. The vascular tree distal to the angioplasty site should be examined after dilation or recanalization, either under fluoroscopy or by formal filming, and before removal of the guidewire to exclude embolic occlusion of previously patent distal vessels. Most distal emboli are asymptomatic; however, if clinically significant they can be aspirated through a nontapered catheter, or surgical embolectomy must be performed (20).

Injudicious administration of large amounts of contrast material may result in renal failure, complicating an otherwise successful angioplasty procedure. This usually occurs in patients with diabetes or borderline renal failure for whom diagnostic angiography, angioplasty, and follow-up angiograms are done at the same setting. Maintaining good hydration and separating the diagnostic from the interventional angiogram by 1 or 2 days can significantly decrease the incidence of this serious

complication. New nonionic contrast material may be less nephrotoxic than that which is now in use and may also help to reduce postangioplasty renal failure.

When complications of thrombus or embolism occur, the angiographer must carefully outline the arterial anatomy proximal and distal to the thrombus or embolus before terminating the angiographic procedure. Thus, if surgery is required, the surgeon will have a preoperative angiogram demonstrating the complication and the status of the proximal and distal vessels. Low-dose fibrinolytic therapy has been used successfully to manage thrombotic complications of angioplasty with a high rate of success since the thrombus is relatively fresh and not yet organized (21) (Fig. 17A–C).

CLINICAL RESULTS

Numerous studies have been performed to assess the results of transluminal angioplasty of the femoral and popliteal vessels for indications ranging from intermittent claudication to limb salvage (22–29). In

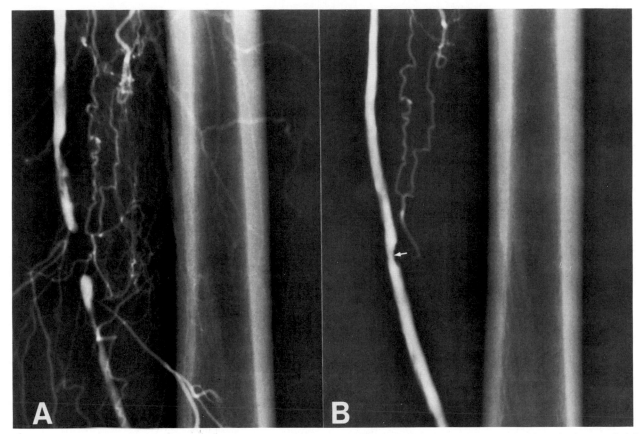

Figure 16.   Control (**A**) and follow-up (**B**) angiograms after recanalization of short-segment occlusion of left superficial femoral artery. Some residual stenosis (arrow) remains; however, since catheter and guidewire were withdrawn proximal to lesion, recrossing was not attempted.

a large cooperative study among 12 different centers published by Gruentzig and Zeitler in 1978 and involving 1184 patients, primary success rate was 74 percent with an overall complication rate of 10 percent. Clinical findings, type of dilation catheter used, stage of disease, and type of obstruction were not specific. Over 20 percent of the patients had occlusions greater than 10 cm in length (22). More recently, with better patient selection, tools, and techniques, results have improved and complications have been reduced (27). For ideal lesions, i.e., short femoropopliteal stenoses, Zeitler et al. have had success rates of 91–94 percent. Similarly, with occlusions less than 10 cm long, the rate of success of recanalization varies from 74 to 91 percent, depending on the stage of ischemic disease (27). In the same series of 1524 patients the complication rate was 3.2 percent, with only 0.13 percent requiring surgery (27). Angioplasty and/or recanalization in patients with end stage vascular disease being considered for amputation was successful in relieving symptoms in 73 percent, none of whom were acceptable candidates for surgical reconstruction (24,25).

In comparing the results of transluminal therapy to femoropopliteal bypass surgery, the type of graft material used must be taken into consideration. For saphenous vein grafts, the cumulative patency rate after 1 and 2 years is very similar to the patency rate of angioplasty. However, when prosthetic grafts are used, the 1- and 2-year patency rates of surgical therapy are significantly lower than that of angioplasty (14,26). Dotter has demonstrated patency of a popliteal recanalization lasting 9 years (30) (Fig. 18A–C). Therefore, if the saphenous vein is not going to be used, not available, or being saved for possible use in other vascular beds (e.g., coronary arteries), transluminal therapy is the treatment of choice (14,26).

FUTURE PERSPECTIVES

Despite the skepticism and resistance that marked the first 20 years of transluminal angioplasty, its

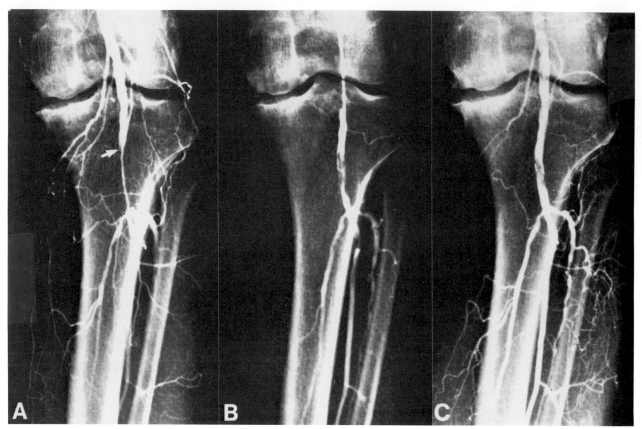

Figure 17.   **A.** Angiograms of a 59-year-old female who developed acute ischemic symptoms of left foot 3 hours after apparently successful left superficial femoral angioplasty. **A.** Embolus (arrow) occluding left popliteal artery. **B.** Small catheter has been advanced so its tip is buried in thrombotic material. **C.** After 4 hours of low-dose (5000 U per hour) streptokinase infusion, embolus has almost completely lysed.

future is bright. In a recent study, Doubilet and Abrams demonstrated that for iliac and femoropliteal angioplasty, the potential benefits are great in terms of dollars, limbs, and lives. On the basis of their projections, if only 40 percent of all patients with iliac or femoral lesions acceptable for angioplasty in 1980 underwent transluminal therapy, even including those procedures that were unsuccessful immediately or had late occlusions, $82 million, 5006 limbs, and 352 lives would have been saved (31). With hospital and medical costs undergoing intense scrutiny and current implementation of DRGs, the cost-effectiveness of transluminal angioplasty compared to traditional therapy will be an important factor. Outpatient angioplasty is a practical step in further reducing hospital costs associated with the therapy of obstructive vascular disease (32). Since the patency rates of femoropopliteal angioplasty and bypass grafting with saphenous veins are roughly

similar and given the fact that 50 percent of patients undergoing femoropopliteal bypass grafts will be dead within 5 years and 75 percent within 10 years, is it not entirely reasonable that they be treated with the method having the least pain, morbidity, mortality, medical cost, and income loss (2,14,26,31)? Furthermore, unsuccessful angioplasty almost never jeopardizes future surgical therapy or results in a change in the type of surgical procedure that would have been done had angioplasty not been attempted.

With the development of new angioplasty balloons and better understanding of the pharmacology of platelet aggregation and fibrinolytic drugs, improved results and decreased complications of transluminal angioplasty are inevitable. As Dotter once wrote, ''This (book) was intended to provide you with state-of-the-art information on transluminal angioplasty. With ingenuity, innovation, and luck, it will soon be out of date'' (33).

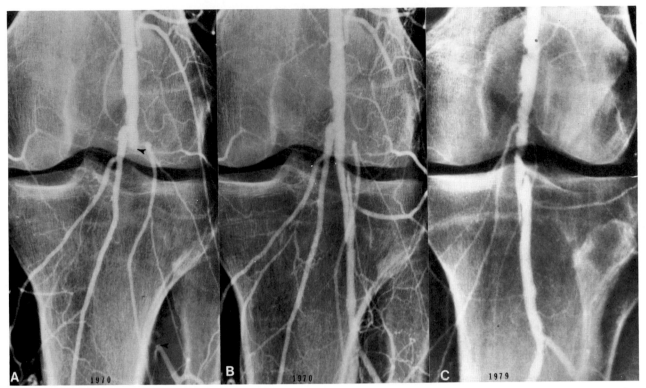

Figure 18.   A 9-year patency of popliteal recanalization. **A.** Control angiogram reveals left popliteal artery occlusion extending into calf vessels. **B.** Immediately and **(C)** 9 years after recanalization artery is patent. (Courtesy of Charles T. Dotter, M.D.)

REFERENCES

1. Boyd AM: The natural course of atherosclerosis of the lower extremities. Angiology 11:10–14, 1960.
2. Deweese JA, Rob CG: Autogenous venous grafts ten years later. Surgery 82:775–784, 1977.
3. Dotter CT, Judkins MP: Transluminal treatment of arteriosclerotic obstruction. Circulation 30:654–670, 1964.
4. Zeitler E, Shoop W, Zahrow E: The treatment of occlusive arterial disease by transluminal angioplasty. Radiology 99:19–26, 1971.
5. Gruentzig AR, Hopff H: Perkutane Rekanalisation chronischer artellier Verschlusse mit einem neuen Dilatationskatheter: Modification der Dotter-technik. Dtsch Med Wochenscher 99:2502–2505, 1974.
6. Abbott WM: Percutaneous transluminal angioplasty: Surgeon's view. AJR 135:917–920, 1980.
7. Neiman HL, Bergan JJ, Yao JST, Brandt TD, Greenburg M, O'Mara CS: Hemodynamic assessment of transluminal angioplasty for lower extremity ischemia. Radiology 143:639–643, 1982.
8. Gruentzig AR, Kumpe DA: Technique of percutaneous transluminal angioplasty with the Gruentzig balloon catheter. AJR 132:547–552, 1979.
9. Sos TA, Sniderman KW: Percutaneous transluminal angioplasty. Semin Roentgenol 16:26–41, 1981.
10. Motarjeme A, Keifer JW, Zuska AJ: Percutaneous transluminal angioplasty and case selection. Radiology 135:573–581, 1980.
11. Motarjeme A, Keifer JW, Zuska AJ: Percutaneous transluminal angioplasty of the deep femoral artery. Radiology 135:613–617, 1980.
12. Bachman DM, Casarella WS, Sos TA: Percutaneous iliofemoral angioplasty via the contralateral femoral artery. Radiology 130:617–621, 1979.
13. Freiman DB, Ring EJ, Oleaga JA, Berkowitz H, Roberts B: Transluminal angioplasty of the iliac, femoral, and popliteal arteries. Radiology 132:285–288, 1979.
14. Greenfield AJ: Femoral, popliteal, and tibial arteries: Percutaneous transluminal angioplasty. AJR 135:927–935, 1980.
15. Waltman AC: Percutaneous transluminal angioplasty: Iliac and deep femoral arteries. AJR 135:921–925, 1980.
16. Abele JE: Technical considerations: Physical properties of balloon catheters, inflation devices, and pressure measurement devices, in Castaneda WR (ed): *Transluminal Angioplasty.* New York, Thieme-Stratton, 1983, pp 20–27.
17. Sos TA, Sniderman KW, Beinart C: Gruentzig catheter with a 10 cm long balloon. Radiology 141:825–826, 1981.

18. Yune HY, Klatte EC: Circumferential tear of percutaneous transluminal angioplasty balloon catheter. AJR 135:395, 1980.

19. Laerum F, Castaneda WR, Amplatz KA: Complications of transluminal angioplasty, in Castaneda WR (ed): *Transluminal Angioplasty*. New York, Thieme-Stratton, 1983, pp 41–44.

20. Sniderman KW, Bodner L, Saddekni S, Srur M, Sos T: Percutaneous embolectomy by transcatheter aspiration. Radiology 150:357–361, 1984.

21. Katzen BT, Van Breda A: Low dose streptokinase in the treatment of arterial occlusions. AJR 136:1171, 1981.

22. Gruentzig AR, Zeitler E: Cooperative study of results of PTR in twelve different centers, in Zeitler E, Gruentzig AR, Schoop W (eds): *Percutaneous Vascular Recanalization*. Berlin, Springer-Verlag, 1978.

23. Colapinto RF, Harries-Jones EP, Johnston KW: Percutaneous transluminal angioplasty of peripheral vascular disease: A two-year experience. Cardiovasc Intervent Radiol 3:213–218, 1980.

24. Lu CT, Zarins CK, Yang CF, Sottiurai V: Long-segment arterial occlusion: Percutaneous transluminal angioplasty. AJR 138:119–122, 1982.

25. Lu CT, Zarins CK, Yang CF, Turcotte JK: Percutaneous transluminal angioplasty for limb salvage. Radiology 142:337–341, 1982.

26. Martin EC, Fankuchen EI, Karlson KB, Dolgin C, Collins RH, Voorhees AB Jr, Casarella WJ: Angioplasty for femoral artery occlusion: Comparison with surgery. AJR 137:915–919, 1981.

27. Zeitler E, Richter EI, Roth FJ, Schoop W: Results of percutaneous transluminal angioplasty. Radiology 146:57–60, 1983.

28. Probst P, Cerny P, Owens A, Mahler F: Patency after femoral angioplasty: Correlation of angiographic appearance with clinical findings. AJR 140:1227–1232, 1983.

29. Katzen BT: Percutaneous transluminal angioplasty for arterial disease of the lower extremities. AJR 142:23–25, 1984.

30. Dotter CT: Transluminal angioplasty: A long view. Radiology 135:561–564, 1980.

31. Doubilet P, Abrams HL: The cost of underutilization: Percutaneous transluminal angioplasty for peripheral vascular disease. N Engl J Med 310:95–102, 1984.

32. Manashil GB, Thunstrom BS, Thorpe CD, Lipson SR: Outpatient transluminal angioplasty. Radiology 147:7–8, 1983.

33. Dotter CT: Introduction in Castaneda, WR (ed): *Transluminal Angioplasty*. New York, Thieme-Stratton, 1983, p 2.

6

Percutaneous Transluminal Angioplasty of Rarer Categories of Vascular Disease

ERIC C. MARTIN, M.D.

INTRODUCTION

The major vessels in the body that cause ischemic symptoms have all been dilated percutaneously at one time or another. This text discusses the major categories of common percutaneous angioplasties in separate chapters; this chapter describes the rarer categories of vascular disease in which percutaneous angioplasty has been applied in relatively small numbers of cases. These rarer vascular conditions are discussed under the following separate subheadings: the extracranial vessels, the mesenteric vessels, the aorta itself, and the venous system, excluding the femoral vein grafts.

THE SUBCLAVIAN AND BRACHIAL ARTERIES

Dilations of the carotid and vertebral artery are discussed in Chapter 7. The brachial and subclavian arteries may conveniently be discussed together in this section.

Not infrequently subclavian artery stenoses are asymptomatic, presumably because collateral chan-

nels provide sufficient flow to the affected target area. Furthermore, the use of the upper extremity requires less physical exertion than use of the lower extremities. Exercise may be avoided, or symptoms may not be severe enough to cause patient complaints. Indeed, claudication of the upper extremity, if one may use the phrase, most frequently follows brachial artery occlusion as a result of cardiac catheterization by the Sones technique. In this instance, the occlusion is more distal and the resulting collateral flow less adequate. Nevertheless, there is no doubt that patients may complain of exertional arm pain resulting from subclavian artery stenosis or occlusion, and most frequently the stenosis is distal to the origin of the vertebral artery. Such clinical symptoms justify revascularization by bypass surgery (1) or transluminal angioplasty.

The other major symptom complex associated with subclavian artery stenosis is the *subclavian steal syndrome*, which results from a stenosis or occlusion proximal to the origin of the vertebral artery (2,3). The extremity is supplied from the contralateral side, through the circle of Willis, via the ipsilateral vertebral artery in a retrograde fashion. In subclavian steal syndrome, physical exertion of the affected upper extremity would cause posterior cerebral ischemia by shunting oxygenated blood from the brain to the upper extremity. This occasionally leads to vertebrobasilar insufficiency with drop attacks, disordered gait, and alterations of balance or to episodes of cortical blindness.

Etiology

Four of the etiologies of subclavian artery obstruction have been recognized and treated by percutaneous angioplasty.

Arteriosclerotic stenoses most frequently occur either at the origin of the subclavian artery or just distal to the origin of the vertebral artery and respond well to balloon dilation. Very proximal stenoses may progress to occlusion over a long segment, up to the inflow from the vertebral artery. Motarjeme reported the only attempted dilation of a complete occlusion in the literature, although his attempt was unsuccessful (4).

Takayasu's disease less commonly affects the brachiocephalic vessels and may be either a short- or long-segment stenosis. It may progress to complete occlusion over a long segment.

Fibromuscular disease less commonly affects the brachiocephalic vessels, but we have encountered one patient with a long-segment stenosis distal to the vertebral artery.

More distal lesions may be secondary to the external compression of the thoracic outlet syndrome. The most common agents are the scalenus anticus muscle and fibrous connections from a cervical rib. We expect that angioplasty would not correct these conditions because the causes of the problems are extravascular in origin.

Indications for Transluminal Dilation

The usual risks of dilation prevail in these conditions as in other organ systems, but the complications may be more significant than in the iliofemoral system; consequently they influence the indications for angioplasty in the upper extremities. Although the risk of embolization is low, even a small embolus may be consequential. Embolic occlusion of the brachial artery at its trifurcation may cause ischemia or even a Volkman's contracture. Although a proximal embolization is probably amenable to embolectomy, it is not possible to recover an embolus from the digital arteries, which are small-caliber distal vessels. One would have to rely on streptokinase, provided the embolus were a clot.

More consequential is a cerebral embolic event.

It has been argued that angioplasty is safe in the subclavian steal syndrome because collateral flow is from the vertebral artery to the subclavian artery. However, when the balloon is deflated after a successful angioplasty, antegrade flow returns, and with it the possibility of embolization and its disastrous consequences. To the best of our knowledge this result has not been reported, and Bachman's report that 30–45 seconds elapse before flow reversal takes place is encouraging (5).

The risk of rupturing a vessel applies here as in any other type of angioplasty. In the proximal subclavian artery this would result in a major intrathoracic hemorrhage. The origin of the vessel could perhaps be occluded by balloon inflation for a temporary hemostasis, but again, to the best of our knowledge, this complication has not been reported.

A third, hypothetical problem is that of dilating a complete occlusion at the site of a bifurcation between subclavian and vertebral arteries. Ring has emphasized the high incidence of contralateral embolization when dilating a common iliac occlusion (6), and it may be that a similar risk accompanies the dilation of a complete occlusion just distal to the vertebral artery origin. Perhaps this complication would arise only if dilation were performed retrogradely from the brachial route, but we would choose not to dilate a complete occlusion of the subclavian in this situation because of possible vertebral artery embolization.

Angioplasty Technique

Prior to dilation procedure, the stenosis will be defined angiographically with an arch aortogram and perhaps with a selective study performed with a "headhunter" catheter. The choice of guidewires and guiding catheters for a transfemoral approach depends on the etiology of the stenosis, as well as the location of the lesion.

The basic technique would be to use a headhunter guiding catheter and to cross the stenosis with a small, soft guidewire, perhaps a 0.028 inch 150 cm J wire or even a Ring A wire. An exchange would then be made over a 300 cm 0.035 inch Rosen J wire and dilation performed with a balloon of appropriate size chosen to be 1 mm larger than the vessel being dilated. Tighter stenoses might require a Newton wire, if one could pass through the stenotic lumen, or even a movable core J wire in order to follow better. If, however, the stenosis is at the origin of the subclavian artery, it may be difficult to catheterize the vessel selectively, necessitating an axillary, or high brachial puncture and a retrograde crossing of the lesion. Such an approach has the advantage that the catheter and wire are well seated in the artery before crossing the lesion is attempted so that the risk of subintimal dissection is decreased. A further advantage is that embolic material would travel down the aorta rather than up into the head so that a small embolus would do less damage.

Because of the increased embolic risk, we would perform all dilations with the patient on acetylsalicyclic acid (aspirin), 325 mg daily, administering 2000 U of heparin intra-arterially at the time the lesion was crossed. We would continue aspirin for 6 months after a successful dilation.

Clinical Results

In 1980 we reported on a successful dilation of the right subclavian artery in a patient with Takayasu's disease (7). She suffered from exertional pain and had a 60 mmHg pressure differential between the two arms that was confirmed by crossing the stenosis. The gradient was eliminated by the dilation, and symptoms abated. She has remained asymptomatic for several years (Fig. 1).

In the same year, Novelline reported successful dilations of two stenoses, one distal to the left vertebral artery and a second of a high right brachial artery, both due to atherosclerosis (8). Also in 1980, Bachman reported the first successful percutaneous dilation for the subclavian steal syndrome and thereby the first successful dilation performed proximal to the vertebral artery (5). In 1982, Hodgins reported another patient with Takayasu's disease in whom the stenosis was proximal to the left vertebral artery and was successfully dilated (9). Motarjeme produced the largest series in the literature in 1982, describing seven patients, five with a stenosis proximal to the vertebral artery origin and two in whom it was distal (4). Five of these patients were cured by dilation, and one was improved. The seventh patient had a complete occlusion that could not be crossed.

It is of interest that in the first 10 reports, the subclavian stenosis was on the left in 9 patients, with half being dilated for the subclavian steal syndrome and half for arm claudication. Since that time we have dilated one long-segment stenosis of the right subclavian artery due to fibromuscular disease. Dilation was unsuccessful and resulted in a complete occlusion, which required a right carotid to axillary bypass graft, the standard operation for these lesions (1).

There have been no reports of embolization to the cerebral circulation as a result of percutaneous dilations of the subclavian artery. The results of angioplasty in the subclavian artery are probably comparable to those in the femoral artery, although the

risks involved are certainly higher in the subclavian artery, and the indications must therefore be more stringent. Since fewer patients warrant dilation in the subclavian artery, a significant series has yet to be accrued. However, we see no reason for the results to be less successful in the subclavian artery than in other vascular systems.

THE MESENTERIC VESSELS

In the mesenteric circulation, dilation has been performed predominantly for mesenteric angina and postprandial syndrome, often accompanied by cramping pain and diarrhea. Patients are reluctant to eat because of the pain, and the consequent weight loss may be substantial (10,11). Angiographic findings include stenosis or occlusion of more than one mesenteric artery, with the third inadequately collateralizing the territory of the others. Most of these patients have arteriosclerotic disease, which is frequently widespread with associated peripheral vascular disease, ischemic heart disease, or cerebrovascular disease. The traditional surgical treatment is an aortomesenteric bypass graft, which carries a significant mortality in such patients and a less than ideal success rate (12).

A trial of angioplasty is preferable in such patients; however, the clinical picture and even the angiographic anatomy can be confusing. Not infrequently, one sees young patients with fibromuscular disease or Takayasu's disease who have occlusion of the superior mesenteric and celiac arteries. Often these individuals are totally asymptomatic, even though the bowel is perfused from the inferior mesenteric artery alone. In patients with mesenteric ischemia, collateralization may be less than adequate, and it is generally considered that at least two mesenteric vessels be compromised before clinical symptoms occur.

The symptom complex may not, however, be clearcut, and the clinical diagnosis may be difficult. Some of these patients may be examined angiographically and only a single stenosis be found. Under these circumstances, is it justifiable to attempt a therapeutic trial of angioplasty? Certainly there are reports in the angiographic literature of dilatations being successful and relieving symptoms when only one vessel is involved (13,14).

A further confusing issue is the *median arcuate ligament syndrome,* whereby the median arcuate ligament of the diaphragm compresses the celiac axis from above so that it appears indented on a lateral aortogram (Fig. 2). The compression is relieved with inspiration but accentuated on expiration. Over the years, the motion of the diaphragm causes trauma

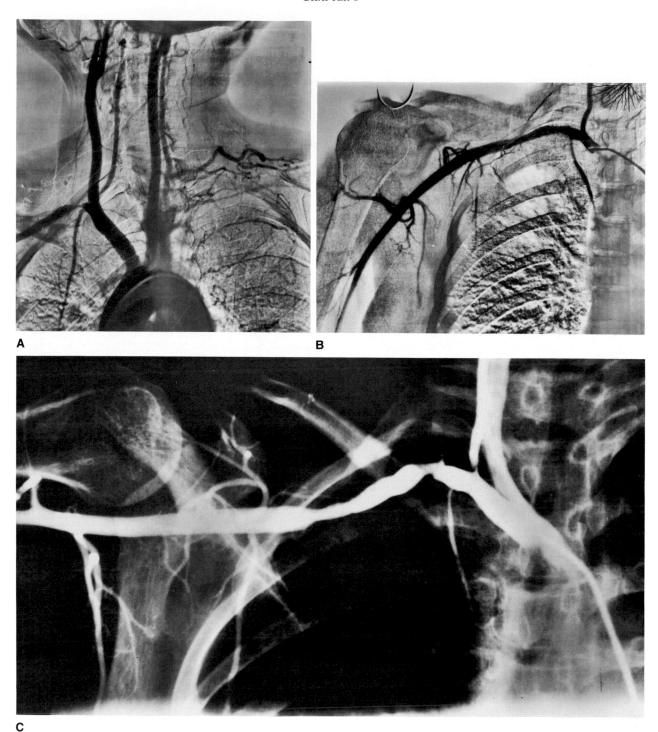

Figure 1. **A.** Tight stenosis at the origin of the right subclavian artery; the vertebral artery is also narrowed. Left subclavian artery is completely occluded, consistent with Takayasu's disease. **B.** Appearances after successful dilation and elimination of the pressure gradient. **C.** Follow-up study 18 months later. Vertebral artery stenosis appears to have progressed, and the origin of internal mammary artery shows a new stenosis. More distal irregularities that were not dilated on the first occasion have also progressed. Four years after first dilation, patient has a good pulse clinically.

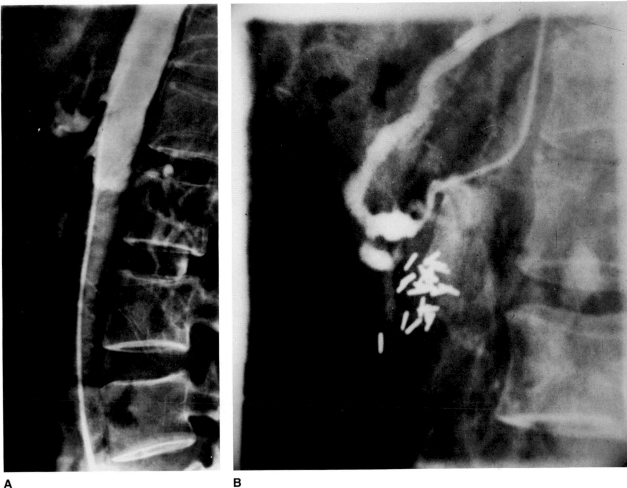

A   **B**

Figure 2.   A. Lateral aortogram of a patient with a median arcuate ligament syndrome. Stenosis is now fixed. The patient had abdominal pain for 8 years. **B.** Appearance after dilation with a 7 French 6 mm balloon, with procedure performed from the left axilla, shows no change in the appearances or pressure gradient.

and fibrosis around the artery and eventually develops a fixed stenosis. There is no logical reason for such a lesion to cause mesenteric ischemia, yet these patients, predominately females, have long histories of abdominal pain. Almost invariably they have been investigated on numerous occasions, and, perhaps as a result of being told nothing is wrong, they usually have had psychiatric treatment for many years because they persist in complaining of abdominal pain. They are often relieved of pain by dividing the median arcuate ligament and freeing up the celiac axis, perhaps by the disruption of the celiac plexus (15,16).

Other causes of abdominal pain secondary to mesenteric ischemia include vasculitis of the distal vessels in patients who have fibromuscular disease in a more proximal level of the same artery (17). The small-artery stenosis secondary to vasculitis is not a good indication for transluminal angioplasty.

Angioplasty Technique

A lateral projection of the aortogram is the ideal way of visualizing the mesenteric vessels in the median arcuate ligament syndrome. Films should be obtained both in deep inspiration and full expiration. If the aortogram is negative in the lateral projection, an AP view of the aorta should be obtained to look for a vasculitis. Selective studies may well be necessary to exclude obvious differentials, such as carcinoma of the pancreas, which also presents a clinical picture including abdominal pain and weight loss.

Only one of the celiac or the superior mesenteric arteries may be dilated, since one-vessel dilation is sufficient to alleviate the ischemic symptoms in mesenteric occlusive disease. We choose the easier of the two vessels, preferably the superior mesenteric artery because dilation of the celiac artery in the median arcuate ligament syndrome is usually unsuccessful.

If the angle of origin of the superior mesenteric artery is favorable, we would catheterize the artery directly with a cobra or a Simmons' catheter. We heparinize with 2000 Units and exchange over a Rosen J wire for an appropriate balloon catheter chosen to be 1 mm larger in diameter than the vessel being dilated. We prefer to cross the lesion with a Newton guidewire before making the balloon catheter exchange. Polyvinyl catheters follow better than polyethylene catheters if the angle is more acute than anticipated. In the majority of patients, the angle of origin of the superior mesenteric artery is too acute to dilate without kinking the balloon during inflation. We would, therefore, choose to perform most dilations from the axillary artery.

Clinical Results

In November 1980, Novelline reported the successful dilation of an arteriosclerotic superior mesenteric artery in a patient with mesenteric angina. This same patient also had the median arcuate ligament syndrome. The first attempt to dilate the celiac artery was not successful, but after successful dilation of the superior mesenteric artery, the patient was cured of her symptoms (8). In the same month Saddekni reported a patient operated upon for the median arcuate ligament syndrome who developed a restenosis 3 months later; in that patient percutaneous balloon dilation was successful (18). Birch reported two successful dilations in mesenteric ischemia (14). In 1982 Golden reported seven patients who were successfully dilated and received extensive follow-up (17). All of Golden's cases had mesenteric angina with two vessel involvements. In six patients the stenoses were due to arteriosclerosis and in one patient to fibromuscular disease (Fig. 3). In six patients the procedure was performed from the axilla. The superior mesenteric artery was dilated in all patients, with a primary success in six. All six patients were completely relieved of their symptoms, with follow-up extending to 28 months in one patient. Single case reports provide comparable data (13,19,20).

It appears that the median arcuate ligament syndrome is seldom amenable to dilation, a finding that concurs with our own experience. We would certainly attempt dilation at the time of the diagnostic study, choosing a polyethylene balloon and attempting dilation with a reinforced Olbert balloon, but would not be hopeful of the outcome unless the patient had had previous surgery (18).

THE HEPATIC ARTERY

Intra-arterial infusion of chemotherapeutic agents, particularly in the liver, has found favor because of the increased drug concentration delivered to the tumor. However, one report suggested that up to 40 percent of these arteries will thrombose or occlude as a result of catheter trauma, prolonged catheterization, vasculitis, or direct irritation from vinca alkaloids, adriamycin, or mitomycin C (21). Reports from M. D. Anderson Hospital described negotiating these occluded segments with a guidewire and dilating the occlusions by passage of a catheter to enable perfusion to be repeated. They reported negotiating 9 of 10 patients successfully without causing perforation (22).

The technique is of limited general value, but in those institutions where intra-arterial infusion is widely practiced, it would be a useful procedure.

THE ABDOMINAL AORTA

Angioplasty of the abdominal aorta will be considered under two separate subheadings: angioplasty in (1) atheromatous disease and in (2) the middle aortic syndrome.

Atherosclerotic Disease of Aorta

Atheromatous disease of the aorta is particularly common, and aortic stenosis or occlusion not rare. In most patients it is associated with iliac disease and, when symptomatic, is treated by an aortofemoral bypass graft, a procedure that has a high success rate (23). However, in patients for whom the risk of major aortofemoral surgery is judged too high, percutaneous balloon angioplasty of the aorta may be considered. We would choose to dilate patients who have only short-segment lesions.

Angioplasty Technique

Once it is established that the aortic lesion is responsible for symptoms, we approach the aortic lesion from each iliac artery with two 9 mm balloons, side by side (Fig. 4). This technique was developed because 9 mm balloons were the largest balloon size

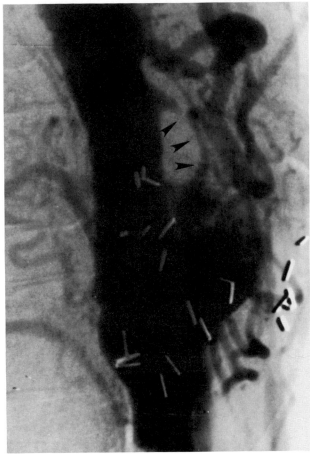

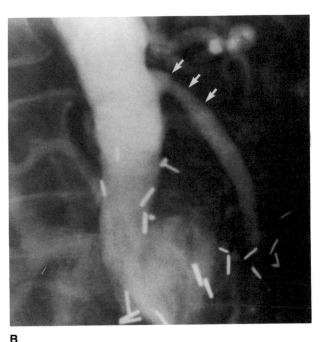

A   **B**

Figure 3.   A. Patient with hyperlipidemia and extensive atherosclerotic disease who had a previous bypass graft with the iliac limbs led to each renal artery. Arrows point to a tight long-segment stenosis of superior mesenteric artery. **B.** A follow-up arteriogram 15 months later, showing a normal superior mesenteric artery (arrows). (Reprinted from Ref. 17. With permission.)

available. A single 9 mm balloon is too small in diameter for aortic angioplasty, but two 9 mm balloons are adequate for aortic dilations.

Indeed, Kumpe reported using three balloons in one patient to achieve a more ideal dilating circumference (24). Now, however, balloons up to 20 mm in diameter are available, and the technique of angioplasty may, perhaps, be modified.

Nevertheless, although midaortic stenoses several centimeters above the bifurcation may be dilated with a single large-diameter balloon, we prefer two balloons side by side, the so-called kissing balloon technique, for dilating lesions at the aortic bifurcation. We have performed dilation in such circumstances with guidewires placed high up in the aorta to keep the catheter tips away from the aortic wall. If guidewires are not used, the tips of the di-

lating catheters, usually 1 cm from the balloon, may be pushed against the contralateral aortic wall and may cause an intimal dissection.

Since the aorta is a high-flow artery, we have not heparinized patients but recommend 6 months of aspirin therapy of 325 mg per day. One could argue, however, that if dilation is performed with two guidewires in place, heparin should be used; if it is, we administer 2000 U intra-arterially at the beginning of the procedure, and particularly in low-flow situations.

Clinical Results

Only isolated case reports may be found in the literature (25–27), the largest series being three patients reported by Kumpe (24). All were successful.

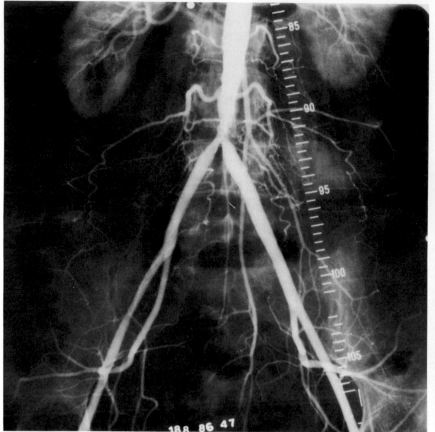

A

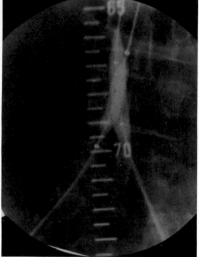

B

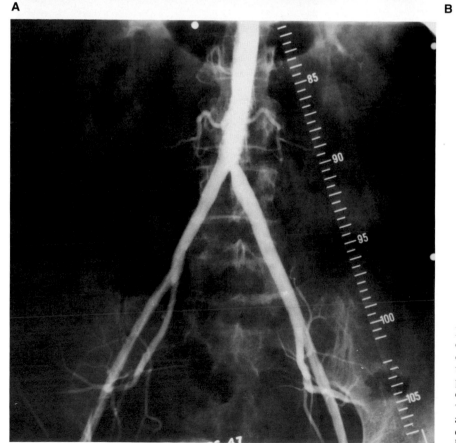

C

Figure 4. **A.** Tight stenosis of abdominal aorta involving the origin of each common iliac artery. Note the lumbar collaterals. **B.** Dilation performed with two 9 mm polyethylene balloons; note that guidewires are in place. **C.** Good result after dilation with bilateral removal of pressure gradients. Lumbar collaterals are smaller and do not contribute to antegrade flow.

Most reports have described the use of two balloons because the majority were dealing with bifurcation stenoses. We have dilated four patients with a primary success in all and have always used two balloons over guidewires.

We would not choose to dilate an aortic occlusion but rather would perfuse with streptokinase in an attempt to dissolve thrombus and to identify a potentially dilatable stenosis.

Aortic Coarctation and Middle Aortic Syndrome

Aortic coarctation and middle aortic syndrome comprise a heterogeneous collection of diseases. *Middle aortic syndrome* is an aortic narrowing commonly associated with renal artery involvements. Aortic narrowing may also be caused by fibromuscular disease (although we have not seen such a case) and by aortic hypoplasia in the supravalvular aortic stenosis syndrome. Such a lesion was described by Lock (42) in the thoracic aorta and was successfully dilated. We have seen it in the abdominal aorta but have not dilated it. We would also include under the term artery *abdominal aortic coarctation* a congenital form of the disease as opposed to an acquired aortitis, although excluding an aortitis as a cause of aortic coarctation may be difficult, and the differentiation may be suspect. Radiation aortitis may cause significant aortic stenosis; we have not dilated such a lesion. Radiation-caused arterial narrowing has been successfully dilated in the iliac artery (18). Retroperitoneal fibrosis may cause aortic narrowing; suc-

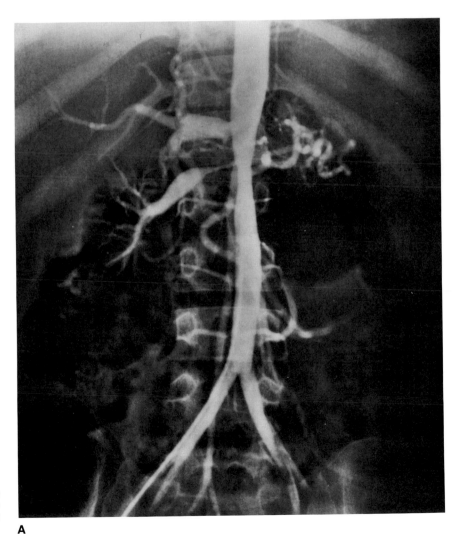

Figure 5. **A.** Takayasu's disease with a middle aortic syndrome; severe stenosis of right renal artery and complete occlusion of left.　　**A**

cessful dilation of such a lesion has been reported by Haynes (28).

Clinical Results

We have dilated one patient with a middle aortic syndrome secondary to Takayasu's disease and a patient in whom it was secondary to neurofibromatosis; both had associated renal artery stenosis (7). The patient with Takayasu's disease had left renal artery occlusion and a tight right renal artery stenosis. The left renal artery occlusion was opened but not successfully dilated. The right-sided stenosis did not yield to the dilating balloon. At the end of the procedure an attempt to dilate the abdominal aorta met with no success (Fig. 5). Castanada-Zuniga has claimed successful dilation of renal artery in Takayasu's disease (29), and we have successfully dilated a subclavian artery but know of no reports on the abdominal aorta. The patient with neurofibromatosis had bilateral renal artery stenoses, which were unsuccessfully dilated, as was the middle aortic lesion (Fig. 6).

In 1983, Nanni successfully dilated an abdominal aortic coarctation in a 12-year-old patient by using an 8 mm balloon; the resulting lumen remained patent over a year's follow-up. Since aortitis was not documented in this patient, the coarctation was assumed to be congenital (30).

We have treated a similar patient, aged 7, in whom the etiology was similarly unclear. There was neither evidence of Takayasu's disease nor of neurofibromatosis, although the patient may have been too young to demonstrate the pathologic stigmata, and we again assumed it to be a congenital coarctation. Flow to the distal aorta was supplied through

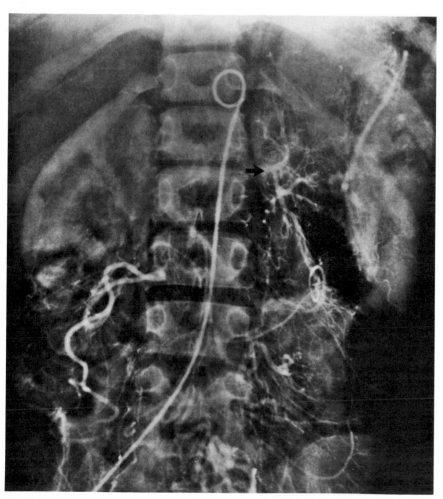

B

Figure 5 (Continued). **B.** Reconstitution of left renal artery (arrow). Occlusion of some ileal vessels and disease in the hepatic artery are seen.

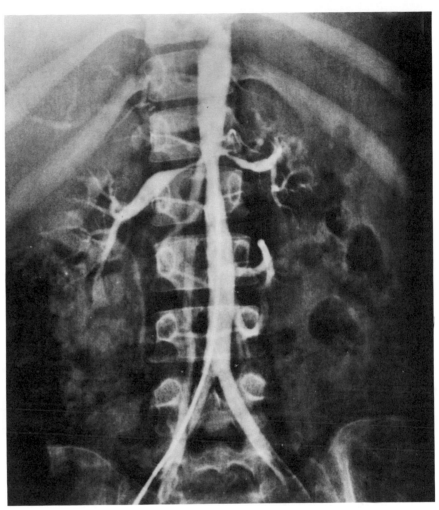

Figure 5 *(Continued).*   C. Attempt to dilate the aorta was unsuccessful. Continuity has been restored to the left renal artery with no change in pressure gradient and no cosmetic improvement of right renal artery. (Reprinted from Ref. 7. With permission.)

C

the superior mesenteric circulation and extensive posterior collaterals (Fig. 7). The complete occlusion was crossed with a 0.025 inch straight guidewire and followed by a number 5 French Van Andel catheter, which was advanced through the obstruction. An exchange was then made for a number 5 French 5 mm balloon catheter, which was inflated in the occluded segment. When the balloon was deflated the femoral pulses were noted to be restored; therefore, a number 8 French 8 mm balloon catheter was substituted. Two inflations of the 8 mm balloon for 30 seconds each never completely removed the constriction, and within seconds of deflation the femoral pulses diminished. An injection at the end of the procedure showed very little change in the appearances.

   In young children, catheter technology limits the size of the balloon available; we would not choose to use a number 8 French catheter in a child less than 6 or 7 years of age. Our impression had been that such aortic stenoses are not amenable to dilation, although Nanni's report is certainly encouraging. Success may certainly be achieved in aortic stenoses due to retroperitoneal fibrosis, radiation, and, of course, atheroma.

VENOUS DILATATIONS

The graft dilations are more commonly performed in the coronary artery and iliofemoral systems. This section will discuss percutaneous balloon angioplasties in (1) angio-access fistulas (2) mesenterocaval shunts for portal hypertension, and (3) the inferior vena cava.

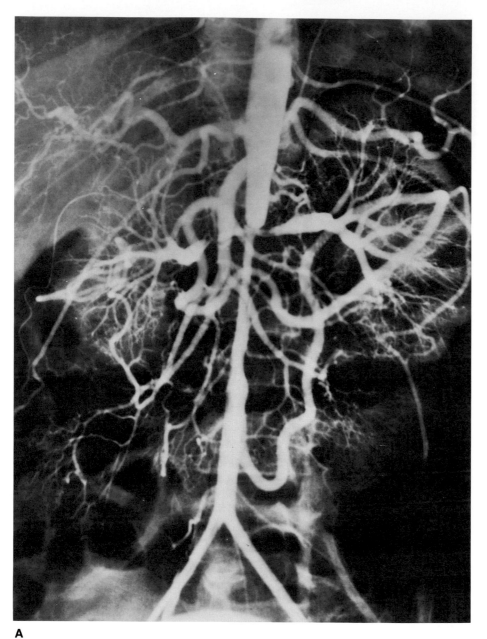

A

Figure 6. A. Neurofibromatosis with severe long-segment stenosis of abdominal aorta starting at the level of the renal arteries.

Angio-Access Fistulas

The most common arteriovenous fistula is the Brescia-Cimino radial artery to cephalic vein fistula at the wrist, which gives access to the veins of the forearm for chronic hemodialysis. Such patients face three main problems resulting from these fistulas: (1) arterial inflow stenosis that results in failure and ultimate thrombosis, (2) venous outflow stenosis that produces high dialysis pressures and therefore lengthy and inadequate dialysis, and (3) pseudoaneurysm formation from repeated venipuncture. When arteriovenous fistulae fail, they may be replaced by direct arteriovenous grafts, vein graft, bovine carotid artery, or expanded polytetrafluoroethylene (PTFE; Gore-Tex) graft. These suffer the same three problems, with venous outflow disease predominating (31). Venous outflow stenosis appears to be due to

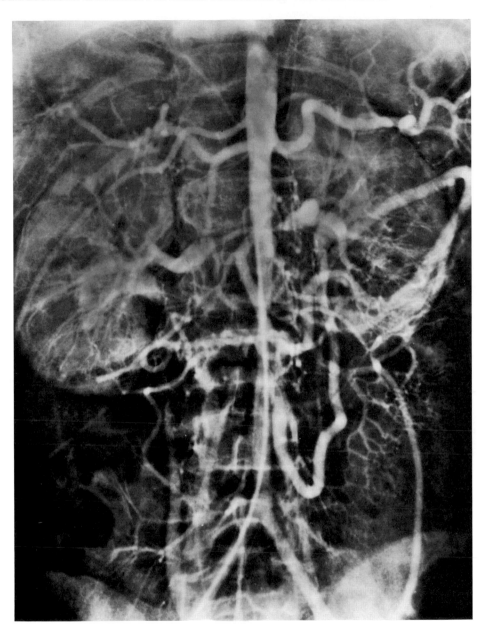

Figure 6 *(Continued)*. **B.** After dilation of each renal artery and abdominal aorta by a 5 mm balloon there is some improvement in the aortic lumen at the level of the renal arteries, although the pressure gradient was not removed. (Reprinted from Ref. 7. With permission.)

B

a neointimal peel that causes circumferential thrombus formation, often complicated by perivascular fibrosis resulting from the repeated venipunctures.

Problems of angio-access fistulas and shunts require preliminary angiographic documentation. Arteriovenous fistulas may be demonstrated by arteriography performed from the brachial artery, or by catheterization from the groin. Direct puncture of arteriovenous shunts demonstrates the venous out-

flow; if a cuff is inflated around the upper arm, the refluxing contrast adequately opacifies the arterial inflow. Short-segment stenoses appear amenable to dilation. If possible, we dilate arterial inflow stenoses from the venous side. A retrograde approach also optimizes the dilation of venous outflow stenosis. But it is acceptable to take an antegrade approach through the venous portion of an arteriovenous fistula or through direct graft puncture in a

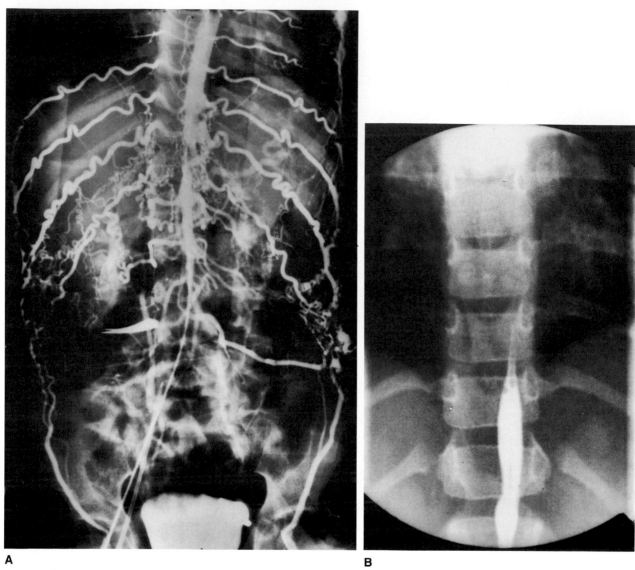

A   **B**

Figure 7.   **A.** An aortogram on a patient apparently without Takayasu's disease or neurofibromatosis; complete aortic occlusion, with the distal aorta fed retrogradely by markedly hypertrophied intercostal collaterals. **B.** The waist in the 8 mm balloon could not be completely removed.

shunt. We prefer not to approach from the arterial side but commonly do. Under such circumstances, we would give heparin, priscoline, xylocaine, and, if necessary, intra-arterial nitroglycerin to counteract spasm. Since perivascular fibrosis frequently makes dilation of venous outflow stenoses difficult, we have used large balloons of 8 and 9 mm in diameter and prefer polyethylene or Olbert balloons. We have also noticed that dilation is painful in such cases. Initially, we heparinized but no longer do so

except on the arterial side. Hemostasis may be achieved after successful dilation, and problem bleeding has not been reported.

Most dilations in the literature have been performed on the venous side, the majority in arteriovenous shunts rather than fistulas. The first report was from Heidler in 1978 (32), and in 1980 we reported on the dilation of three arteriovenous fistulas (7). In one patient we dilated both the arterial inflow and the venous side (Fig. 8) and in the other two,

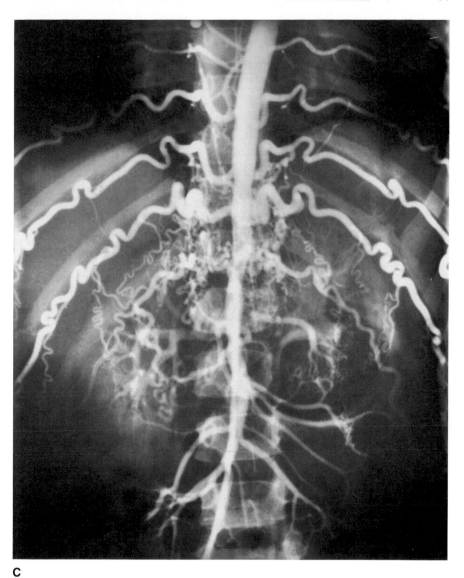

Figure 7 *(Continued)*. **C.** Aorta is now patent to the level of the superior mesenteric artery, but distal aorta still fills from collaterals. Femoral pulse diminished within seconds of balloon's deflation.   **C**

the venous outflow (Fig. 9). We utilized a retrograde approach in all patients for the venous side but needed an antegrade puncture for the arterial inflow. Lawrence in 1981 dilated venous outflow stenoses in six patients with arteriovenous shunts with success in three (33). In 1982 Probst was successful in dilating four out of five arteriovenous fistulas with stenosis at the arterial as well as the venous end (34). The largest report is from Gordon in 1982, who was successful in 15 out of 16 dilations (35). From his report the distribution between grafts and fistulas is not clear, but 15 of the 16 dilations were performed at the venous outflow. Summarizing these results, 20

out of 31 dilations were successful, a lower primary success rate than in the ileofemoral system. Furthermore, the follow-up is poor, and the long-term results appear disappointing.

Nevertheless the technique is worth attempting because it can be performed on outpatients and may prolong the lives of grafts or fistulas. The situation in which it is performed is dismal in any case as the average life expectancy of a fistula is only 1.6 years (36), and patients require 0.6 operations per year on their angio-access fistulas (37). Our impression is that dilation in arteriovenous fistulas is more successful than in shunts.

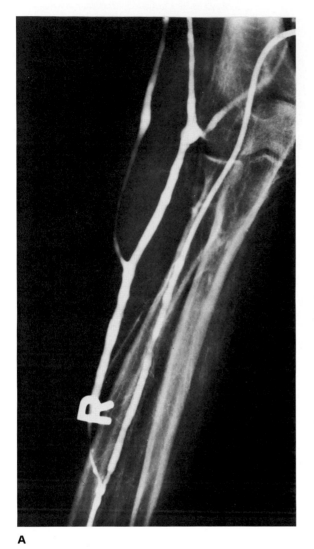

A

B

Figure 8. **A.** Modified Brescia-Cimino arteriovenous (AV) fistula; inflow stenosis on arterial side and relatively short segment venous occlusion just below the misplaced label. **B.** After dilation, arterial inflow was dilated by 5 French 5 mm balloon catheter from arterial side and the venous stenosis by a retrograde puncture of the arterialized vein. Both stenoses are improved cosmetically; some extravasation remains at venous puncture site. Associated spasm soon disappeared.

Mesenterocaval Shunts

In November 1980, Cope reported successful dilation of four patients with synthetic polyester shunts between the superior mesenteric vein and the inferior vena cava (38). Three of the patients had total occlusions, and one shunt was stenotic. He described probing the anterior wall of the cava with a guidewire until the Dacron prosthesis was entered and then sequentially dilating until a 9 mm balloon could be accommodated. Cope also used the access gained to embolize residual varices and suggested that such shunts should be studied intermittently so that prophylactic dilation could be undertaken to

maintain shunt patency. Cope postulated that stenosis starts on the mesenteric side, where low flow produces layering of fibrin until the graft occludes. In the same paper he reported the successful dilation of a recently thrombosed (36 hours) portocaval shunt. The shunt has remained patent after dilation for 16 months.

Bredfeldt also reported the successful dilation of a stenotic portocaval shunt in which the stenosis occurred in the first week after surgery (39). Novelline described a Warren shunt that had stenosed 3 years after surgery and was successfully dilated (8).

Certainly reoperation in such patients carries a

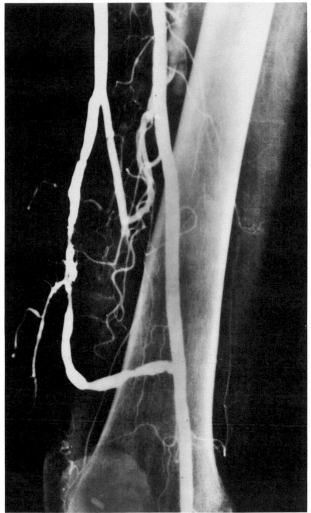

A

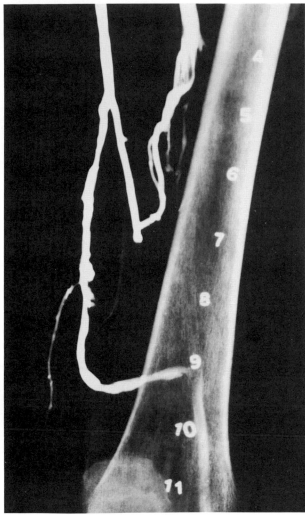

B

Figure 9. **A.** AV fistula between femoral artery and saphenous vein, with stenoses in midportion. **B.** Stenoses approached from right femoral vein retrogradely, with a good cosmetic result. Fistula functioned well until patient's death 9 months later. (Reprinted from Ref. 7. With permission.)

significant mortality, and it may be lifesaving to open and dilate the anastomoses angiographically. Cope has suggested that anastomotic sites be marked with metallic markers and that regular venous studies be performed on patients so that prophylactic dilation can be undertaken. It is obviously an excellent technique but still in its infancy (Fig. 10).

Budd-Chiari Syndrome and Inferior Vena Cava

In 1981, Meier reported on the dilation of a web that was causing a Budd-Chiari syndrome by obstructing the main hepatic vein as it entered the cava (40). The web was dilated with two 8 mm balloons, one from each femoral vein, and although the stenosis recurred at 2½ months, it was redilated successfully.

Web obstruction of the inferior vena cava causing the Budd-Chiari syndrome has been reported previously, but a web obstructing the hepatic veins appears to be unique.

In 1982, Yamada reported dilating five patients with inferior vena cava obstruction due to a web causing the Budd-Chiari syndrome (41). All five patients were successfully treated and remained so

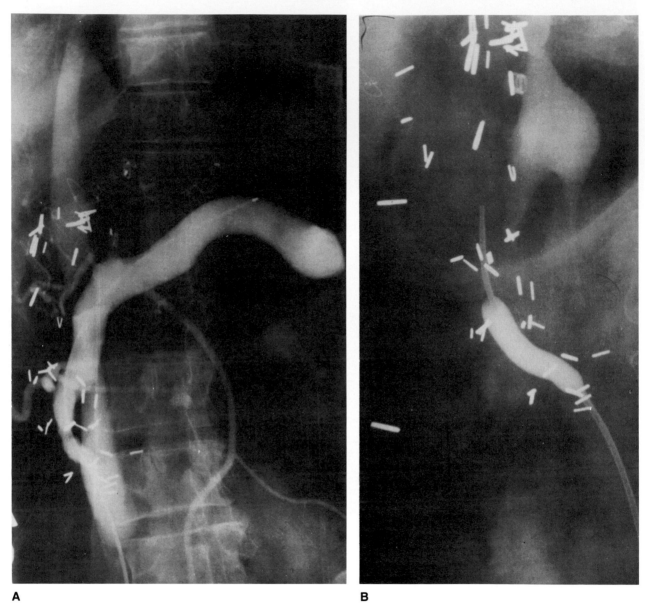

A   **B**

Figure 10. **A.** Injection after catheterizing a mesenterocaval shunt; long-segment stenosis at the level of the anastomosis with a 35 mmHg gradient. **B.** Dilation balloon (9 mm) in position; gradient was reduced to 11 mmHg. (Courtesy of Dr. Ernest J. Ring, San Francisco.)

over a long follow-up (Fig. 11). The technique he used involved four balloons, two placed in each femoral vein and inflated simultaneously in the inferior vena cava. He was using 9 mm balloons to achieve a 30–40 mm caval diameter; with the newer large balloons, two 20 mm balloons, one from each femoral vein, should suffice.

CONCLUSION

This chapter has described percutaneous transluminal balloon dilation for a collection of heterogeneous categories of vascular conditions. The percutaneous techniques for these rarer conditions were borrowed from the basic percutaneous techniques

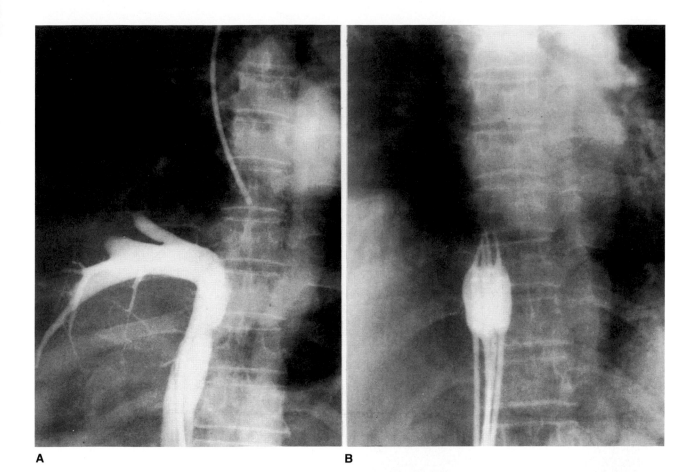

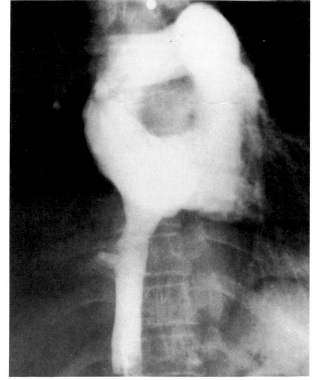

Figure 11. **A.** Complete occlusion of inferior vena cava by a web; patient suffers from a Budd-Chiari syndrome; this combination is more common in Japan. **B.** Four 8 mm balloon catheters inflated in the lesion, two catheters introduced from each femoral vein. **C.** Appearances after dilation; patient was clinically improved. (Courtesy of Dr. R. Yamada of Wakayama, Japan.)

A

B

C

of the iliofemoral system and modified for specific needs of individual categories of the rarer conditions. Since it is difficult to build up any high level of skills with these uncommon conditions, they should probably be attempted by experienced angiographers only, not because the techniques are necessarily more difficult, but because an ad hoc approach is called for in such circumstances. It will take some time before meaningful series are accrued and adequate follow-up results known in these categories of rarer vascular conditions.

REFERENCES

1. Beebe HG, Stark R, Johnson ML, Hill LD: Choices of operation for subclavian vertebral artery disease. Surgery 139:616–623, 1980.
2. Herring M: The subclavian steal syndrome—a review. Surgery 43:222–228, 1977.
3. Fields WS, Lemak NA: Joint study of extracranial arterial occlusion VII subclavian steal—a review of 168 cases. JAMA 222:1139–1143, 1972.
4. Motarjeme A, Keifer JW, Zuska AJ: Percutaneous transluminal angioplasty of the brachiocephalic vessels. AJR 138:457–462, 1982.
5. Bachman DM, Kim RM: Transluminal dilatation for subclavian steal syndrome. AJR 135:995–996, 1980.
6. Ring EJ, McLean GK: *Interventional Radiology: Principles and Techniques.* Boston, Little, Brown, 1980, p 190.
7. Martin EC, Diamond NG, Casarella WJ: Percutaneous transluminal angioplasty in non-atherosclerotic disease. Radiology 135:27–33, 1980.
8. Novelline RA: Percutaneous transluminal angioplasty: Newer applications. AJR 135:983–993, 1980.
9. Hodgins GW, Dutton JW: Subclavian and carotid angioplasties for Takayasu's disease. J Can Assoc Radiol 33:205–207, 1982.
10. Morris GC, DeBakey ME: Abdominal angina: Diagnosis and treatment. JAMA 176:89–92, 1961.
11. Stoney RJ, Ehrenfeld WK, Wylie EJ: Revascularization methods in chronic visceral ischemia caused by atherosclerosis. Ann Surg 186:468–476, 1977.
12. Reul GJ, Wukash DC, Dandiford EM, Chiarillo L, Hallman GL, Cooley DA: Surgical treatment of abdominal angina—review of 25 patients. Surgery 75:682–689, 1974.
13. Castanada-Zuniga WR, Gomes A, Weenes C, Ketchum D, Amplatz K: Transluminal angioplasty in the management of mesenteric angina. Fortschr Rontgenstr 137:330–332, 1982.
14. Birch SJ, Colapinto RF: Transluminal dilatation in the management of mesenteric angina—a report of two cases. J Can Assoc Radiol 33:46–47, 1982.
15. Snyder MA, Mahoney EB, Rob CG: Symptomatic celiac artery stenosis due to constriction by the neurofibromatous tissue of the celiac ganglion. Surgery 61:372–380, 1967.
16. Stoney RJ, Wylie EJ: Recognition and surgical management of visceral ischemic syndromes. Ann Surg 164:714–721, 1966.
17. Golden DA, Ring EJ, McLean GK, Frieman DB: Percutaneous angioplasty in the treatment of abdominal angina. AJR 139:247–249, 1982.
18. Saddekni A, Sneiderman KW, Hilton S, Sos TA: Percutaneous transluminal angioplasty of non-atherosclerotic lesions. AJR 135:975–982, 1980.
19. Furrer J, Gruentzig A, Kugelmeier J, Goebel N: Treatment of abdominal angina with percutaneous dilatation of an arterial mesenteric superior stenosis. Cardiovasc Intervent Radiol 3:43–44, 1980.
20. Uflacker R, Goldany MA, Constant S: Resolution of mesenteric angina with percutaneous transluminal angioplasty of a superior mesenteric artery stenosis using a balloon catheter. Gastrointet Radiol 5:367–369, 1980.
21. Oberfield RA, McGaffrey JA, Pulio J, Klauss ME, Hamilton T: Prolonged and continuous percutaneous intraarterial hepatic infusion chemotherapy in advanced metastatic liver adenocarcinoma from colorectal primary. Cancer 44:414–423, 1979.
22. Soo CS, Chuang VP, Wallace S, Carrasco H: Catheterization of thrombosed arteries for intraarterial cancer treatment. AJR 140:607–609, 1983.
23. Brewster DC, Darling RC: Optimum methods of aortoiliac reconstruction. Surgery 84:739–748, 1981.
24. Kumpe DA: Percutaneous dilatation of an abdominal aortic stenosis. Radiology 141:536–537, 1981.
25. Grollman JH, Del Vicario M, Mittal AK: Percutaneous transluminal abdominal aortic angioplasty. AJR 134:1053–1054, 1980.
26. Tegtmeyer CJ, Wellons HA, Thompson RN: Balloon dilatation of the abdominal aorta. JAMA 244:2636–2637, 1980.
27. Velasquez G, Castanada-Zuniga WR, Formanek MD, Zollikofer C, Barreto A, Nickloff D, Amplatz J, Sullivan A: Nonsurgical aortoplasty in Leriche syndrome. Radiology 143:359–360, 1980.
28. Haynes GI, Simon J, West RJ, Hamer JD: Idiopathic retroperitoneal fibrosis with occlusion of the abdominal aorta treated by transluminal angioplasty. Br J Surg 69:432–433, 1982.
29. Castanada-Zuniga WR, Formanek A, Lillehei C, Tadavarthy M, Amplatz K: Nonsurgical treatment of Takayasu's disease. Cardiovasc Intervent Radiol 4:245–248, 1981.
30. Nanni GS, Hawkins IF, Alexander JA: Percutaneous transluminal angioplasty of and abdominal aortic coarctation. AJR 140:1239–1241, 1983.
31. Butt KMH, Freidman EI, Kuntz SL: Angioaccess. Curr Probl Surg 13:1–67, 1976.
32. Heidler R, Zeitler E, Gessler U: Percutaneous transluminal dilatation of stenosis behind AV fistulas in hemodialysis patients, in Gruentzig A, Zeitler E,

Schoop W, (eds): *Percutaneous Vascular Recanalization.* New York, Springer-Verlag, 1978.

33. Lawrence PF, Miller FJ, Mineau DE: Balloon catheter dilatation in patients with failing arteriovenous fistulas. Surgery 89:439–442, 1981.

34. Probst P, Marla F, Kreneta A, Descoeudres C: Percutaneous transluminal dilatation for restoration of angio access in chronic hemodialysis patients. Cardiovasc Intervent Radiol 5:257–259, 1982.

35. Gordon DH, Glanz S, Butt KM, Adamsons RJ, Koenig MA: Treatment of stenotic lesions in dialysis access fistulas and shunts by transluminal angioplasty. Radiology 143:53–58, 1982.

36. Giacchino JL, Geis P, Buckingham JM, Vertuno LL, Bansal VK: Vascular access: Long term results: New techniques. Arch Surg 114:403–409, 1979.

37. Mandel SR, Martin PL, Blumoff RL, Mattern WD: Vascular access in a university transplant and dialysis program. Arch Surg 112:1375–1381, 1977.

38. Cope C: Balloon dilatation of closed mesocaval shunts. AJR 135:989–993, 1980.

39. Bredfeldt JE, Pingoud EG, Groszmann M, Tilson MD, Conn HO: Balloon dilatation of stenotic portocaval anastomosis. Hepatology 1:448–454, 1980.

40. Meier WL, Waller RM, Sones PJ: Budd-Chiari web treated by percutaneous transluminal angioplasty. AJR 137:1257–1258, 1981.

41. Yamada R, Sato M, Kawabata M, Nakatsura H, Nakamura K: Successfully treated segmental obstruction of the hepatic IVC by means of transluminal angioplasty using Gruentzig balloon catheters. Presented at the Radiological Society of North America Chicago, 1982.

42. Lock JE, Bass JL, Amplatz K, Furlsman BP, Castaneda-Zuniga WR: Balloon dilatation angioplasty of aortic coarctation in infants and children. Circulation 68:109–116, 1983.

7

Percutaneous Transluminal Angioplasty of Carotid and Vertebral Arteries

ANTON N. HASSO, M.D.
C. ROGER BIRD, M.D.

INTRODUCTION

The use of balloon catheters to dilate stenotic lesions in the cephalic vessels started with vascular surgeons who wanted an improved intraoperative method to relieve nonulcerative fibrodysplastic lesions within the internal carotid arteries (1). Graduated Bakes biliary dilators (2–5 mm in diameter) were customarily inserted through a carotid arteriotomy to treat fibromuscular dysplasia (FMD) in the high cervical portion of the internal carotid artery. However, because of an unacceptable complication rate caused by the rigid metallic dilators, Fogarty or other balloon catheters were used as a substitute for the Bakes dilators (2,3). Concomitant developments in percutaneous techniques elsewhere in the human vascular system led us and others to employ transluminal angioplasty for fibromuscular dysplasia of the internal carotid artery (4–7).

A few cases of transluminal angioplasty in the internal or common carotid artery for arteriosclerotic occlusive disease have been reported (3,8–10). Generally the procedure is done on high risk patients who could not tolerate a surgical end-arterectomy.

There is great hesitancy to use percutaneous transluminal angioplasty for stenotic lesions in the supra-aortic vessels because of the potential for showering emboli distally into the cranial circulation. Most authors rightfully believe that the readily accessible carotid bifurcation lesions are safely treated by end-arterectomy during which the carotid artery can be clamped and allowed to back-bleed in order to wash out any debris through the arteriotomy incision.

On the other hand, stenotic lesion of fibromuscular dysplasia of the internal carotid artery may be entirely in the medial wall of the vessel, without involvement of the intima. This is particularly apparent in the most common form of fibromuscular dysplasia in which the disease is primarily a medial hyperplasia (6,11).

This chapter describes our experience with transluminal angioplasty for fibromuscular dysplasia involving the cervical internal carotid artery and occlusive disease involving the external carotid artery. A review of reports of other authors who have performed percutaneous transluminal angioplasty in the supra-aortic vessels such as the subclavian and vertebral arteries is included.

PATIENT SELECTION AND CARE

Patients having neurological symptoms such as transient ischemic attacks (TIAs), amaurosis fugax, drop attacks, or dizziness with angiographically documented stenotic fibromuscular dysplasia are candidates for percutaneous transluminal angioplasty. The model lesions are in the high cervical portions of the internal carotid artery which are often inaccessible without high neck dissection and its known risks of injury to the hypoglossal and/or vagus nerves. Patients with associated kinks, loops, or coils are best treated surgically, since it may be possible to resect a portion of the redundant vessel while relieving the stenotic lesion.

There must be a unanimity of opinion among the referring physician, vascular surgeon or neurosurgeon, and radiologist that the patient's symptoms are ascribable to fibromuscular dysplasia and that there is the possibility of meaningful improvement following the procedure. Noninvasive hemodynamic studies, such as oculoplethysmography and cerebral flow studies can provide additional information regarding the hemodynamic significance of the stenosis.

Patients with complete occlusion of the internal carotid artery frequently have ipsilateral focal cerebral or retinal ischemic episodes. Some cases may have concomitant critical stenosis of the ipsilateral external carotid artery origin. It is now recognized that transient ischemic attacks or amaurosis fugax in these patients may be caused by the occlusive disease in the external carotid vessel (7,12–13). The cerebral and/or retinal ischemia is due to transient blockage or emboli transmitted via the well-known transophthalmic collaterals developing in patients with severe carotid occlusive disease or, rarely, due to direct origin of the ophthalmic artery from the external carotid branches (14). Percutaneous transluminal angioplasty is indicated in these patients in order to (1) increase transophthalmic collateral blood flow to the brain and (2) decrease the likelihood of recurrent retinal ischemic events.

Angioplasty treatment of an external carotid artery stenosis is also useful in patients undergoing a superficial temporal artery to middle cerebral artery transcranial anastomosis to bypass an occluded internal carotid artery or stenotic carotid or middle cerebral artery inaccessible to direct surgery. Successful percutaneous transluminal angioplasty of the external carotid artery will likely increase blood flow through the superficial temporal branch of the external carotid artery and aid in keeping the anastomosis patent (13,15–16).

Arteriosclerotic lesions of the subclavian or vertebral arteries are rarely treated by direct surgical end-arterectomy. Both of these vessels require a thoracotomy for access to their origins with additional inherent surgical risks. Furthermore, the vertebral artery is difficult to end-arterectomize inasmuch as the vessel is small and highly susceptible to spasm. Subclavian artery stenosis is often managed by one of several extrathoracic bypass grafts such as caroticosubclavian, axilloaxillary, or femoroaxillary procedures. Patients in these two categories with smooth stenotic lesions have been considered candidates for percutaneous transluminal angioplasty. Several authors have reported successful angioplasties of the subclavian and vertebral arteries with a low incidence of complications (8,18–22).

Intraoperative transluminal balloon angioplasty may be performed as an alternative to direct end-arterectomy. The diseased artery is surgically exposed and a balloon catheter is positioned in the stenotic segment with the aid of a fluoroscope. An angioplasty is accomplished after which the vessel is allowed to back bleed. The procedure is particularly useful when percutaneous transluminal angioplasty is not desirable or not possible such as in the basilar artery (23).

Before internal carotid artery angioplasty, the procedure including risks and benefits is carefully explained to the patient and informed consent is obtained. We usually give the patient the choice of an intraoperative balloon angioplasty or a percutaneous transluminal angioplasty via a transfemoral approach, emphasizing the advantage of back bleeding with the former and the advantage of being awake with continuous monitoring of neurological signs and symptoms during the latter. The risks including transient cerebral ischemia (about 10 percent) and stroke with permanent deficits (about 1 percent) are discussed with the patient. The risks are relatively minor for external carotid angioplasty, probably no greater than the risks of conventional angiography.

Patients undergoing percutaneous transluminal angioplasty are placed on antiplatelet therapy with aspirin (300 mg PO, tid) and/or Persantine (dipyridamole, Geigy), (50 mg PO, tid) for 2 or 3 days prior to the procedure. Premedication usually consists of Valium (diazepam, Roche), 10 mg PO, dexamethasone, 10 mg PO, and atropine 0.4 mg IM. It is important that patients not be oversedated so that their neurological status can be closely monitored throughout the procedure.

Antiplatelet therapy with aspirin and/or dipyridamole is continued for at least 6 months following balloon angioplasty. Otherwise, the patients are treated in the same manner as patients experiencing conventional angiography.

TECHNIQUE

Following local anesthesia, a 5 or 7 French diagnostic catheter is introduced percutaneously into the common femoral artery and advanced into the common carotid artery on the involved side. A straight catheter with a gentle curve at the tip (HT-1, Cook) is used in most cases. In patients who have tortuous vessels or an elongated aortic arch, a catheter with a reverse curve (Simmons 3 or 4, Cook) is occasionally needed.

Once the common carotid artery has been selectively catheterized, an arteriogram is performed to precisely locate the stenosis in the internal or external carotid arteries. It is important to localize the stenosis fluoroscopically and identify a bony landmark or place lead markers on the skin to later aid in positioning the balloon.

The balloon size used in the internal carotid artery is 2–4 cm in length by 4–5 mm in diameter (Figure 1) and in the external carotid artery 1.5–2 cm by 4 mm. The ideal balloon size is determined by the length of the stenosis and diameter of the uninvolved part of the artery.

The balloon should always be placed through an introducer sheath in the groin which facilitates a smooth, atraumatic insertion and withdrawal through the arterial wall. Prior to insertion, the balloon is inflated with dilute contast while held upside down and then deflated to remove any trapped air bubbles.

After the arteriogram, the diagnostic catheter is exchanged for the 100 cm long 5 or 7 French polyethylene Gruentzig balloon catheter over a 240–260 cm long guidewire. During the exchange, it is important to observe the guidewire tip on the fluoroscope and prevent movement out of the common carotid artery.

Next, a floppy tipped guidewire (Bentson, Cook) is inserted and advanced. If there is difficulty selecting a branch of the common carotid artery, forming a gentle curve at the tip of the guidewire will usually help. The wire is then carefully advanced through the stenosis. If there is any resistance, a rotating movement of the wire during advancement will usually allow a smooth passage. The wire should never be forced. In the internal carotid artery, the wire should not be advanced further than the petrous carotid canal. In the external carotid artery, it should not be advanced further than the origin of the maxillary artery. Special care must be taken when manipulating the guidewire in the external carotid artery to prevent spasm.

Once the guidewire has been positioned, the balloon catheter is advanced and the balloon is placed at the site of previously identified stenosis. Hyperextension of the neck helps prevent wedging of the

Figure 1. Diagrammatic representation of the balloon catheter and inflating syringe.

catheter in the internal carotid artery. We routinely use continuous pressure monitoring via a transducer hooked up to the catheter. Slight dampening of the pressure waves is tolerated in the internal carotid artery. However, a completely flat wave form requires insertion of a smaller catheter.

Rotation of the neck during positioning in the external carotid artery helps in advancing the balloon. The balloon should be positioned so the stenosis is in its midportion to avoid migration during inflation. Balloons with radiopaque markers at each end make positioning much easier and quicker.

The balloon is then inflated under fluoroscopic control with dilute contrast material for approximately 5 seconds and then deflated. Care must be taken during inflation to avoid overdistension and possible rupture of the balloon. This may be done by a pressure gauge attached to the catheter. The inflation may be repeated, if necessary.

Once the stenosis has been relieved, as evidenced by fluoroscopic observation and relief of any significant pressure gradient, the catheter is withdrawn into the proximal internal carotid artery or common carotid artery and a postdilatation arteriogram is performed. It is best not to reposition the balloon in the stenotic area following dilatation because of increased chance of raising an intimal flap, producing a dissection or detaching emboli from the ruptured plaque or site of fibromuscular dysplasia.

After the postdilatation arteriogram, the balloon catheter is withdrawn while maintaining negative pressure to the balloon. Manual compression is used to maintain hemostasis in the groin.

The procedure for the subclavian artery has to be slightly modified, since the vessel is larger than the carotid artery and requires a 7 French catheter with a 6 mm diameter balloon or a 9 French balloon catheter (8,20). Generally, the guidewire is left anchored in position across the stenosis while the balloon is inflated, which limits migration of the balloon beyond the area of stenosis (20). The balloon must remain inflated for 10 seconds or longer in order to obtain the desired results of dilatation.

Percutaneous transluminal angioplasty of the vertebral artery requires a 5 French balloon catheter. No pressure monitoring is used since even this small catheter may cause damped pressure waves before and after dilatation and cannot be used as an accurate indicator of dilatation results (19). Moderate dampening of the pressure is usually withstood by the patient since the contralateral vertebral artery continues to supply the brain. A patient with a lesion in a solitary vertebral artery or bilateral critical lesions should be considered an extremely high risk for angioplasty.

Patients with atherosclerosis are generally given a loading dose of heparin (5–10,000 I.U.) through the catheter before or after the stenosis has been crossed. Some authors recommend this regimen plus a continuous heparin drip for patients with fibrodysplastic disease of the internal carotid artery. We have not given heparin for balloon angioplasty of fibromuscular dysplasia, since these patients have an increased risk of dissection and occlusion or fistula formation (23). An intimal tear causing dissection or an intramural hematoma might be aggravated by heparinization. More importantly, a cerebral pale (thrombotic or embolic) infarct may be converted to a red (hemorrhagic) infarct with heparinization.

ANGIOGRAPHIC AND CLINICAL RESULTS

We have successfully treated eight patients with fibromuscular dysplasia of the internal carotid artery (Table 1). In seven patients the transfemoral approach was used and in one patient an intraoperative approach was chosen. Technical success, as evidenced by increase in the luminal diameter by at least 50 percent was achieved in all of the cases (Figures 2 and 3). All patients had complete or significant resolution of cerebral ischemic symptoms. The cases with audible neck bruits also had relief of ischemic signs following the procedure.

One patient (VH, Case 4, Table 1) developed a right hemispheric ischemic episode during the angioplasty which was documented angiographically as an embolus to the right middle cerebral artery. The symptoms resolved within 20 minutes without any long-term sequela. The patient had a successful angioplasty (Figure 3) following the ischemic event. A subsequent cranial CT scan showed no evidence of cerebral infarction.

Most patients developed small fissures in the intimal walls of the internal carotid artery after angioplasty (Figures 2 and 3). This is likely caused by splitting of the intima and media during expansion of the balloon (Figure 4) which leads to an increase in the luminal diameter of the vessel.

Nine patients with critical origin stenosis of the external carotid artery have been treated (Table 2). Eight of the nine had origin occlusion of the ipsilateral internal carotid artery and one had critical stenosis of the cavernous part of the ipsilateral internal carotid artery. Decrease of the stenosis to less than 50 percent of the luminal diameter was attained initially in eight of nine patients (Figures 6 and 7). In one case the stenosis was successfully dilated during a repeat procedure. None of the nine patients experienced neurological symptoms during angioplasty. One patient had spasm in the external carotid artery just distal to the stenosis on the postdilatation arteriogram (Figure 5).

Reports on the percutaneous balloon angioplasty of internal or common carotid artery for occlusive disease have been sparse for the reasons discussed earlier. No more than a handful of cases have been attempted and no double blind statistical comparisons vis-a-vis conventional end-arterectomy vs angioplasty are available. Preliminary data suggests an approximately 20 percent complication rate (4,5) which is well above the usual norm for end-arterectomy.

Percutaneous transluminal angioplasty of the vertebral artery may be of value in patients suffering from vertebrobasilar insufficiency. Proximal stenotic lesions without calcific plaques or ulcerations have been successfully treated in 11 out of 13 patients with no significant complications (23). Reports on subclavian artery angioplasty in nearly 20 cases are equally encouraging, particularly in patients with a subclavian steal syndrome (8,18–21).

TABLE 1 Clinical and Angiographic Findings: Internal Carotid Artery

Case, Initials Age, Gender	Initial Signs and Symptoms	Arteriographic Findings	Angioplasty Methods	Technical Success	Complications	Clinical Results	Follow-Up Period
#1, AW 54, F	Rt amaurosis fugax bruit rt carotid	FMD beading rt ICA	Transfemoral	+	None	No bruit, symptoms resolved	28
#2, CL 52, F	Lt amaurosis fugax bruit lt carotid	FMD beading lt ICA	Transfemoral	+	None	No bruit, symptoms resolved	22 mo
#3, JR 61, F	Transient rt cerebral ischemia	FMD with 75% web, stenosis rt ICA	Transfemoral	+	Minor spasm ICA	Symptoms initially resolved 10 mos later, 2 TIA's	24 mo
#4, VH 71, F	Transient dysphasia, weakness and numbness of rt face	FMD beading rt ICA	Transfemoral	+	Embolus to rt MCA with TIA quickly resolved	Symptoms resolved	30 mo
#5, CL 51, F	Sudden rt cerebral ischemia, bruit rt carotid	FMD beading rt ICA	Intraoperative	+	None	No bruit, symptoms resolved	28 mo
#6, AW 53, F	Minor stroke residual, lt cerebral TIS's, lt amaurosis fugax	FMD beading lt ICA	Transfemoral	+	None	Decreased frequency and severity symptoms	Angio, 6 wk later, no change
#7, JS 53, F	Pulsatile ringing in lt ear, lt carotid bruit	FMD with web stenosis, lt ICA	Transfemoral	+	None	No bruit, symptoms resolved	Angio, 18 mo later, no change
#8, MA 75, F	Dizziness, recurrent lightheadedness	FMD beading bilateral ICA, stenosis rt ICA	Transfemoral	+	None	No bruit, symptoms resolved	15 mo

Abbreviations: rt = right, lt = left, FMD = fibromuscular dysplasia, ICA = internal carotid artery, mo = month, TIA = transient ischemic attack, MCA = middle cerebral artery, angio = angiogram

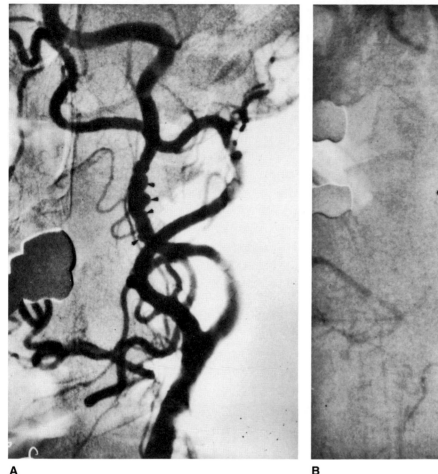
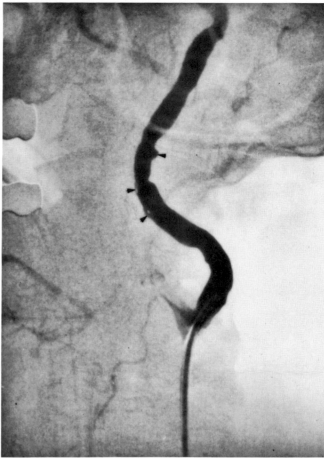

A
B

Figure 2. CL/A 52-year-old female. Predilation AP left common carotid (A), and postdilation AP left internal carotid (B) arteriograms. **A.** Typical "string of beads" appearance of the cervical internal carotid artery (arrows). **B.** Following dilatation there is increased caliber to vessel lumen. Small "fissures" are noted in the vessel wall (arrows).

FOLLOW-UP RESULTS

There is evidence that durable lumen patency can be expected following surgical treatment for occlusive disease due to fibromuscular dysplasia of the internal carotid artery. Some cases of fibromuscular dysplasia treated by open graduated internal dilatation have remained widely patent on follow-up angiography several years later. Clinically, our cases have remained under control for up to 30 months (Table 1). One patient had a follow-up angiogram 18 months after angioplasty with no change in the previously documented appearance of the internal carotid artery.

The natural history of fibromuscular dysplasia in the cephalic vessels has not been well documented, but could be progressive as it appears to be in the renal arteries. One study showed increased disease in two of six patients who had repeated angiograms 2 to 4 years later (24). It must be mentioned that percutaneous transluminal angioplasty only relieves a stenotic lesion and does not cure fibromuscular dysplasia. Since the disease may continue to advance, long-term follow-up is needed.

The patients with external carotid artery stenosis who later underwent a superficial temporal artery to middle cerebral artery anastomosis (Cases 1, 2, 5, 8, and 9, Table 2) had subsequent angiograms in order to document patency of the transcranial anastomosis. The postoperative studies showed no change in the angioplasties during the 7–10 day follow-up period. The experience of other authors is similar with generally good results in the external carotid vessel (13,17).

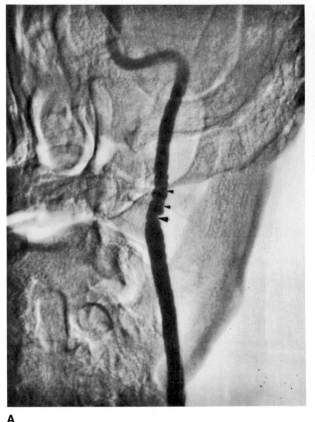

A

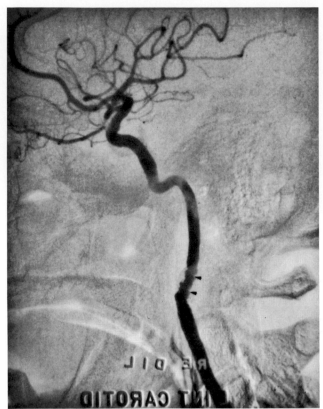

B

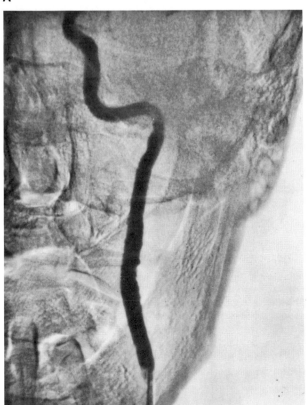

C

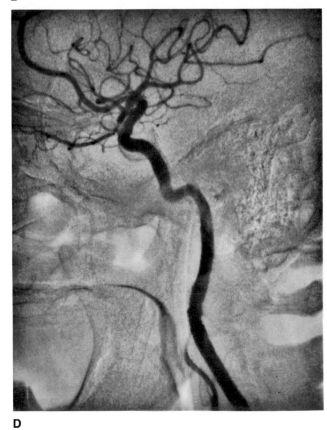

D

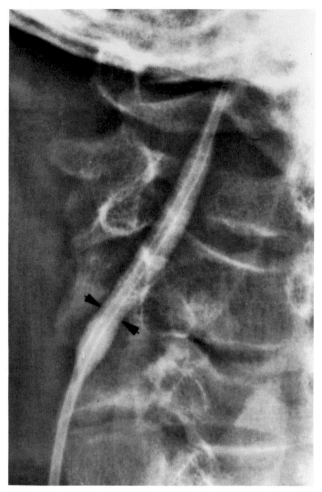

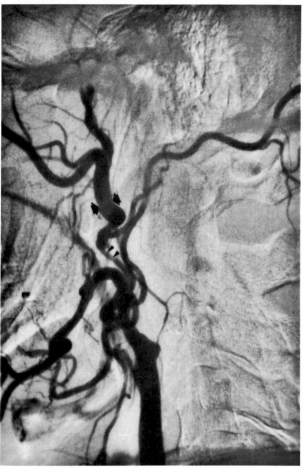

Figure 4. Film taken during dilatation with the balloon catheter inflated shows "waisting" of the balloon at the site of stenosis.

Figure 5. HO/A 71-year-old male with amaurosis fugax. Right common carotid arteriogram (lateral view) following dilation of the external carotid artery origin. Note the spasm in the external carotid artery just distal to the origin (small arrows). Note normal luminal diameter beyond the area of spasm (arrow heads).

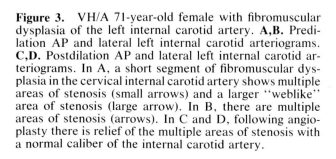

Figure 3. VH/A 71-year-old female with fibromuscular dysplasia of the left internal carotid artery. **A,B.** Predilation AP and lateral left internal carotid arteriograms. **C,D.** Postdilation AP and lateral left internal carotid arteriograms. In A, a short segment of fibromuscular dysplasia in the cervical internal carotid artery shows multiple areas of stenosis (small arrows) and a larger "weblike" area of stenosis (large arrow). In B, there are multiple areas of stenosis (arrows). In C and D, following angioplasty there is relief of the multiple areas of stenosis with a normal caliber of the internal carotid artery.

There is only limited follow-up available on the reported cases of vertebral and subclavian percutaneous transluminal angioplasty (8,18–21). No cases have returned for recurrent disease and no doubt the results will be similar to that in vessels of equivalent size elsewhere in the human vascular tree.

CONCLUSION

Despite dramatic advancements in peripheral and coronary arterial percutaneous transluminal angioplasty, progress in the cephalic vessels has been slow and hesitant. In the opinion of the authors,

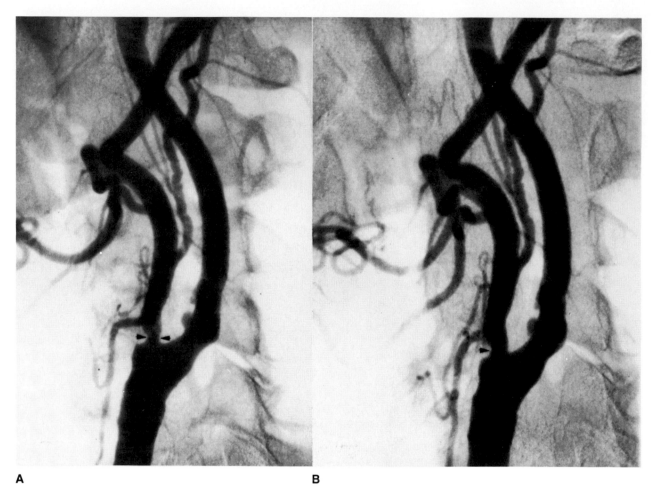

A   B

Figure 6. GW/A 65-year-old male with left cerebral TIA's. **A.** Lateral view, left common carotid arteriogram demonstrating moderate stenosis of the external carotid artery origin (arrows) as well as an ulcerated plaque of the internal carotid artery origin. **B.** Following dilation there is an increase in the luminal diameter of the external carotid artery. Internal carotid artery is unchanged in appearance.

percutaneous transluminal angioplasty should not be performed routinely on patients with atherosclerotic occlusive disease involving the internal or common carotid arteries. If the patient has a high operative risk and if the patient cannot tolerate medical treatment with anticoagulants (previous cerebral hemorrhage, cerebral aneurysms, etc.), then there may be a role for percutaneous transluminal angioplasty.

Patients with symptomatic fibromuscular dysplasia of the internal carotid artery are excellent candidates for angioplasty because the disease is usually limited to the media or adventitia of the vessel. Our preliminary experience suggests an acceptable complication rate, although the data is admittedly

meager. Long-term patency needs to be determined, possibly by a relatively noninvasive procedure such as digital subtraction angiography.

External carotid artery percutaneous transluminal angioplasty can be considered the treatment of choice since the risks are minimal. The benefits occur primarily in patients who have TIA's or amaurosis fugax in the distribution of an occluded internal carotid artery and/or prior to a transcranial anastomosis.

Percutaneous transluminal angioplasty of the vertebral artery may be of value in patients suffering from vertebrobasilar insufficiency. It is unlikely that the procedure will gain widespread use since an iatrogenic infarct within the vertebrobasilar system is potentially catastrophic. Percutaneous transluminal

TABLE 2   Clinical and Angiographic Findings: External Carotid Artery

Case, Initials Age, Gender	Initial Signs and Symptoms	Arteriographic Findings	Technical Success	Complications	STA-MCA Anastomosis	Clinical Results
#1, LM 54, M	Amaurosis fugax right eye	90% stenosis rt ECA, occluded rt ICA	+	None	Yes	Symptoms resolved
#2, GS 57, M	Lt cerebral TIA's transient dysphasia	75% stenosis lt ECA, occluded lt ICA	+	None	Yes	Symptoms resolved
#3, HO 71, M	Amaurosis fugax right eye	70% stenosis rt ECA, occluded rt ICA	+	Spasm ECA	No	Amaurosis resolved
#4, BC 69, F	Amaurosis fugax left eye	70% stenosis lt ECA, occluded rt ICA	+	None	No	Amaurosis resolved
#5, CD 56, F	Lt cerebral TIA's amaurosis fugax lt eye	80% stenosis lt ECA, occluded lt ICA	Initial-0 repeat-+	None	Yes	Symptoms improved
#6, MG 68, F	Partially resolved lt cerebral infarct	75% stenosis lt ECA, occluded lt ICA	+	None	No	Symptoms improved
#7, HS 57, F	Lt cerebral TIA's transient dysphasia	75% stenosis lt ECA, occluded lt ICA	+	None	No	No improvements
#8, GW 65, M	Lt cerebral TIA's	75% stenosis lt ECA, critical stenosis lt cavernous portion ICA	+	None	Yes	Symptoms resolved
#9, CB 69, M	Amaurosis fugax lt eye	90% stenosis lt ECA, occluded lt ICA	+	None	Yes	Symptoms resolved

Abbreviations: rt = right, lt = left, ECA = external carotid artery, ICA = internal carotid artery, TIA = transient ischemic attack

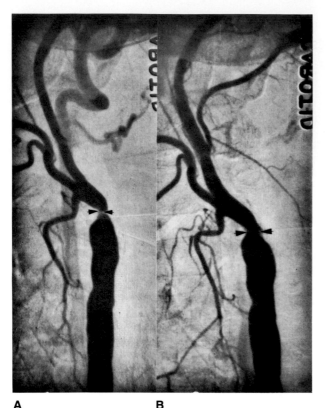

A   **B**

Figure 7. CB/A 69-year-old male with amaurosis fugax
O.S. **A.** Right common carotid arteriogram (lateral view)
shows marked stenosis of the external carotid artery origin (arrows) as well as an occluded internal carotid artery.
B. Following dilation there is marked improvement in the
caliber of the external carotid artery origin which is now
near normal.

angioplasty of the subclavian artery is often successful and beneficial in the appropriate clinical situation.

A potential new application for carotid artery percutaneous transluminal angioplasty is in patients with
rapid internal carotid artery restenosis following end-
arterectomy (25). The mechanism for this form of
postoperative stenosis is thought to be fibrous
myointimal hyperplasia which consists of smooth
muscle cell and fibroblastic proliferation occurring
in the vessel wall without lipid-laden or foam cells
seen in atherosclerosis (26). There are no intimal
plaques, which makes this entity favorable for angioplasty with minimal possibility for embolic complications. One patient with this disease process has
recently been reported to have undergone percutaneous transluminal angioplasty successfully (25).

REFERENCES

1. Upson J, Raza ST: Fibromuscular dysplasia of internal carotid arteries. NY State J Med 76:972–974, 1983.
2. Appleberg M: Graduated internal dilatation in the
   treatment of fibromuscular dysplasia of the internal
   carotid artery. SA Med J 51:244–246, 1977.
3. Effeney DJ, Ehrenfeld WK, Stoney RJ, Wylie EJ:
   Why operate on carotid fibromuscular dysplasia? Arch
   Surg 115:1261–1265, 1980.
4. Bockenheimer SAM, Mathias K: Percutaneous transluminal angioplasty in arteriosclerotic internal carotid
   artery stenosis. AJNR 4:791–792, 1983.
5. Bockenheimer SAM, Mathias K: Percutaneous transluminal angioplasty in supra-aortic artery disease.
   Medicamundi 28:87–89, 1983.
6. Corrin LS, Sandok BA, Houser OW: Cerebral ischemic events in patients with carotid artery fibromuscular dysplasia. Arch Neurol 38:616–618, 1981.
7. Countee RW, Vijayanathan T, Chavis P: Recurrent

retinal ischemia beyond cervical carotid occlusions.
J Neurosurg 55:532–542, 1981.
8. Damuth HD Jr., Diamond AB, Rappoport AS, Ren-
   ner JW: Angioplasty of subclavian artery stenosis
   proximal to the vertebral origin. AJNR 4:1239–1242,
   1983.
9. Dublin BA, Baltaxe HA, Cobb CA III: Percutaneous
   transluminal carotid angioplasty in fibromuscular dysplasia. Case report. J Neurosurg 59:162–165, 1983.
10. Wiggli U, Gratzl O: Transluminal angioplasty of ste-
    notic carotid arteries: Case reports and protocol. AJNR
    4:793–795, 1983.
11. Hasso AN, Bird CR, Zinke DE, Thompson JR: Fi-
    bromuscular dysplasia of the internal carotid artery:
    Percutaneous transluminal angioplasty. AJNR
    2:175–180, 1981.
12. Schuler JJ, Flanigan P, DeBord JR, Ryan TJ, Cas-
    tronuovo JJ, Lim LT: The treatment of cerebral is-

chemia by external carotid artery revascularization. Arch Surg 118:567–572, 1983.

13. Vitek JJ: Percutaneous transluminal angioplasty of the external carotid artery. AJNR 4:796–799, 1983.

14. Weinberg PE, Patronas NJ, Kim KS, Melen O: Anomalous origin of the ophthalmic artery in a patient with amaurosis fugax. Arch Neurol 38:315–317, 1981.

15. Mullan S, Duda EE, Patronas NJ: Some examples of balloon technology in neurosurgery. J Neurosurg 52:321–329, 1980.

16. Bird CR, Hasso AN: Transluminal angioplasty of the carotid artery, in Castaneda-Zuniga, WR: *New Applications of Transluminal Angioplasty*. New York, Thieme-Stratton, Inc, 1983, pp 154–161.

17. Vitek JJ, Morawetz RB: Percutaneous transluminal angioplasty of the external carotid artery: preliminary report. AJNR 3:451–546, 1982.

18. Bachman DM, Kim RM. Transluminal dilatation for subclavian steal syndrome. AJR 135:995–996, 1980.

19. Motarjeme A, Keifer JW, Zuska AJ: Percutaneous transluminal angioplasty of the vertebral arteries. Radiology 139:715–717, 1981.

20. Motarjeme A, Keifer JW, Zuska AJ: Percutaneous transluminal angioplasty of the brachiocephalic arteries. AJNR 3:169–174, 1982.

21. Novelline RA: Percutaneous transluminal angioplasty: new applications AJR 135:983–988, 1980.

22. Reddy SVR, Karnes WE, Earnest F IV, Sundt TM Jr: Spontaneous extracranial vertebral arteriovenous fistula and fibromuscular dysplasia. J Neurosurg 54:399–402, 1981.

23. Sundt TM Jr, Smith HC, Campbell JK, Vlietstra RE, Cucchiara RF, Stanson AW: Transluminal angioplasty for basilar artery stenosis. Mayo Clinic Proc 55:673–680, 1980.

24. So EL, Toole JF, Dalal P, Moody DM: Cephalic fibromuscular dysplasia in 32 patients. Clinical findings and radiological features. Arch Neurol 38:619–622, 1981.

25. Tievsky AL, Druy EM, Mardiat JG: Transluminal angioplasty in post surgical stenosis of the extracranial carotid artery. AJNR 4:800–802, 1983.

26. Palmaz JC, Hunter G, Carson SN, French SW: Postoperative carotid restenosis due to neointimal fibromuscular hyperplasia. Radiology 148:699–702, 1983.

Intraoperative Balloon Angioplasty of the General Vascular System

THOMAS J. FOGARTY, M.D.
JAMES C. FINN, M.D.

INTRODUCTION

The first published report on clinical trial of transluminal angioplasty was by Dotter and Judkins (1). The technique utilized the sequential passage of tapered teflon coaxial catheters over a guidewire positioned through a stenotic artery under fluoroscopy. Although such a technique could be applied successfully in the radiology suite, the passage of a guidewire or rigid tube through an atheromatous plaque in the operating room without fluoroscopic control has a significant risk. Lack of trained personnel and unfamiliarity with the concept and instrumentation limited the acceptance of angioplasty by the surgeon. Certain surgeons, however, noted the potential value of this technique as an adjunct to standard bypass revascularization procedures and developed a staged technique in conjunction with their radiology counterparts (2,3).

Patients who had unilateral iliac artery occlusion and significant contralateral iliac stenosis (associated with severe multisystem disease) were considered candidates for angioplasty procedure. Some patients underwent percutaneous dilation of the iliac artery stenosis in preparation for use of the artery as a donor in a femorofemoral bypass a few days later. The percutaneous approach obviated entering the abdominal artery and crossclamping a diseased aorta and allowed the use of regional anesthesia in selected cases. Because of its dramatic initial success, this staged technique was later extended to lower-risk patients who required enhancement of iliac artery inflow to facilitate distal artery reconstructions. In these patients, both the extent and risk of the reconstructive procedure were reduced with patency rates that were comparable with more extensive surgical procedures alone (4).

Limitations of vessel access and maximum diameter of the coaxial dilators of original Dotter design were some of the technical obstacles. During this period, the technology of the balloon catheter evolved. Dotter himself recognized these drawbacks (5,6) and introduced a caged-balloon dilating catheter (6,7) similar to that employed in Europe by Porstmann (8). Consisting of a thick-walled latex balloon reinforced by a section of longitudinally slit number 8 French plastic outer tube, this catheter permitted a lower relative silhouette upon introduction into the artery and a threefold increase in diameter upon balloon inflation. Though more effective than coaxial tube dilators, this caged-balloon catheter was thrombogenic, and its use was confined to iliac vessels.

Some years after Dotter's developments, Gruentzig and Hopff (9,10) introduced a new dilating catheter based on a nonelastomeric balloon of relatively fixed maximum diameter. Having a smoother tip (reducing thrombogenicity) and greater variety of balloon diameters and catheter sizes, this Gruentzig-Hopff design represented a considerable improvement over previously used dilation catheters.

Staged procedures involving dilation of proximal iliac stenoses with subsequent bypass of distal femoral or contralateral iliac obstructions in both high and low-risk patients were now being performed with the Gruentzig balloon catheter (11–13), with a reported dilation complication rate of 5 percent (14). Although the initial patency rates of the stenoses dilated with the Gruentzig catheter were comparable to those of surgically reconstructed vessels, the combination of angioplasty and bypass revascularization performed at two separate times exposed the patient to the specific risks of two major procedures rather than one.

The first report of adjunct intraoperative angioplasty of the peripheral vascular system performed by surgeons appeared in the discussion of the July 1980 article by Beebe (13). In this paper, Fogarty reported an initial experience with the Gruentzig type coaxial balloon catheter, as well as successful application of a new extrusion balloon catheter in 43 additional cases.

The precedent for coronary artery dilation (15) by surgeons was also being established with encouraging results and has since been extensively reported (16–22). Intraoperative use of balloon catheters allowed easy access to the distal coronary lesions, which were difficult to access by percutaneous transluminal coronary angioplasty (PTCA), and such procedure improved distal runoff.

In June 1981, Lowman (23) published a study on the intraoperative use of the Gruentzig balloon by vascular surgeons. Balloon angioplasty of the peripheral arterial system was felt to confer several advantages over similar percutaneous uses in patients who would require reconstructive operations in any event. Peripheral embolization had been reported in approximately 5 percent of patients (24) with percutaneous angioplasty of peripheral arteries. This rate could be reduced by arterial "flushing" through the arteriotomy site during intraoperative dilation (23). Inevitable and serious complications such as thrombosis or rupture of vessels could be addressed with greater effectiveness and directness if they occurred in an operative site rather than in the radiology suite during percutaneous procedure. Furthermore, the occurrence of serious wound hematomas from percutaneous manipulation would be eliminated. Also, combining the angioplasty and surgery in the means of intraoperative angioplasty would avoid the psychological and physical stress of subjecting the patient to two separate procedures and would reduce the length of the hospital stay. Finally, technical success rates in the operating room should, in theory, be comparable with those obtained in percutaneous approaches.

Many lesions were found to be easily accessible to the surgeon with the Gruentzig type balloon, particularly if they were near or within the operative site or located in relatively nontortuous vessels. However, problems arose when the lesions to be dilated were far from the operative site or located within or beyond tortuous vessels and difficult bifurcations. Blind passage of guidewires or catheters through such vessels could lead to complications, including perforation or thrombosis secondary to subintimal dissection (23,24). To overcome these deficiencies, some institutions reported dilations by vascular radiologists using C-arm fluoroscopy units during operative procedures (25,26). However, the need for specialized (and often uncomfortable to the surgeon) operating tables and equipment (25), the manipulation of an unsterilizable fluoroscopic unit around the operative field, the need for additional personnel within the sterile region, and the lengthening of the operative time posed serious obstacles to the practical implementation of this approach.

In November 1981, Fogarty reported on the intraoperative dilation of 66 vessels utilizing a new concept in balloon catheters (27). Linear extrusion balloons are extruded across the lesion rather than being slid across as are the Gruentzig type balloon catheters. The balloon unrolls onto the luminal wall of the vessel with no longitudinal movement in relation to the arterial wall (Fig. 1). This property of the balloon gives it certain qualities that are advantageous to the surgeon. For one, the balloon finds its way through both tortuous vessels and tight lesions without the aid of a guidewire. Furthermore, the *shear force* (the force of frictional contact between catheter body or balloon and vessel wall) generated is minimal as the balloon negotiates through the lesion (Fig. 2), thus eliminating a major potential cause of plaque embolus, perforation, or subintimal passage (24,28–37). This catheter has greatly simplified the technique of intraoperative angioplasty, obviating the need for guidewire application. It also follows that lesions located outside the perimeter of the operative exposure, and those located beyond a tortuous segment are accessed more easily. With experience, the need for routine intraoperative fluoroscopy is diminished as well.

Adjunctive intraoperative angioplasty at our institution is employed in those situations in which the magnitude of the operation can be reduced or the effectiveness of the individual graft can be enhanced. Examples of such procedures are dilation of iliac arteries to increase inflow to femoropopliteal

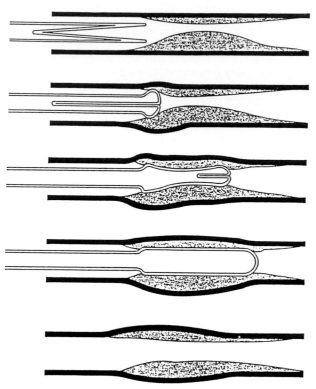

Figure 1. Fundamental mechanism of linear extrusion. Top: undeployed balloon. Second: initial contact with stenosis. Middle: extrusion across stenosis. Note that the balloon only partially dilates the lesion during its deployment. Fourth: full dilation. Bottom: lesions after dilation.

and femorofemoral grafts and dilation of popliteal and tibioperoneal arteries to increase outflow from grafts bypassing more proximal lesions (Fig. 3).

Since Fogarty's article, there have been several reports of angioplasty, utilizing coaxial type balloon dilation catheters, as an adjunct to surgery (26,38,39). However, the technique of adjunctive transluminal angioplasty employing the linear extrusion catheter will be the primary subject of this chapter.

DYNAMIC DIFFERENCES BETWEEN COAXIAL AND EXTRUSION BALLOON CATHETERS

The linear extrusion concept relies on principles very different from those of the coaxial balloon system (Fig. 4). Coaxial balloons are presently the most commonly used catheter for angioplasty. Because the balloons are attached in a colinear fashion on a catheter body, the minimum profile that can be achieved upon introduction is dependent on the min-

imum diameter of catheter body that can be used and the bulk added by the collapsed balloon. This in turn limits the severity of stenosis that can be crossed without generating excessive shear force that will cause the attendant complications. Furthermore, since coaxial balloons must be directly located across the stenosis by a sliding movement, the degree of tortuosity of vessel or lesion that can be traversed is limited by the combined rigidity of catheter body and deflated balloon. Guidewires placed under direct fluoroscopic control represent the safest technique for properly placing coaxial systems.

Linear extrusion relies on the unrolling motion of an extruding balloon element, much like that of a carpet unrolling onto a floor. The balloon has been inverted into a catheter body. When the catheter is inflated with fluid, the balloon extrudes forward with no relative motion between vessel wall and the skin of adjacent balloon. Furthermore, the extrusion balloon, with only the bulk of fluid and collapsed unextruded balloon skin inside, can negotiate very tight strictures with a minimum of resistance. Thus, the balloon finds its way through the path of least resistance, passing tortuous vessels or lesions without the need of guidewires or fluoroscopic guidance.

The issue of the shear force between luminal wall and balloon is important. Excessive shear force has been directly correlated histologically with intimal damage and stripping by embolectomy catheters (29,30). Clinical studies present evidence that persistent forceful advancement of balloon catheters into arterial stenoses may produce intimal dissection, flap formation, and embolization (24,31–37) due to shear force. It thus appears likely that excessive shear force may be important in intimal flap formation, possible subsequent intimal dissection or thrombosis, and generation of atherosclerotic emboli during angioplasty. Measurements made by our group show that the shear forces generated by linear extrusion balloon in both human exvivo iliac samples and experimental models are significantly lower than those generated by either coaxial balloon catheters or Dotter coaxial dilators. The severity, length, compliance, vessel angulation, and catheter advancement rate were all controlled in the experimental models during these tests (Figs. 5–7). Furthermore, there is evidence that long-term complications, particularly the accelerated formation of atherosclerotic plaque (30,40), may ensue from intimal damage produced by balloon catheters. It is of interest that the mechanical damage to the endothelium produced by balloon catheters is used to create accelerated atherosclerotic disease in both hyperlipidemic and nonhyperlipidemic experimental animals (30,41–47).

Shear force also plays a role, along with catheter

SHEAR FORCE CONCEPT

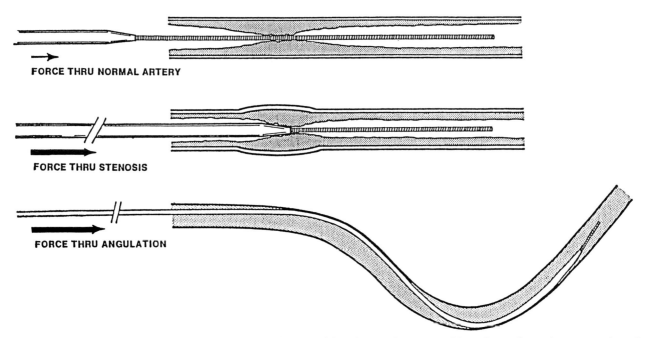

Figure 2. Relative magnitudes of shear force during coaxial catheter placement. Shear force depends on severity of stenosis, length, compliance, vessel angulation, and advancement rate.

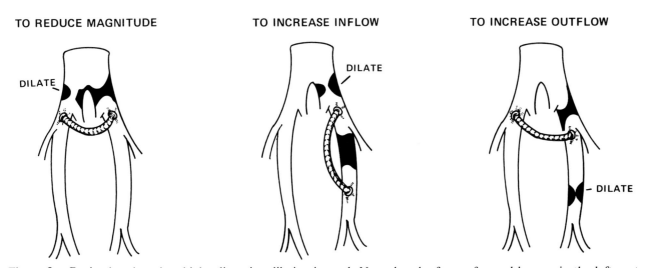

Figure 3. Basic situations in which adjunctive dilation is used. Note that the femorofemoral bypass in the leftmost diagram is placed in lieu of an aortofemoral graft.

COAXIAL DILATOR   COAXIAL BALLOON CATHETER   LINEAR EXTRUSION CATHETER

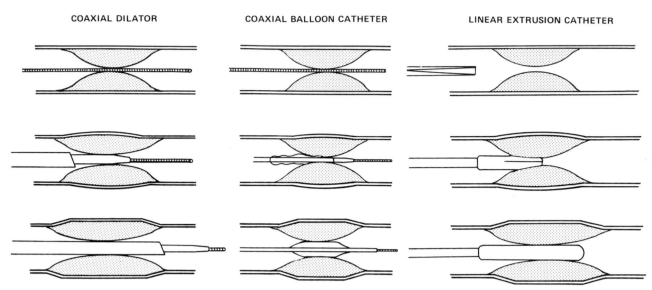

Figure 4. Comparison of basic mechanisms of angioplasty catheters. There is a large difference in shear force between the extrusion catheter and the other two systems.

rigidity (as measured by the flexural modulus), in precluding catheter placement through tortuous, tight, or long vessels and lesions (1,24,31,34,35,48,49). Furthermore, the physical rigidity of the catheter body and its tip may contribute to the complications mentioned in the clinical literature, including intimal dissection, flap formation, embolization, and perforation (24,31–37). Preliminary measurements in our laboratory indicate that the forces required for the tip of either a guidewire or a catheter to dissect an exvivo human arterial wall are directly proportional to the size of the tip and to the angle between the catheter and the vessel wall (both due to the greater surface area of contact and thus decreased stress concentration). Perforation forces required were also found to vary directly with catheter tip size but were found to be relatively independent of incident angle. This characteristic may be due to the pronounced deflection of the arterial wall caused by the imping-ing catheter that accompanies the high forces re-quired for perforation, thus reducing the influence of incident angle. It should be noted that the buck-ling strength of a catheter body can be substantially enhanced when it is supported by a tube such as an artery; thus its column strength, which determines how much force can be applied by the catheter tip, is greater in an artery than it is in free air.

The linear extrusion balloon reduces the problem of negotiating tortuous, tight, or long strictures by decreasing shear force as compared to coaxial bal-loon catheters. Because the linear extrusion balloon can have a much lower column strength than a rigid coaxial catheter body during extrusion through ves-sels and lesions, it may help reduce the problem of perforation or dissection, although it does not elim-inate it. Furthermore, because the balloon is un-rolling and not translating in relation to a vessel lumen or lumen of stenotic segment, the linear ex-trusion balloon tends to deflect more easily around curves and tortuosities.

The column strength of the linear extrusion bal-loon depends upon its inflation pressure. This pres-sure may vary during deployment, depending on the severity of the lesion stenosis or vessel tortuosity that must be negotiated by the balloon. When re-sistance to forward movement of the balloon in-creases, so does the inflation pressure force nec-essary to move the unfolding balloon past the resistance, thus increasing its column strength at that point in time. Once the resistance has been passed, the pressure requirement will drop once again.

One important potential problem with the linear extrusion balloon is the nondirectability of its for-ward movement. If a lesion lies beyond a bifurcation or branching that the extruding balloon must pass before reaching the lesion, the balloon may enter the wrong branch. This can be particularly danger-ous if the branch entered has a diameter smaller than the maximum diameter of the balloon and the in-tended target branch has a larger diameter that matches the balloon size. In this case, balloon en-trance into the branch may cause avulsion of the

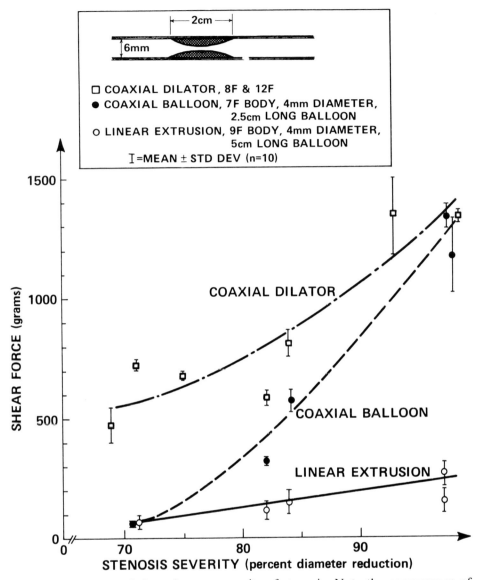

Figure 5. Relation of shear force to severity of stenosis. Note the convergence of coaxial balloon and coaxial dilators in the range of high-degree stenosis. The tests were made on 11 necropsy specimens of atherosclerosis. Significant differences ($P < .025$) exist between these three design models.

vessel from its orifice. In fact, the author has experienced such a case in the coronary system. The problem may be avoided either by locating the catheter tip beyond the bifurcation or branching or, where this is not possible, by utilizing a two-step procedure. A smaller linear extrusion balloon matched to the diameter of the small branch is first extruded through the lesion. With the lesion partially dilated,

a second, larger balloon is extruded through the lesion in the artery of larger diameter.

The dilation mechanism of the linear extrusion balloon catheter is identical to that for the coaxial dilation catheter. In a study conducted by our lab, the major portion of luminal increase (87–93 percent) was found to be due to disruption of the atherosclerotic plaque and vessel wall, with extrusion of fluid content from the plaque accounting for only approximately 6–12 percent of the luminal increase (Figs. 8,9) (50). As in other studies (51–59), we have found calcified atherosclerotic plaque to be essentially incompressible.

SELECTION OF PATIENT AND TARGET VESSEL

In light of the fact that the primary role of adjunct intraoperative angioplasty is to minimize the magnitude of the vascular surgery and to enhance the effectiveness of individual grafts, intraoperative dilations must be anticipated and planned and the operative strategy designed around them. Therefore, both patient and vessel selections are made prior to surgery and based on a variety of diagnostic procedures, including noninvasive assessment using Doppler or pulse volume recording with calculation of the leg-arm indices. Angiographic documentation of the location, severity, and length of the diseased segment of the artery provides details for formulating and planning the intended vascular surgery. These diagnostic tests are performed in conjunction with

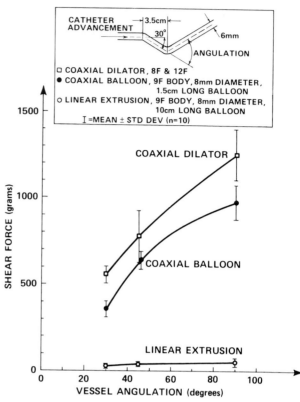

Figure 7. Relation of shear force to vessel angulation. The shear force generated through an angulated vessel is significantly greater in the other two designs than that produced by extrusion catheter ($P < .025$). The shear force generated in an angulated vessel is determined by the stiffness of the catheter body or the balloon.

standard clinical evaluations of the patients. Particular evaluations must assess potential contraindications to surgery and general anesthesia or to treatment with anticoagulants. In our experience, most presenting patients are moderately debilitated long-term smokers with multisystem disease. The systemic nature of atherosclerotic vascular disease often makes assessment of cerebrovascular and coronary systems necessary when the primary revascularization is planned for the peripheral vascular system.

Generally, total occlusion, cul-de-sac lesions, and lesions at critical bifurcations (e.g., at the aortic bifurcation) are considered specific contraindications to intraoperative angioplasty. Furthermore, lesions containing fresh thrombus and ulcerative lesions that appear to be a source of peripheral emboli such as in the blue toe syndrome are also avoided. It is recommended not to attempt lesions over 10 cm in length with extrusion catheters. It should be noted, however, that tandem lesions that can be

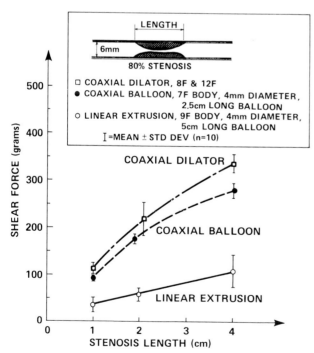

Figure 6. Relation of shear force to stenosis length. Across the different lengths of stenotic segment, extrusion catheter produces significantly less ($P < .025$) shear force than the other two designs.

FACTORS INVOLVED IN DILATATION

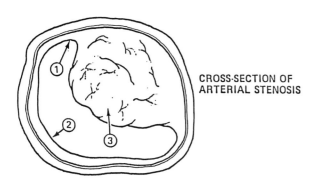

CROSS-SECTION OF
ARTERIAL STENOSIS

1. DISRUPTION OF PLAQUE-ARTERY JUNCTION ⎫ 86.8–93%
2. DISRUPTION OF ARTERY ⎬
3. PLAQUE FLUID EXTRUSION (6 – 12%)

Figure 8. Cross-section of arterial stenosis.

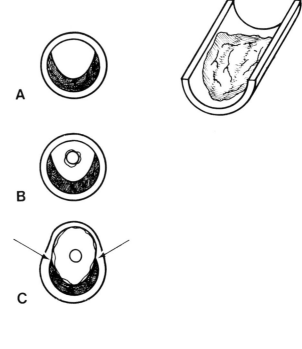

A

B

C

D

E

reached in a single extrusion are not a contraindication and can be dilated by a method described in the section Operative Techniques Using the Extrusion Balloon Catheter.

Common lesions often suitable for dilatation include those that significantly restrict inflow to a graft, significantly restrict outflow from a graft, or specifically allow reduction of the extent of the vascular surgery.

OPERATIVE TECHNIQUES USING THE EXTRUSION BALLOON CATHETER

Currently used balloons range from 4 to 8 mm in diameter in 1-mm increments and are either 5 or 10 cm in length. All catheter bodies are 100 cm long. The maximum recommended volume and pressure of inflation vary with the diameter and length of the balloon, and are specified in each catheter package. All catheters have an infusion port leading to an infusion orifice located just behind the catheter tip and exiting from the side of the body. This port is suitable for injection of radiopaque contrast media.

The technique for intraoperative dilation using the linear extrusion balloon involves two basic phases: *calibration* and *dilation*. Although this procedure may be used without fluoroscopic guidance, fluoroscopic visualization during early experience with this system may provide valuable learning experiences.

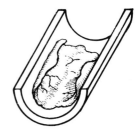

Figure 9. An important factor in dilating a stenosis is separation between the plaque and the vessel wall. Arrows indicate points of initiation of separation. The compliant part of arterial wall not covered by the plaque or separated from plaque attachment can now distend to a greater circumferential length under physiological pressure, no longer bound by the more rigid plaque.

Operative calibration prior to dilation confirms angiographic findings and provides an accurate measurement of the distance to the stenosis from the arteriotomy, as well as the diameter of the adjacent uninvolved arterial segment. A Fogarty peripheral balloon calibrator is introduced through the arteri-otomy site in a partially inflated state. The infla-tion medium should be sterile fluid, and the system should be purged of air prior to use. The calibrator is then advanced to the level of the stenosis, at which point resistance to further advancement is encoun-tered (Fig. 10). The calibrator should never be forced or repeatedly probed during this procedure. The dis-tance to the arteriotomy is then read off the depth markings on the calibrator body (placed every 10 cm). Figure 10 is noted because it indicates proper placement of the dilation catheter. The diameter of the adjacent uninvolved arterial segment is next as-sessed by slowly inflating the calibrator balloon while gently moving the catheter back and forth. Resist-ance to movement confirms that the balloon has engaged the arterial wall. The size of this balloon represents the intraluminal diameter of the artery proximal to the target lesion. The volume of liquid injected is noted, and the balloon is deflated and removed. Outside the patient, the balloon is rein-flated with the volume previously noted, and a bal-loon calibrator scale is used to measure its diameter (Fig. 11). This estimate of the uninvolved segment diameter is noted, and the Fogarty-Chin dilation catheter with a balloon diameter closest to this size is chosen. During calibration of arterial diameter, care should be taken not to over inflate the calibrator balloon as this may cause damage of the arterial segment (Fig. 12). Furthermore, the dilation cath-eter, which is selected on the basis of the calibrator measurement, may be too large if the calibrator is overinflated.

The dilation catheter has three ports: infusion,

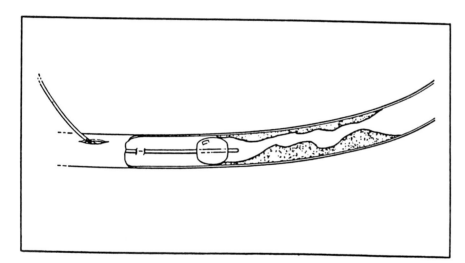

Figure 10. Calibration of the distance between arteriotomy and stenotic segment, as well as of the diameter of the parent artery.

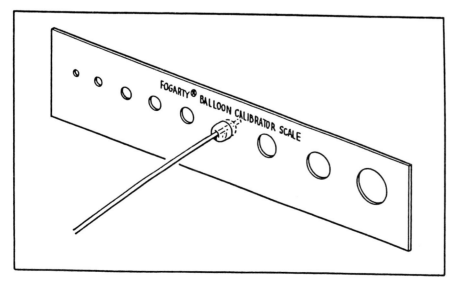

Figure 11. Measuring diameter of reinflated calibrator balloon. This calibration is an estimation of the diameter of uninvolved artery proximal to the target lesion.

vent, and inflation (Fig. 13). After air has been purged from the inflation and infusion lines, an adjustable sliding silicone marker on the catheter body is set to a distance from the body tip 2 cm less than the length measured by the calibration catheter. The dilation catheter is next advanced through the arteriotomy until the adjustable marker reaches the incision site (Fig. 14). The smallest syringe that will accommodate the average recommended injection volume should be used to extrude the balloon. Sterile saline or dilute contrast medium is used as the inflation fluid. Fluid injection into the inflation port results in extrusion of the balloon when the vent gate valve is closed. Balloon pressures can be monitored during extrusion. It will be found that high pressures, even exceeding the maximum recom-

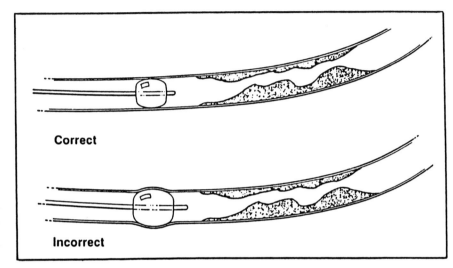

Figure 12. Overinflation of the calibrator balloon should be avoided to prevent damage to the arterial wall and overestimation of the artery diameter.

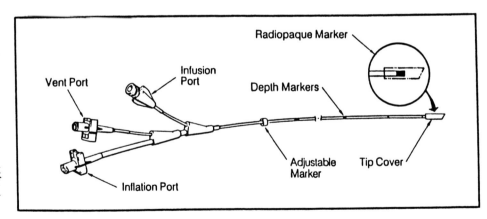

Figure 13. The connecting ports and depth markers of Fogarty-Chin peripheral dilation catheter.

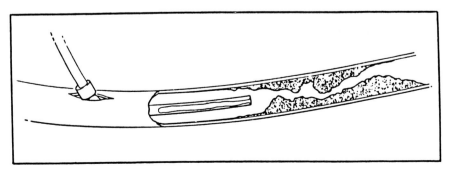

Figure 14. Placement of Fogarty-Chin catheter. Note that the catheter tip is located 2 cm proximal from the leading edge of the lesion.

mended inflation pressure for the balloon, may be experienced prior to the balloon eversion (Fig. 15). This initial rise will not damage the balloon and should not alarm the surgeon. A mechanical assist inflation device or a small-bore syringe may be used in such instances to help overcome these high initial pressures. A sudden drop in pressure will be noted upon balloon exit from the catheter tip. Balloon contact with the stenosis will be noted as a second rise in pressure, the magnitude of which will depend on the severity of the stenosis. In particularly severe or tortuous stenoses, where substantial resistance (as noted by pressure) to balloon passage (as noted by volume) is encountered, a technique of inflation-deflation of the balloon has been found to reduce the peak pressures required to negotiate the stenosis

or tortuosity (Fig. 16). The inflation-deflation maneuver is accomplished by repetitive inflation to high pressure (holding the pressure for 1–2 s), followed by full aspiration on the syringe. This is continued until it is evident that the balloon has negotiated the restricted section of the lesion. The pressure will again drop upon balloon exit from the stenosis and will remain low until full balloon extrusion. It should be noted that tandem lesions along an artery can be dilated by simply extruding the balloon through all lesions located within the maximum length of the balloon. In this case, the pressure will sequentially rise and fall each time a stenosis is passed.

During the initial phase of balloon contact with the leading edge of a stenotic lesion, the catheter body may tend to back out of the arteriotomy site.

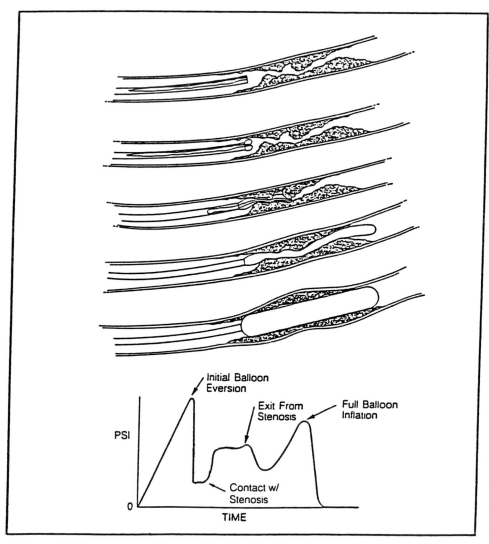

Figure 15. Inflation pressure recording during an extrusion. Note that actual variations in pressure during extrusion depend on the severity of each stenosis encountered.

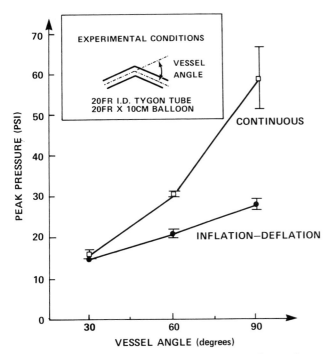

Figure 16. The relation of peak pressure and vessel angle. Note that the inflation-deflation maneuver substantially reduces required inflation pressures in an angulated vessel. Bars indicate one standard deviation.

In this event, the catheter body can be braced by clamping and holding a right angle clamp on the lateral surface of the silicone adjustable marker (Fig. 17). Care must be taken not to clamp any part of the catheter body itself.

When the balloon is fully extruded, fluid injection is continued until either the desired degree of dilatation is achieved or the maximum recommended pressure is reached. If the lesion resides within the exposed segment of an artery in the operative field, one may observe a bulging and blanching of the diseased segment that reflects dilation and disruption of the arterial wall, as well as the plaque. If one

cannot directly visualize the artery, and one is not using fluoroscopy, the volume of fluid injected can be used as a gauge of inflation diameter. When the entire volume is injected, one can be sure the balloon is at or near its nominal maximum diameter. The inflation gate valve is now closed and the balloon left inflated for as long as desired. The catheter may be left in place as an arterial occlusion device during the remainder of the reconstructive procedure, particularly when it is used to dilate donor artery lesions.

Prior to completing the final anastomosis to the arteriotomy site used as access for dilation, the balloon is removed after it is actively deflated by aspiration with the syringe. Generous flushing with blood, when possible, is employed to wash any air or atherosclerotic debris that might have been loosened from the luminal wall during dilation.

Intraoperative assessment of the efficacy of dilation may be achieved by making a postdilation arteriogram or by observing the balloon inflated with contrast medium within the dilated segment either under fluoroscopy or by radiography. Direct recalibration of the dilated segment itself with a bougie or balloon calibrator should never be performed since the risk of dislodging emboli or damaging the arterial wall is too great.

In light of the fact that vascular surgeons generally have an anatomical orientation, it should be pointed out that pursuit of a perfect anatomical appearance is not the goal of dilation. Isolated stenoses must be particularly severe (greater than 90 percent for aortic or iliac lesions) for there to be a clinically significant reduction in blood flow or distal pressure (60–62). Furthermore, because resistance across a stenosis is related exponentially by the fourth power to the radius of the residual lumen (63), small increases in lumen size will result in large gain in reduction of lesion resistance. For example, a twofold increase in luminal diameter would result in a sixteen-fold reduction in resistance against flow. In reality, the pressure gradient usually drops along with a more modest increase in flow. Even in smaller

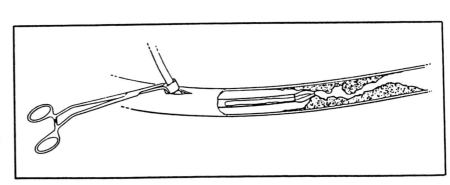

Figure 17. Bracing the dilation catheter with a right angle clamp. Note that only the silicone marker, not the catheter body, is clamped.

arteries, an increase in lumen size to only 50 percent of normal diameter eliminates the pressure gradient and returns flow to normal (27).

Because of the nearly incompressible nature of plaque (50–59), attempts at restoration of normal luminal diameter may cause overdilation and rupture. Thus, full restoration of physiological function while avoiding unnecessary mural damage can be accomplished with only partial enlargement of the residual lumen diameter.

The precise role of preoperative antiplatelet therapy is as yet unclear. It is the practice of this group to pretreat all patients with 5 gr of aspirin 1 day prior to surgery and to maintain them postoperatively on 5 gr of aspirin and 75 mg of dipyridamole daily for an indefinite period of time.

CLINICAL RESULTS

The first reported clinical trial employing the linear extrusion catheter involved 24 dilations at this institution (27). Table 1 presents the specific peripheral lesions dilated. It should be noted that this initial trial involved five renal, three carotid, one subclavian, and one axillary dilation, in addition to dilation of the great vessels of the lower extremity. Intraoperative recalibration was carried out in 19 of the 24 dilated vessels and demonstrated an increased lumen size in 18 of these 19 vessels. Improvements in vessel diameter as measured by this technique ranged from 30 to 166 percent over the baseline with a mean increase of 78 percent. Postoperative angiograms were obtained in nine patients at periods ranging from 2 months to 3 years after the initial procedure. Continued success was demonstrated in eight

of the nine angiographic studies. Ankle-arm indices were used to provide immediate postoperative and long-term follow-up, when applicable. In 9 of 10 individuals in whom noninvasive studies were conducted, improvement was found during a follow-up period ranging from 3 months to 3 years.

The single known operative complication from this group followed dilation of a 90 percent midsuperficial femoral lesion. Embolization to the adductor tendon level was demonstrated by postoperative arteriograms. No subsequent repair was performed. The ankle-arm index, however, showed a small but significant improvement.

The overall success rates of a multicenter study involving 12 institutions and 82 dilations of peripheral arteries are presented in Table 2. The protocol used in this study is essentially identical to the guidelines elucidated in the section Operative Techniques Using the Extrusion Balloon Catheter. As can be seen, an 81 percent primary success rate (61/75) was achieved as documented by flow or pressure gradient measurements, intraoperative angiography, or by an increase in the ankle-arm index of greater than 0.10. Secondary success, as determined by the follow-up at 3 months by angiography (when the patient was symptomatic), ankle-arm indices, and clinical signs (such as skin color, temperature, and pulse return), is distinguished into two categories. Of the 48 patients for whom follow-up data are available, 38 were considered primary successes, and 10 were considered primary failures. Of those 38 individuals who were considered initially successful, 95 percent (36/38) were considered successful at 3 months. Of the 10 individuals classified as initial failures, 3 were found to fit the criteria of secondary success at 3 months. Thus, an overall follow-up success rate of 81 percent (39/48) was determined for this study.

Five major immediate complications were experienced during this study. Three of these complications occurred during procedures necessitated by mismatches between the balloon diameter and the lumen size of the artery, which were violations of the study protocol. In one case, the tibioperoneal trunk was ruptured during a double-vessel dilation in which the popliteal artery was effectively dilated. No subsequent repair of the ruptured vessel was required. In the second case, a dissection of the posterior tibial artery occurred during a double dilation of the trifurcation region in which an effective dilation of the anterior tibial artery was achieved. No subsequent repair was performed, and the patient was found to be clinically well at 3-month follow-up, with an ankle-arm index of 1.0. The third was a dissection of the right popliteal artery. A below-the-knee graft was placed, and the patient recovered satisfactorily with no additional surgery.

TABLE 1 Vessels Dilated in Original Study

Vessel	Number
Renal	
Fibromuscular	3
Atherosclerotic	2
Iliac	4
Superficial femoral	4
Carotid fibromuscular	3
Mid–deep femoral	3
Popliteal	2
Tibioperoneal	1
Subclavian	1
Axillary	1
Total	24

Source: Ref. 27.

TABLE 2A   **Multicenter Study Results**

Effective Sample Size		*Primary (48-hour) Success Analysis*	
82	Total cases	61	Primary successes 81%
4	Out of protocol	18	By flow/gradient data
3	Insufficient information	31	By intraoperative angiography
75	Cases for Primary success analysis	11	By arm/leg index increase $\geq$ .10
		11	By direct calibration

Of the two complications that occurred within protocol, one was an embolization into the common femoral artery during an external iliac dilation. The patient satisfactorily recovered after additional surgery. The second was a thrombosis of the external iliac artery during dilation of the iliac. Satisfactory recovery was achieved after an iliofemoral bypass graft was placed.

Six patients required below-the-knee amputations at varying intervals following surgery. All had severe peripheral vascular disease and gangrenous changes prior to surgery. Five of the six were considered initial successes immediately after the surgery, but all were excluded from the follow-up study once it became evident that their gangrene could not be reversed.

Two deaths occurred during follow-up periods. One patient died 3 weeks following operation as the result of a myocardial infarction. The other died 5 days postoperation after multiple hypotensive episodes and stroke. This second patient had uncontrollable hypertension prior to surgery and progressive renal failure with a 99 percent stenosis of the right renal artery, which was successfully dilated.

Since the initial report by Fogarty (27), 51 additional lesions have been dilated at this institution. The success of these dilations has been studied by using a protocol similar to that of the multicenter study. The composite primary success rate of this institution encompassing both the initial and subsequent series of 75 total dilations, is 87 percent, with a continued patency of the dilated vessels of 95 percent during the follow-up period. These follow-up success rates are at 3 months for the last 51 dilations and at varying intervals beyond 2 months for the first 24 dilations. No significant complications have occurred in this group since the initial study.

CONCLUDING PERSPECTIVES

As emphasized by Beebe (13), angioplasty should not stand apart from the options of conventional therapy in an isolated or competitive way, rather it should be considered by the vascular surgeon among the possible alternatives for appropriate patients.

Prior to the advent of the linear extrusion catheter, intraoperative dilation without fluoroscopic assistance posed a significant technical challenge to the surgeon and a significant risk of complication to the patient. Staged adjunctive angioplasty, however, exposes the patient to the risks of percutaneous angioplasty (35,64–66), along with the added psychological, physical, and economic burdens of two separate procedures (25) and a lengthened hospital stay (13). Adjunctive angioplasty performed under fluoroscopic control by a vascular radiologist during the operative procedure does eliminate these problems but creates new obstacles. These include the need for specialized and often uncomfortable operating tables and equipment (25), the manipulation of an unsterilizable fluoroscopic unit around the operative field, additional personnel within the sterile region, and additional operative time. Adjunctive intraoperative dilation performed with the linear extrusion catheter, without guidewire placement or the necessity of fluoroscopic control, reduces to a minimum the extent, time, and complexity of the adjunctive procedure and maximizes the versatility of balloon dilation during vascular surgery.

Reducing the magnitude of a surgical operation is important in making available reconstructive pro-

TABLE 2B   **Multicenter Study Results**

Secondary (3-Month)* Success Analysis			
Primary Classification	Secondary Classification	Frequency	Secondary Success Rate
Successful	Successful	36	36/38 = 95%
Successful	Unsuccessful	2	
	Subtotal	38	
Unsuccessful	Successful	3	
Unsuccessful	Unsuccessful	7	
	Subtotal	10	
	Overall		39/48 = 81%

*Because of lack of clinical information all cases did not have 3-month follow-up.

cedures to patients with many risk factors. Extra anatomical procedures, such as femorofemoral (67) and axillofemoral (68) bypass, can be facilitated by angioplasty when donor or recipient arterial obstruction is the prohibiting factor. Furthermore, evidence suggests that long-term patency rates following angioplasty for localized stenoses are now comparable to those achieved with reconstructive surgery (35,65,66,69–72). Therefore, it may be possible to reduce the required number of grafts by use of angioplasty to correct localized stenoses during an extensive reconstructive procedure.

Enhancing efficacy and survivability of reconstructive grafts is also an important function of adjunctive angioplasty. Most patients with truly disabling claudication or rest pain, for example, have multilevel involvement of the arterial tree (23). In this setting, a graft will often be fed by a significantly obstructed donor artery or be drained by an obstructed recipient artery that would not routinely be bypassed but would nevertheless compromise the effectiveness of the bypass surgery (13,73–76). Angioplasty in this circumstance can be used to increase the graft flow efficiently by dilating either the

inflow or outflow tract. Furthermore, since a certain minimal flow through a graft is necessary to maintain graft patency, atherosclerotic disease in both inflow and outflow channels can contribute significantly to graft failure (77,78). In one study (78), femorofemoral grafts with up to a 50 percent obstruction of the donor iliac artery experienced a graft patency rate of 90 percent. However, those with greater than 50 percent stenosis of the donor iliac uniformly failed. In addition, those individuals with significant outflow disease and no obstruction to inflow had a 59 percent graft failure rate. These grafts failed either from direct hemodynamic causes or thrombosis. Therefore, when graft flow is increased by dilation of inflow or outflow channels, survivability of the grafts can also be expected to increase.

The future of intraoperative angioplasty lies in increased understanding by vascular surgeons of the mechanisms, clinical pitfalls, and potential benefits of this adjunctive procedure. Greater clinical experience with this technique is required to elucidate more fully its role in vascular reconstructive surgery.

REFERENCES

1. Dotter CT, Judkins MP: Transluminal treatment of arteriosclerotic obstruction: Description of a new technique and a preliminary report of its application. Circulation 30:654–670, 1964.
2. Porter JM, Eidemiller LR, Dotter CT, Rosch J, Vetto RM: Combined arterial dilatation and femorofemoral bypass for limb salvage. Surg Gynecol Obstet 137:409–412, 1973.
3. Eidemiller LR, Porter JM, Rosch J, Dotter CT, Krippaehne WW: Surgical treatment of bilateral iliac artery occlusive disease in high-risk patients. Am Surg 40:511–517, 1974.
4. Porter JM, Eidemiller LR, Hood RW, Wesche DH, Dotter CT, Rosch J: Transluminal angioplasty and distal arterial bypass. Am Surg 43:695–702, 1977.
5. Dotter CT, Frische LH, Judkins MP, Mueller R: The "nonsurgical" treatment of iliofemoral arteriosclerotic obstruction. Radiology 86:871–875, 1966.
6. Dotter CT, Rosch J, Anderson JM, Antonovic R, Robinson M: Transluminal iliac artery dilatation: Nonsurgical catheter treatment of atheromatous narrowing. JAMA 230:117–124, 1974.
7. Dotter CT, Rosch J, Porstmann W, et al: Caged balloon catheter system for iliac artery dilatation. Presented before the American Heart Association, 46th Scientific Session, Atlantic City, N.J., November, 1973.
8. Porstmann W: Ein Neuen Korsett Balloon Katheter zur Transluminalen Rekanalisation nach Dotter unter besonderer Berucksichtigung von Obliterationen an den Beckenarterien. Radiol Diagn 14:239–244, 1973.
9. Gruentzig A, Hopff H: Perkutane Rekanalisation chronischer arterieller Veschlusse mit einem Neuen

Dilatations katheter. Dtsch Med Wochenschr 99:2502–2505, 1974.
10. Gruentzig A, Kumpe DA: Technique of percutaneous transluminal angioplasty with the Gruentzig balloon catheter. AJR 132:547–552, 1979.
11. Alpert JR, Ring EJ, Freiman DB, Oleaga JA, Gordon RG, Berkowitz HD, Roberts B: Balloon dilation of iliac stenosis with distal arterial surgery. Arch Surg 115:715–717, 1980.
12. Roberts B, Gertner MH, Ring EJ: Balloon-catheter dilation as an adjunct to arterial surgery. Arch Surg 116:809–812, 1981.
13. Beebe HG, Stark R, Freeny PC: Indications for transluminal angioplasty: A surgical view. Am J Surg 140:31–39, 1980.
14. Alpert JR, Ring EJ, Freiman DB, Oleaga JA, Gordon R, Berkowitz HD, Roberts B: Treatment of iliac artery stenosis by balloon catheter dilatation. Surg Gynecol Obstet 150:481–485, 1980.
15. Wallsh E, Franzone AJ, Clauss RH, Bruno MS, Steichen F, Stertzer SH: Transluminal coronary angioplasty during saphenous coronary bypass surgery. Ann Surg 191:234–237, 1980.
16. Katz R, Leiboff R, Aaron B, Mills M, Wasserman A, Ross A: Intraoperative retrograde left anterior descending balloon angioplasty for reperfusion of jeopardized proximal branches. (Abstract) Circulation 64(suppl IV):90, 1981.
17. Mills NL, Doyle D: Does operative transluminal angioplasty (OTA) extend the limits of coronary bypass procedure? (Abstract) Circulation 64(suppl IV):292, 1981.

18. Villemot JP: Preoperative dilatation of coronary arteries. Nouv Presse Med 11:2559–2562, 1982.
19. Roberts AJ, Feldman RL, Conti CR, et al: Preliminary experience with intraoperative transluminal balloon catheter dilatation and coronary artery bypass grafting for the treatment of symptomatic diffuse coronary artery disease. Ann Thorac Surg 34:504–514, 1982.
20. Wallsh E, Franzone AJ, Weinstein GS, Alcan K, Clavel A, Stertzer SH: Use of operative transluminal coronary angioplasty as an adjunct to coronary artery bypass. J Thorac Cardiovasc Surg 84:843–848, 1982.
21. Jones EL, Ring SB: Intraoperative angioplasty in the treatment of coronary artery disease. J Am Coll Cardiol 1(3):970–971, 1983.
22. Mills NL, Ochsner JL, Doyle DP, Kalchoff WP: Technique and results of operative transluminal angioplasty in 81 consecutive patients. J Thorac Cardiovasc Surg 86:689–696, 1983.
23. Lowman BG, Queral LA, Holbrook WA, Estes JT, Dagher FJ: Transluminal angioplasty during vascular reconstructive procedures. Arch Surg 116:829–832, 1981.
24. Katzen BT, Chang J, Knox WG: Percutaneous transluminal angioplasty with the Gruentzig balloon catheter: A review of 70 cases. Arch Surg 114:1389–1399, 1979.
25. Motarjeme A, Keifer JW, Zuska AJ: Percutaneous transluminal angioplasty as a complement to surgery. Radiology 141:341–346, 1981.
26. Waltman AC, Greenfield AJ, Novelline RA, et al: Transluminal angioplasty of the iliac and femoropopliteal arteries. Arch Surg 117:1218–1221, 1982.
27. Fogarty TJ, Chin A, Shoor PM, Blair GL, Zimmerman JJ: Adjunctive intraoperative arterial dilatation: Simplified intrumentation technique. Arch Surg 116:1391–1398, 1981.
28. Dobrin PB: Balloon embolectomy catheters in small arteries: I. Lateral wall pressures and shear forces. Surgery 90:177–185, 1981.
29. Jorgensen RA, Dobrin PB: Balloon embolectomy catheters in small arteries: IV. Correlation of shear forces with histologic injury. Surgery 93:798–808, 1983.
30. Chidi CC, Depalma RG: Atherogenic potential of the embolectomy catheter. Surgery 83:549–557, 1978.
31. Cowley MJ, Vetrovec GW, Wolfgang TC: Efficacy of percutaneous transluminal coronary angioplasty: Technique, patient selection salutary results, limitations and complications. Am Heart J 101:272–280, 1981.
32. Staple TW: Modified catheter for percutaneous transluminal treatment of arteriosclerotic obstructions. Radiology 91:1041–1043, 1968.
33. Van Andle GJ: *Percutaneous transluminal angioplasty: The Dotter Procedure*. Amsterdam, Excerpta Medica, 1976.
34. Gruentzig A, Senning A, Seigenthaler WE: Nonoperative dilatation of coronary artery stenosis: Percutaneous transluminal coronary angioplasty. N Engl J Med 301:61–68, 1979.
35. Spence R, Freiman DB, Gatenby R, et al: Long term results of transluminal angioplasty of the iliac and femoral arteries. Arch Surg 116:1377–1385, 1981.
36. Zeitler E: Complications in and after PTR, in Zeitler R, Gruentzig A, Schoop W, (eds): *Percutaneous Vascular Recanalization*. Berlin, Springer-Verlag, 1978, pp 120–125.
37. Kumpe DA, Jones DN: Percutaneous transluminal angioplasty: Radiologic viewpoint. Appl Radiol 11:29–40, 1982.
38. Kadir S, Smith GW, White RI, et al: Percutaneous transluminal angioplasty as an adjunct to the surgical management of peripheral vascular disease. Ann Surg 195:786–795, 1982.
39. Howell HS, Ingram CA, Parham AR, et al: Transluminal angioplasty of the iliac artery combined with femorofemoral bypass. South Med J 76:49–51, 1983.
40. Ross R, Glomset JA: The Pathogenesis of atherosclerosis. N Engl J Med 295:376–377, 1976.
41. Baumgartner HR: Eine neue Method zur Erzeugung von Thromben durch Gesielte Uberdehung der Gefusswand. Z Gesamte Exp Med 137:227, 1963.
42. Friedman M, Byers S: Aortic atherosclerosis intensification in rabbits by prior endothelial denudation. Arch Pathol 79:345, 1965.
43. Insul W Jr, Chidi CC: Model of arterial injury by balloon de-endothelialization: Reappraisal, improvement, and application to measurement of endothelial growth, in Shettler G, et al (eds): *Atherosclerosis: Proceedings of the Fourth International Symposium*, New York, Spring-Verlag, 1976, p 273.
44. Katocs AS, Largis EE, Will LW, et al: Sterol deposition in aortae of normocholesteremic and hypercholesteremic rabbits subjected to aortic de-endoth elialization. Artery 2:38, 1976.
45. Lee WM, Lee RT: Advanced coronary atherosclerosis in swine produced by combination of balloon catheter injury and cholesterol feeding. Exp Mol Pathol 23:491, 1975.
46. Minick CR. Stemerman MB, Insull W Jr: Effect of regenerated endothelium on lipid accumulation in the arterial wall. Proc Natl Acad Sci USA 74(4):1724, 1977.
47. Moore S: Thromboatherosclerosis in normolipidemic rabbits. A Result of continued endothelial damage. Lab Invest 29:478, 1973.
48. Katzen BT, Chang J, Lukowsky GH, Abramson EG: Percutaneous transluminal angioplasty for treatment of renovascular hypertension. Radiology 131:53–58, 1979.
49. Greenfield AJ: Femoral, popliteal, and tibial arteries: Percutaneous transluminal angioplasty. AJR 135:927–935, 1980.
50. Kinney TB, Chin AK, Rurik GW, Finn JC, Shoor PM, Hayden WG, Fogarty TJ: The physical mechanisms of transluminal angioplasty: A mechanical pathophysiological correlation. Radiology 153:85–89, 1984.
51. Castaneda-Zuniga WR, Formanek A, Tadavarthy M, Vlodaver Z, Edwards JE, Zollikofer C, et al: The mechanisms of balloon angioplasty. Radiology 135:565–571, 1980.
52. Pasternak RC, Baughman KL, Fallon JT, Block PC: Scanning electron microscopy after coronary transluminal angioplasty of normal canine coronary arteries. Am J Cardiol 45:591–593, 1980.
53. Block PC, Baughman KL, Pasternak RC, Fallon JT:

Transluminal angioplasty: Correlation of morphologic and angiographic findings in an experimental model. Circulation 61:778–785, 1980.

54. Hoffman MA, Balloon JT, Greenfield AJ, Waltman AC, Athanasoulis CA, Block PC: Arterial pathology after percutaneous transluminal angioplasty. AJR 137:147–149, 1981.

55. Block PC, Myler RK, Stertzer SH, Fallon JT. Morphology after transluminal angioplasty in human beings. N Engl J Med 307:382–385, 1981.

56. Chin AK, Fogarty TJ, Kinney TB, Hayden WG, Shoor PM, Rurik GW: Pathophysiologic bases for transluminal dilatation. Surg Forum 32:323–325, 1981.

57. Castaneda-Zuniga WR, Amplatz K, Laerum F, Formanek K, Subley R, Edwards J, et al: Mechanics of angioplasty: An experimental approach. Radiographics 1:1–14, 1981.

58. Saffitz JE, Totty WG, McClennan BL, Gilula LA: Percutaneous transluminal angioplasty. Radiological pathological correlation. Radiology 141:651–654, 1981.

59. Zarins CK, Lu CT, Gewertz BL, Lyon RT, Rush DS, Glagov S: Arterial disruption and remodeling following balloon dilatation. Surgery 92:1086–1095, 1982.

60. May AG, Oeberg LV, DeWeese JA, Rob CG: Critical arterial stenosis. Surgery 54:250–259, 1963.

61. Haimovici H. Escher DJW: Aortoiliac stenosis. AMA Arch Surg 72:107, 1956.

62. Wylie EJ, McGuinnes JS: The recognition and treatment of arteriosclerotic stenosis of major arteries. Surg Gynecol Obstet 97:425, 1953.

63. Berne RM, Levy MN: *Cardiovascular Physiology*. St. Louis: C. V. Mosby, 1977, p 59.

64. Connolly JR, Kwaan JHM, McCart PM: Complications after percutaneous transluminal angioplasty. Am J Surg 142:60–66, 1981.

65. Freiman DB, Spence R, Gatenby R, et al: Transluminal angioplasty of the iliac and femoral arteries. Follow-up results without anticoagulation. Radiology 141:347–350, 1981.

66. VanAndel GJ: Transluminal iliac angioplasty: Long term results. Radiology 135:607–611, 1980.

67. Vetto RM: The treatment of unilateral iliac artery obstruction with a transabdominal subcutaneous femoropopliteal graft. Surgery 52:342–345, 1962.

68. Louw JH: Splenic to femoral and axillary-to-femoral bypass grafts in diffuse arteriosclerotic occlusive disease. Lancet 1:1401–1402, 1963.

69. Schmidke I, Fettler E, Schoop W: Late results of percutaneous catheter treatment (Dotter's technique) in occlusion of the femoropopliteal arteries, stage II, in Zeitler E, Greuntzig A, Schoop W (eds): *Percutaneous Vascular Recanalization*. New York, Springer-Verlag, 1978, pp 96–110.

70. Malone JM, Moore WS, Goldtone J: The natural history of bilateral aortofemoral bypass grafts for ischemia of the lower extremities. Arch Surg 110:1300–1306, 1975.

71. Inahara T: Eversion endarterectomy for aortoiliofemoral occlusive disease. A 16 year experience. Am J Surg 138:196–204, 1979.

72. Brewster DC, Darling RC: Optimal methods of aortoiliac reconstruction. Surgery 84:739–748, 1978.

73. Archie JP: Objective improvement after aortofemoral bypass for exercise ischemia. Surg Gynecol Obstet 149:374–376, 1979.

74. Bone GE, Hayes AC, Slaymaker EE, Barnes RW: The value of segmental limb blood pressures in predicting results of aorto-femoral bypass. Am J Surg 132:733, 1976.

75. Garrett WV, Slaymaker EE, Heintz SE, Barnes RW: Intraoperative prediction of symptomatic results of aortofemoral bypass from changes in ankle pressure index. Surgery 82:504, 1977.

76. Strandness DE: Functional results after revascularization of the profunda femoris artery. Am J Surg 119:240, 1970.

77. Brief DK, Alpert J, Parsonnet V: Crossover femorofemoral grafts: Compromise or preference: A reappraisal. Arch Surg 105:889, 1972.

78. Flanigan DP, Pratt DG, Goodreau JJ, et al: Hemodynamic and angiographic guidelines in selection of patients for femorofemoral bypass. Arch Surg 113:1257–1262, 1978.

9

Percutaneous Transluminal Angioplasty in Congenital Heart Disease

Eric C. Martin, M.A. (OXON)

INTRODUCTION

Since Rashkind first performed a balloon atrial septotomy in 1966 (1), there has been considerable interest in interventional radiology in congenital heart disease, spurred primarily by Rashkind himself (2), but it is only very recently that angioplasties have been performed.

PULMONARY VALVULOPLASTY

In 1979, Semb reported on a balloon valvulotomy in a neonate with tricuspid regurgitation caused by Ebstein's disease and severe pulmonary stenosis (3). A Berman catheter maneuvered across the right ventricular outflow tract revealed a gradient of 20 mmHg with a right ventricular pressure that was two-thirds systemic. The balloon was then inflated with carbon dioxide and pulled down into the right ventricle, thereby reducing the outflow gradient to 6 mmHg. The patient's condition rapidly improved, the systolic murmur disappeared, and he required no treatment over the following year.

Balloon dilation of the pulmonary valve has been considered for many years, but Semb's report brought it closer to reality. Kan, Anderson, and White, encouraged by this work, developed a technique for balloon angioplasty in animals (4), and in the same year Kan, White, et al. reported the successful treatment of an 8-year-old child (5). The gradient across the pulmonary valve was 48 mmHg and the right ventricular pressure 60 mmHg, which was two-thirds systemic. Balloon dilation reduced the gradient to 20 mmHg, and the right ventricular pressure fell to 28 mmHg. On recatheterization at 4 months the appearances were essentially unchanged, with only a 20 mmHg gradient remaining.

In the same report they commented on four other patients in whom an immediate reduction in right ventricular pressure had been achieved with valvuloplasty, but no follow-up was available.

In a 1983 abstract presented to the American College of Cardiology, Kan updated this experience to six patients (6). The mean pressure gradient before valvuloplasty was 66 mmHg and after valvuloplasty 28 mmHg; and no patient developed the clinical finding of pulmonary insufficiency on follow-up. The six patients ranged in age from 3 months to 24 years, and the most dramatic result was in a child of 6 months whose gradient dropped from 118 mmHg to 22 mmHg. The child was dilated twice, once at 3 months and once at 6 months. Although one 8-year-old had the gradient reduced from 45 mmHg to 14 mmHg, it appears that patients with lesser gradients, i.e., those with mild to moderate pulmonary valve stenosis, have inferior responses: one patient's gradient was reduced only from 37 mmHg to 35 mmHg.

Pepine, in 1982, reported on a 59-year-old woman with pulmonary valve stenosis whose gradient was reduced from 140 mmHg to 40 mmHg and whose right ventricular pressure fell from a systemic level to 65 mmHg with balloon valvuloplasty. These pressures were maintained at the 10th day when catheterization was repeated (7).

Angioplasty Technique

A specially modified angioplasty catheter is required for pulmonary valvulotomy.* It is polyethylene and of the usual coaxial manufacture but contains two side ports for rapid evacuation of the balloon. Obviously, when the balloon is inflated the right-sided cardiac output is occluded, and rapid inflation and deflation are essential. The catheters have larger than usual balloon sizes ranging from 10 to 20 mm with a consequent low inflation pressure of around 3 atmospheres. It is important to size the pulmonary annulus by measurement, either from the cine angiogram or from the videotape, correcting for magnification. The catheter, with its known French size, may be used as a scale. Placing a centimeter ruler on the output phosphor at the level of the sternum not only acts as a convenient reference for the annulus size, but also provides a fluoroscopic reference point for the position of the pulmonary valve.

Kan and White's technique involves positioning one catheter in the right ventricle and a second across the pulmonary valve to be used in exchanging for the valvulotomy catheter. A third catheter, or at least a sheath, is placed in the femoral artery to monitor systemic pressure. Any appropriate catheter is used to cross the valve and an exchange made by using a 200 or 300 cm, 0.9 mm (0.035 inch) Rosen J wire positioned in either the left or right pulmonary artery. The chosen number 9 French balloon catheter is then passed over the guidewire and the balloon positioned at the level of the pulmonary valve. The balloon markers are easily visualized, and we have found a ruler on the lateral output phosphor to be of assistance. Additionally, we have used a single frame from a tape of a previous right ventricular cine angiocardiogram, filmed in the lateral view and projected on to the anteroposterior (AP)

*Meditech, Inc., Watertown, Macssahusetts.

monitor as a still frame for reference. The balloon is then inflated to 3 atmospheres, and a "waist" should appear in the middle of the balloon. If not, the balloon is rapidly deflated before full inflation has occurred and the balloon then repositioned. Inflation continues until the constriction expands suddenly, and then the balloon is rapidly deflated with the systemic pressure being monitored throughout. On occasion, inflation may have to be terminated because of a falling systemic pressure. If the systemic pressure is falling quickly, it may be necessary to advance the balloon over the guidewire, once it is partially deflated, to allow forward flow to resume. For this reason the procedure is performed with a guidewire in place. The right ventricular pressure is useful for assessing the gradient after dilation and for deciding whether or not to reinflate the balloon.

During dilation we pay most attention to the systemic pressure. We have also noted a change in the pulmonary artery pressure trace to be useful in monitoring the results of dilation. With the catheter in place across the pulmonary valve the pulmonary artery pressure is frequently damped; after successful dilation a phasic quality returns.

Our experience is limited to three patients, all with only moderate stenoses and right ventricular pressures the order of 50 mmHg, in whom only moderate results were achieved (Fig. 1). Although it does not appear particularly satisfactory to lower the gradient by only 10 or 15 mmHg, nevertheless, reducing the RV pressure to 35 or perhaps 50 mmHg eliminates the need for surgery.

It appears from the results so far that the technique is most successful in patients with tight stenoses and that the procedure is unsuccessful only in those patients who have dysplastic valves. We feel that consent for balloon pulmonary valvulotomy should be obtained for all patients with pulmonary valve stenosis and that valvuloplasty should be attempted at the time of the initial catheterization except in patients with dysplastic valves. Only after a significant series will a decision on its value be validated. Follow-up is eagerly awaited.

PERIPHERAL PULMONARY ARTERY STENOSIS

In 1980, we reported on a patient with pulmonary atresia who had a Dacron Hancock graft from the right ventricle to each pulmonary artery (8). Tight anastomotic stenoses were demonstrated angiographically between the distal ends of the Y graft and the pulmonary arteries with approximately 100-mm gradients. Dilation was performed on each side with an overstretched 9 mm polyvinyl balloon cath-

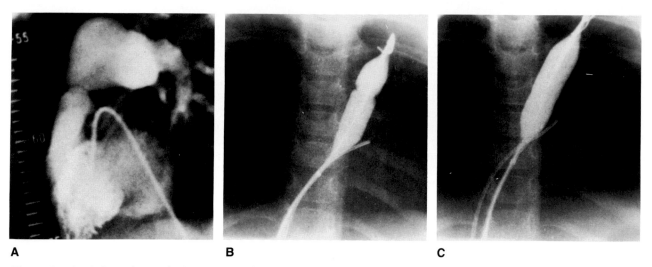

Figure 1.   A. A lateral ventriculogram showing the typical appearance of pulmonary valve stenosis. The ruler on the output phosphor is a useful guide to the position of the pulmonary valve. **B.** The valvulotomy balloon partially inflated, showing "waist." Note the additional right ventricular catheter. The guidewire has pulled back from the left pulmonary artery. **C.** During inflation, the waist has disappeared. (Dilation was performed in conjunction with Dr. Carl Steeg.)

eter on 2 successive days; despite a change in the phasic quality of the distal pulmonary artery pressures, the right ventricular pressure did not change (Fig. 2). The patient was then operated upon and had the graft replaced but died postoperatively. At

surgery each pulmonary artery accepted a 9 mm dilator. Autopsy indicated a severe pulmonary valvular stenosis in the Hancock valve that had not been appreciated during the preoperative catheterization. Since that time we have performed three

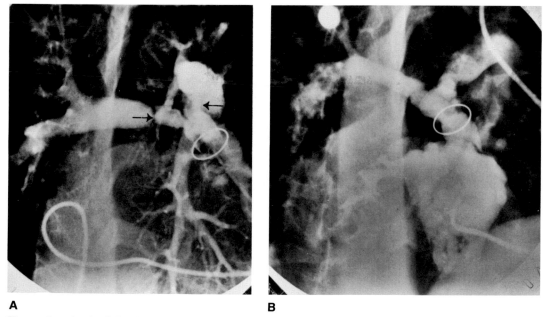

Figure 2.   A. An injection in the graft from the right ventricle to each pulmonary artery, showing tight anastomotic strictures (arrows). **B.** A right ventriculogram after dilation. The stenoses have been removed, with no change in the right ventricular pressure. Valvular stenosis in the Hancock valve is not readily appreciated. (Reprinted from Ref. 8.)

other dilations, all using polyvinyl balloons, in patients with peripheral pulmonary artery stenoses. One patient, again, had a Hancock graft from the right ventricle to the pulmonary artery with an anastomotic stenosis of the left pulmonary artery and damage to the main right pulmonary artery as a result of a previous Waterston shunt. The constriction in the balloon could not be removed at the site of the anastomotic stricture, but the balloon expanded fully in the narrowed segment of the Waterston anastomosis, again without significant change in the main pulmonary artery or right ventricular pressures (Fig. 3). A third patient had a long-segment stenosis, again as a result of a Waterston shunt; dilatation had no significant hemodynamic effect, although the balloon expanded fully (Fig. 4). The fourth patient had a congenitally hypoplastic pulmonary artery; although the balloon expanded easily, there was again no change in pressure. The balloon chosen was the diameter of the normal pulmonary artery.

In an attempt to clarify the problem of pulmonary artery branch stenosis, Lock created pulmonary artery stenoses in 27 newborn lambs by an interrupted suture technique similar to his model of experimental coarctation (9) (Fig. 5). Nine lambs were long-term survivors and were dilated with a modified Gruentzig balloon catheter with apparent success. Polyethylene catheters were used; the balloon sizes were not specified, although it is assumed that they were of the order of 8 mm in diameter. Dilation was associated with a decrease in the systolic gradient from a mean of 35 mmHg to 8 mmHg and with an increase in the diameter of the pulmonary artery from 4.6 to 7.6 mm. The percentage of total flow to the dilated lung appeared to rise from 19 percent to 45 percent when measured with radioactive microspheres. Four of the lambs were catheterized 4 months later, and the average gradient remained below 10 mmHg despite the growth that had occurred during that interval. Histology showed linear tears in the intima, completely healed at 2 months. No morbidity could be attributed to the procedure.

The study may be criticized for its choice of an experimental model of pulmonary artery stenosis. Lock used a suture stenosis, which we know from Gruentzig's original work in the coronary arteries in 1976 may easily be dilated (10). It may not, however, be representative of hypoplastic pulmonary arteries, nor of other congenital stenoses. Rather, it is more akin to the anastomotic strictures that we know to have been successfully dilated in other locations (8,11,12) although less successfully in the pulmonary arteries.

Encouraged by these experimental results, Lock applied the technique to hypoplastic and stenotic pulmonary arteries in five patients with tetralogy of Fallot who had already been operated upon and who had branch pulmonary artery stenosis (13). In all patients the right ventricular pressure was greater than two-thirds systemic; in all patients an increase in diameter was demonstrated angiographically, as well as were a decreased gradient and an increase in the percentage of blood flow to the lung on the dilated side. The right ventricular pressures fell from a mean

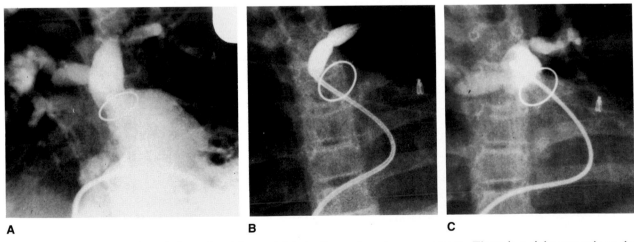

A   **B**   **C**

Figure 3.   **A.** A Hancock graft from the right ventricle to the main pulmonary artery. There is a tight stenosis at the origin of the left pulmonary artery and a stenosis in the middle of the right pulmonary artery. (The Gianturco coil is the result of an unsuccessful attempt at embolization of a collateral.) **B.** The balloon inflated in the left pulmonary artery. The waist could not be eliminated. **C.** The appearances after dilation, with essentially no cosmetic improvement and pressures unchanged.

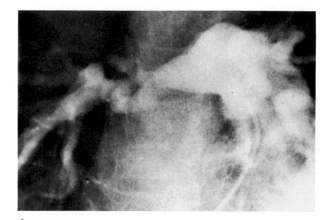

A

B

Figure 4. **A.** The stenosis in the right pulmonary artery is visible as a result of the Waterston anastomosis. **B.** After dilation the appearance are minimally improved. There was a slightly more phasic quality to the pressure trace distally and the right ventricular pressure was unchanged.

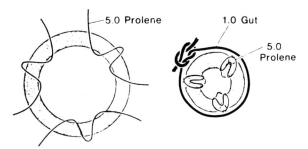

Figure 5. Lock's technique for creating experimental pulmonary artery stenosis. (Reprinted from Ref. 9.)

of 104 mmHg to a mean of 80 mmHg. The diameter of the narrowed segment increased from a mean of 3.7 mm to a mean of 6.8 mm, and the percentage of blood flow increased from 41 to 52 percent. There was no morbidity. Follow-up angiograms in three patients indicated persistent cosmetic improvement.

A closer examination of the paper, however, is confusing. All the patients appeared to have two gradients, one across the outflow tract and the other between the branch pulmonary artery and the main pulmonary artery. The relative severity of each is not clear from the numbers given, making interpretation very difficult. Nor is the systemic pressure recorded; so that the significance of the mean 20 mmHg fall in the right ventricular pressure is unclear. The branch pulmonary artery and main pulmonary artery gradients are almost uninterpretable since the RV outflow gradients range from 25 to 95 mmHg, and the systemic pressures are unavailable. However, the cosmetic results illustrated in one patient are most dramatic. Three children have been restudied, apparently demonstrating successful results; however, no pressure data are given.

It is clear that a considerable amount of work remains to be done to assess which patients are amenable to dilation of pulmonary artery branch stenoses. Anastomotic strictures may be dilatable if the balloon can be expanded, and this may be dependent on the age of the anastomosis. In other stenoses, expansion of the balloon does not necessarily produce a good hemodynamic result, but obviously it would be a particularly valuable technique if successful, and Lock's work on overdilating hypoplastic lesions is most encouraging.

COARCTATION OF THE AORTA

Sos successfully dilated a postmortem specimen of coarctation in 1979 (14), and his success was confirmed by Lock in an excised human specimen (15). Lock suggested, from the histological appearance, that higher than usual dilating pressures of the order of 8 atmospheres, would be required for successful in vivo dilatation, but fortunately the mechanics of balloon angioplasty require higher pressures for smaller balloons (16). Currently the largest balloon mounted on a number 5 French catheter is 5 mm; when manufactured from polyethylene, this size easily takes 8 atmospheres of pressure. The equipment for dilating coarctation of the aorta is therefore available. Theoretically, however, one would not expect the anatomy of coarctation to respond very favorably to percutaneous dilation. Typically, a shelf opposite the ductus creates an eccentric lesion. If the anterior aortic wall is elastic, or at least compliant, then dilation should fail. Marked-over dilation, to compensate for the elasticity, obviously increases the risk of aortic rupture.

In 1982, Lock developed an animal model of coarctation in which excision of a wedge from the posterior aortic wall created an eccentric shelf similar to the anatomy of coarctation (17) (Fig. 6). Thirteen lambs in which artificial coarctations were created underwent balloon dilatations using 7F balloons of 6–12 mm in diameter. The mean gradient was reduced from 37 (± 16) mmHg to 12 (± 6) mmHg. A trace of extravasation appeared in one animal on contrast injection, but otherwise there were no complications. After sacrifice, the histology showed intimal tears on the anterior wall with intramedial hemorrhage and separation of the muscle fibers. Six lambs were restudied 6–12 months later, and the gradient remained low despite their growth. After sacrifice, histology demonstrated evidence of medial thinning, but no aneurysm formation or atheroma, and the intima was intact.

This experimental study raises a number of questions. First, it is possible to tear the aorta sufficiently to extravasate contrast. Intimal tears and medial stretching occur in atheromatous disease but appear not to be a cause for concern. Rather, they suggest a successful angioplasty, but we are not yet sure that they are so benign in coarctation. Furthermore, it is possible that the dilated segment will go on to aneurysm formation or even early atheroma, but as yet there is no evidence.

Second, if percutaneous dilation is to be performed, it must be successful. The literature suggests that coarctation corrected before the age of 5 years removes the risk of late hypertension and per-

haps therefore the late cardiovascular complications of coarctation of the aorta (18–20). A partially successful dilation may not so affect the natural history.

Third, the technique employed by Lock and Castaneda involved numbers 8 and 9 French catheters. It is commonly accepted that such catheters are too large for a neonate, and the risk of trading a moderate result of coarctation dilation for an ischemic limb appears unacceptable.

It was encouraging, therefore, when Sperling reported on two infants, 3 weeks and 11 months in age, treated with number 5 French, 5 mm balloon catheters (21). The cosmetic results illustrated in their paper appeared satisfactory, and the gradients were reduced from 50 mmHg to 8 mmHg systolic and from 23 mmHg to 8 mmHg systolic, respectively. However, the good result did not persist in the first patient, and the indication for dilation in the second patient appears to have been marginal.

In 1983 Lock followed up his animal work and reported on two infants with coarctation who were dilated at 1 week and 4 weeks of age respectively (22). He used number 5 French catheters with 4 and 5 mm balloons and inflated for 30–60 seconds once only. He used a balloon diameter 2.5 times the coarctation diameter. Neither patient was helped by the procedure, and both then had surgery. One patient had a diminished femoral pulse as a result of angioplasty.

In the same paper Lock also reported dilating a hypoplastic aortic segment in a 16-year-old with the supravalvular aortic stenosis syndrome. An excellent result was achieved for a few hours, although the stricture then recurred as if it were elastic. Parenthetically, we have had a similar experience in a

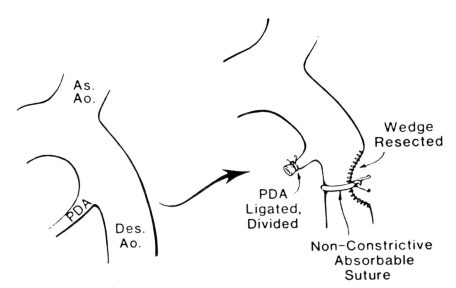

Figure 6. Lock's technique for creating an experimental model of coarctation. (Reprinted from Ref. 17.)

renal artery in a patient with the supravalvular aortic stenosis syndrome.

Our only experience in primary coarctation of the aorta involves one infant less than 1 week of age in whom dilation was performed with a number 5 French, 5 mm balloon catheter. The balloon expanded easily and without resistance, but the coarcted segment appeared elastic. An angiogram following the procedure showed no change nor did the gradient. It would appear, therefore, from the preliminary data that primary coarctation of the aorta is seldom amenable to dilation.

RECOARCTATION OF THE AORTA

The surgical repair of coarctation of the aorta in infancy results in a significant incidence of recoarctation, varying from 5 percent to as much as 50 percent in one report of excision and end-to-end reanastomosis (23,24). Dissatisfaction with these results has led to alternative repairs, such as the subclavian turndown operation (25,26), but balloon angioplasty would obviously be an attractive alternative to reoperation. Extensive experience in the angioplasty of vein grafts and renal transplants has demonstrated that surgical anastomoses are dilatable (8,11,12,27) but not always, as we have seen, in the heart.

Castaneda-Zuniga developed an experimental model to examine the feasibility of dilating recoarctation of the aorta (28). The surgery was performed in the abdominal aorta of dogs, but the model is an equivalent of postoperative coarctation. Thirteen dogs had coarctations created, and in six animals it was performed by wedge resection, transection, and reanastomosis, a model of the normal operative procedure (Fig. 7). Five of the animals subsequently underwent angioplasty with 9 mm balloon catheters, with a resulting mean increase in diameter of 0.8 mm and a mean drop in systolic gradient of 12 mmHg from an initial mean of 24 mmHg. The histology showed stretching and thinning of the media and some areas of intimal splitting. In the control animal a marked fibroblastic reaction involved the media and adventitia as a result of the surgery. The authors concluded that recoarctation was not a suitable lesion for balloon dilation, probably because of the strong circumferential scar tissue. They further postulated that because the adventitia is no longer intact, the risk of rupture may be higher than in normal arteries.

Despite the claims of a lower incidence of recoarctation following the subclavian turndown operation, we have had the opportunity to dilate two such restenoses. In a 1-year-old patient we used a

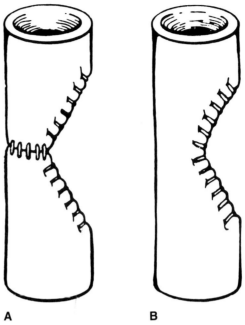

A   **B**

Figure 7. Castaneda-Zuniga's technique for creating an experimental model of postoperative coarctation of the aorta. Wedge resection with and without reanastomosis. (Reprinted from Ref. 28.)

number 5 French, 5 mm polyvinyl catheter and dilated a hypoplastic transverse arch in addition to the recoarctation (Fig. 8). In the 6-year-old, the restenosis was dilated with a number 7 French, 8 mm polyethylene balloon catheter (Fig. 9). Both patients had a return of normal femoral pulses after the procedure, and one was left with a mild residual gradient but only a moderate cosmetic result. The other had the gradient completely removed. They have not yet been recatheterized, but both have had normal femoral pulses clinically up to 3 months follow-up.

Since the subclavian turndown is essentially an onlay patch and therefore eccentric, it may not dilate easily unless there is concentric, circumferential perivascular fibrosis. We are not, therefore, sanguine about the results of the dilation of recoarctation when due to a subclavian turndown operation.

Despite their experimental results, Lock et al. reported on five patients who had dilation of postoperative coarctation (22). Two patients had subclavian flap operations. Both had an immediate reduction of gradient; from 74 mmHg to 33 mmHg in one patient and from 30 mmHg to 10 mmHg in the other, with increases in diameter from 2.2 to 2.4 mm in the first patient and from 1.8 to 2.8 mm in

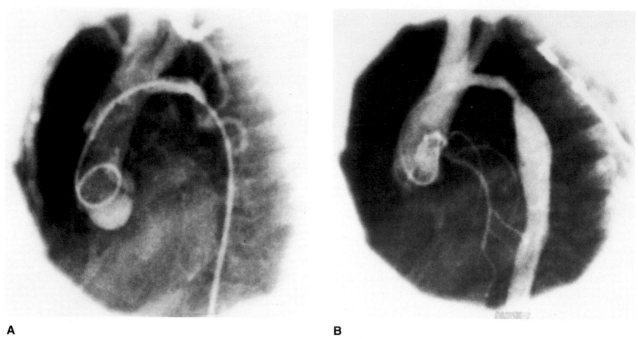

A   B

Figure 8. **A.** A severe coarctation after a subclavian turndown operation. The catheter (number 5 French) is occlusive. There are also marked hypoplasia of the transverse arch, and 80 mmHg systolic pressure gradient. **B.** After dilation with a 5 mm balloon, the gradient was reduced to 20 mmHg. There has been no dilation of the hypoplastic arch. Some overstretching distal to the coarctation appears.

the second. However, the favorable results persisted for only a day or two. The three patients with end-to-end anastomosis did better. The first patient, aged 6 years, had the gradient reduced from 67 to 49 mmHg and the diameter increased from 3.5 to 5.4 mm. The second, aged 9 years, had the gradient reduced from 42 to 18 mmHg and the diameter increased from 4.3 to 7.3 mm, and the third patient, aged 22 years, had the gradient reduced from 43 to 0 mmHg and the diameter increased from 8.5 to 13.5 mm. The balloons used had diameters of 8, 12, and 20 mm and were chosen to be approximately 2.5 times the diameter of the coarcted segment. All three patients maintained good distal pulses clinically. The first patient was then redilated and the gradient reduced to 8 mmHg, which appears to have persisted clinically for 3 months.

We feel, at the moment, that angioplasty in coarctation and recoarctation should be explored only in a few centers and that the procedure will probably be employed only for patients with recoarctation and end-to-end anastomoses. But further results are awaited and this impression is distinctly premature.

MUSTARD BAFFLE

In 1982 we attempted to dilate the obstructed upper limb of a Mustard baffle in a 3½-month-old child who had a 18 mmHg gradient between the superior vena cava (SVC) and the right atrium (Fig. 10). We used a number 5 French, 5 mm balloon catheter, and the balloon expanded easily without a "waist" being visible. The pressure gradient fell transiently, but it was not reproducible throughout the rest of the catheterization. The patient's clinical situation was complex, including some superimposed pulmonary venous compromise, but although no immediate clinical improvement occurred, there was a gradual improvement over the next few months. Nine months later, the patient did not require reoperation. We had assumed that the stenosis would be cicatricial at the site of the anastomosis, but rather it proved to be elastic and the result is difficult to assess. Certainly, the procedure was performed as an emergency in an infant requiring ventilatory support and now reoperation is no longer being considered. It is, however, difficult to place the results of the dilation in context.

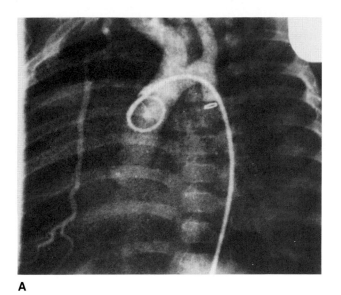

A

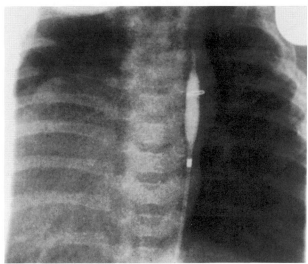

B

Figure 9. A. Arch aortogram some years after a coarctation repair. The clip may have been on a PDA. The internal mammary artery is hypertrophied and collateralizes distally. The catheter is occlusive. (The patient has situs inversus and complex congenital heart disease.) **B.** The balloon is inflated. The waist has disappeared. **C.** The appearance after dilatation. Note that the contrast column is at the level of the diaphragm while the internal mammary artery is filled down only to the level of the coronary arteries.

C

SUPERIOR VENA CAVA

A not dissimilar case was described by Rocchini in 1982 (29). Theirs was a child with total anomalous pulmonary venous connection to the right atrium whose pulmonary veins were baffled into the left atrium, and the patient was operated upon on two more occasions before developing a superior vena caval syndrome clinically. At catheterization the gradient between the SVC and the right atrium was 30 mmHg. The stenosis was dilated with a 6 mm balloon catheter, and a fair cosmetic result was obtained, decreasing the gradient from 30 mmHg to 16

mmHg. Four hours later the patient deteriorated and died shortly thereafter. It appeared at autopsy that the SVC had contained organized thrombus, which one would expect to respond to dilation but presumably is not a very common finding. The death was apparently unrelated to the procedure.

PULMONARY VEIN STENOSIS

In 1982, Driscoll reported on the dilation of three patients with pulmonary vein stenosis, a rare lesion almost invariably associated with other congenital

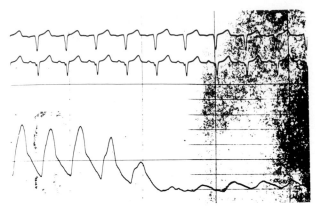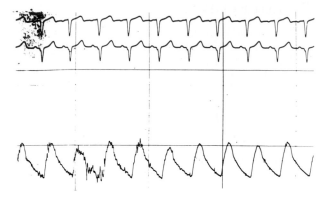

D

Figure 9 (*Continued*).   D. The pressure tracing before (left) and after (right) dilation. Balloon was pulled back across the coarctation.

heart disease (30). The first patient, aged 17 months, had dilation of both the upper left and the left lower pulmonary veins, lowering the pressure from 63 to 10 mmHg in the upper pulmonary vein and from 33 to 18 mmHg in the lower pulmonary vein. Following this, the systemic saturation increased from 65 to 97 percent, but 2 weeks later, the patient deteriorated clinically and died during surgery. A 5 mm catheter had been used. The second patient, aged 2 months, had a similar reduction from 65 to 25 mmHg in the pulmonary vein when dilated with a 6 mm balloon catheter but died 36 hours later of sepsis. It

is not stated whether or not the sepsis was related to the dilation, but it is presumed not to have been. The third patient, aged 7 months, had a moderately successful hemodynamic result reducing the pressure from 68 to 45 mmHg in the dilated pulmonary vein, but understandably no clinical improvement was achieved. In one patient, autopsy showed chronic medial hypertrophy of the vein at the junction with the left atrium, and severe overdilation might be successful. Certainly, the findings in this series do not lead one to much enthusiasm, but the results are, as yet, preliminary.

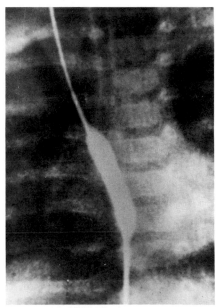

A   **B**

Figure 10.   A. An injection in the right atrium, showing reflex up the stenotic superior limb of the Mustard baffle. **B.** A 4 mm balloon inflated in the stenosis.

BLALOCK-TAUSSIG SHUNTS

Not infrequently Blalock-Taussig shunts become stenotic at the point of insertion into the pulmonary arteries, and patients become more cyanotic as this occurs. In those individuals who have complex lesions, unsuitable for total correction or even for a Fontan procedure, dilation of the stenotic Blalock-Taussig shunt is attractive. We have attempted such dilations in three shunts in two patients, without hemodynamic success, and on each occasion we used polyvinyl balloons.

The first patient was 16 years old with pulmonary atresia and a ventricular septal defect but no continuity of the pulmonary arteries across the midline. Separate Blalock-Taussig shunts supplied each pulmonary artery. On the right were a long-segment stenosis in the subclavian artery and an anastomotic stenosis; on the left the stenosis was purely anastomotic. Dilation was performed with number 7 French 6 mm balloons that inflated with difficulty and never completely removed the constriction. The surgery had been performed many years before. Recently, Banner et al. have observed that in ureteric anastomoses dilation may be expected to be successful only if the anastomosis is less than 6 months old and very rarely if it is much older (31). This is probably a reflection not of the ureter itself, but of the cicatricial fibrosis, and it is tempting to correlate this finding with the anastomoses in the patient described. The second patient was 8 years old, again with complex anatomy, and was dilated with a 6 mm polyvinyl balloon also without significant hemodynamic improvement.

On each occasion the lesion was crossed with a cobra catheter and an exchange made over a 0.035 inch Rosen J wire without administration of heparin. Because of the fibrotic nature of the anastomosis and the difficulty in inflation, we have assumed that polyethylene might perform better and an Olbert balloon better still but have not yet had the opportunity to test this hypothesis. Again the technique is potentially very valuable, but its efficacy is far from proved.

CONCLUSION

Interventional radiology in congenital heart disease is an exciting field, encompassing the embolization of collaterals, shunt closure in various sites, and angioplasty, but it is not yet clear what contribution angioplasty will make. Balloon valvulotomy of the pulmonary valve appears to be a successful procedure although it is unlikely that it is superior to open heart repair. Nevertheless, it is as effective as closed repairs, and the long-term results may well be sufficiently adequate to replace surgery except in dysplastic valves. Determining its value will, however, depend on accumulated long-term results. Angioplasty in other aspects of congenital heart disease is less easy to assess. In the arterial system, balloon technology limits the largest balloon mounted on a number 5 French catheter to 5 mm. Larger catheter shafts are probably contraindicated in the youngest patients, and 5 mm may not be an adequate size. On the venous side, and on the arterial side, where larger catheters may be used, the results are still ambiguous. Rupture has been reported and is the complication most feared. Whether or not severe overdistention leads to late atheroma or aneurysm formation is not yet known.

It may, of course, not be true that improved balloons will enhance the results. Certainly polyethylene balloons have a better working curve than polyvinyl balloons, and the new reinforced Olbert balloons modified to take much higher pressures seem more attractive, although they may not culminate in better results.

One source of confusion is the etiology of the stenoses being dilated. Anastomotic strictures are dilatable, certainly in vein grafts and occasionally in other vessels, but we know very little about congenital stenoses or hypoplasia. There are no adequate experimental models for these strictures, and it is often not clear exactly what one is dilating at the time of the procedure. Clearly, on theoretical grounds, one would expect hypoplastic lesions to be purely elastic, but Lock's work on overdilating hypoplastic lesions is encouraging (13).

Although it is important that angioplasty be explored in a few centers, ultimately, apart from valvular pulmonary stenosis, it will probably have only a limited role, specifically in patients with complex congenital heart disease. It may well be that dilation of recoarctation of the aorta will emerge to stand beside percutaneous pulmonary valvulotomy as a procedure of proven worth.

REFERENCES

1. Rashkind WJ, Miller WW: Creation of an atrial septal defect without thoracotomy: A palliative approach to complete transposition of the great arteries. JAMA 196:991–992, 1966.
2. Rashkind WJ: Transcatheter treatment of congenital heart disease. Circulation 67:711–716, 1983.
3. Semb GKH, Tjonneland S, Stake G, Aabyholm G: "Balloon valvulotomy" of congenital pulmonary valve

stenosis with tricuspid insufficiency. Cardiovasc Radiol 2:239–241, 1979.

4. Kan JS, Anderson J, White RI: Experimental basis for balloon valvuloplasty: A new method for treating congenital pulmonary valve stenosis. N Engl J Med 307:540–542, 1982.

5. Kan JS, White RI, Mitchell SE, Gardner TJ: Percutaneous balloon valvuloplasty: A new method for treating congenital pulmonary valve stenosis. N Engl J Med 307:540–542, 1982.

6. Kan JS, White RI, Mitchell SE, Gardner TJ: Transluminal balloon valvuloplasty for the treatment of congenital pulmonary valve stenosis; abstracted J Am Coll Cardiol 1(2):588, 1983.

7. Pepine CJ, Gessner IH, Feldman RL: Percutaneous balloon valvuloplasty for pulmonic valve stenosis in the adult. Am J Cardiol 150:1442–1445, 1982.

8. Martin EC, Diamond NR, Casarella WJ: Percutaneous transluminal angioplasty in non-atherosclerotic disease. Radiology 135:27–33, 1980.

9. Lock JE, Niemi T, Einsiz S, Amplatz K, Burke B, Bass JL: Transvenous angioplasty of experimental branch pulmonary artery stenosis in newborn lambs. Circulation, 64:886–893, 1981.

10. Gruentzig AR, Turina IM, Schneider JS: Experimental percutaneous dilatation of coronary artery stenosis. Circulation 54:81–85, 1976.

11. Sniderman KW, Sos TA, Sprayregen S, Saddekni S, Cheigh JS, Tapia L, Veith FJ: Percutaneous transluminal angioplasty in renal transplant arterial stenosis for relief of hypertension. Radiology 135:23–26, 1980.

12. Alpert JR, Ring EJ, Berkowitz HD, Frieman DB, Oleaga JA, Gordon R, Roberts B: Treatment of vein graft stenosis by balloon catheter dilatation. JAMA 242:2769–2771, 1979.

13. Lock JE, Castaneda-Zuniga WR, Fuhrman BP, Bass JL: Balloon dilatation angioplasty of hypoplastic and stenotic pulmonary arteries. Circulation 67:962–967, 1983.

14. Sos T, Sniderman KW, Rettek-Sos B, Strupp A, Alonso DR: Percutaneous transluminal dilatation of thoracic aorta post mortem. Lancet 2:970–971, 1979.

15. Lock JE, Castaneda-Zuniga WR, Bass JL, Foker JE, Amplatz K, Anderson RW: Balloon dilatation of excised aortic coarctations. Radiology, 143:689–691, 1982.

16. Abele JE: Balloon catheters and transluminal dilatation—technical considerations. AJR 135:901–906, 1980.

17. Lock JE, Niemi T, Burke DA, Einsiz S, Castaneda-Zuniga WR: Transcutaneous angioplasty of experimental aortic coarctation. Circulation 66:1280–1285, 1982.

18. Liberthson R, Pennington DG, Jacobs ML, Daggett WN. Coarctation of the aorta: Review of 234 patients and clarification of management problems. Am J Cardiol 43:835–840, 1979.

19. Clarkson PM, Nicholson MR, Barratt-Boyes BG, Neukze JM, Whitlock RM. Results after repair of coarctation of the aorta beyond infancy: A 10–28 year follow up with particular reference to late systemic hypertension. Am J Cardiol 51:1481–1488, 1983.

20. Maron BJ, Humphries JO, Rowe RD, Mellites RD: Prognosis of surgically corrected coarctation of the aorta: 20 years post operative appraisal. Circulation 47:119–126, 1973.

21. Sperling DR, Dorsey TJ, Rowen M, Gazzaniga AB: Percutaneous transluminal angioplasty of congenital coarctation of the aorta. Am J Cardiol 51:562–564, 1983.

22. Lock JE, Bass JL, Amplatz K, Fuhrman BP, Castaneda-Zuniga WR: Balloon dilatation angioplasty of aortic coarctations in infants and children. Circulation 68:109–116, 1983.

23. Williams WG, Shindo G, Trusler GA, Dische MR, Olley PN: Results of repair of coarctation of the aorta during infancy. J Thorac Cardiovasc Surg 79:603–608, 1980.

24. Harkmann AF, Goldring D, Hernandez A, Behrer MR, Schad N, Ferguson T, Buford T: Recurrent coarctation of the aorta after successful repair in infancy. Am J Cardiol 29:405–410, 1970.

25. Waldhausen JA, Nahrwold DL: Repair of coarctation of the aorta with subclavian flap. J Thorac Cardiovasc Surg 51:532–533, 1966.

26. Pierce WS, Waldhausen JA, Verman W, Whitman V: Late results of the subclavian flap procedures in infants with coarctation of the thoracic aorta. Circulation 58(suppl 1):1.78–1.82, 1978.

27. Ring EJ, McLean GK: *Interventional Radiology: Principles and Techniques*. Boston, Little, Brown, 1981, pp 227–239.

28. Castaneda-Zuniga WR, Lock JE, Vlodaver Z, Rusnak B, Rysavy JP, Herrera M, Amplatz K: Transluminal dilatation of coarctation of the abdominal aorta. Radiology 143:693–697, 1982.

29. Rocchini AP, Cho KJ, Byrum C, Heidelberger K. Transluminal angioplasty of superior vena cava obstruction in a 15-month-old child. Chest 82:506–508, 1982.

30. Driscoll DJ, Hesslein PS, Mullins CE: Congenital stenosis of individual pulmonary veins: Clinical spectrum and unsuccessful treatment by transvenous balloon dilatation. Am J Cardiol 49:1767–1772, 1982.

31. Banner MP, Pollack HM, Ring EJ, Wein AJ: Catheter dilatation of benign ureteral strictures. Radiology 147:427–433, 1983.

10

Vascular Pathophysiology of Transluminal Angioplasty

ANDREW CRAGG, M.D.
KURT AMPLATZ, M.D.

INTRODUCTION

When angioplasty was first introduced in 1964 by Dotter and Judkins, its effectiveness was attributed to compression and redistribution of the "soft" atheromatous intima (1). Since that time extensive research has demonstrated that the pathophysiology of angioplasty is much more complex. The morphologic alterations produced in the vessel wall by dilation also give rise to perhaps equally important physiological changes.

An understanding of the mechanism of angioplasty is important in order that patient selection, technique, and pharmacologic intervention can be applied judiciously. In this chapter, the pathophysiologic changes which occur in the dilated artery are discussed, as well as their clinical significance. Changes in the arterial wall produced by angioplasty can be divided into two general categories: morphologic and physiologic (Figure 1).

MORPHOLOGY

In the original explanation of the mechanism of angioplasty, it was postulated that the atheromatous plaque was puttylike in consistency and would

undergo a "cold flow" along the wall of the vessel (Figure 2). By redistributing the atheromatous material along the wall of the vessel, an enlarged lumen was created without distention of other layers of the vessel wall. This process results in an elongation of the atheroma which is never seen in clinical practice. Therefore, to better understand the mechanism by which angioplasty produces its effect, we initiated a series of experimental studies using both atherosclerotic cadaver arteries and dogs (2).

Abdominal aortas and coronary, renal, superior mesenteric, and iliac arteries were harvested from cadavers. Iliac arteries and aorta were also dilated in normal dogs in order to assess the long-term histopathologic effects of angioplasty. Dilation was carried out in each preparation three times using Gruentzig balloon catheters inflated to 4 atmospheres for 1 minute.

Based on these initial experiments, a new theory was proposed to explain the morphologic success of angioplasty (Figure 3). This theory has since been confirmed in atherosclerotic animal models as well as in human specimens obtained at autopsy following in vivo dilation (3–6).

Atherosclerotic intimal plaques which are responsible for most stenotic vascular lesions are solid or infrequently semiliquid in composition and both

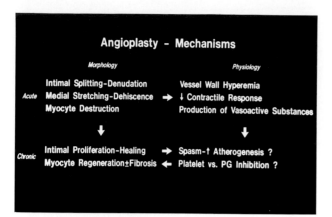

Figure 1. Pathophysiologic arterial wall changes after angioplasty can be divided into two general categories: morphologic and physiologic. Acute morphologic effects, such as intimal splitting, intimal-medial dehiscence, and medial stretching probably account for a large part of the immediate enlargement in luminal diameter. Other changes, including endothelial denudation and myocyte destruction, incite local inflammation which is manifested by altered vessel physiology. Physiologic changes include vessel wall hyperemia, depressed contraction, and release of vasoactive inflammatory substances. The use of anti-inflammatory medications such as aspirin after angioplasty may inhibit the production of these substances whose effects may span the gamut of vascular physiology from the regulation of thrombosis and vessel tone to the development of intimal hyperplasia and atherosclerosis.

types are, therefore, incompressible. Although it may theoretically be possible to squeeze very soft atheromatous material along the arterial wall by inflating a balloon, redistribution of material is impossible with the much more common fibrous or calcified plaques. In a now huge clinical material, this phenomenon has never been demonstrated although it is theoretically possible.

Endothelial Denudation

One of the earliest and most consistent consequences of any intravascular trauma, including angioplasty, is endothelial denudation (7). Vascular endothelium is important since it provides a nonthrombogenic interface between the blood and the vessel wall, in part by production of the antiaggregatory prostaglandin and prostacyclin. After angioplasty, the deendothelialized surface is thrombogenic and quickly becomes covered by a carpet of platelets. The risk of sudden thrombotic occlusion of the dilated artery due to platelet thrombi exists but is a very uncommon occurrence in clinical prac-

tice. Pharmacologic intervention, therefore, is directed toward decreasing the rate and amount of platelet deposition following dilation. Aspirin, by interfering with arachidonate metabolism, may decrease platelet aggregation at the site of dilation. Some studies have shown that platelet adhesion, a separate phenomenon from aggregation, may be unaffected by aspirin (8). Low molecular weight dextran has been shown to be effective in decreasing platelet adhesion (7).

Intimal Splitting

The first event which appears to occur following balloon inflation is stretching of the more normal portions of the vessel wall. If dilation is insufficient, the artery will return to its original size without morphologic alteration other than endothelial denudation. If dilation is continued, however, as occurs clinically the first gross morphologic alteration

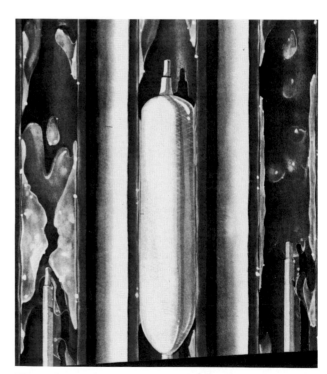

Figure 2. In the original explanation of angioplasty's effect, the atheromatous intima was described as being compressed and redistributed with little deformation of the surrounding arterial wall. Since most plaques are solid or fluid filled, however, compaction by the dilating balloon is negligible. Redistribution of the sclerotic plaque along the vessel wall also does not appear to occur to any significant extent.

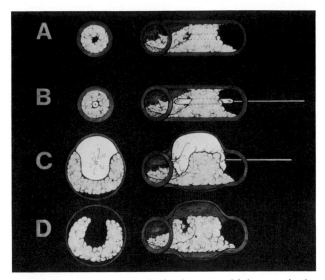

Figure 3. Acute morphologic changes which occur in the atherosclerotic arterial wall as a result of dilation are depicted here. As the balloon expands within the stenotic lumen, fractures occur in the intimal plaque. With greater dilation, these fracture lines extend between the intima and media. As the encasing effect of the intima is relieved, the media is stretched resulting in a permanent increase in vessel diameter. (Reprinted with permission from Ref. 2.)

produced is splitting, or fracturing, of the atheromatous plaque (Figure 4). If moderate dilation is performed, only a small cleft may be created in the plaque. Frequently, this may be sufficient to produce a significant increase in the arterial luminal diameter. In the clinical situation, however, it appears that more significant morphologic changes are usually produced.

Intimal-medial Dehiscence

As the diameters of the vessel lumen and the balloon become more disparate and number of inflation and pressures are increased, more significant fracturing of the plaque is likely to occur. As the split extends to the media, actual separation of the intima and media may occur. Intimal-medial dehiscence is a common occurrence in clinical angioplasty. This phenomenon has been observed both experimentally and clinically in arteries which have come to pathologic examination (Figure 5A).

Histologically, the intimal cleft appears to occur most often at the junction of a focal eccentric plaque and the more normal vessel wall (9). These clefts extend between the intima and media for a variable

distance around the circumference of the vessel. Complete intimal-medial dehiscence, which would be manifested by distal embolization of the plaque, is fortunately a rare clinical occurrence. Embolization of atherosclerotic debris has not been demonstrated to any significant extent experimentally (6,10).

Radiographically, intimal-medial dehiscence is indicated by a narrow extraluminal collection of contrast material at the site of dilation (Figure 5B). This represents the subintimal collection of contrast material in the cleft created by dilation. In almost all instances, intimal-medial dehiscence is a benign consequence of angioplasty. Rapid healing of the vessel occurs by neointima formation usually with preservation of the newly enlarged lumen.

It is possible that by freeing the relatively normal, compliant media from the encasing effect of the atheromatous intima the artery can better adapt to blood flow demands and pressure and actually increase its diameter. This may be one explanation for the apparent increase in arterial diameter weeks or months after angioplasty. Others have postulated that remodeling of the plaque occurs as a result of phagocytosis which results in a rounding off of the intimal fracture and an increase in the arterial diameter (11). It is our opinion that this explanation is very unlikely.

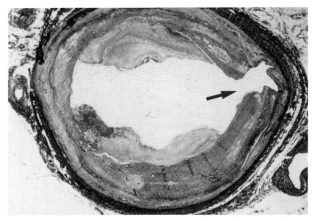

Figure 4. Intimal splitting. Human femoral artery. Specimen obtained 1 week after in vivo dilation. An intimal fracture (arrow) is easily seen extending through the intima and partially involving the underlying media. In addition, thrombus is seen on the opposite wall at the site of another tear. With minimal dilation, no other morphologic changes may be produced in the arterial wall. These fractures may heal with "remodeling" of the surrounding plaque which results in persistent widening of the newly enlarged lumen. (Illustration courtesy of Dr. J. Schneider and Dr. C. Zollikofer, University of Zurich, Zurich, Switzerland.)

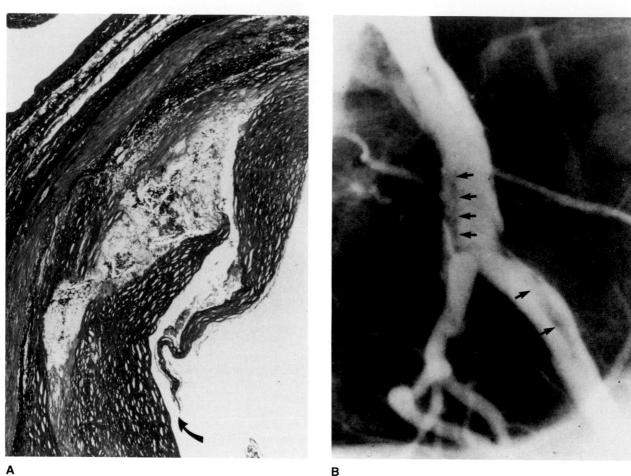

A

B

Figure 5. Intimal-medial dehiscence. **A.** Human femoral artery. Following in vitro dilation, a large cleft has been created between the intima and media. Intimal-medial dehiscence occurs frequently in clinical angioplasty; however, complete separation of intima and media, manifested by distal embolization of the plaque, is seen only rarely. (Illustration courtesy of Dr. J. Schneider and Dr. C. Zollikofer, University of Zurich, Zurich, Switzerland.) **B.** Postangioplasty aortogram demonstrates a longitudinally oriented radiolucency (arrows) separating a narrow collection of contrast material from the main arterial lumen. This is the characteristic appearance of subintimal dissection and is the angiographic counterpart of the intimal-medial dehiscence which is seen histologically. This radiographic picture almost always resolves without sequelae. (Reprinted with permission from Castaneda-Zuniga WR, et al: Radiographics 1:1–14, 1981.)

Medial Stretching

If dilation is continued after the plaque has been separated from the arterial media, significant stretching of the media can be produced without vessel rupture. This effect has been demonstrated in normal vessels experimentally (Figure 6). The normal media is very compliant and can withstand distention of 25 percent to 50 percent over its original diameter without permanent deformation. If a critical point of overstretching is reached, the artery does not resume its original diameter and becomes dilated. In the normal vessel, this point appears to occur when the balloon size exceeds the normal arterial diameter by approximately 50 percent. In clinical situations, balloon sizes are generally chosen to be equal to the original arterial diameter. When inflated in the stenotic vessel, however, the incompressible atherosclerotic plaque acts as a fulcrum resulting in severe overstretching of the arterial media at the plaque. Thus a permanent mechanical deformation of the media may be part of all angioplasties after the intima has been split.

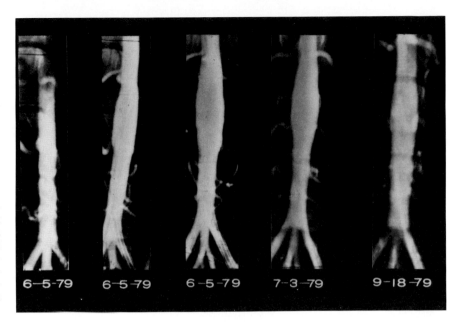

Figure 6. Medial stretching. Canine abdominal aorta. Permanent deformation of the arterial media is possible if it is suffiently stretched during angioplasty. This phenomenon may be important in producing a lasting increase in luminal diameter. (Reprinted with permission from Ref. 2.)

Medial Necrosis

The effect of medial stretching is quite apparent at the ultrastructural level. Carotid arteries in eight dogs were dilated approximately 50 to 100 percent over their original diameter. The animals were sacrificed at several intervals during the first 48 hours and at weekly intervals for 4 weeks. Specimens were obtained for both light and electron microscopy. Figure 7 demonstrates the progression from medial damage to repair which occurs at the cellular level. In the first 48 hours after angioplasty the smooth muscle cells of the arterial media become pyknotic and disintegrate. Macrophages invade the media to clear the cellular debris and are followed closely by invading fibroblasts which assist in the repair of the damaged vessel. In many instances this repair may be more or less complete, however, if sufficient damage to the vessel wall has been produced, fibrosis of the media may occur. The long-term effect of this process is not known but the late occurrence of vessel rupture following angioplasty has not been described. Most of the reparative changes which occur following angioplasty are probably complete, however, within 4 to 6 weeks.

Intimal Proliferation

Repair of the damaged intimal layer may occur by intimal proliferation. It is important clinically as a potential cause of restenosis. As organization of platelet thrombi occurs at the site of dilation, regrowth of endothelium occurs. This usually begins at the margins of the denuded area and proceeds toward the center. Neointima formation after endothelial denudation provides a new nonthrombogenic surface. Frequently, however, a hyperplastic endothelial response is produced (Figure 8). This intimal proliferation may be confined to the microscopic level or it may be massive resulting in gross restenosis of the newly widened vascular lumen.

The long-term clinical implications of this phenomenon are potentially twofold. First, intimal proliferation may be responsible for the smoothing out and remodeling of the fractured plaque which occurs after dilation. An exaggerated response of the vessel may then result in long-term restenosis. Second, the phenomenon of endothelial denudation, intimal proliferation, and accelerated atherogenesis are well known in experimental animals. Intimal proliferation is believed to be part of the process which results in accelerated atherogenesis after vessel trauma. Interestingly, platelet-derived factors may mediate the development of intimal proliferation (12). Thus, the rationale for long-term aspirin administration may extend beyond the prevention of platelet aggregation and thrombosis to include also inhibition of intimal proliferation. In the next section we will attempt to better define the relationship between the physiologic alterations produced by dilation and the morphologic changes we have just discussed.

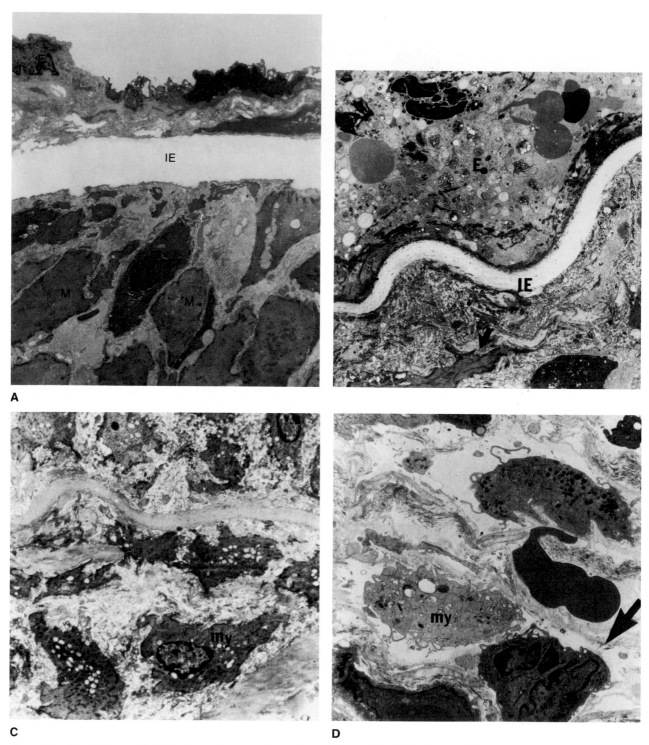

Figure 7. Medial necrosis. Electron microscopic sections of the media of a dog's carotid artery document the extensive myocyte destruction and subsequent repair which occurs after dilation. **A.** Normal electron micrograph. Normal endothelium (E), internal elastica (IE), and medial smooth muscle cells (M) can be seen. **B.** Immediate postdilation. Red blood cells, platelets, and fibrin are deposited on the denuded endothelial surface (E). The internal elastica is stretched and has an undulating appearance. Myocytes in the media also show early signs of degeneration (arrow). **C.** After 48 hours. Extensive cellular vacuolization and accumulation of necrotic debris in media. **D.** After 1 week. Macrophages (arrow) invade media to digest necrotic myocytes (m) and begin the process of repair.

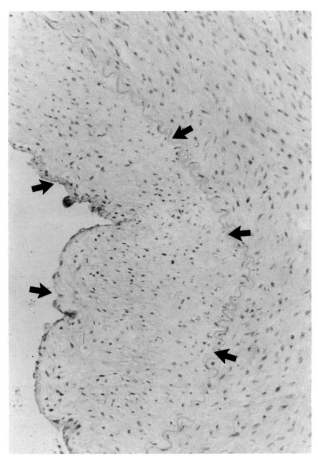

Figure 8. Intimal proliferation. The endothelium of the normal canine carotid artery is one cell thick. Following angioplasty, a hyperplastic intimal response (arrow) occurs as is illustrated in this case. Intimal proliferation may be one cause of restenosis after angioplasty. Recent evidence indicates that arachidonic acid metabolites released either from platelets or the damaged vessel wall may stimulate intimal hyperplasia. One benefit of long-term aspirin administration, therefore, may be the reduction of intimal proliferation by interrupting the conversion of arachidonate to the prostaglandins. (Courtesy of Dr. C. Zollikofer, University of Zurich, Zurich, Switzerland.)

PHYSIOLOGIC ALTERATIONS AFTER ANGIOPLASTY

The physiologic changes which occur in the vessel wall following angioplasty are undoubtedly very complex. We have recently examined several aspects of postangioplasty physiology which may be important not only in understanding how angioplasty works but also in improving the long-term success of the procedure.

Aspirin is used routinely following angioplasty to decrease platelet aggregation and thus the risk of arterial thrombosis. Some investigators feel that aspirin may also inhibit intimal proliferation which occurs at the site of endothelial denudation (13). As we will see, aspirin may also affect other physiologic responses of the artery to angioplasty. To illustrate this point, we can review several recent studies which indicate that significant metabolic changes occur in the vessel wall as a result of angioplasty.

Vasa Vasorum

The vasa vasorum exist in the adventitia and provide nutrient blood flow to the media of most larger systemic arteries. Interruption of the vasa vasorum has been implicated in both aneurysm formation and accelerated atherogenesis. Since these same two complications have been attributed to angioplasty, we investigated the effect of balloon dilation on the vasa vasorum of the dog (14).

In a preliminary set of experiments, the morphology of the vasa vasorum was studied by in situ injection of a latex preparation (Figure 9A). Control and dilated arteries were obtained and examined grossly and microscopically to determine if the vasa vasorum were indeed interrupted by angioplasty. No morphologic alterations could be demonstrated in the vasa vasorum after significant overstretching of the vessel wall. Vasa were still seen penetrating the outer layers of the media and no extravasation of latex was demonstrated. Only normal vessels were dilated, however. In atherosclerotic vessels the inner portions of the media are more dependent on nourishment from the vasa vasorum since the thickened, atheromatous intima inhibits the normal diffusion of oxygen from the blood vessel lumen. It is thus still possible that vasa vasorum, which penetrates the inner layers of the atherosclerotic vessel, are interrupted by dilatation, contributing in part to the medial necrosis that occurs.

In another series of dog studies, a single angioplasty protocol was used to examine a number of different physiologic alterations after angioplasty. In the first experiment, blood flow in the vasa vasorum was measured to determine whether local vessel autoregulation is affected by dilation. Second, vessel contraction was measured both in vivo and in vitro to define the effect of angioplasty on vascular tone. Finally, the production of several important vasoactive prostaglandin metabolites was measured to better define the role of these substances in postangioplasty spasm, thrombosis, and atherogenesis.

Pentobarbital-anesthetized mongrel dogs were used for all studies. In each dog, one or both carotid arteries were dilated using nonexpandable polyethylene balloons with inflated diameters 25 percent

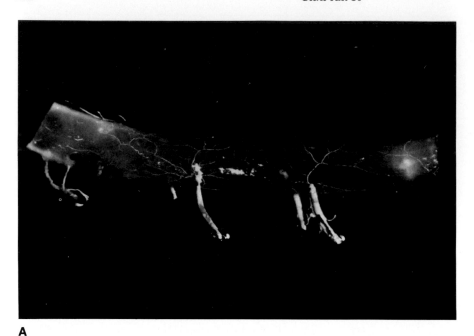

A

Effect of Aspirin on Vessel Wall Blood Flow
60 Minutes After Angioplasty

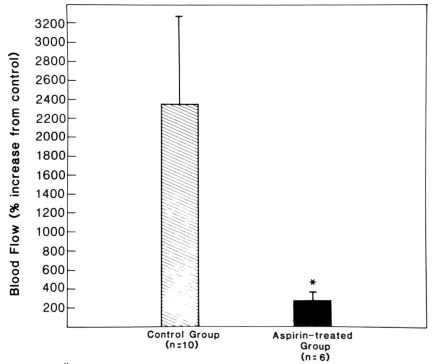

B

Figure 9. Vasa vasorum. **A.** The vasa vasorum of this canine aorta have been filled with a special latex preparation to demonstrate the rich perivascular network that supplies most larger systemic arteries. Interruption of the vasa vasorum has been shown to cause medial necrosis and may lead to late aneurysm formation. Disruption of the vasa vasorum after angioplasty was not demonstrated in normal vessels, although it is possible that disruption of the more extensive network of vasa in the atherosclerotic vessel could contribute to the medial necrosis which has been shown to occur after angioplasty. (Reprinted with permission from Ref. 14.) **B.** Blood flow through the vasa vasorum is significantly affected by angioplasty. A persistent hyperemic response can be demonstrated which is also significantly attenuated by aspirin. This suggests that aspirin may decrease local inflammation again by interfering with local prostaglandin production. (Reprinted with permission from Ref. 15.)

to 100 percent larger than the original arterial diameter. Oversized balloons were used to simulate the degree of medial stretching produced clinically. Carotid arteries were chosen because of their percutaneous accessibility and large size.

Vessel Wall Hyperemia

To better understand the metabolic alterations produced in the vessel wall by angioplasty, blood flow in the vasa vasorum was measured using the radioactive microsphere technique (14). Sequential injections of different radioactive-labeled microspheres (15 μm diameter) were made up to 6 hours after angioplasty in mechanically ventilated close-chested dogs. Samples of the dilated vessel wall were obtained and counted in a gamma scintillation spectrometer. Blood flow was calculated using the reference sample technique. In a second set of animals, aspirin (10 mg/kg) was administered prior to dilation (15).

Blood flow in the vasa vasorum up to 6 hours after angioplasty was persistently elevated 10- to 30-fold over control levels. This vessel wall hyperemia may reflect the degree of mechanical damage which occurs in the vessel wall. Perhaps most interesting is the potential mechanism of its occurrence. In the presence of aspirin, the hyperemia is significantly attenuated (Figure 9B). This suggests that aspirin may inhibit local inflammation after angioplasty by interfering with prostaglandin metabolism. Aspirin is known to act in part by inhibiting the enzyme cyclo-oxygenase which catalyzes the formation of the prostaglandins in both platelets and blood vessels.

Prostacyclin (PGI$_2$), a well-known vasodilating prostaglandin and inhibitor of platelet aggregation, has now been shown to be released acutely after angioplasty (16). Thus, release of prostacyclin from damaged endothelial cells or over a longer period from thrombin stimulation of intact endothelium adjacent to the site of dilation may be one explanation for a persistent postangioplasty vessel wall hyperemia.

Vessel wall hyperemia may be a reflection of the vessel's attempt to repair itself by release of vasoactive substances which decrease platelet aggregation and mediate local vasodilation. The inhibition of this phenomenon by aspirin may or may not have long term clinical implications. Importance has been placed recently, however, on the ability to differentially inhibit platelet aggregation while sparing production of "good" prostaglandins such as prostacyclin, which causes local vasodilation and inhibits platelet aggregation. At present, it appears that very low doses of aspirin, approximately 1 mg/kg/day, may achieve this result.

Altered Vessel Tone

A second physiologic manifestation of altered vessel wall morphology after angioplasty is an acute decrease in the vascular contractility. Two sets of experiments were performed to evaluate this phenomenon. In one set of animals, the in vivo contractile response of the dilated carotid artery was assessed by administration of vasopressin, which results in contraction of the media. In those vessel segments in which significant overstretching of the media occurred, contraction was found to be absent (Figure 10A) (17). The vessel adjacent to the site of dilation, however, responded in a normal fashion.

In another set of animals, immediate and long-term effects on contraction were assessed in vitro by suspending dilated carotid arterial rings (3 mm width) from an isometric force transducer, which was placed in an organ bath of oxygenated buffered Krebs solution (18). Isometric contractile force was then measured with varying concentrations of norepinephrine.

The experiment was repeated in the presence of indomethacin (3 mg/mL) which acts similarly to aspirin by inhibiting prostaglandin production. As expected, in the first 6 hours after dilation contraction was significantly depressed relative to the control level (Figure 10B). In those animals studied from 1 day to 3 months after dilation, however, it was found that vessel tone had returned toward normal. Indomethacin was found to improve contraction in all vessels; however, the improvement was most dramatic in the subgroup of short-term dilated arteries. In this group the isometric tension generated by the dilated artery was approximately three times greater in the presence of indomethacin. Depressed vessel contractility after angioplasty has now been confirmed by others (19).

These findings suggest that medial damage, as was demonstrated morphologically, probably accounts for a large part of the depressed vascular contractility after angioplasty. However, both the fact that vessel contraction improved within 24 hours after dilation and the fact that contraction could be augmented in the acute phase by indomethacin suggest that humoral factors may also affect the balance of contraction and dilation after angioplasty. This study, therefore, provides additional indirect evidence to support the concept that aspirin or aspirinlike drugs either suppress the production of a vasodilating substance, such as prostacyclin, or augment the production of a vasoconstrictor.

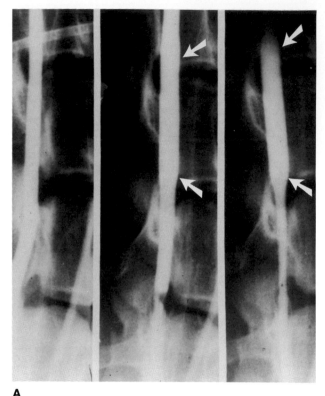

A

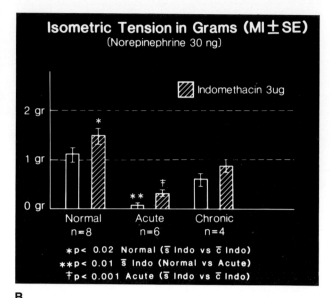

B

Figure 10. Altered vessel tone. **A.** The dilated canine carotid artery fails to contract at the same time that the normal artery responds dramatically to vasoconstricting stimuli. *Left*: control; *center*: immediately after dilation; *right*: vasoconstriction. Disruption of medial architecture probably accounts for a large part of this response. (Reprinted with permission from Ref. 14.)

Figure 10 (*Continued*). **B.** This same phenomenon is demonstrated in vitro by measuring the isometric tension generated by a normal and dilated arterial rings. Contraction is depressed acutely less than 6 hours after dilation but is improved by indomethacin and returns toward normal over a period of days. These findings suggest, again, that the acute release of humoral vasodilating substances may contribute to depressed vessel contractions after significant medial stretching has occurred.

One group of recently discovered biologic compounds, the hydroperoxy acids, are known to be patent vasoconstrictors (20). These substances, which are produced in vascular tissue from arachidonic acid—the precursor of the prostaglandins—have also been linked to the atherogenic process (21). In order to better understand the physiologic basis of potential angioplasty-induced phenomena such as vasospasm, intimal hyperplasia, and accelerated atherogenesis, the production by the dilated artery of both prostacyclin and a hydroperoxy acid metabolite by the dilated artery was measured.

Production of Vasoactive Substances (22)

Sixty minutes following dilation of a single carotid artery in eight dogs, the animals were heparinized and both carotid arteries were removed. The in vitro

conversion of carbon-14 (^{14}C) -arachidonic acid (AA) to 6-keto PGF$_{1d}$ (prostacyclin-PGI$_2$) PGE$_2$ (another vasodilatory prostaglandin), and 12L-hydroperoxy-5,8,10,14-eicotetraenoic acid (12-HPETE) was determined using thin layer radiochromatography.

Interestingly, angioplasty resulted in an acute decrease in the production of PGI$_2$ (70 percent) and PGE$_2$ (44 percent). This decrease, although measured acutely, may in fact represent the chronic state in which endothelial denudation results in depressed PGI$_2$ and PGE$_2$ production until endothelial and medial repair occurs. In the acute phase in vivo, however, endothelial damage results in prostacyclin release. As was stated previously, this phenomenon, in conjunction with prostacyclin production by thrombin-stimulated adjacent endothelium, may explain transient phenomena such as vessel wall hyperemia or altered tone. It is very likely in the long-term, however, that local prostacyclin production is depressed.

The production of the hydroperoxy acid 12-HPETE was significantly elevated (104 percent over control) after dilation. That this substance is produced by the vessel wall in response to trauma is a new finding with several potentially important clinical implications.

First, since 12-HPETE is known to be a potent vasoconstrictor, it is possible that in cases in which sufficient medial stretching is not performed, production of 12-HPETE could result in local spasm immediately after angioplasty. Sudden closure of the dilated coronary artery occurs in 3 percent to 8 percent of the cases. In addition, the hydroperoxy acids have now been shown to induce smooth muscle cell migration, a well-recognized part of the atherosclerotic process (21). Its effect on intimal proliferation is not known. The implications of these studies therefore, may extend beyond an explanation of the pathophysiology of angioplasty to establish a biopharmacologic basis for atherogenesis following vascular trauma.

SUMMARY

Angioplasty results in a number of morphologic changes in the vessel wall. Intimal splitting, medial stretching, and necrosis all occur acutely depending on the extent of dilation. Repair of the damaged vessel occurs rapidly and in most cases produces an excellent morphologic result. Intimately connected with the morphologic changes produced by angioplasty are certain physiologic alterations. Among these, disruption of prostaglandin metabolism may be important in the etiology of spasm, inflammation, increased atherogenesis, and thrombosis. By understanding the effect of angioplasty on vessel wall morphology and physiology we may be able to intervene pharmacologically to produce a more satisfactory long-term result.

REFERENCES

1. Dotter CT, Judkins MP: Transluminal treatment of arteriosclerotic obstruction: Description of a new technique and a preliminary report of its application. Circulation 30:654–670, 1964.
2. Castaneda-Zuniga WR, Formanek A, Tadavarthy M, et al: The mechanism of angioplasty. Radiology 135:565–571, 1980.
3. Block PC, Baughman KL, Pasternak RC, Fallon JT: Transluminal angioplasty: Correlation of morphologic and angiographic findings in an experimental model. Circulation 61:778–785, 1980.
4. Saffitz JE, Totty WG, McClennan BL, Gilula LA: Percutaneous transluminal angioplasty. Radiological-pathological correlation. Radiology 141:651–654, 1981.
5. Block PC, Myler RK, Stertzer S, Fallon JT: Morphology after transluminal angioplasty in human beings. N Engl J Med 305:382–385, 1981.
6. Sanborn TA, Faxon DP, Waugh D, Small DM, Haudenschild C, Gottsman SB, Ryan TJ: Transluminal angioplasty in experimental atherosclerosis. Analysis for embolization using an in vivo perfusion system. Circulation 66:917–922, 1982.
7. Pasternak RC, Baughman KL, Fallon JT, Block PC: Scanning electron microscopy after coronary transluminal angioplasty of normal canine coronary arteries. Am J Cardiol 45:591–598, 1980.
8. Bick RL, Adams T, Schmalhorst WR: Bleeding times, platelet adhesion and aspirin. Am J Clin Pathol 65:69–72, 1976.
9. Laerum F, Castaneda-Zuniga WR, Rysavy J, Moore R, Amplatz K: The site of wall rupture in transluminal angioplasty: An experimental study. Radiology 144:760–770, 1982.
10. Block PC, Elmer D, Fallon JT: Release of atherosclerotic debris after transluminal angioplasty. Circulation 65:950–952, 1982.
11. Block PC: Percutaneous transluminal coronary angioplasty. AJR 135:955–959, 1980.
12. Ross R: Atherosclerosis: A problem of the biology of arterial wall cells and their interaction with blood components. Arteriosclerosis 1:293–311, 1981.
13. Hagen P, Wang Z, Mikat EM, Hacket DB: Antiplatelet therapy reduces aortic intimal hyperplasia distal to small diameter vascular prostheses (PTFE) in nonhuman primates. Ann Surg 195:328–339, 1981.
14. Cragg AH, Einzig S, Rysavy A, Castaneda-Zuniga WR, Borgwardt B, Amplatz K: The vasa vasorum and angioplasty. Radiology 148:75–80, 1983.
15. Cragg A, Einzig S, Rysavy J, Castaneda-Zuniga W, Borgwardt B, Amplatz K: Effect of aspirin on angioplasty-induced vessel wall hyperemia. AJR 140:1233–1238, 1983.
16. Probst R, Pachinger O, Sinzinger H, Kaliman J: Release of prostaglandins after percutaneous transluminal coronary angioplasty. Circulation 68:144 (Suppl III), 1983.
17. Casteneda-Zuniga WR, Laerum F, Rysavy J, Rusnak B, Amplatz K: Paralysis of arteries by intraluminal balloon dilation. Radiology 144:75–76, 1982.
18. Zollikofer CL, Cragg AH, Einzig S, et al: Prostaglandins and angioplasty: An experimental study in canine arteries. Radiology 149:681–685, 1983.
19. Consigny PM, Tulenko TN: Acute effects of angioplasty on vascular smooth muscle. Radiology 149(P):241 (Abstr), 1983.
20. Asano M, Hidaka H: Contractile response of isolated rabbit aortic strips to unsaturated fatty acid peroxides. J Pharmacol Exp Ther 208:347–353, 1979.
21. Nakao J, Ooyama T, Chang W, Murota S, Osimo H: Platelets stimulate aortic smooth muscle cell migration in vitro. Atherosclerosis 43:143–150, 1982.
22. Cragg A, Einzig S, Castenada-Zuniga W, Amplatz K, White JG, Rao GR: Vessel wall arachidonate metabolism after angioplasty: Possible mediators of postangioplasty vasospasm. Am J Cardiol 51:1441–1445, 1983.

11

Current Technology of Angioplasty Catheters and Accessories

WILLIAM H. HARTZ, M.D.
GORDON K. McLEAN, M.D.

THE CURRENT STATE OF TECHNOLOGY FOR ANGIOPLASTY CATHETERS AND ACCESSORIES

The technology of balloon catheters and ancillary equipment is rapidly evolving. In this chapter, we discuss the current variety and applications of balloon catheters and related accessories.

General Engineering Principles of Balloon Catheters

Force of Dilation

One of the most important considerations in opening stenoses or occlusions is the dilating force applied directly to the arterial wall. This force is dependent on multiple factors, which have been well described by Abele (1,2). The radial force acting against the arterial wall varies with balloon pressure, area of stenosis, balloon diameter and length, and degree of stenosis.

For a balloon of any given diameter, an increase in the inflating pressure produces an increased dilating force in a direct linear relation. This is true

only if the diameters of the stenosis and the balloon remain constant. At a fixed inflation pressure, as the lesion begins to dilate, the radial force applied to a stenosis decreases. Thus, continued increases in pressure are necessary if a constant dilating force is to be applied to the lesion.

The pressure needed to dilate a lesion varies both with lesion length and degree of stenosis. Of these two determinants, the degree of stenosis is the more critical. Consequently, more force is necessary to dilate a very narrow, short stenosis than a long arterial segment that is only moderately stenotic.

The diameter of the balloon selected is also an important consideration. Early balloon angioplasty catheters had stated diameters that were not reproducible, even during their initial inflations. Because of this, an indefinite amount of force was expended in stretching the balloon material rather than dilating the stenosis. Clearly, more dilating force is applied directly to the arterial wall if the balloon does not stretch under maximum dilating pressure. If the balloon material is stretchable, a shorter balloon tends to exert more force against the arterial wall than a longer balloon of the same diameter at a uniform inflating pressure. With the current generation of balloon catheters, balloon diameter and length are relatively constant; thus, short balloons are not necessarily more effective instruments for the treatment of very short, tight stenoses.

Hoop stress is a circumferential force applied to the internal wall of the artery as the balloon inflates. This is present because deflated balloons must be wrapped around the catheter shaft and, on inflation, exert a circumferential tearing force. Because of their surface area, larger-diameter balloons exert more hoop stress when dilating than do smaller balloons.

Synthetic Plastics Used in Balloon Construction

The earliest dilating balloons were made of latex and were useful only when applied to relatively soft lesions. These balloons tended to rupture because they expanded in the direction of least resistance. Thus tightly calcified or fibrotic lesions were essentially untreatable. Porstmann's attempt to overcome

the limitations of latex by "caging" the balloon was unsuccessful because of the tendency of the cage to injure the vessel wall (3).

Balloons made of polyvinylchloride (PVC) were enthusiastically received because of their relative stiffness when compared to latex. However, PVC balloons still deformed at higher inflation pressures and could not maintain a reproducible outer diameter within the vessel. The pressure at which balloon deformation occurred was quite close to the pressure at which the balloon would burst.

In recent years, polyethylene (PE) has largely replaced PVC in balloon catheter construction (4,5). Compared to PVC balloons, PE balloons have a greatly decreased tendency to deform at high inflation pressures (Fig. 1). Moreover, the pressure needed to rupture the balloon is higher than that easily obtainable by hand injection. Because of these advantages, the angiographer may exert a greater pressure against the internal arterial lumen without fear of balloon overexpansion or rupture.

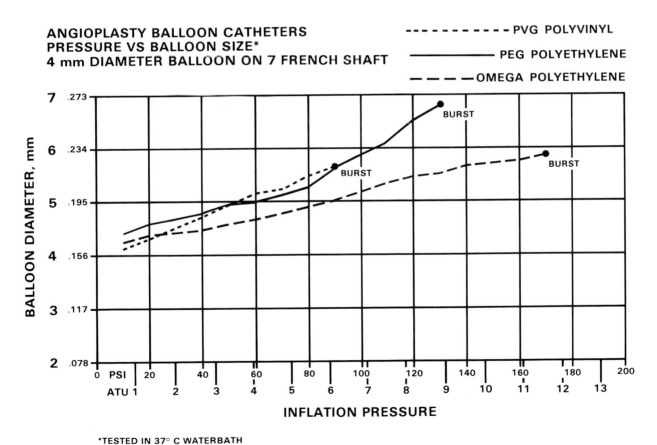

Figure 1. Comparison of polyvinylchloride (PVC) balloons with early (PEG) and recent (OMEGA) polyethylene varieties. Note the significantly increased burst pressure and more uniformly inflated balloon size.

Recently, the plastic Mylar* has been used in the construction of vascular dilating balloons. The most impressive aspect of this plastic is its ability to withstand extremely high inflation pressures without bursting. Currently available Mylar balloons can contain a sustained pressure of 16–18 atmospheres. Since, under normal conditions no more than 10–12 atmospheres can be generated by a hand-held syringe of 5 mL or greater, this provides an extremely wide safety margin. Although Mylar has great tensile strength, because of its thinness and lack of flexibility it tends to tear more easily than PE. Theoretically, this might present a problem when the balloon is passed through or used to dilate calcified and irregular (i.e., sharp) vessels.

Because the earliest PVC balloons had a tendency to overdistend and to burst easily, the so-called reinforced, or dual component, balloons were developed. These balloons are made of either polyurethane or PVC and have external reinforcement with nylon filaments. The balloon is bonded to the end of the catheter and, because of the reinforcement, cannot overdistend. In addition the balloon shape is much more uniform. Higher atmospheric pressures can be used to inflate the balloon without the fear of rupture, especially in the smaller-diameter balloons. These balloons do not fold on themselves in the deflated state and are closely applied to the catheter shaft.

The dual component balloon was advocated for use in occluded vessels or in stenotic vessels that would not dilate with conventional balloons. Greater lateral wall pressure can be applied with this reinforced balloon than with earlier PVC catheters. These balloons are also very useful in dilating cutaneous tracts for nephrostomies and T tubes. There are no folds, wrinkles, or "umbrella effects" when the catheter is withdrawn.

Because of the development of the stronger plastics, notably Mylar and PE, the reinforced balloon has become less useful for intravascular work. Problems with balloon rupture and overdistention are being eliminated by these newer plastics, and nylon reinforcement is no longer necessary.

Catheter and Guidewire Specifications

Balloon Catheter Shape

Currently available balloon angioplasty catheters have a variety of shapes and tip configurations. The most generally useful balloons are those mounted on straight shafts (Fig. 2). These are used in femoral,

*Dupont

popliteal, tibial, iliac, subclavian, and many renal dilations. Balloons are available on preshaped catheters such as Simmons's (Fig. 3), cobra, and Levin (Fig. 4) configurations (6). These preformed catheters are familiar to the majority of vascular radiologists and may be quite useful in certain situations. The Simmons's configurations are useful for entering vessels originating at extremely acute angles from the abdominal aorta. More open angles such as those often present in the renal arteries can be entered with cobra or Levin shaped balloon catheters.

Balloon Sizes

Polyethylene and PVC balloons of 4–20 mm are available. Dual component and Mylar balloons are, at present, available in a more limited array of sizes. This size range is appropriate for all vessels currently being dilated.

The length of the balloon was previously an important consideration with PVC balloons. The current generation of PE and Mylar plastic make length less important, since these balloons do not overdistend. As a general rule, the balloon should be at least 2 cm longer than the length of stenosis to provide the necessary overlap when placing the balloon across the lesion under fluoroscopic guidance. Polyethylene and PVC balloons can be obtained in lengths up to 10 cm for use with diffusely diseased and narrowed superficial femoral or external iliac arteries (Fig. 5) (7).

Tip length may be a significant consideration when dilating stenoses near small branch vessels (e.g., distal renal artery lesions). On most current balloon catheters, the inflated balloon does not begin to widen on the shaft until 1 cm from the tip. Balloons placed at 2 and 3 cm from the catheter tip are also available. A recent balloon design, which has a gradual taper over the leading 2 cm, has the advantage of progessively enlarging a tight stenosis when the balloon catheter will not cross the lesion. Thus the balloon can be advanced slowly across a high-grade stenosis with several sequential dilations.

Syringes and Pressure Monitors

When a fixed force is applied to the plunger of an inflating syringe, the pressure generated in the syringe will vary inversely with the diameter of the plunger. Even with extreme caution, PVC and PE balloons can be easily ruptured when inflated with a 1-mL tuberculin syringe. Accordingly, most angiographers use a 10 mL or larger syringe for balloon inflation. The readily available disposable plastic syringes are acceptable, even though they may oc-

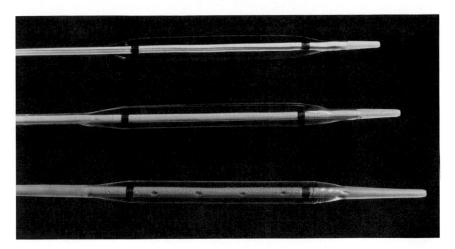

Figure 2. Evolution of balloon catheters. **Top.** Polyvinylchloride balloon bonded to the entire catheter length, with inflation through a groove on the outer surface of the catheter. **Middle.** Older polyethylene, still bonded to the entire length of catheter, but more resistant to bursting. **Bottom.** Latest design with a two-lumen catheter (sideholes inflate balloon) and balloon bonded only to end of catheter. Note the tight taper of the tip and ends of the balloon.

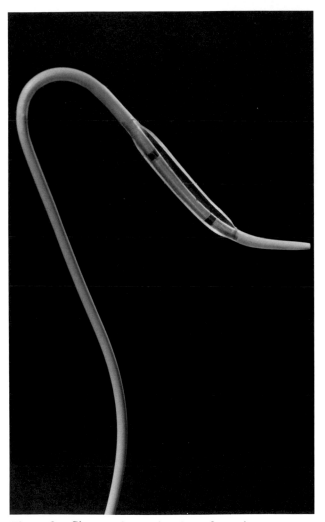

Figure 3. Simmons's number 2 configuration.

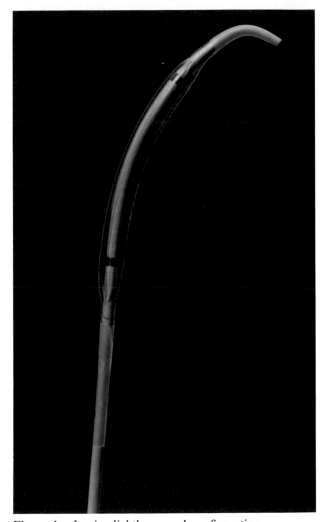

Figure 4. Levin slightly curved configuration.

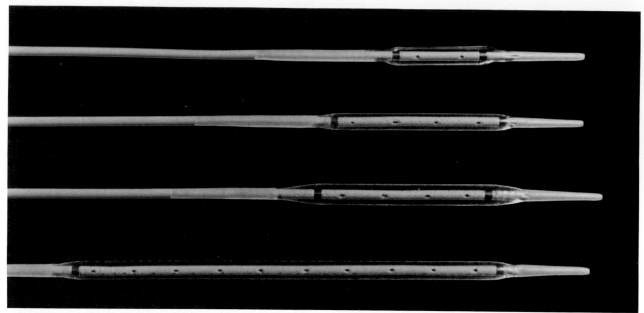

Figure 5. Examples of various balloon lengths and sizes. **Top to bottom**: 2 cm length with 5 mm diameter; 4 cm length, 4 mm diameter; 4 cm length, 5 mm diameter; 10 cm length, 4 mm diameter.

casionally bend or collapse under pressure. Glass syringes should be avoided when using direct hand injection, because the glass may break, possibly lacerating the hand.

Pressure gauges are readily available from most catheter companies and may be easily interfaced between the balloon inflation port and the inflation syringe. These gauges are useful in preventing overinflation and balloon rupture when using older PVC or PE balloons. However, with the newer generation of balloon catheters it is almost impossible to reach the bursting pressure with hand injection.

Many angiographers do not use pressure monitoring devices because they feel balloon bursting is not a primary concern. If a "waist" is still present at the maximum rated balloon pressure, additional force is usually applied until the waist disappears or the balloon ruptures. Although case reports associating balloon rupture with vessel damage have appeared in the literature, the vast majority of balloon ruptures is innocuous (8). Since the balloon is inflated with a pure hydraulic system, breaching of the balloon should result in a "square wave" decompression and not explosive expansion. It should be recalled that the larger balloons have a lower burst pressure than smaller balloons made of the same material and that pressure monitors may be useful in this setting (Fig. 6).

Guidewires

Standard Teflon coated or noncoated alloy stainless steel wires used in general angiography are employed for angioplasty. These wound spring wires with either fixed or movable cores have a wide variety of lengths and tip configurations.

The wire should provide a soft, flexible tip with a 2–15 cm taper to a more rigid midportion in order for the balloon catheter to follow easily through the puncture site and across the lesion. An exception to this rule occurs when passing a guidewire and a catheter together through a total obstruction. With this technique, a 3 or 1.5 mm J guidewire "leads" a straight catheter through a total occlusion (9). The guide is simply folded back over the catheter, and in fact no flexible tip leads.

Torque control guidewires are recent innovations that facilitate passage through eccentrically diseased and tortuous vessels. These are available with a variety of distal curves, which are selected according to the specific anatomy of the lesion.

Since inadvertent back and forth motion of the curve often occurs when exchanging diagnostic for balloon catheters, the tight (1.5 mm radius) Rosen wire with its short (2 cm) distal taper is often preferred for popliteal and renal angioplasty (10). It cannot be used in very small vessels, however. The

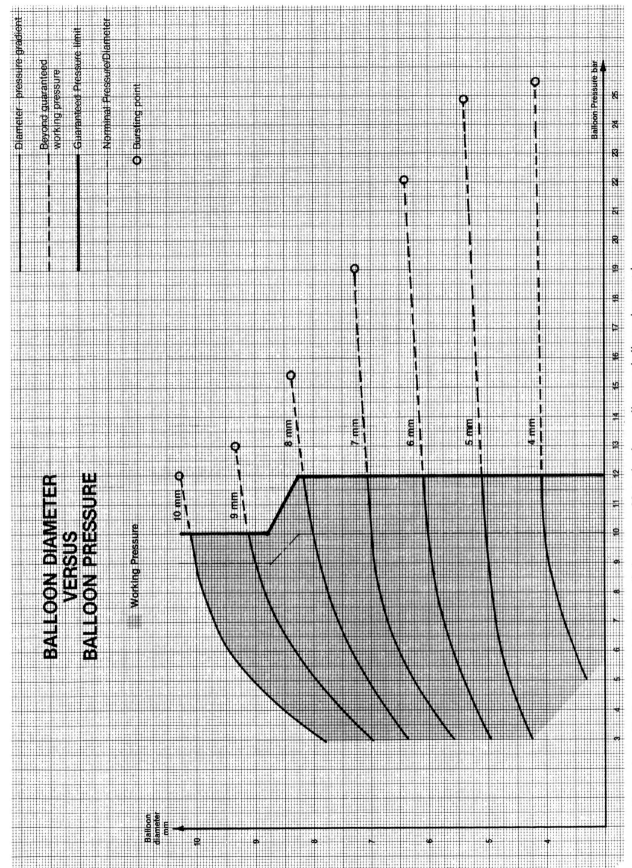

Figure 6. Comparison of balloon diameter with applied pressure. Note that larger diameter balloons burst at lower pressures.

gically created arteriovenous fistulas (35,36). Balloon size is dictated by the size of the shunt or draining vein. Very stiff balloons must be used since the area of dilation includes both rigid graft and extremely distensible draining vein. The balloon should have a good tip taper because introduction through the graft itself or through scarred surrounding tissues may be quite difficult.

Dilation of Graft-Related Stenoses

Stenoses occurring at the juncture of a native vessel with a graft (whether synthetic or saphenous vein) are approached as any peripheral lesion (37). It is imperative that extremely stiff balloons be used since the overdistention of a soft (i.e., PVC) balloon may seriously damage the native vein.

Lesions that occur within vein grafts may also be dilated very successfully. Careful sizing is indicated since these veins may be overdistended and torn, leading to false aneurysm formation and consequent loss of the graft. It is possible to introduce balloons directly through a graft placed subcutaneously (38,39). This may be done in a synthetic graft (e.g., axillofemoral bypass grafts) or through native saphenous vein grafts. An oblique puncture of the graft is recommended since this allows for more effective hemostasis at completion of the procedure. A good taper on the distal portion of the dilating balloon is essential for atraumatic introduction. Following removal of the balloon the graft must be compressed gently for an extended period of time to avoid extravasation.

Venous grafts between the systemic and portal venous system have also been dilated (40). The opening of a portacaval or mesocaval shunt may effectively prolong a decompressive procedure performed for portal hypertension. Sizing of these grafts is extremely important since the anastomosis is between two relatively fragile veins. Balloons from 6 to 10 mm will generally produce an adequate lumen.

Recommendations for Specific Catheter Systems

Polyvinylchloride Catheters

Polyvinylchloride (PVC) balloons were those first available for modern balloon angioplasty. Although used successfully in many thousands of dilations, they have largely been superseded by more rigid plastics. PVC balloon catheters may still be used for routine dilations of the iliac and femoral arteries, but their tendency to overdistend makes size selection somewhat difficult. In retrospect, it appears that

PVC balloon catheters were routinely selected at a size that was actually smaller than the vessel to be dilated, but this undersizing was compensated for by the balloon's tendency to overinflate. PVC catheters should never be used when a lesion lies between vessels of disparate size. The balloon's tendency to overinflate in the larger vessel often precludes successful dilation of the smaller vessel. More important, these balloons should never be used in fragile vessels; vein grafts, cervicocephalic vessels, and renal vessels should never be dilated with PVC catheters since they may be easily damaged.

One of the principal advantages of the early PVC catheters was that the flexible nature of this plastic allowed the balloon catheter to follow the guidewire into the lesion. The stiffness of the carrier catheter that plagued many early PE balloons has now largely been overcome.

Polyethylene Balloon Catheters

Polyethylene (PE) balloons have largely replaced the PVC balloons in routine angioplasty. A variety of PE formulations exist, varying in their physical characteristics according to different treatment by the manufacturers. Sizing of PE balloons to the vessel is generally quite accurate. The stiffness that plagued early PE balloon catheters has generally been overcome by current manufacturing methods.

Mylar Balloon Catheters

Mylar balloons are currently available in some angioplasty catheters. This plastic is extremely rigid and will withstand pressures of up to almost 20 atmospheres. These balloons offer the advantage of almost instantaneous inflation to the rated diameter and fairly rigid conformation to this diameter right up to the point of failure. Some early Mylar balloons suffered from fragility and were easily torn on introduction, but this problem has generally been resolved.

Dual Component Balloon Catheters

Several angioplasty catheter systems that use a dual component approach are available. An inner driving balloon applies force to a rigid outer balloon that fixes the inflated diameter. Although quite resistant to bursting, these balloons have several disadvantages, principally cost and complexity. Some dual component balloons also exert a longitudinal force on the vessel wall as the balloon is inflated; theoretically, this places a nonessential and potentially hazardous stress on the diseased portion of the vessel wall.

Two-Centimeter versus Four-Centimeter Balloons

For general angioplasty of the extremities, 4-cm balloons offer some advantages. The longer balloon length makes precise centering of the lesion in the balloon center less critical. Centering is important to keep the balloon from "wedging" itself out of the lesion. Short balloons may be used to advantage in renal arteries when it is desirable to apply dilating force only to a very limited segment of the vessel.

Ten-Centimeter Balloons

Ten-centimeter-long balloon catheters are used in the superficial femoropopliteal arteries. The rationale for this balloon is that the diffuse disease in these vessels warrants generalized dilation of the entire vessel, rather than application to particular or isolated segments.

Preshaped Balloon Catheters

Both the cobra and Simmons's shapes are available for dilating renal and visceral stenoses. These catheters avoid the need for exchange once the vessel has been catheterized, thus minimizing the risk of guidewire and catheter damage to the vessel. However, since they are not truly torque-controlled, they do not offer the safety of the diagnostic catheters on which they are based. In certain applications, (c.g., axillary approach to renal artery stenoses) they seem to be the safest alternative.

Guidewires

Virtually any guidewire may be used as an exchange guide for a balloon dilation catheter. Generally, heavy-duty guides are preferred since they minimize the tendency of the catheter to buckle in the vessel or at the groin as it is introduced through a tight lesion. Straight tip guides are preferred when the vessel to be dilated is small. Guidewires whose tip is formed into a tight J configuration are preferred when larger vessels are dilated. Variable stiffness guides offer a compromise in which the tip of the wire is floppy and the more proximal portion of the guide is stiffened by the introducing mandril.

Inflation Systems

A variety of mechanical inflating devices with accompanying pressure gauges are available. Pressure gauges in themselves are useful when the skills of angioplasty are first being acquired since they give the angiographer a method of quantifying the pressure applied to the syringe. The safety factor that these devices afford is less essential than it was several years ago, since the new, stiffer plastics have minimized the danger of inadvertent balloon overinflation and consequent rupture. Mechanical inflation devices also may be used to quantify precisely the amount of pressure applied to the balloon. Many angiographers feel that this feature is not particularly helpful since the end point of dilation is generally the fluoroscopically observed appearance of the balloon and vessel, rather than a predetermined intraluminal pressure. Some mechanical inflating devices can apply much greater pressure to the balloon than can be achieved by hand injection alone. However, such high pressures are rarely necessary for intravascular dilation.

FUTURE TECHNOLOGY

Balloon catheter research and development have exploded in recent years. Many companies are competing to produce the "ideal balloon catheter" as well as its necessary accessories. Future refinements will center on decreasing the external diameter of the catheter shaft, increasing the strength and resistance of the balloon, and increasing the ability of these catheters to follow selective guidewires.

REFERENCES

1. Abele JE: Balloon catheters and transluminal dilatation: Technical considerations. Am J Roentgenol 135:901–906, 1980.
2. Abele JE: Technical considerations: Physical properties of balloon catheters, inflation devices, and pressure measurement devices, in Castaneda-Zuniga WR (ed): *Transluminal Angioplasty.* New York, Thieme-Stratton, 1983, pp 20–27.
3. Porstmann W: Ein neuer Korsett-Ballon Katheter zur transluminalen Rekanalisation nach Dotter unter besonderer Beruecksichtigung von Obliterationen an den Beckenarterien. Radiol Diagn (Berl) 14:239–244, 1973.
4. Athanasoulis CA: Percutaneous transluminal angioplasty: General principles. Am J Roentgenol 135:893–900, 1980.
5. Gerlock AJ Jr, Regen DM, Shaff MI: An examination of the physical characteristics leading to angioplasty balloon rupture. Radiology 144:421–422, 1982.

6. Levin DC, Murray P, Harrington DP: New curved catheter for renal angioplasty. Am J Roentgenol 138:359–360, 1982.

7. Sos TA, Sniderman KW, Beinart C: Gruentzig catheter with a 10 cm long balloon. Radiology 141:825–826, 1981.

8. Zeitler E: Complications in and after PTR, in Zeiter E, Gruentzig A, School W (eds): *Percutaneous Vascular Recanalization*. Berlin, Springer-Verlag, 1978, pp 120–125.

9. Frieman DB, McLean GK, Oleaga JA, Ring EJ: Percutaneous transluminal angioplasty, in Ring EJ, McLean GK (eds): *Intraventional Radiology: Principles and Techniques*. Boston, Little, Brown, 1981, pp 117–244.

10. Rosen RJ, McLean GK, Oleaga JA, Freiman DB, Ring EJ: A new exchange guide wire for transluminal angioplasty. Radiology 140:242–243, 1981.

11. Levin DC, Harrington DP, Bettmann MA, Garnic JD, Torman H, Murray P, Boxt LM, Geller SC: Equipment choices, technical aspects and pitfalls of percutaneous transluminal angioplasty. Cardiovasc Intervent Radiol 7:1–10, 1984.

12. Shaver RW, Soong J: Angioplasty through aortofemoral graft: Use of catheter-introducer sheath. Am J Roentgenol 138:168–169, 1982.

13. Brunner V, Gruentzig A: Vascular surgery and transluminal dilatation/recanalization: Complementary procedures for the reconstruction of peripheral occlusive diseases, in Zeitler E, Gruentzig A, School W (eds): *Percutaneous Vascular Recanalization*. Berlin, Springer-Verlag, 1978, pp 160–166.

14. Wierny L, Plass R, Porstmann W: Long-term results in 100 consecutive patients treated by transluminal angioplasty. Radiology 112:543–548, 1974.

15. Simonetti G, Rossi P, Passareillo R, Faraglia V, Spartera C, Pistolese R, Fiorani P: Iliac artery rupture, a complication of transluminal angioplasty. Am J Roentgenol 140:989–990, 1983.

16. Frieman DB, Ring EJ, Oleaga JA, Berkowitz H, Roberts B: Transluminal angioplasty of the iliac, femoral, and popliteal arteries. Radiology 132:285–288, 1979.

17. Ring EJ, Frieman DB, McLean GK, Schwarz W: Percutaneous recanalization of common iliac artery occlusions: An unacceptable complication rate? Am J Roentgenol 139:587–589, 1982.

18. Lu C-T, Zarins CK, Yang C-F, Turcotte JK: Percutaneous transluminal angioplasty for limb salvage. Radiology 142:337–341, 1982.

19. Grollman JH Jr, Del Vicario M, Mittal AK: Percutaneous transluminal abdominal aortic angioplasty. Am J Roentgenol 134:1053–1054, 1980.

20. Castenda-Zuniga WR, Lock JE, Vlodaver Z, Rusnak B, Rysavy JP, Herrera M, Amplatz K: Transluminal dilatation of coarctation of the abdominal aorta: An experimental study in dogs. Radiology 143:693–697, 1982.

21. Tegtmeyer CJ, Wellons HA, Thompson RN: Balloon dilatation of the abdominal aorta. JAMA 244:2636–2638, 1980.

22. Cicuto KP, McLean GK, Oleaga JA, Frieman DB, Grossman RA, Ring EJ: Renal artery stenosis: Anatomic classification for percutaneous transluminal angioplasty. Am J Roentgenol 137:599–601, 1981.

23. Colapinto RF, Stronell RD, Harries-Jones EP, Gildiner M, Hobbs BB, Farrow GA, Wilson DR, Morrow JD, Logan AG, Birch SJ: Percutaneous transluminal dilatation of the renal artery: Follow-up studies on renovascular hypertension. Am J Roentgenol 139:727–732, 1982.

24. Grable GS, Smith DC: The use of the Simmons "sidewinder" catheter in percutaneous transluminal angioplasty of the renal arteries. Radiology 137:541–543, 1980.

25. Tegtmeyer CJ, Ayers CA, Wellons HA: The axillary approach to percutaneous renal artery dilatation. Radiology 135:775–776, 1980.

26. Grossman RA, Dafoe DC, Shoenfeld RB, Ring EJ, McLean GK, Oleaga JA, Freiman DB, Naja A, Perloff LJ, Barker CF: Percutaneous transluminal angioplasty treatment of renal transplant artery stenosis. Transplantation 34:339–343, 1982.

27. Gerlock AJ Jr, MacDonell RC Jr, Smith CW, Muhletaler CA, Parris WCV, Johnson HK, Tallent MB, Richie RE, Kendall RI: Renal transplant arterial stenosis: Percutaneous transluminal angioplasty. Am J Roentgenol 140:325–331, 1983.

28. Golden DA, Ring EJ, McLean GK, Freiman DB: Percutaneous transluminal angioplasty in the treatment of abdominal angina. Am J Roentgenol 139:247–249, 1982.

29. Roberts L Jr, Wertman DA Jr, Mills SR, Moore AV Jr, Heaston DK: Transluminal angioplasty of the superior mesenteric artery: An alternative to surgical revascularization. Am J Roentgenol 141:1039–1042, 1983.

30. Galichia JP, Bajaj AK, Vine DL, Roberts RW: Subclavian artery stenosis treated by transluminal angioplasty: Six cases. Cardiovasc Intervent Radiol 6:78–81, 1983.

31. Kobinia GS, Bergman H Jr: Angioplasty in stenosis of the innominate artery. Cardiovasc Intervent Radiol 6:82–85, 1983.

32. Motarjeme A, Keifer JW, Zuska AJ: Percutaneous transluminal angioplasty of the brachiocephalic arteries. Am J Roentgenol 138:457–462, 1982.

33. Motarjeme A, Keifer JW, Zuska AJ: Percutaneous transluminal angioplasty of the vertebral arteries. Radiology 139:715–717, 1981.

34. Hasso AN, Bird CR, Zinke DE, Thompson JR: Fibromucular dysplasia of the internal carotid artery: percutaneous transluminal angioplasty. Am J Roentgenol 2:175–180, 1981.

35. Probst P, Mahler F, Krneta A, Descoeudres C: Percutaneous transluminal dilatation for restoration of angioaccess in chronic hemodialysis patients. Cardiovasc Intervent Radiol 5:257–259, 1982.

36. Gordon DH, Glanz S, Butt KM, Adamsons RJ, Koenig MA: Treatment of stenotic lesions in dialysis access fistulas and shunts by transluminal angioplasty. Radiology 143:53–58, 1982.

37. Mitchell SE, Kadir S, Kaufman SL, Chang R, Williams GM, Kan JS, White RI Jr: Percutaneous transluminal angioplasty of aortic graft stenoses. Radiology 149:439–444, 1983.

38. Zajko AB, McLean GK, Freiman DB, Oleaga JA, Ring EJ: Percutaneous puncture of venous bypass grafts for transluminal angioplasty. Am J Roentgenol 137:799–802, 1981.

39. Alpert JR, Ring EJ, Berkowitz HD, Freiman DB, Oleaga JA, Gordon R, Roberts B: Treatment of vein graft stenosis by balloon catheter dilation. JAMA 242:2769–2771, 1979.

40. Cope C: Balloon dilatation of closed mesocaval shunts. Am J Roentgenol 135:989–993, 1980.

PART
II

CORONARY ANGIOPLASTY

12

Clinical Choice of Coronary Revascularization: Coronary Angioplasty versus Bypass Surgery

KENNETH R. JUTZY, M.D.
WILLIAM H. WILLIS, JR., M.D.
G. DAVID JANG, M.D.

INTRODUCTION

Cardiovascular disease remains the number one cause of death in the United States with over 1 million deaths per year, despite the steady decrease in the mortality rate over the past decade. Myocardial infarction accounts for 600,000 of these deaths (1). Approximately 350,000, or a little over 60 percent of the deaths resulting from myocardial infarction, occur before the patient arrives at the hospital. The estimated health care cost for treating cardiovascular disease in 1983 was approximately $57 billion. In 1981 approximately 160,000 bypass surgeries were performed, at an estimated cost of $3.25 billion (1). The projected cost of treating cardiovascular disease in the year 2000 with currently available methods is estimated at $366 billion (2).

Advances in treatment of coronary artery disease in the last two decades have paralleled those in the diagnostic methods of ischemic artery disease. The introduction of selective coronary arteriography in the mid-1960s made it possible to ascertain the pres-

ence and severity of coronary artery narrowing in vivo (3–7). The concurrent development of noninvasive studies of the physiologic function of the heart, such as exercise stress testing and nuclear isotope imaging, has enabled investigators to correlate physiologic function with anatomic information (8–13). Once coronary heart disease could be diagnosed safely and accurately, new treatment regimens were possible. The widespread use of nitrate preparations starting in the 1960s, beta blockers in the 1970s, and calcium antagonists in the 1980s allowed a broad choice of effective antianginal therapy. Of various surgical modalities attempted, coronary artery bypass surgery emerged as the safest and most clinically effective way of alleviating symptoms (14). In 1977, the first report of nonsurgical revascularization using percutaneous transluminal coronary angioplasty was reported (15).

In general, coronary artery revascularization is felt to be appropriate in patients with angina pectoris refractory to medical therapy, in those with severe angina who are intolerant to medications, or in those with coronary artery disease in the high-risk anatomic locations.

Percutaneous transluminal coronary angioplasty (PTCA) has evolved into an acceptable technique for revascularization in a selected group of patients. The focus of this chapter will be on the selection of patients for revascularization in general and for percutaneous transluminal coronary angioplasty in particular.

CURRENT CONCEPTS OF TREATING CORONARY ARTERY DISEASE

Medical Regimen

It has long been recognized that certain factors are epidemiologically linked to a high incidence of coronary artery disease. In the male population, the five leading factors in order of decreasing importance are age, smoking history, hypercholesterolemia, systolic hypertension, and an abnormal electrocardiogram. Among women, the factors in order of decreasing importance are age, systolic hypertension, abnormal electrocardiogram, hypercholesterolemia, and cigarette smoking (16). Because of the complexity and variety of risk factors involved, it has been very difficult to prove, by population studies, that altering these risk factors decreases the development or progression of coronary artery disease. Some studies, however, have shown a statistical correlation between reduction of risk factors and decrease in coronary artery disease (17–22).

Risk factor reduction includes cessation of smoking, control of hypertension, and reduction of hyperlipidemia by dietary or pharmacologic methods. The role of cholesterol in ischemic heart disease and the importance of cholesterol reduction in males have been reaffirmed (23), but while the value of a regular exercise program in the prevention of coronary artery disease remains widely touted, it is still difficult to prove with controlled studies (24).

Organic nitrates remain the most commonly used drugs in treating patients with angina. Nitrates relax vascular smooth muscle with resultant dilatation, more prominent in venous than arterial circulation (25–27). Nitrates also dilate coronary arteries (28) and redistribute blood flow to the ischemic areas of myocardium, particularly the subendocardium (29). Many different forms of organic nitrates are available at the present time. Nitroglycerin has sublingual, oral, ointment, and transdermal forms. Isosorbide dinitrate, similar in its physiologic action to nitroglycerin, is available in sublingual, chewable,

and oral preparations. Pentaerythritol tetranitrate and erythrityl tetranitrate are also marketed in oral forms. Sublingual nitrates are effective in treating acute episodes of angina pectoris (30). Long-acting forms have been shown effective in preventing angina pectoris and in allowing increased exercise capacity free of angina (31). The most common side effects are headache and hypotension.

Beta-adrenergic blocking medications have been used in the treatment of angina since the late 1960s and early 1970s. The mechanism of action appears to be a combination of decrease in inotropism and a decrease in heart rate (32,33). Clinically available agents in the United States include atenolol, metoprolol, nadolol, pindolol, propranolol, and timolol. Beta blockers have proved effective in allowing increased exercise capacity without angina. In addition, studies involving timolol maleate (34) and propranolol hydrochloride (35) have shown a small but statistically significant decrease in mortality in individuals with prior myocardial infarction when compared to control groups. Side effects from beta blockage include bradycardia, hypotension, exacerbation of asthma, generalized fatigue, and worsening of left ventricular function.

Calcium antagonists were first introduced in the United States in the early 1980s. Their mechanism of action is relaxation of smooth muscle by inhibition of slow inward calcium currents, resulting in vasodilation of coronary arteries and systemic arterioles and venules (36). These medications have been effective with both variant and chronic stable angina (37–39). Currently available agents in the United States include diltiazem hydrochloride, nifedipine, and verapamil hydrochloride. Side effects include dizziness, headache, peripheral edema (more common with nifedipine), constipation (more common with verapamil hydrochloride), and increase in atrioventricular block (more common with verapamil and diltiazem).

In recent years much effort has been directed toward finding effective antiarrhythmic drugs for the treatment of intractable atrial and ventricular rhythm disturbances. Many of these drugs, either available commercially or under investigation, have been effective in decreasing the incidence of ectopy. Because ventricular ectopy is associated with a higher mortality following a myocardial infarction (40,41), most therapy has focused on such high-risk patients. However, except for a subgroup of well-defined individuals with recurrent ventricular tachycardia, no controlled studies have clearly demonstrated reduction in mortality by such treatment. Studies currently in progress hope to better define the population groups most at risk and specific drugs most useful in reducing mortality.

Revascularization Procedures

Coronary artery bypass surgery has become the most common cardiac surgery in the United States. It is usually performed by using the patient's own saphenous vein. The vein is attached end to side to the ascending aorta and then side to side or end to side to the affected coronary artery beyond the site of stenosis. Alternatively, the internal mammary artery can be freed from the chest wall and attached end to side with the coronary artery beyond the site of stenosis. Some studies have shown an increase in graft patency when the latter procedure is used (42–44). Using current techniques, mortality for elective coronary bypass surgery by experienced surgeons varies from 1 to 2 percent, with about a 5 percent perioperative myocardial infarction rate (45–47). Graft patency at 1 year has been reported to be 75–90 percent (48). Symptoms of angina pectoris are abolished or significantly improved in 70–95 percent of patients receiving bypass surgery (49).

Percutaneous coronary angioplasty has evolved as an alternative to bypass surgery for coronary revascularization in selected patients. With current techniques, the primary success rate at centers with highly experienced staff is approximately 90 percent, with a 2–5 percent emergency surgery rate, a 2–3 percent myocardial infarction rate, and a mortality of around 1 percent (50). Restenosis at 1 year of 17 to 47 percent has been reported (51–57). Most of these patients have a repeat angioplasty with a higher success rate than that of first-time patients.

CLINICAL DOCUMENTATION OF MYOCARDIAL ISCHEMIA

Symptoms and Signs

Classical angina pectoris is described as a retrosternal chest pain or discomfort (1) with characteristic qualities of pressure, tightness, squeezing, heaviness, or aching; (2) located in the midline anywhere from the epigastrium to pharynx, sometimes with radiation to the arms, neck, or jaw; (3) lasting 0.5–10 minutes; (4) brought on by exertion, cold environment, postprandial state, or emotional stress; (5) relieved by rest or sublingual nitroglycerin. The prevalence of coronary artery disease is 90 percent in adult males who have classical angina pectoris (58). Symptoms of angina can be stratified by grading of efforts (59) to elicit angina pectoris according to the Canadian Cardiovascular Society (CSS) (Table 1). Stable angina pectoris implies a condition in which the eliciting factors are relatively constant over several months to years. Unstable angina pectoris implies a condition in which pain is more frequent, elicited by less exertion, more prolonged, or new in onset.

Exercise Electrocardiogram

Exercise electrocardiogram recording before, during, and after graded exercise can be useful in detecting the presence and to a certain degree the extent of coronary artery disease. Early onset of ST segment depression, long persistence of ST depression following exercise, and a horizontal or downsloping configuration of the ST segment are all strongly associated with extensive coronary artery disease (8,9). In addition, certain clinical findings, such as the occurrence of chest pain early in the test and development of hypotension, are also associated with increased risk of future cardiac events (10,11). The sensitivity of exercise electrocardiography to detect coronary artery disease is approximately 64 percent and the specificity 89 percent (12). The sensitivity of exercise electrocardiograms is related to the number of diseased coronary arteries, falling from 85 percent in patients with three vessel disease to 44 percent in those with single vessel disease. Patients whose exercise electrocardiograms cannot be properly interpreted include those with baseline electrocardiographic abnormalities such as bundle branch block, left ventricular hypertrophy, or digitalis effect; those who are incapable of achieving 85 percent or more of the maximum predicted heart rate because of noncardiac limitations; and those who are on cardiac medications preventing achievement of near maximal heart rate.

TABLE 1 Grading of Angina of Effort by the Canadian Cardiovascular Society

1. "Ordinary physical activity does not cause angina": For example, walking and climbing stairs. Angina with strenuous, rapid, or prolonged exertion at work or recreation.
2. "Slight limitation of ordinary activity": Walking or climbing stairs rapidly, walking uphill, walking or stair climbing after meals, or in cold, or in wind, or under emotional stress, or only during the few hours after awakening. Walking more than two blocks on the level and climbing more than one flight of ordinary stairs at a normal pace and in normal conditions.
3. "Marked limitation of ordinary physical activity": Walking one to two blocks on the level and climbing one flight of stairs in normal conditions and at a normal pace.
4. "Inability to carry on any physical activity without discomfort—anginal syndrome may be present at rest."

Coronary Perfusion Scans

Exercise thallium 201 myocardial perfusion scans compare images obtained immediately after exercise and those following a rest period of reperfusion. Persistent perfusion defects imply infarction, and defects during exercise that normalize at reperfusion imply ischemia in the affected area. The average sensitivity of this method for detecting coronary artery disease is approximately 83 percent with a specificity of 90 percent (12,13).

In gated exercise radionuclide ventriculogram, technetium 99 is used to obtain measurements of ejection fraction and regional wall motion at rest and during gated exercise. Abnormalities suggestive of coronary artery disease include a failure of the ejection fraction to rise during exercise and the presence of a new regional wall motion abnormality during exercise. Sensitivity and specificity are both approximately 90 percent with this method (12).

Cardiac Angiography

Left ventriculography is useful in assessing regional and global wall motion to detect areas of prior infarction or severe ischemia (60). Accentuation of apparently akinetic areas is possible by inducing postextrasystolic beats or by administering nitroglycerin (61,62). Quantitative assessment of coronary stenosis from coronary arteriogram is usually made by estimation by an experienced observer, although several computer-assisted quantitative methods are now available. In general, reduction of flow reserve does not occur until 50 percent diameter stenosis is present; reduction of flow sufficient to cause symptoms does not occur until 70 percent diameter stenosis is present (63). Recent hyperemic flow studies utilizing digital subtraction methods before and after contrast injection confirm the significance of these different degrees of narrowing (64). Open chest studies with flow meters during bypass surgery may be useful in determining the ultimate significance of narrowing seen in the diagnostic arteriogram (65).

SELECTION CRITERIA OF PATIENTS FOR CORONARY ANGIOPLASTY

Single Vessel Disease

Patients being considered for percutaneous transluminal coronary angioplasty (PTCA) should have a clinical history compatible with ischemic heart disease, with symptoms of sufficient magnitude to justify revascularization (66). Ideally, they should have further documentation of ischemia by treadmill or radionuclide perfusion scans. Coronary arteriography should confirm the presence of single vessel disease of at least 60–70 percent diameter stenosis (67). Since it is well known that cardiac ischemia can occur in the absence of symptoms, patients with markedly positive treadmill or thallium 201 scans and significant angiographic stenosis may also be candidates (68). A small subgroup of patients consists of those with major angiographic stenosis but insignificant symptoms and negative noninvasive studies. These individuals are often the focus of considerable debate, some physicians arguing for revascularization to prevent myocardial infarction and others pointing to the low mortality rate in controlled studies in single vessel disease (69). In general, because of the low mortality associated with single vessel disease, most physicians reserve percutaneous revascularization for those with medically refractory symptoms (70). The final decision, however, should also take into consideration the amount of myocardium supplied by the diseased vessel and the patient's age.

The ideal angiographic lesion is a discrete, concentric, focal lesion located in the proximal part of an artery (66). The left anterior descending in its proximal portion is the most easily approachable and was the major constituent of most early angioplasty series. The circumflex and right coronary arteries were initially more problematic, but with the introduction of steerable systems, various guidewires, and different guiding catheters, they have become less of a problem (Fig. 1). Also, with the present angioplasty systems most major artery side branches are accessible for dilation (Fig. 2). Further discussion regarding these will be presented later. Features such as eccentricity, associated vessel calcification, and distal location of the lesion have recently been shown to be only relative contraindications.

The success of vein graft dilation depends on the site of stenosis (71). The location with the best long-term patency rate following angioplasty is the distal anastomosis site of the vein graft to the native coronary artery (Fig. 3). The proximal anastomotic site and the body of the vein graft have shown good primary success rates but up to 50 percent restenosis rate. In general, it is felt that the more recent the bypass surgery, the safer dilation becomes. Older vein grafts, those in place for longer than 1–2 years, have a higher incidence of friable atheromatous deposits and a greater risk of distal embolization during angioplasty (72).

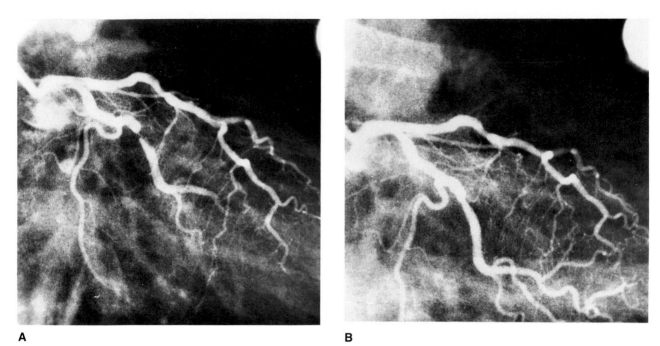

A   **B**

Figure 1. Sixty-five-year-old male admitted with unstable angina. **A.** Before angioplasty. **B.** Immediately after angioplasty.

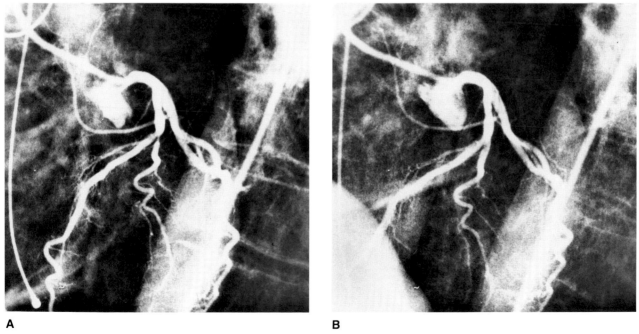

A   **B**

Figure 2. Fifty-seven-year-old woman with exertional angina. **A.** Original diagonal stenosis before angioplasty. **B.** After angioplasty.

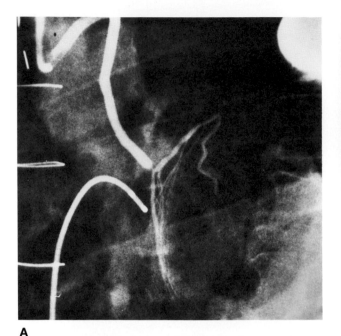

A

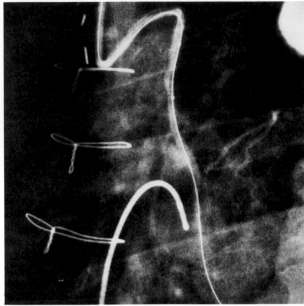

B

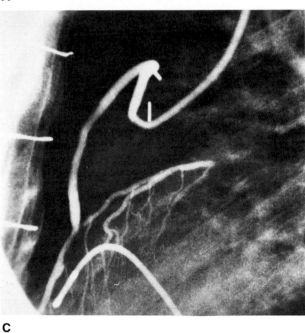

C

Figure 3. Fifty-six-year-old male 3 months after bypass surgery with a distal graft stenosis. **A.** Before angioplasty. **B.** Left venous bypass graft guiding catheter and balloon catheter. **C.** After angioplasty.

Single Vessel Dilation in Multivessel Disease

Occasional patients are seen with multivessel disease in which only one stenosis appears to cause the majority of symptoms or the other involved vessels are not appropriate for revascularization. Such

vessels could include those with prior occlusion and total infarction in the area supplied (Figs. 4 and 5). Other examples would be non-flow-limiting disease or flow-limiting disease in vessels too small for either angioplasty or bypass surgery. A further example would be a patient with multivessel disease and prior bypass surgery with only one vessel now unpro-

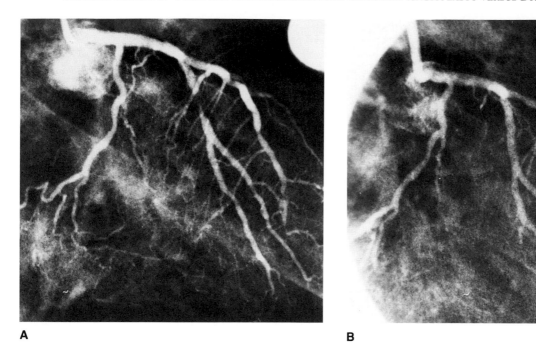

A   B

Figure 4. Sixty-two-year-old male with prior occlusions of his distal anterior descending, distal left circumflex, and proximal right coronary arteries. **A.** Principal diagonal stenosis, before angioplasty. **B.** After angioplasty.

tected by a bypass graft (Fig. 6). The safety of such dilations depends to a large degree on the coronary distribution and the residual ventricular function. If prior infarction in a different vessel has occurred and if abrupt closure of the vessel involved in the angioplasty were to result in a loss of myocardium of greater than 35–40 percent of left ventricular mass, the high risk of mortality in such a patient might contraindicate angioplasty in favor of bypass surgery.

Multivessel Dilations in Multivessel Disease

Each stenosis in multivessel disease should meet the selection criteria suggested for single vessel disease. Thus, each should be accessible to the balloon catheter and of such a degree as to cause significant symptoms. Further, the stenosis should not have the contraindications listed in the discussion following (Fig. 7). The most severe or most hemodynamically important lesion should be addressed first and the remaining lesions subsequently. When sequential stenoses occur in the same vessel, usually the first stenosis is dilated first so that distal pressures can be accurately assessed. The skill or experience of the operator becomes of primary importance when dealing with multivessel disease, since the time in-

volved in each dilation is multiplied by the number of lesions attempted. The hemodynamic importance of each vessel requiring angioplasty must be kept in mind. If abrupt closure of any of the proposed lesions would cause severe and likely fatal impairment of ventricular function, the advisability of angioplasty in such patients should be reconsidered. Multivessel dilations can be performed either consecutively during the same angiographic procedure or stepwise on different days. The advantage of performing all dilations during the same procedure is the saving of time and expense involved. The primary disadvantage is that if abrupt reclosure of an earlier dilatation were to occur while the physician was working on the second or third lesion that also became occluded, very severe hemodynamic impairment would result.

Partial versus Complete Revascularization

In selecting patients for angioplasty, the concept of complete revascularization should be kept in mind. The highly symptomatic patient should have all vessels with significant stenosis revascularized if at all possible. Thus, successfully dilating one vessel while another large vessel is occluded but supplied by collaterals without infarction may leave the patient with

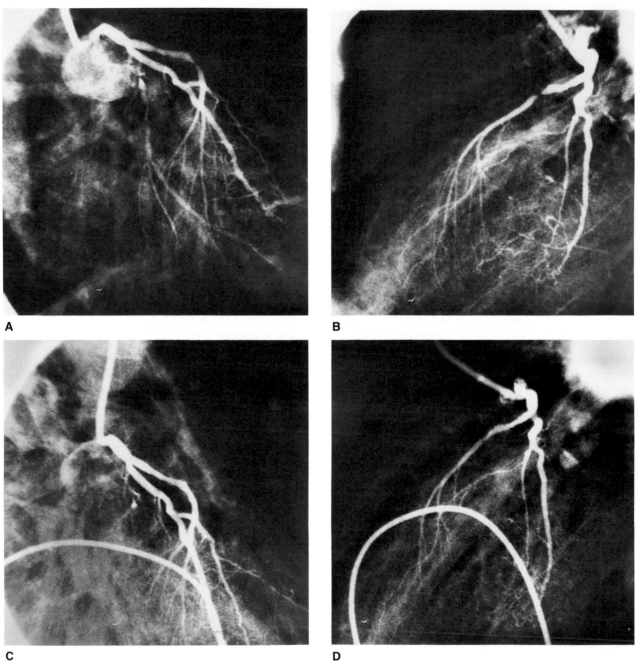

A

B

C

D

Figure 5. Fifty-six-year-old male with previous anterior descending and circumflex occlusions but with a first septal stenosis causing unstable angina. **A,B.** RAO and LAO preangioplasty views. **C,D.** RAO and LAO views after angioplasty with relief of symptoms.

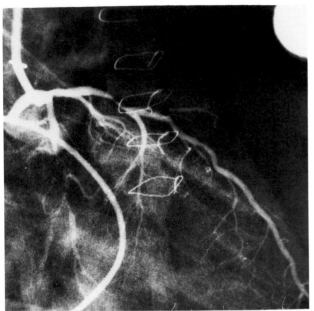
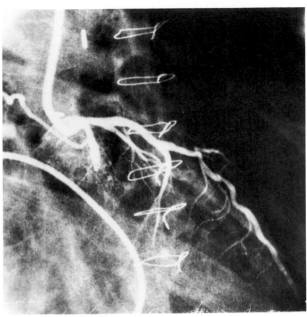

A   **B**

Figure 6. Seventy-three-year-old male 10 years after bypass graft surgery with development of a stenosis in the anterior descending artery not protected by a graft. **A.** Before angioplasty. **B.** After angioplasty.

significant symptoms. In general, a good rule of thumb is to avoid angioplasty in patients who could be more completely revascularized by bypass surgery. Some patients may have diffuse, distal disease or small vessels that would not be suitable candidates for surgery. In that case, dilation of other vessels may be helpful in reducing the amount of angina and the amount of medications required to control the angina (Fig. 8).

Contraindications

Because of the high mortality associated with abrupt closure or late reclosure of the left main coronary artery, dilation of "unprotected" left main coronary arteries is strongly contraindicated (73). The term *unprotected* refers to a left main coronary artery without adequate blood flow distally from other sources. It may be entirely appropriate to dilate a left main coronary artery where either the circumflex or the anterior descending artery has a widely patent bypass graft that perfuses the distal vessels (Fig. 9).

Patients who are not surgical candidates or are poor surgical risks because of cardiac or noncardiac factors are usually not considered for angioplasty. In spite of recent advances in techniques, a small

risk of abrupt closure during or immediately after angioplasty remains. Patients who are not surgical candidates could then be at high risk for myocardial infarction or death. There are occasional cases, however, in which patients are so severely limited by angina that they would be willing to accept the risk of myocardial infarction and even possible death if angioplasty failed (Fig. 10). Such a decision would have to be made by the patients and their families with a full understanding of the risk. The final decision should involve both the medical and surgical teams.

Long, eccentric, ulcerated, or extensively calcified lesions are relatively contraindicated (74) (Fig. 11). All of these factors are associated with a slightly higher risk of failure or complications of angioplasty but are not absolute contraindications. In each case the potential benefits are weighed against the slight increased risk of the procedure.

Bifurcation Lesions

When an important side branch arises near the site of stenosis, careful angiographic visualization must be used to ascertain the exact relation of the side branch to the stenosis. If the side branch arises from

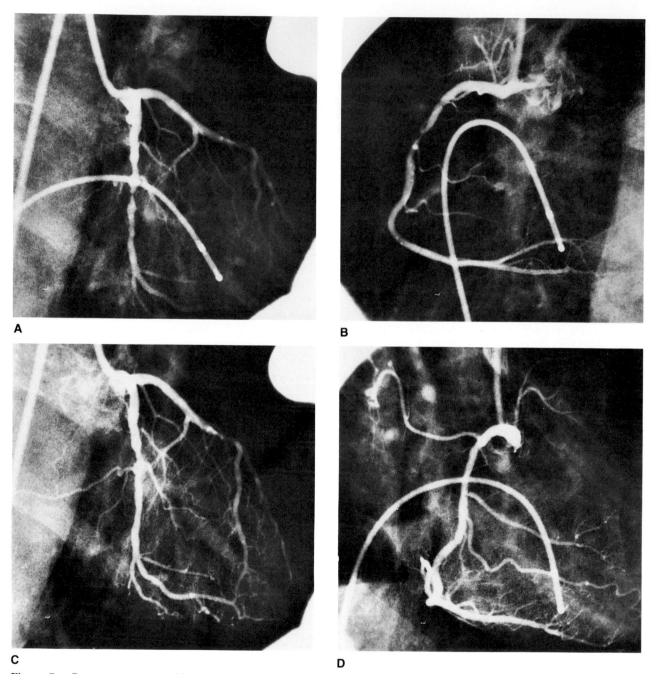

Figure 7. Seventy-one-year-old male with exertional angina. **A.** Left coronary artery: before angioplasty. **B.** Right coronary artery: before angioplasty. **C.** Left coronary artery: after angioplasty. **D.** Right coronary artery: after angioplasty.

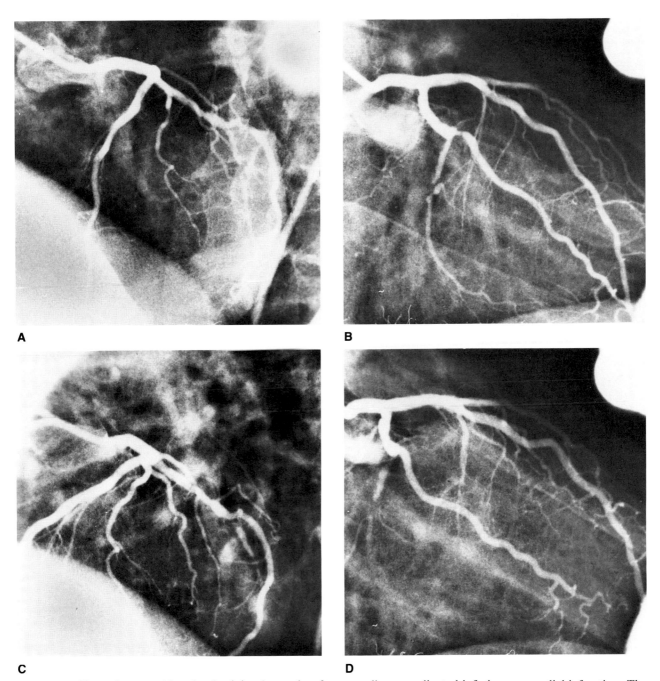

Figure 8.   Sixty-six-year-old male physician 2 months after a small uncomplicated inferior myocardial infarction. The diagonal and superior obtuse marginal arteries were dilated, but the guidewire would not advance into the inferior obtuse marginal artery. **A,B.** Preangioplasty views of the left coronary artery. **C,D.** Postangioplasty views of the left coronary artery.

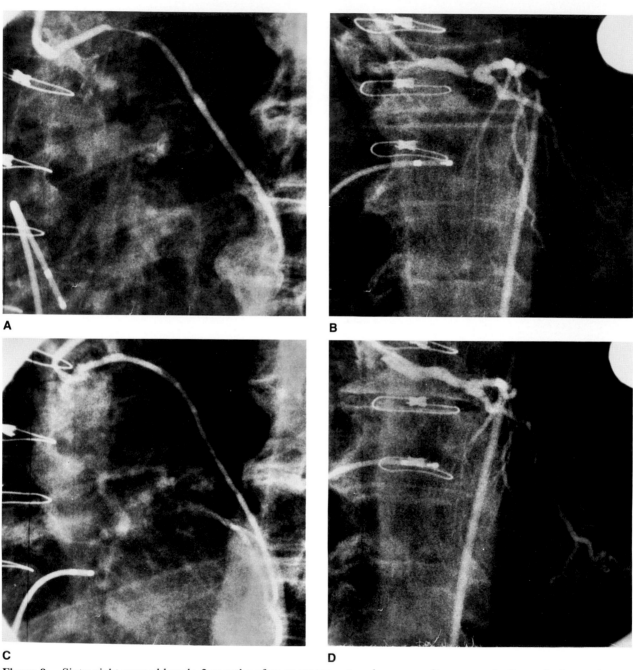

A

B

C

D

Figure 9. Sixty-eight-year-old male 2 months after coronary artery bypass graft surgery. **A.** Preangioplasty view of the graft to the obtuse marginal artery. **B.** Preangioplasty view of the native left coronary artery. The graft to the anterior descending artery was occluded. **C.** Postangioplasty view of the graft to the obtuse marginal artery. **D.** Postangioplasty view of the left coronary artery. Both angioplasty sites were widely patent at restudy 6 months later.

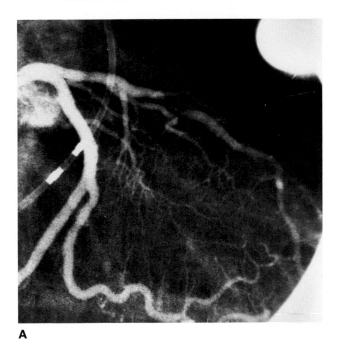
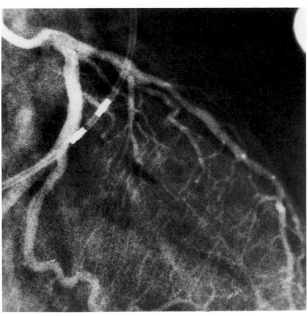

A   B

Figure 10. Sixty-three-year-old female with severe chronic congestive pulmonary disease and disabling angina. **A.** Preangioplasty left coronary anterior descending stenosis. **B.** Postangioplasty view.

the stenosis or has severe proximal disease, the chance of occlusion of the side branch during dilation of the primary vessel is approximately 15 percent (Fig. 12). If the side branch is included in the area covered by the balloon during inflation but is not involved in the stenosis itself, the chance of side branch occlusion is only 1–2 percent (75) (Fig. 13). The involved side branch lesions, if located in large vessels, may be dilated while a separate complete angioplasty system is positioned in the aorta with the guidewire across the main artery and the side branch separately dilated (76).

Complete Occlusions

Complete occlusions may be acute, subacute, or chronic. Not infrequently a high-grade stenosis progresses to a complete occlusion when there is some delay between the diagnostic study and the angioplasty procedure. If the complete occlusion occurs in this setting, it can be successfully recanalized in 70 percent of cases (77,78) (Fig. 14). Usually the occlusion can be easily crossed by the guidewire and successfully dilated. When the occlusion occurs in the setting of an acute myocardial infarction of only a few hours' duration, the occlusion can be treated by thrombolysis, and if there is a severe

residual stenosis after thrombolysis, the occlusion may be dilated (Fig. 15). The occlusion may be crossed with the guidewire and balloon initially and thrombolysis administered as indicated. Presently, data to determine the best protocol are too scant; however, the first-mentioned approach, thrombolysis followed by dilation, is currently most widely used (79,80).

FOLLOW-UP PATIENT CARE

Medication

The role of medications following successful angioplasty is to reduce the chance of the reclosure or restenosis. Additional medications may be required in selected cases to treat coronary artery disease not amenable to revascularization. The current recommendation for antiplatelet regimen is aspirin in doses from 100 to 1300 mg per day. The addition of other medications with antiplatelet activities, such as dipyridamole or sulfinpyrazone, is advocated by some. Because patients with coronary artery spasm have a higher incidence of restenosis (80), many physicians routinely administer calcium antagonists after angioplasty in hopes of preventing restenosis brought about by this mechanism. An isolated 1983

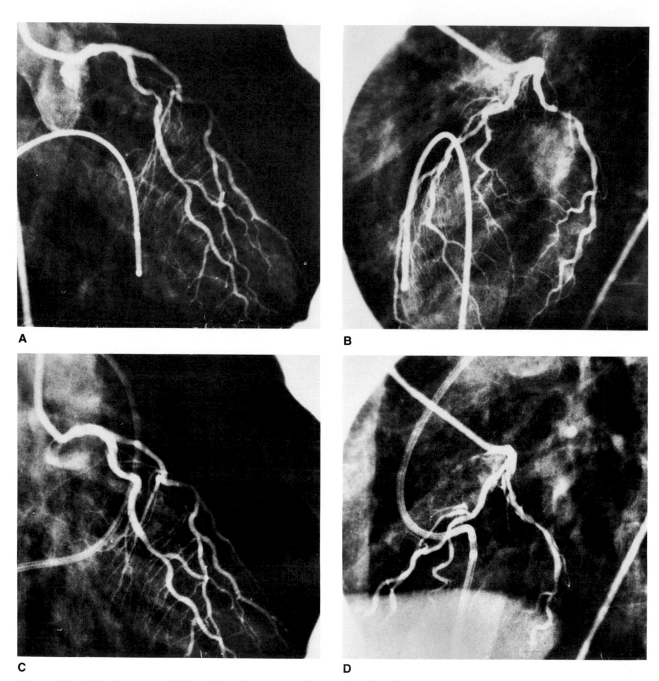

A

B

C

D

Figure 11. Fifty-five-year-old female with progressive angina. On fluoroscopy there was extensive left coronary calcification. **A,B.** Preangioplasty views. **C,D.** Postangioplasty views. The angioplasty site shows a good result but proximal dissection from the site that involved the left main artery and necessitated bypass graft surgery.

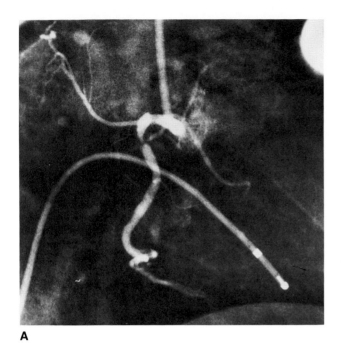
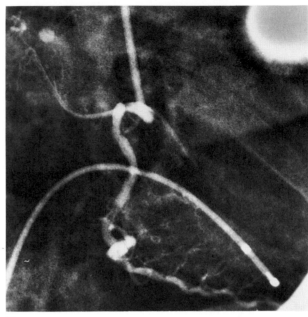

A   **B**

Figure 12.   Seventy-year-old female with exertional angina and a right coronary stenosis with involvement of the first right ventricular branch. **A.** Before angioplasty. **B.** After angioplasty; note poor filling of right ventricular branch.

study combining aspirin, isosorbide dinitrate, and verapamil hydrochloride has produced the lowest restenosis rate: 17 percent at 1 year (52). No controlled studies at the present time verify the efficacy of any of these medications in preventing restenosis.

Follow-up Parameters

Data from most centers suggest a 25–35 percent average rate of restenosis, with the majority occurring in the first 2–6 months (45,52,53). Thus, close attention is necessary during this time to identify patients in whom restenosis occurs. Since almost all patients have significant reduction in angina following successful angioplasty (54), recurrence of cardiac ischemic symptoms is a good clinical marker of restenosis. In patients with abnormal preangioplasty treadmill, electrocardiograms, or perfusion scans, every effort should be made to repeat those tests soon after angioplasty to document the clinical efficacy of the procedure and to provide a baseline for comparing studies at a later date. Since it has been shown that a certain number of patients may develop a restenosis without a significant change in symptoms (54,55), these noninvasive studies assume a greater importance in each patient's evaluation (80). Repeat angiography is currently the most

accurate way of determining the exact rate of restenosis in any given population because in some patients restenosis occurs without symptoms (54,55). For practical management of patients in a clinical setting, however, it would seem reasonable to reserve coronary arteriogram for those individuals having significant recurrence of symptoms or showing marked ischemia on noninvasive studies (56). A small percentage of patients with recurrence of symptoms do not show restenosis at subsequent angiography but do show progression of disease in other vessels that account for the symptoms (Fig. 16).

Follow-up Schedule

In patients without recurrent symptoms, noninvasive testing such as exercise electrocardiography or nuclear perfusion studies should be performed at 1–2 weeks, 3–6 months, 12 months, and then yearly thereafter. Patients with modifiable risk factors should be encouraged to reduce them whenever possible. Persons with recurrent stable symptoms should have noninvasive studies when symptoms recur and subsequent angiography to define the reason for symptom recurrence.

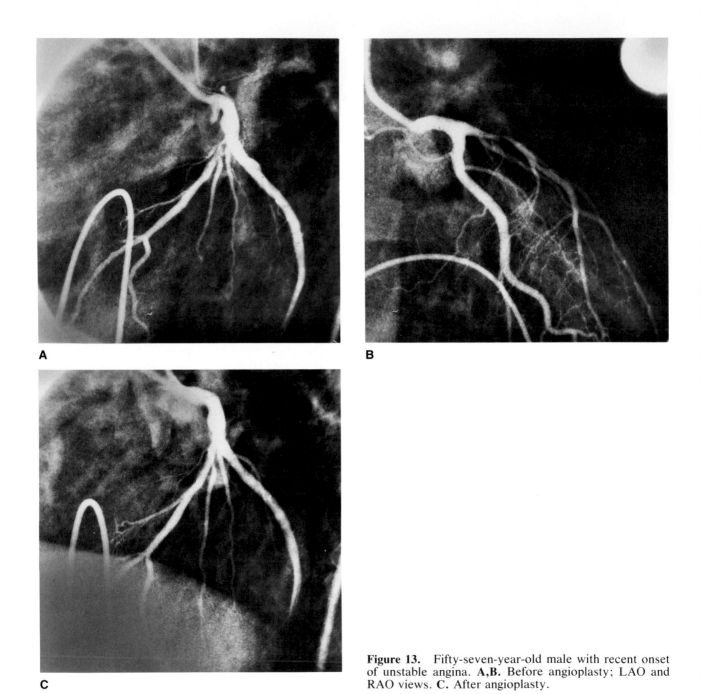

A

B

C

Figure 13. Fifty-seven-year-old male with recent onset of unstable angina. **A,B.** Before angioplasty; LAO and RAO views. **C.** After angioplasty.

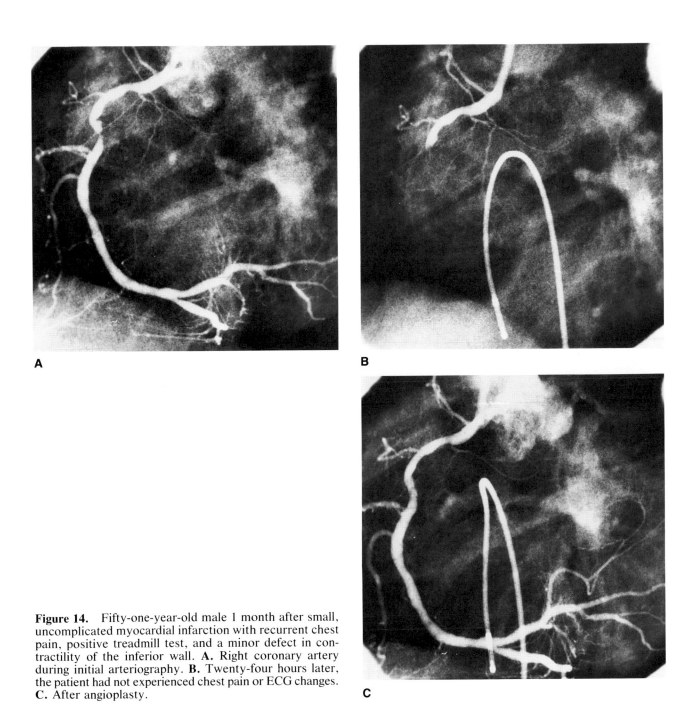

Figure 14. Fifty-one-year-old male 1 month after small, uncomplicated myocardial infarction with recurrent chest pain, positive treadmill test, and a minor defect in contractility of the inferior wall. **A.** Right coronary artery during initial arteriography. **B.** Twenty-four hours later, the patient had not experienced chest pain or ECG changes. **C.** After angioplasty.

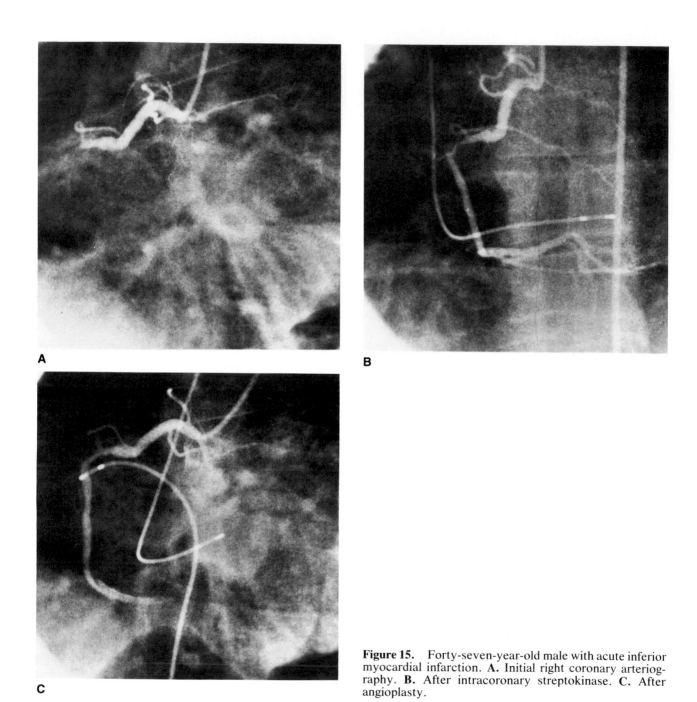

A

B

C

Figure 15. Forty-seven-year-old male with acute inferior myocardial infarction. **A.** Initial right coronary arteriography. **B.** After intracoronary streptokinase. **C.** After angioplasty.

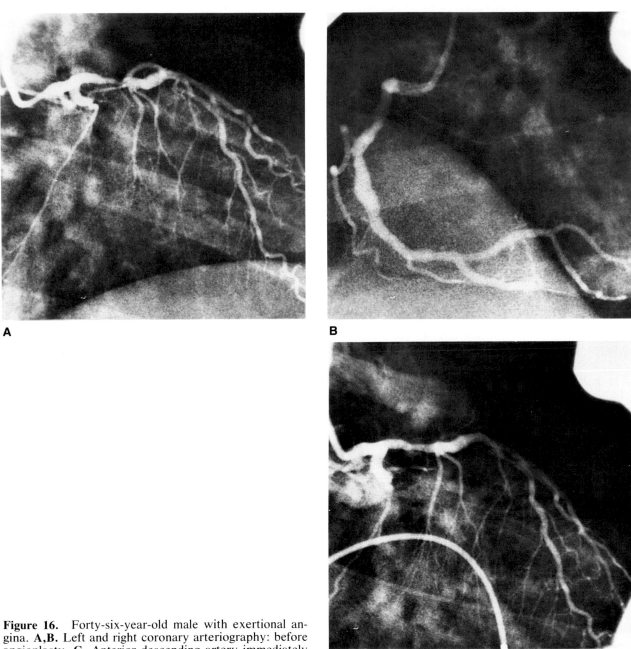

A

B

C

Figure 16. Forty-six-year-old male with exertional angina. **A,B.** Left and right coronary arteriography: before angioplasty. **C.** Anterior descending artery immediately after angioplasty.

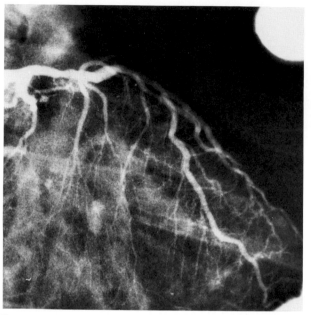

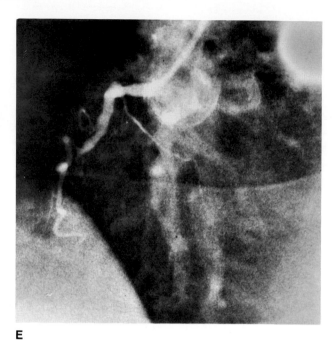

D   **E**

Figure 16 *(Continued)*.   **D,E.** Left and right arteriography 1 year after angioplasty, showing continued patency of the anterior descending artery at the angioplasty site but interval occlusion of the right coronary artery.

Management of Crossover Patients

Initial PTCA Success to Subsequent Bypass Surgery

In patients with initial angioplasty success whose symptoms recur (Fig. 17), the majority have restenosis and can be managed with a second angioplasty if clinically indicated (57). Individuals having significant ischemic symptoms who are not able to undergo a successful repeat angioplasty, those who develop restenosis following a second or third angioplasty, or those who develop multivessel disease not amenable to angioplasty are best managed by bypass surgery at that time. Although there is some controversy over how many times a single vessel should undergo angioplasty with subsequent restenosis before the patient receives surgery, most physicians limit angioplasty of a single lesion to two or three times and will refer to surgery after a second or third restenosis.

Initial Bypass Surgery to Subsequent PTCA

Some patients who have had coronary artery bypass surgery redevelop symptoms due to early or late failure of the grafts (Fig. 17). Recurrent symptoms can also develop because of development of disease in nonbypassed vessels or in vessels distal to the bypass grafts. If the symptoms and vessels involved are appropriate for percutaneous revascularization, many such patients can be managed by angioplasty.

Recrossover of Patients

As patients are followed for long periods, many have multiple crossovers between angioplasty and bypass surgery groups. Some patients with initially successful angioplasty who later require bypass surgery may again require angioplasty because of partial bypass graft failure. Likewise, some individuals with initially successful bypass surgery who receive subsequent angioplasty for partial graft failure may eventually need repeat bypass surgery because of progression of disease in other vessels or because of long-term angioplasty failure. Thus, it becomes obvious that patients in either treatment group require a careful long-term follow-up with cooperation between the cardiologist and the cardiovascular surgeon.

CURRENT STATE OF CORONARY REVASCULARIZATION

With presently available techniques of angioplasty, the procedure has evolved into a safe, angiograph-

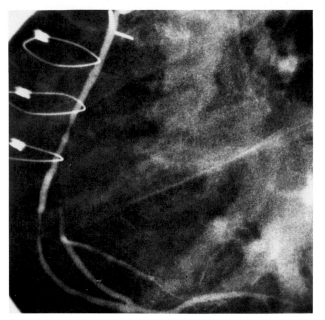

A

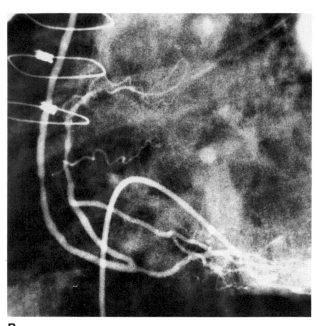

B

Figure 17. Sixty-two-year-old physician with a recurrent right graft stenosis 2 months after angioplasty. **A.** Before repeat angioplasty. **B.** After repeat angioplasty.

ically demonstrable, and clinically effective method of achieving coronary revascularization in carefully selected patients (50). As the techniques and clinical experience improve, the percentage of patients considered angioplasty candidates will increase.

The major advantages of angioplasty over bypass surgery include lower cost (81–84), decreased hospital stay (85,86), and decreased duration of disability (81,85,86). The major disadvantages of angioplasty include the narrowness of the subset of eligible patients out of the entire population of patients with coronary artery disease to whom the procedure can be applied, loss of some direct control over the anatomy of the heart when complications occur, inability to address concomitant cardiac pathology such as valvular disease or ventricular aneurysms, and inaccessibility of some coronary lesions (Table 2). With present techniques, the initial complication rates and recurrences of symptoms following successful procedures are similar for coronary angioplasty and bypass surgery.

As angioplasty techniques and clinical skills improve, the indications for the procedure in relation to bypass surgery will increase. Thus, more and more patients with multivessel disease and complicated coronary pathology considered too difficult to attempt in the past (77,78) will undergo angioplasty procedures. Although this may help to decrease the cost of clinical care, caution needs to be exercised

in pursuing such a course. As the complexity and severity of coronary disease increases, potential complications also increase (73). Mortality in the face of acute occlusion will obviously be higher in the patient with severe three vessel disease and poor ventricular function than in the one with single vessel disease and good ventricular function.

Management of complications that occur during bypass surgery is facilitated by the use of cardiopulmonary bypass, a technique not available to the angiographer. Thus, the risk from total occlusion during the procedure must always be kept in mind,

TABLE 2   Angioplasty versus Bypass Surgery for Revascularization

Advantages of Angioplasty	Advantages of Bypass Surgery
Lower cost	Wider subset of amenable patients
Decreased hospital stay	Availability of cardiopulmonary bypass to deal with complications
Decreased duration of disability	Ability to repair concomitant cardiac lesions
Less invasiveness	

and patients who would likely survive long enough to be transferred to the operating room for subsequent bypass surgery should be selected. Another problem with multivessel angioplasty is restenosis. Although the current restenosis rate of 25–35 percent may be acceptable in single vessel disease, the problem may be magnified correspondingly when multiple vessels are dilated in a single patient.

FUTURE OF RESEARCH AND IMPROVEMENT

In spite of recent advances in angioplasty technique, the problem of complications during angioplasty requiring emergency bypass surgery remain significant. The reported emergency bypass rate of 2–5 percent at centers with highly experienced personnel is quite low, but it is still too high for angioplasty without available surgical backup. Although certain broad correlations have been observed between certain anatomic landmarks and risk of acute dissection or occlusion, such as lesions in curves or ulcerated lesions, there still is no accurate way to predict a reasonable probability of who will develop an acute complication during coronary angioplasty. When acute occlusions do occur, a significant percentage of such patients suffer at least partial myocardial necrosis in the involved area, as manifested by electrocardiogram (ECG) and enzyme changes (73). Thus, another valuable area for investigation would be developing methods to preserve or perfuse the myocardium during acute occlusion until bypass surgery could be performed.

Restenosis at the site of successful angioplasty continues to be a significant problem, occurring in 17–47 percent of patients in spite of treatment with antiplatelet agents following angioplasty (51–57). Because the pathophysiology of successful angioplasty, acute intimal disruption, and the healing process following angioplasty are not completely understood, it remains unclear why some patients develop new stenoses at the angioplasty site while others maintain or improve the residual lesion after angioplasty. Various studies in progress are assessing variables such as balloon size, maximum pressure, inflation time, associated vessel spasm, shape or length of stenosis, vessel size and postangioplasty medication, and so forth, in hopes of finding solutions to the problem or of predicting the probability of restenosis in specific subsets of patients.

As expertise in angioplasty technique increases, both the number of patients attempted and the number of physicians performing the procedure will also increase. Equipment involved in the technique will improve in terms of balloon catheter size, guidewire size, steerability, balloon profile and pressure thresholds, and imaging techniques to visualize the stenotic vessel. Other percutaneous revascularization techniques continue to be investigated; they include laser angioplasty, extrusion balloons, fiberoptic angioscopic visualization of vessels before and after dilations, and techniques for removing the thrombotic material instead of spatially rearranging it. Other major advances to be expected could include noninvasive or invasive imaging techniques that will help determine which patients are the best candidates for revascularization.

REFERENCES

1. American Heart Association: *Heart Facts*. Dallas, 1983.
2. Mushkin SJ, Smelker M, Wyss D, Vehorn CL: Cost of disease and illness in the United States in the year 2000. Public Health Rep 93:494, 1978.
3. Sones FM Jr, Shirey EK: Cine coronary arteriography. Mod Concepts Cardiovasc Dis 31:735, 1962.
4. Rickets HJ, Abrams HL: Percutaneous selective coronary cine arteriography. JAMA 181:620, 1962.
5. Judkins MP: Selective coronary arteriography: A percutaneous transfemoral technic. Radiology 89:815, 1967.
6. Amplatz K, Formanek G, Stanger P, Wilson W: Mechanics of selective coronary artery catheterization via femoral approach. Radiology 89:1040, 1967.
7. Bourassa MG, Lesperance J, Campeau L: Selective coronary arteriography by the percutaneous femoral artery approach. Am J Roentgenol 107:377, 1969.
8. Dagenais GR, Rouleau JR, Christen A, Fabia J: Survival of patients with a strongly positive exercise electrocardiogram. Circulation 65:452, 1982.
9. Goldschlager N, Selzer A, Cohn K: Treadmill stress tests as indicators of presence and severity of coronary artery disease. Ann Intern Med 85:277, 1976.
10. Irving JB, Bruce RA, DeRouen TA: Variations in and significance of systolic pressure during maximal exercise (treadmill) testing: Relation to severity of coronary artery disease and cardiac mortality. Am J Cardiol 39:841, 1977.
11. Cole JP, Ellestad MH: Significance of chest pain during treadmill exercise: Correlation with coronary events. Am J Cardiol 41:227, 1978.
12. Gibson RS, Beller GA: Should exercise electrocardiographic testing be replaced by radioisotope methods? in Rahimtoola SH (ed): *Controversies in Coronary Artery Disease*. Philadelphia, Davis, 1983, pp 1–31.
13. Iskandrian AS, Wasserman LA, Anderson GS, Hakki

H, Segal BL, Kane S: Merits of stress thallium-201 myocardial perfusion imaging in patients with inconclusive exercise electrocardiograms: Correlation with coronary arteriograms. Am J Cardiol 46:553, 1980.

14. Garrett HE, Dennis EW, DeBakey M: Aorto-coronary bypass with saphenous vein graft: Seven year follow-up. JAMA 223:792, 1973.

15. Gruentzig AR, Myler RK, Itanna ES, Turina MI: Transluminal angioplasty of coronary artery stenosis, abstracted. Circulation 56:84, 1977.

16. Kannel WB, McGee D, Gordon T: A general cardiovascular risk profile: The Framingham Study. Am J Cardiol 38:46, 1976.

17. Wilhelmsson C, Vedin JA, Elmfeldt D, Tibblin G, Wilhelmsen L: Smoking and myocardial infarction. Lancet 1:415, 1975.

18. Report of the Surgeon General: Smoking and Health. Publication 79-50066. U.S. Depart. of Health, Education, and Welfare, 1979.

19. Friedman GD, Siegelaub AB: Changes after quitting cigarette smoking. Circulation 61:716, 1980.

20. Friedman GD, Petitti DB, Bawol RD, Siegelaub AB: Mortality in cigarette smokers and quitters. Effect of base-line differences. N Engl J Med 304:1407, 1981.

21. Five-year findings of the Hypertension Detection and Follow-up Program: I. Reduction in mortality of persons with high blood pressure, including mild hypertension. JAMA 242:2562, 1979.

22. Report of the Management Committee: The Australian therapeutic trial in mild hypertension. Lancet 1:1261, 1980.

23. Lipid Research Clinics Program: The Lipid Research Clinics Coronary Primary Prevention Trial Results. JAMA 251:351–365, January 1984.

24. Fox SM III, Naughton JP, Hasket WL: Physical activity and the prevention of coronary heart disease. Ann Clin Res 3:404, 1971.

25. Brachfeld N, Bozer J, Gorlin R: Action of nitroglycerin on the coronary circulation in normal and in mild cardiac subjects. Circulation 19:697, 1959.

26. Gorlin R, Brachfeld N, MacLeod C, Bopp P: Effect of nitroglycerin on the coronary circulation in patients with coronary artery disease or increased left ventricular work. Circulation 19:705, 1959.

27. Mason DT, Braunwald E: The effects of nitroglycerin and amyl nitrite on arteriolar and venous tone in the human forearm. Circulation 32:755, 1965.

28. Likoff W, Kasparian H, Lehman JS, Segal BL: Evaluation of coronary vasodilators by coronary arteriography. Am J Cardiol 13:7, 1964.

29. Bache RJ, Ball RM, Cobb FR, Rembert JC, Greenfield JC Jr: Effects of nitroglycerin on transmural myocardial blood flow in the unanesthetized dog. J Clin Invest 55:1219, 1975.

30. Abrams J: Nitroglycerin and long-acting nitrates. N Engl J Med 302:1234, 1980.

31. Markis JE, Gorlin R, Mills RM, Williams RA, Schweitzer P, Ransil BJ: Sustained effect of orally administered isosorbide dinitrate on exercise performance of patients with angina pectoris. Am J Cardiol 43:265, 1979.

32. Parmley WW: Beta blockers in coronary artery disease. Cardiovasc Rev Rep 2:655, 1981.

33. Koch-Weser J, Frishman WH: Beta-adrenoceptor antagonists: New drugs and new indications. N Engl J Med 305:500, 1981.

34. The Norwegian Multicenter Study Group: Timolol-induced reduction in mortality and reinfarction in patients surviving acute myocardial infarction. N Engl J Med 304:801, 1981.

35. Beta Blocker Heart Attack Study Group: The Beta-Blocker Heart Attack Trial. JAMA 246:2073, 1981.

36. Braunwald E: Mechanism of action of calcium-channel-blocking agents. N Engl J Med 307:1618, 1983.

37. Sherman LG, Liang CS: Nifedipine in chronic stable angina: A double blind placebo-controlled crossover trial. Am J Cardiol 51:706, 1983.

38. Leon MB, Rosing DR, Bonow RO, Lipson LC, Epstein SE: Clinical efficacy of verapamil alone and combined with propranolol in treating patients with chronic stable angina pectoris. Am J Cardiol 48:131, 1981.

39. Wagniart P, Ferguson RJ, Chaitman BR, Achard F, Benacerraf A, Belanguenhagen B, Morin B, Pasternac A, Bourassa MG: Increased exercise tolerance and reduced electrocardiographic ischemia with diltiazem in patients with stable angina pectoris. Circulation 66:23, 1982.

40. Mukharji J, Rude R, Gustafson N, Poole K, Passamani E, Thomas LJ Jr, Strauss HW, Muller JE, Roberts R, Raabe DS Jr, Braunwald E, Willerson JT, et al: Late sudden death following myocardial infarction: Interdependence of risk factors. J Am Coll Cardiol 1:585, 1983.

41. Moss AJ, Bigger JT, Case RB, Gillespie J, Goldstein R, Greenberg H, Krone R, Marcus FI, Odoroff CL, Oliver GC: Risk stratification and prognostication after myocardial infarction, abstracted. J Am Coll Cardiol 1:716, 1983.

42. Lytle BW, Loop FD, Cosgrove DM, Easley K, Taylor PC: Long-term (5–12 year) sequential studies of internal mammary artery and saphenous vein coronary bypass grafts, abstracted. Circulation 68(suppl III): 114, 1983.

43. Campeau L, Enjalbert M, Lesperance J, Bourassa MG, Grondin CM: Comparison of late changes (closure and atherosclerosis at 10 years) in internal mammary artery and saphenous vein coronary artery grafts, abstracted. Circulation 68(suppl III): 114, 1983.

44. Tector AJ, Schmahl TM, Canino VR: The internal mammary artery graft: The best choice for bypass of the diseased left anterior descending coronary artery. Circulation 68(suppl II):214, 1983.

45. Kennedy JW, Kaiser GC, Fisher LD, Fritz JK, Myers W, Mudd JG, Ryan TJ: Clinical and angiographic predictors of operative mortality from the collaborative study in coronary artery surgery (CASS). Circulation 63:793, 1981.

46. Burton JR, FitzGibbon GM, Keon WJ, Leach AJ: Perioperative myocardial infarction complicating coronary bypass: Clinical and angiographic correlations and prognosis. J Thorac Cardiovasc Surg 82:758, 1981.

47. Kouchoukos NT, Oberman A, Kirklin JW, Russell RO Jr, Karp RB, Pacifico AD, Zorn GL: Coronary bypass surgery: Analysis of factors affecting hospital mortality. Circulation 62(suppl I):84, 1980.

48. Rahimtoola SH: Coronary bypass surgery for chronic angina—1981: A perspective. Circulation 65:225, 1982.

49. Frick MH, Harjola PT, Valle M: Persistent improvement after coronary bypass surgery: Ergometric and angiographic correlations at 5 years. Circulation 67:491, 1983.

50. Gruentzig AR, Meier B: Percutaneous transluminal coronary angioplasty: The first five years and the future. Int J Cardiol 2:319, 1983.

51. Renkin J, David PR, Dangoisse V, Lesperance J, Bourassa MG: Coronary angiographic results 6 and 18 months after successful percutaneous transluminal angioplasty in 53 consecutive patients, abstracted. Circulation 68(suppl III):314, 1983.

52. Kaltenbach M, Scherer D, Kober G: Long-term results of coronary angioplasty, abstracted. Circulation 68(suppl III):95, 1983.

53. Holmes D, Vliestra R, Smith H, Kent K, Bentivoglio L, Block P, Dorros G, Gosselin A, Gruentzig AR, Myler RS, Simpson J, Stertzer SH, Williams DO, Bourassa M, Vetroved G, Kelsey S, Detre K, Passamani E, Van Raden M, Mock M: Restenosis following percutaneous transluminal coronary angioplasty (PTCA): A report from the NHLBI PTCA Registry, abstracted. Circulation 66(suppl III):95, 1983.

54. Jutzy KR, Berte LE, Alderman EL, Ratts J, Simpson JB: Coronary restenosis rates in a consecutive patient series one year post successful angioplasty, abstracted. Circulation 66(suppl II):331, 1982.

55. Ewels CJ, Rosing DR, Kent KM: Restenosis following transluminal coronary angioplasty, abstracted. Circulation 68(suppl III):96, 1983.

56. Scholl JM, Chaitman BR, David PR, Dupras G, Brevers G, Val PG, Crepeau J, Lesperance J, Bourassa MG: Exercise electrocardiography and myocardial scintigraphy in the serial evaluation of the results of percutaneous transluminal coronary angioplasty. Circulation 66:380, 1982.

57. Meier B, King SB III, Gruentzig AR, Douglas JS, Hollman J, Ischinger T, Galan K, Tankersley R: Repeat coronary angioplasty, abstracted. Circulation 68(suppl III):96, 1983.

58. Diamond GA, Forrester JS: Analysis of probability as an aid in the clinical diagnosis of coronary artery disease. N Engl J Med 300:1350, 1979.

59. Campeau L: Grading of angina pectoris (letter). Circulation 54:522, 1976.

60. Helfant RH, Bodenheimer MM, Banka VS: Asynergy in coronary heart disease: Evolving clinical and pathophysiologic concepts. Ann Intern Med 87:475, 1977.

61. Cohn PF: Contractile reserve in coronary heart disease: Detection and clinical importance. Primary Cardiol 7:48, 1981.

62. McAnulty JH, Hattenhauer MT, Rosche J, Kloster FE, Rahimtoola SH: Improvement in left ventricular wall motion following nitroglycerin. Circulation 51:140, 1975.

63. Report of Inter-Society Committee for Heart Disease Resources: Optimal resources for coronary artery surgery. Circulation 46(suppl A):325, 1972.

64. Vogel RA, LeFree M, Bates E: Application of digital techniques to selective coronary arteriography: Use of myocardial contrast appearance time to measure coronary flow reserve. Am Heart J 107:153, 1984.

65. Wright C, Doty D, Eastham C, Laughlin D, Krumm P, Marcus M: Method for assessing the physiologic significance of coronary obstructions in man at cardiac surgery. Circulation 62(suppl I):111, 1980.

66. Levy RI, Mock MB, Willman VL, Frommer PL: Percutaneous transluminal coronary angioplasty, editorial. N Engl J Med 301:101, 1979.

67. Ischinger T, Gruentzig AR, Hollman J, King S III, Douglas J, Meier B, Bradford J, Tankersley R: Should coronary arteries with less than 60 percent diameter stenosis be treated by angioplasty? Circulation 68:148, 1983.

68. Cohn PF: Prognosis and treatment of asymptomatic coronary artery disease. J Am Coll Cardiol 1:959, 1983.

69. Coronary Artery Surgery Study (CASS): A randomized trial of coronary artery bypass surgery survival data. Circulation 68:939, 1983.

70. Hlatky MA, Califf RM, Kong Y, Harrell FE Jr, Rosati RA: Natural history of patients with single-vessel disease suitable for percutaneous transluminal coronary angioplasty. Am J Cardiol 52:225, August 1983.

71. Douglas JS Jr, Gruentzig AR, King SB III, Hollman J, Ischinger T, Meier B, Craver JM, Jones EL, Waller JL, Bone DK, Guyton R: Percutaneous transluminal coronary angioplasty in patients with prior coronary bypass surgery. J Am Coll Cardiol 2:745, 1983.

72. Aueron F, Gruentzig AR: Distal embolization of a coronary artery bypass graft atheroma during percutaneous transluminal coronary angioplasty. Am J Cardiol 53:953, 1984.

73. Dorros G, Cowley MJ, Simpson J, Bentivoglio LG, Block PC, Bourassa M, Detre K, Gosselin AJ, Gruentzig AR, Kelsey SF, Kent KM, Mock MB, Mullin SM, Myler RK, Passamani ER, Stertzer SH, Williams DO: Percutaneous transluminal coronary angioplasty: Report of complications from the National Heart, Lung, and Blood Institute PTCA Registry. Circulation 67:723, 1983.

74. Meier B, Gruentzig AR, Hollman J, Ischinger T, Bradford JM: Does length or eccentricity of coronary stenoses influence the outcome of transluminal dilatation? Circulation 67:497, 1983.

75. Meier B, Gruentzig AR, King SB III, Douglas JS Jr, Hollman J, Ischinger T, Aueron F, Galan K: Risk of side branch occlusion during coronary angioplasty. Am J Cardiol 53:10, 1984.

76. Dervan JP, Baim DS, Cherniles J, Grossman W: Transluminal angioplasty of occluded coronary arteries: Use of a movable guide wire system. Circulation 68(suppl 4):776–784, 1983.

77. Holmes DR, Vliestra RE, Reeder GS, Bresnahar JF, Smith HC, Bove AA, Schaff HV: Angioplasty in total coronary artery occlusion. J Am Coll Cardiol 3:845, March 1984.

78. Hartzler GO, Rutherford BD, McConahay DR: Per-

cutaneous transluminal coronary angioplasty application for acute myocardial infarction. Am J Cardiol 53:117C, 1984.

79. Gold HK, Cowley MJ, Palacios IF, Vetrovec GW, Atkins CW, Block PC, Leinbach RC: Combined intracoronary streptokinase infusion and coronary angioplasty during acute myocardial infarction. Am J Cardiol 53:122C, 1984.

80. David PR, Waters DD, Scholl JM, Crepeau J, Szlachcic J, Lesperance J, Hudon G, Bourassa MG: Percutaneous transluminal coronary angioplasty in patients with variant angina. Circulation 66:695, 1982.

81. Holmes DR Jr, Vliestra RE, Mock MB, Smith HC, Dorros G, Cowley MJ, Kent KM, Hammes LN, Janke L, Elveback LR, Vetrovec GW: Employment and recreation patterns in patients treated by percutaneous transluminal coronary angioplasty: A multicenter study. Am J Cardiol 52:710, 1983.

82. Jang GC, Block PC, Cowley MJ, Gruentzig AR, Dorros G, Holmes DR, Kent KM, Leatherman LL, Myler RK, Sjolander SME, Stertzer SH, Vetrovec GW, Willis WH Jr, Williams DO: Relative cost of coronary angioplasty and bypass surgery in a one vessel disease model. Am J Cardiol 53:52C, 1984.

83. Jang GC, Baskerville AL, Willis WH Jr, Jacobson JG: Cost analysis and eligibility assessment of transluminal coronary angioplasty. Circulation 64(suppl IV):90, October 1981.

84. Reeder GS, Krishan I, Nobrega FT, Naessens J, Kelly M, Christianson JB, McAfee MK: Is percutaneous coronary angioplasty less expensive than bypass surgery? N Engl J Med 311:1157–1162, November 1984.

85. Jang GC, Gruentzig AR, Block PC, et al: Work profile of patients following coronary angioplasty or coronary bypass surgery. Circulation 66(suppl II):122, October 1982.

86. Holms DR, Van Raden MJ, Reeder GS, et al: Return to work after coronary angioplasty: A report from the National Heart, Lung, and Blood Institute PTCA Registry. Circulation 53:48C, June 1984.

13

Transfemoral Approach to Percutaneous Coronary Angioplasty

RICHARD K. MYLER, M.D.

INTRODUCTION

In 1964, Dotter (1) pioneered transluminal angioplasty for peripheral vascular disease; the technique was further developed by Zeitler (2) and modified by Gruentzig (3).

Since the introduction of the transfemoral technique for coronary angiography by Judkins (4) and Amplatz (5) in 1967, this approach has gained wide acceptance in the performance of diagnostic coronary arteriography. The transfemoral approach to percutaneous transluminal coronary angioplasty (PTCA) is derived directly from these two techniques.

Initially, large (number 9.4 French) guiding catheters of solid Teflon were utilized to perform coronary angioplasty. The relatively large size of the guiding catheters and the material from which they were made (Teflon) were necessary to allow the passage of dilation catheters of earlier design through the lumen of guiding catheters. Thus, the femoral approach became a natural choice. Initially, these nontapered guiding catheters were used without the aid of an insertion sheath until such a device was developed to accommodate the more convenient introduction of the guiding catheters. With the evolution of angioplasty catheter technology resulting in smaller French sizes (numbers 8 and 9) and more

sophisticated introducer sheath systems, smaller and safer punctures were made via the transfemoral approach (Fig. 1).

The development of more sophisticated guiding catheters has allowed the success rate and safety of coronary angioplasty performed by the transfemoral approach to improve significantly over the past several years. The advancement in design and technology of the guiding catheter has been accompanied by rapid development of the steerable and low profile balloon dilation catheter systems, leading to further improvement in the success rate and the ability to perform angioplasty in vessels and with lesions that was not possible with earlier nonsteerable systems. Thus, the evolution of the guiding and dilation catheters since 1977 from prototype models to their present sophisticated forms has allowed for an improvement in the success rate, a decrease in complication rate, a shortening of the learning curve, and a widening of patient selection (6–33).

CHARACTERISTICS OF CATHETER DESIGN

The equipment necessary to perform percutaneous transfemoral coronary angioplasty includes a guiding catheter, a dilation catheter, guidewires, an in-

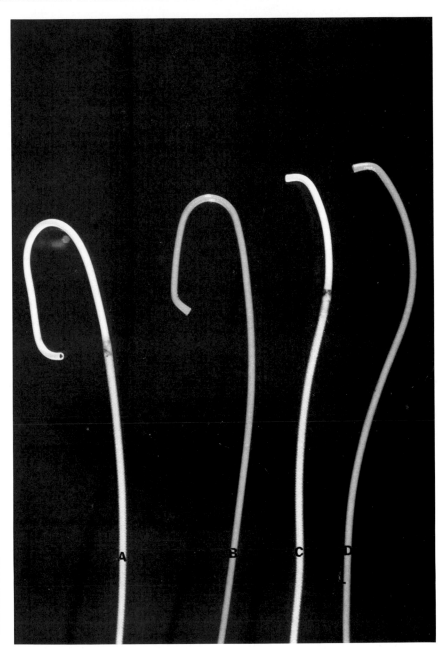

Figure 1. Comparison of original number 9.4 French solid Teflon left (**A**) and right (**C**) Judkins guiding catheters (Schneider Medintag) and current number 8 French left (**B**) and right (**D**) Judkins guiding catheters (USCI).

troducer sheath system, manifolds, and an inflation device (Table 1). The guiding catheters of USCI (United States Catheter and Instrument Company, Billerica, MA) and of Schneider Medintag, Zurich, Switzerland are radiopaque and have an external diameter of French number 8 or 9 (Fig. 2). These catheters are nontapered, without distal side holes; however, a recent guiding catheter by USCI and Interventional Medical (Arlington, MA) is supplied with two small distal side holes to allow flow through the main catheter lumen (Fig. 3). The guiding catheters are made of a bonded composite of three layers (Fig. 4). The outer jacket is of polished polyurethane and provides catheter memory and "stiffness." The middle layer is a wire braid furnishing support and torque control. The inner surface is Teflon because its low coefficient of friction (*lubricity*) allows easy passage of the smaller dilation catheter through its

TABLE 1 Equipment for Coronary Angioplasty

Guiding catheter
Dilation catheter
Steerable wire
Torquer
Y connectors (2)
Manifolds (2)
Insertion introducer system
Pressure gauge
Inflation device
Pacemaker catheter
Heat gun

lumen. Femoral guiding catheters are made with a variety of configurations. Unlike the angiographic catheter counterparts, the angioplasty guiding catheter must engage the coronary artery in a stable and coaxial fashion since its purpose is to guide the dilation catheter into the target coronary artery and allow "back-up support" or "power" for the passage of the dilation catheter across severe stenotic lesions located distally.

The configurations of left coronary guiding catheters are made in the Judkins, Amplatz, and multipurpose shapes. The Judkins catheter, for the left coronary artery, is of French number 8 or 9 with a tip curve configuration that is available in an in-plane orientation or in a 30 percent anterior or 30 percent posterior out-of-plane orientation that can provide a conduit for selective advancement of the dilation catheter into the anterior descending or circumflex arteries, respectively. These Judkins guiding catheter configurations are available with 3.5 to

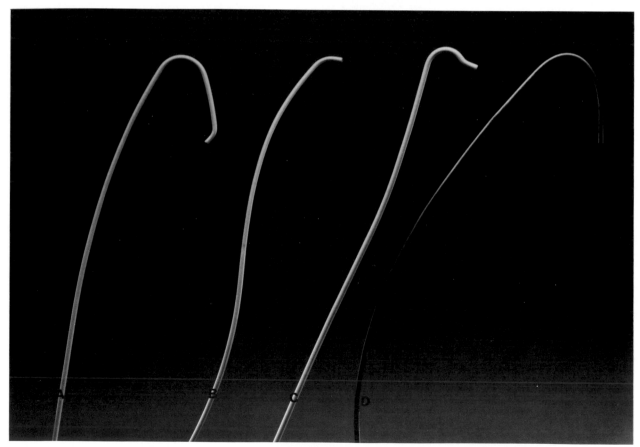

Figure 2.   A. Current (USCI) guiding catheters: left Judkins (**A**), right Judkins (**B**), right Amplatz (**C**), and Stertzer brachial (**D**).

Figure 3. Schematic of perfusion port Judkins style right coronary guiding catheter (USCI).

6 curves (Fig. 5A). The Amplatz left coronary guiding catheter configurations are formed in I and II style curves.

The right coronary catheter's configurations are of Judkins and Amplatz forms, as well as the multipurpose and El Gamal types. The right Judkins catheters are available in 4 and 6 curves and the Amplatz catheters are available in I and II curves (Figures 5B, 6) (Table 2). The right coronary artery orifice, in certain patients, may be small in relation to the tip of the number 9 or even 8 French guiding catheter. Therefore, side holes have been made in the guiding catheter systems offered by USCI, Interventional Medical, and ACS (Advanced Catheter Systems, Mt. View, CA) to allow antegrade blood flow via the side holes into the distal coronary artery, as well as the continued pressure monitoring in the aorta. A new prototype coated number 8 French guiding catheter with a 9 French lumen, yet with a stiffer shaft (for "power"), has been developed and has been termed the "large lumen" or "high flow" 8F guiding catheter (Fig. 7) (Table 3).

The variability of individual coronary anatomy, as well as the individual configuration of saphenous vein graft origins from the ascending aorta, can always be accommodated by modification of the available guiding catheters by employing a heat gun (Fig.

8) that permits reshaping of the guiding catheter to simulate the anatomic relationship of the aorta and the coronary artery ostium. The three layers of the guiding catheter have different melting points (e.g., Teflon melts at a much higher point than does polyurethane). Therefore, the guiding catheter exposure to heat must be brief. Once this is accomplished, the heated portion of the catheter is immediately placed into sterile cold water to "set" the desired curve. Careful inspection of the guiding catheter, after reshaping with the 0.063 inch wire removed, should always be done, if any defect in the catheter is noted, it must be discarded. Reshaping the guiding catheters, as well as other manipulations such as cutting, filing the distal tip, and making side holes in these catheters, although necessary on occasion, are not recommended by the catheter manufacturers. Therefore, if these catheter modifications are performed, they necessitate defensible reasoning and scrupulous care.

In 1976, Gruentzig modified and miniaturized his peripheral angioplasty catheter to perform coronary angioplasty. These catheters were tested initially in canine model and, later, in human cadaver experiments (34–37). Subsequently, this new catheter was utilized intraoperatively in lesions proximal to the arteriotomy during the performance of elective aor-

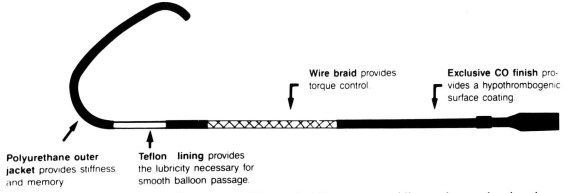

Figure 4. Schematic of Myler/Gruentzig Judkins style left coronary guiding catheter, showing three-layered design common to all styles of femoral guiding catheters (USCI/Schneider).

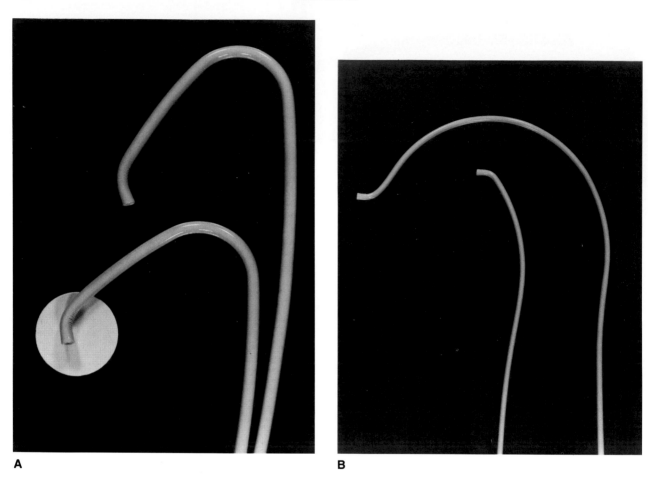

A   B

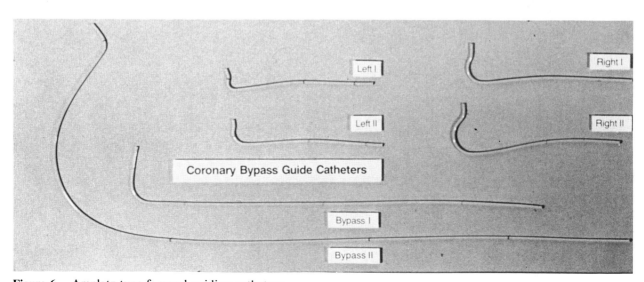

Figure 6.   Amplatz type femoral guiding catheters.

TABLE 2 Guiding Catheters (USCI): Femoral

Catheter Style	Curve Style	Shaft Diameter, F^a	Usable Shaft Length cm	Minimum Tip Opening ID mm
Left				
Judkins (in-plane) (FL)	3.4–6	8–9	100	2.64–2.97
Anterior descending (FLG)	3.5–5	8–9	100	2.64–2.97
Circumflex (FLP)	4–5	8–9	100	2.64–2.97
Amplatz (AL)	I–II	8–9	100	2.64–2.97
Right				
Judkins	4–6	8–9	100	2.64–2.97
Amplatz (AR)	I–II	8–9	100	2.64–2.97
Amplatz	FRS	8–9	100	2.64–2.97
Others				
King Multipurpose		8–9	100	2.64–2.97
El Gamal		8–9	100	2.64–2.97
SVG		8–9	100	2.64–2.97
Bourassa (Schneider)		8–9	100	2.64–2.97

tocoronary bypass surgery, to examine this technique's effectiveness and safety for in vivo atherosclerosis (38). From its earliest form, the dilation balloon catheter has shown rapid evolutionary development, becoming increasingly sophisticated in its technology.

Dilation catheters have always had two lumens. The central lumen is used for pressure measurement and superselective contrast injection, and the second eccentric lumen for balloon inflation-deflation. Initially, it was felt that hemo perfusion via the central lumen would be necessary during balloon inflation, which causes complete temporary occlusion of the target coronary artery. However, subsequent observation in humans showed that perfusion was

not necessary with brief coronary artery occlusion, and more recent experience has revealed that even more prolonged (60 s or more) inflation of the balloon is possible without untoward sequelae.

These prototype balloon catheters evolved to forms having a fixed distal wire and infusion side holes (Fig. 9) and later to the presently most widely used "steerable" dilation catheters (Fig. 10). The steerable dilation catheters have a central lumen for pressure measurement, superselective contrast injection, and passage of a steerable guidewire, and a second eccentric lumen for balloon inflation and deflation. The steerable system has led to a significant improvement of primary success rate, a decrease in the complication rate (by allowing safe recrossing of lesions so long as the wire guide remains distal to the lesion), and the ability to perform angioplasty in patients with complicated coronary anatomy. Tortuous vessels, severe eccentric lesions, tandem

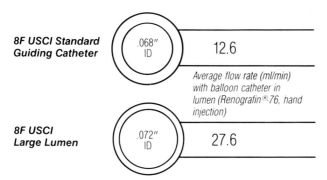

Figure 7. Schematic of cross-section of standard 8F and large lumen (or high-flow) 8F guiding catheter comparing flow rates and internal diameters (ID).

TABLE 3 8F versus 9F Femoral Guiding Catheters

8F*	9F
Smaller arterial puncture	Increased torque control
Less catheter wedging	Better dye delivery
Ability to deep-seat	Better pressure waveform
Less potential for intimal trauma	Stiffer for more power or back-up

*The new "high-flow" 8F combines advantages of both 8F and 9F.

Figure 8. Heat gun (Stanley Corporation) capable of delivering 750°F.

lesions, multivessel lesions, recent total occlusions, and branch lesions requiring the "kissing balloon" technique are all now possible because of the new steerable and low profile dilation catheter systems.

The *low profile steerable* (LPS) dilation catheters are Teflon-coated with balloons manufactured from polyvinyl chloride (PVC), although newer, stronger materials are currently being tested. The length of the balloon is 25 mm, and inflated diameters 2.0–4.0 mm. At each end but within the length of the balloon, two radiopaque platinum markers allow one to position the balloon across the stenosis under fluoroscopy. The shaft diameters of the dilation catheters

vary between numbers 2.3 and 4.3 French sizes, and the length of the catheter shaft is 135 cm (Table 4) (Fig. 11). All dilation catheters, with the exception of ACS's super low profile (0.030 inch or 0.76 mm) catheter, allow guidewire placement and manipulation through the central lumen. The central lumen is for pressure measurement and subselective contrast angiography; both can be achieved even with the steerable wire in place in the central channel of the USCI/Schneider dilation catheter, although not so well with the current ACS system. As indicated, the second lumen is for inflation and deflation of the balloon (Figs. 12–15).

The main advantage of the steerable and low profile systems is the steerable wire itself. These wires vary in length between 175 and 300 cm (the latter being an exchange wire), with shaft outer diameters 0.014 inch (0.356 mm) with tapered central core mandrels and different tip configurations, some more flexible than others (Table 5). The steerable wire is made from a metal alloy coiled over a central mandrel with its distal end tapered to the tip of the coiled wire, or ending 2 or 3 cm from the tip. The former type allows a curve with memory and torque control; the latter is floppy. Newer wires have a forming ribbon one end of which is attached to the tip of the tapered mandrel and the other end to the end of the coiled wire. These newer wires are coated with Teflon for greater trackability (Figs. 16, 17). Selection of the appropriate wire often depends on the dilation catheter selection. For example, a low profile steer-

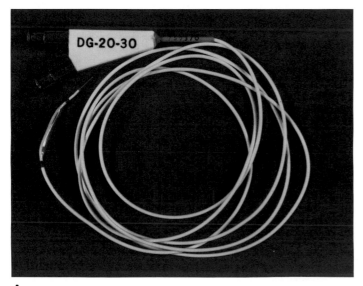

A

B

Figure 9. **A.** Coronary dilation balloon catheter, type DG 20-30. **B.** Close-up of type DG 20-30 (balloon 20 mm long with 3.0 mm inflated diameter). (Reprinted from Ref. 15.)

Figure 10. Coronary balloon catheter (Schneider Medintag) including early style 0.063 inch wire (**A**), steerable (**B**), low profile (**C,F**), double balloon (**D**), and fixed wire (**E**).

able catheter of the USCI 25-20 type may allow only a 0.014 inch (0.356 mm) steerable wire if one is to measure pressures effectively at the distal tip of the catheter. It should be mentioned that dilation catheter and steerable wire technology is in continuing developmental flux. Therefore, new dilation catheters with different shapes and sizes (low profile and otherwise), variable balloon dimensions and mate-rial, new wire configurations, and more advanced capabilities are evolving at the time of this writing.

The steerable wires are introduced through the dilation catheter via a steerable Y connector (yellow) in the USCI system. The dilation catheter enters the guiding catheter through a standard Y connector (clear). The second portal of the yellow Y connector in the dilation catheter allows pressure

TABLE 4   **Steerable Dilation Catheters (USCI): Low Profile**

Inflated Balloon Diameter, mm	Average Deflated Balloon Profile, mm	Balloon Length, mm	Shaft Diameter, F	Shaft Taper, F	Usable Shaft Length, cm
2.0	1.07	25	4.3	2.3	135
2.5	1.22	25	4.3	2.3	135
3.0	1.27	25	4.3	2.3	135
3.5	1.37	25	4.3	2.3	135
4.0	1.42	25	4.3	2.3	135

and contrast injection through it, and the second portal of the clear Y connector attached to the guiding catheter allows pressure measurement and contrast injection through it (Figs. 18–21). In order to introduce the nontapered guiding catheters, numbers 8 and 9 French insertion introducer sheath systems (with or without side port and back bleed diaphragm) are available. The three-piece system consists of a long number 5 French Teflon dilator within a shorter number 8 (or 9) French tapered dilator, which in turn is inside a number 8 (or 9) French thin-walled sheath (Figs. 22, 23). These sheaths have allowed safe and painless introduction and exchange of the guiding catheters.

A variety of manifold systems are available with two, three, or four ports and are utilized in con-

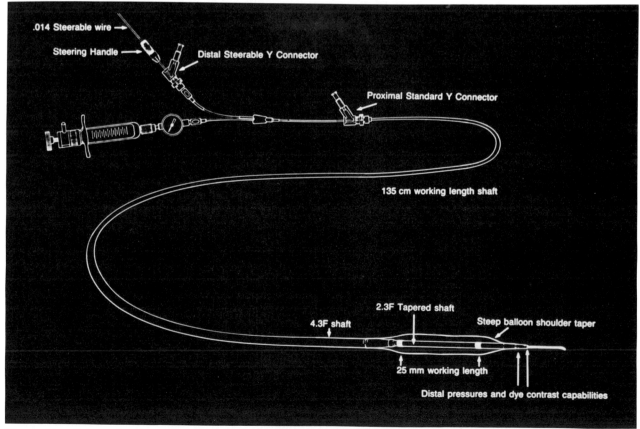

Figure 11.   Schematic of current (USCI) steerable dilation system.

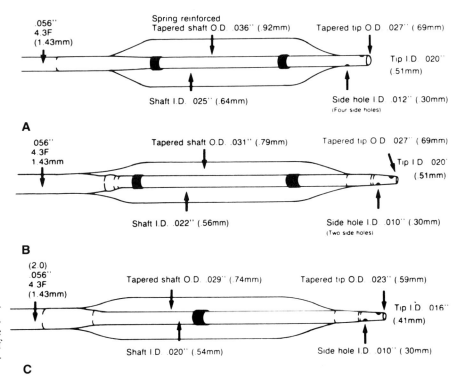

Figure 12. **A.** Schematic of standard steerable dilation catheter. **B.** Schematic of low profile steerable dilation catheter. **C.** Schematic of 2.0 (mm) low profile steerable dilation catheter.

junction with guiding and dilation catheters. These manifolds permit recording of pressure, as well as on-line supply of contrast and saline (Fig. 24). The balloon is inflated witih a calibrated pressure delivery system that is at present a hand-held, screw-type inflation device (USCI and Interventional Medical) that is relatively inexpensive (Figs. 25–28).

Thus, the dilation catheter system has evolved from a primitive prototype of a fixed-wire system to a movable and steerable wire and a low profile dilation catheter system, with a commensurate increase in the success rate and a decrease in the complication rate resulting from these dramatic technical advances.

PATIENT PREPARATION FOR ANGIOPLASTY

Perhaps the most important preangioplasty patient preparation is discussing the goals and risks of the angioplasty procedure. It is recommended that, with rare exception, all patients who undergo coronary angioplasty also be candidates for coronary artery bypass surgery. This requires a presigned informed consent form from the patient for potential emergency coronary bypass surgery should the angio-

plasty procedure fail and an abrupt reclosure of a dilated coronary artery occur.

Of course, careful review of the diagnostic coronary arteriogram prior to the patient's arrival in the hospital and then again just before coronary angioplasty is the primary "selection" method. In addition, patients should understand that in the interval since the initial diagnostic coronary arteriogram was performed, new abnormalities, which may make the patient a less suitable candidate for angioplasty, may have developed. Appropriate radiographic projections in the control coronary arteriogram immediately prior to the planned angioplasty may show new or unsuspected abnormalities (not visualized in the initial study), possibly excluding the patient from consideration for coronary angioplasty (39). Obviously, a dramatic change in the patient's clinical pattern, for example, an evolving myocardial infarction, may also alter plans for the elective coronary angioplasty.

The day before the planned angioplasty, the patient should receive enteric coated aspirin (325 mg bid) or dipyridamole (75 mg tid) and nifedipine (10 mg qid), in addition to nitrates (oral, sublingual, or topical). If the patient is on beta-blockade therapy, it is recommended that this therapy be discontinued or the dosage decreased. On the day of coronary

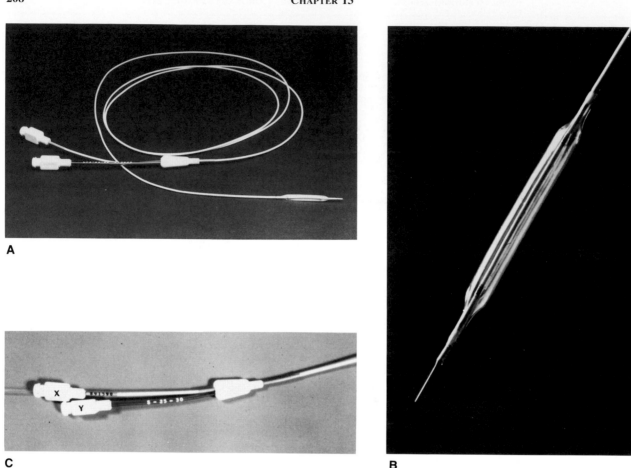

Figure 13. **A.** Steerable coronary dilation balloon catheter (USCI). **B.** Close-up of low profile steerable dilation catheter (Schneider Medintag). Close-up of proximal dual portals for pressure, contrast, wire manipulation (X), and balloon inflation (Y). (Courtesy of Dr. D. Levin.)

angioplasty, in addition to the usual precardiac catheterization medications for analgesia and anxiety, nitrates or calcium channel blockers, as well as aspirin, are indicated. We recommend that two units of packed cells be available on standby in case emergency surgery is necessary. In addition, both groins should be prepped and an intravenous line placed, usually in the patient's left upper extremity. During the angioplasty procedure, the patient is heparinized, usually with 10,000 units intravenously, with additional heparin administered as necessary if the

procedure is prolonged. Intracoronary nitroglycerin and sublingual nifedipine are also used. Intravenous dextran had been recommended (40) to suppress platelet aggregation, which results from endothelial desquamation, splitting of the fibrous cap, and exposure of the necrotic core of the atheroma to circulating blood elements (41–43). Certain untoward side effects (e.g., hypotension) have made dextran less attractive, and in two reports, no difference in platelet deposition (44) or incidence of abrupt reclosure of "dilated" segments after angioplasty in

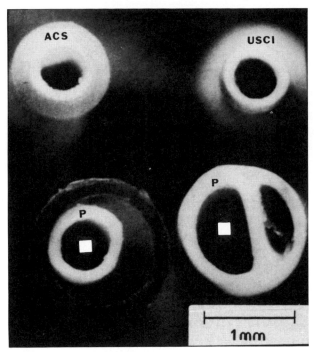

Figure 14. Comparison of ACS (left) and USCI (right) dilation balloon catheters. Cross section at level of distal tip (above) and at balloon (below), showing larger "pressure" (P) lumen (white square) of USCI versus ACS catheter. Thus the 0.356 mm (0.014 inch) steerable wire nearly fills ACS lumen and makes pressure recording more difficult than with the USCI. (Reprinted from Ref. 91.)

TABLE 5   Steerable Wires (USCI)

Configuration	Length, cm	Shaft OD, inches	Tip Configuration
Steerable	175	0.014*	J
Flexible and steerable	175	0.014	Straight
Flexible steerable	175	0.014	J
Very flexible steerable	175	0.014	Straight
Very flexible steerable	175	0.014	J
Exchange wire	300	0.014	Straight

*0.014 inches = 0.356 mm.

A

Figure 15. **A.** Steerable coronary balloon catheters (USCI) with steerable wire (arrows) emerging from tip.

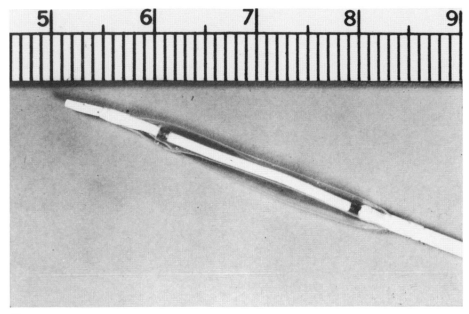

B

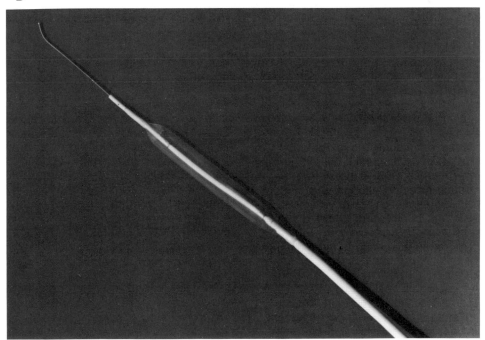

C

Figure 15 (Continued).   B. Close-up of steerable coronary balloon catheters (USCI). (Courtesy of Dr. D. Levin.) **C.** Close-up of low profile steerable balloon catheter with steerable wire.

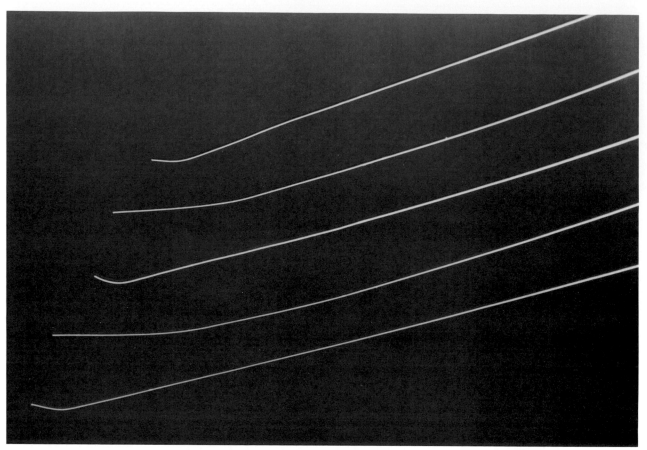

Figure 16. Steerable wires (USCI) for dilation balloon catheters.

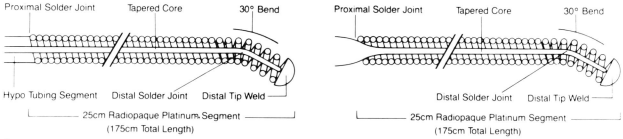

A

Figure 17. **A.** Schematic of 0.016 inch and 0.014 inch standard steerable J guidewire (USCI). **B.** Schematic showing "torque control" manipulation of steerable J guidewire. **C.** Schematic of 0.014 inch flexible steerable straight tip guidewire (USCI). **D.** Schematic of 0.014 inch flexible steerable J tip guidewire (USCI). **E.** Schematic showing advancement of flexible steerable guidewire in tortuous vessel. **F.**Schematic of 0.014 inch very flexible steerable straight tip guidewire (USCI). **G.** Schematic of 0.014 inch very flexible steerable J tip guidewire (USCI). **H.** Schematic showing advancement of very flexible steerable guidewire in tortuous small distal vessel branch. **I.**Schematic comparing new Teflon-coated standard, flexible, and very flexible steerable J tip wires with forming ribbon (USCI).

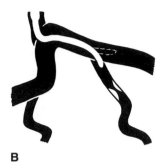

B

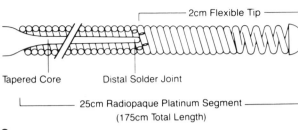

2cm Flexible Tip

Tapered Core   Distal Solder Joint

25cm Radiopaque Platinum Segment
(175cm Total Length)

C

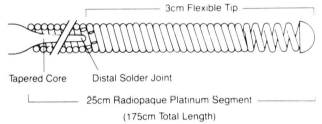

2cm Flexible Tip

Tapered Core   Distal Solder Joint

25cm Radiopaque Platinum Segment
(175cm Total Length)

D

E

3cm Flexible Tip

Tapered Core   Distal Solder Joint

25cm Radiopaque Platinum Segment
(175cm Total Length)

F

3cm Flexible Tip

Tapered Core   Distal Solder Joint

25cm Radiopaque Platinum Segment
(175cm Total Length)

G

H

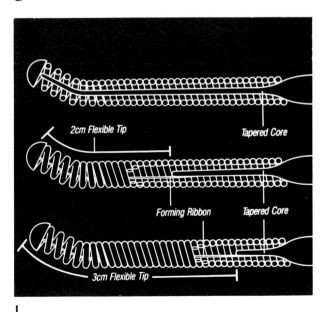

2cm Flexible Tip   Tapered Core

Forming Ribbon   Tapered Core

3cm Flexible Tip

I

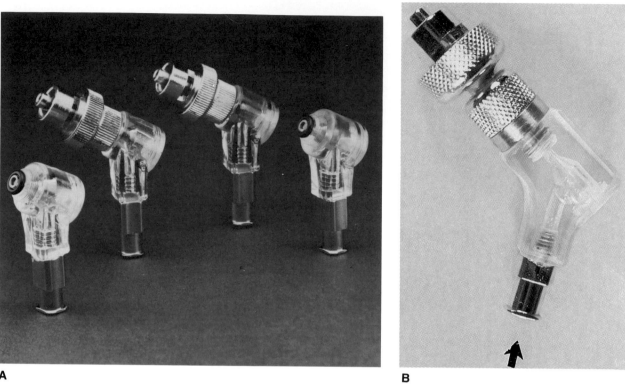

A   **B**

Figure 18. **A.** Standard (clear) and steerable Y connectors and replacement bodies (USCI). **B.** Close-up of Y connector (USCI) with back bleed assembly, also showing side port (arrow) for monitoring pressure, flushing, and contrast media injection.

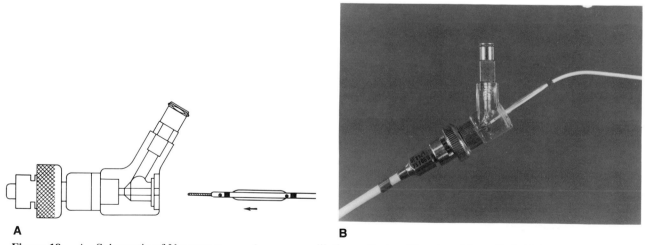

A   **B**

Figure 19. **A.** Schematic of Y connector and coronary dilation catheter. (Reprinted from Ref. 17.) **B.** Introduction of coronary dilation catheter into Y connector (attached to guiding catheter). (Reprinted from Refs. 15 and 17.)

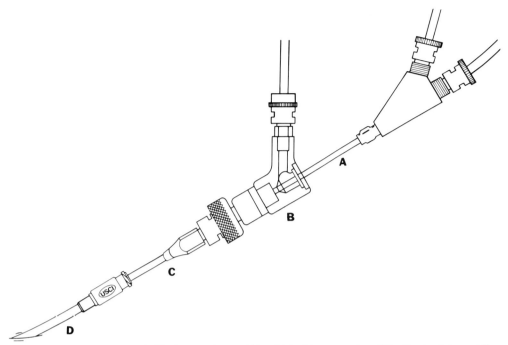

Figure 20. Schematic of dilation catheter (**A**) within Y connector (**B**) attached to guiding catheter (**C**), which is inside sheath (**D**). (Reprinted from Ref. 17.)

humans (45) was noted with or without dextran. Therefore, we have stopped using it.

After angioplasty, the platelet inhibitors (aspirin alone or aspirin and dipyridamole), nitrates, and calcium channel blockers are continued, especially in patients with associated coronary spasm (46). The platelet inhibitors are generally used for 6 months and the nitrates and calcium channel blockers for 1–3 months after coronary angioplasty (Table 6).

PREDILATION ARTERIOGRAM

The importance of a high-quality preangioplasty coronary arteriogram cannot be overemphasized. The specific coronary arteries and their branches must be examined in detail, using specific projections and obliquities to isolate lesions, as well as branches that may be adjacent to or involved with the stenotic lesion(s). Although the coronary anatomy and pathologic characteristics vary considerably from patient to patient, certain views have been noted to be most helpful in exposing lesions in the target coronary arteries. The arteriogram projections for the left anterior descending artery should always include the right anterior oblique, left anterior oblique,

and if possible a lateral projection. It may be advisable to add 15° of caudal or cranial angulation to the 15–30° right anterior oblique projection to visualize the foreshortened proximal lesions and overlapped side branches that are not uncommon in standard right anterior oblique views. In the left anterior oblique projection, the left anterior descending coronary artery is best evaluated from 45 to 60° with an additional cranial angulation between 15 and 30°. The 90° left lateral projection is particularly helpful in evaluating lesions in the midzone of the left anterior descending artery. With proximal stenoses, 10–15° caudal or cranial angulation in the lateral position may be useful. Other angiographic views, including so-called splenic and spider views, may be desirable in certain cases in isolating the lesions under investigation.

TABLE 6 Medications for Angioplasty

Before	During	After
Aspirin	Nifedipine	Aspirin
Dipyridamole	Nitrates	Dipyridamole
Nifedipine	Heparin	Nifedipine
Nitrates		Nitrates

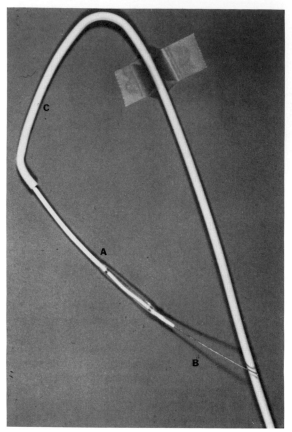

Figure 21. Steerable dilation catheter (**A**) and wire (**B**) emerging from left Judkins guiding catheter (**C**). (Courtesy of Dr. D. Levin.)

The left circumflex coronary artery is well visualized in the right anterior oblique projection between 15 and 30° and is often better visualized with a caudal angulation of 15°. The left anterior oblique projection and left lateral projection also are necessary to expose lesions in the circumflex artery and the relation of these lesions to major coronary branches. The right coronary artery is best visualized in the left anterior oblique views of 45–60° and is also well seen in the right anterior oblique view of 30–45° and left lateral projections. Cranial or caudal angulations can be added to these standard right coronary projections, particularly for distal lesion evaluation.

The basic intent of coronary arteriography prior to transluminal angioplasty is to isolate the specific lesion, its relation to the major branches, and unobstructed views of the coronary anatomy with regard to aortic origin and distal distribution. The angle of origin of the left anterior descending artery from the left main stem may be better appreciated in a left anterior oblique cranial projection. In contrast, the origin of the left circumflex coronary artery may be best appreciated in the right anterior oblique projection, which gives full display of the length of the left main stem and the angulation of the left circumflex takeoff from the left main stem (Fig. 29 A,B).

The use of a right heart catheter with pacemaker lead is advisable for the safety of the patient during angioplasty to prevent bradycardia and asystole and as a reference marker for the location of coronary lesions. With a right heart catheter, for example, the Myler multipurpose pacemaker catheter with the

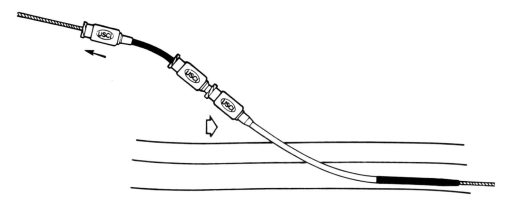

Figure 22. Schematic of three-piece introducer insertion system over safety wire guide. Note approximately 45° introduction angle (open arrow). (Reprinted from Refs. 15 and 17.)

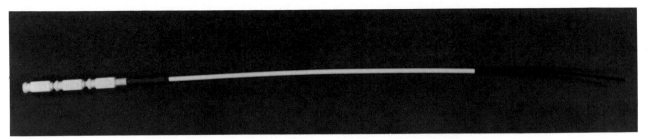

Figure 23. Three-piece introducer system (without side arm.) (Reprinted from Ref. 17.)

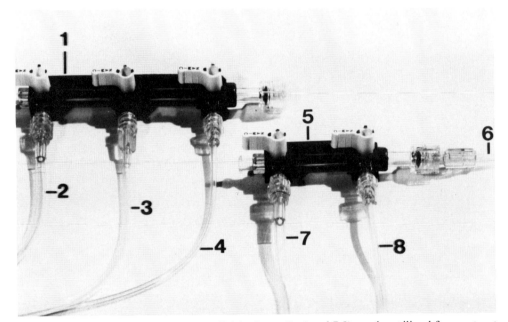

Figure 24. Two- and three-port manifolds. Ports (2–4 and 7,8) can be utilized for contrast and saline injection and pressure measurements from guiding (two-port manifold and dilation (three-port manifold) catheters.

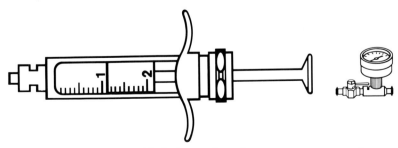

Figure 25. Schematic of inflation syringe (insert: pressure gauge).

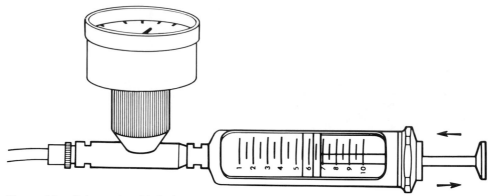

Figure 26.   Schematic of inflation syringe and pressure gauge.

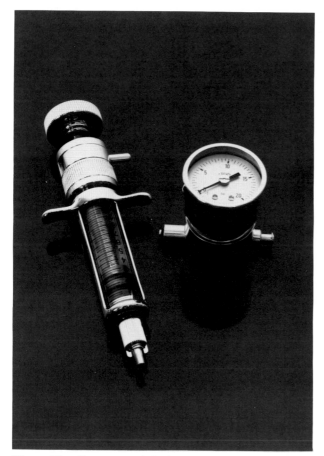

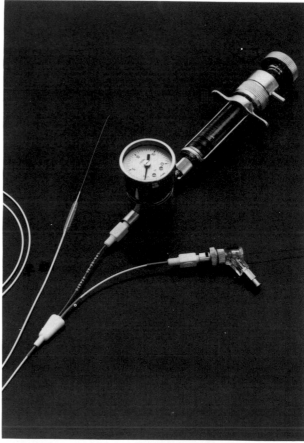

Figure 27.   Screw type inflation device with syringe and a new type of pressure gauge capable of 20-atmospheres of pressure (USCI).

Figure 28.   Screw type inflation device (arrow) and pressure gauge attached to proximal end of steerable coronary dilation catheter.

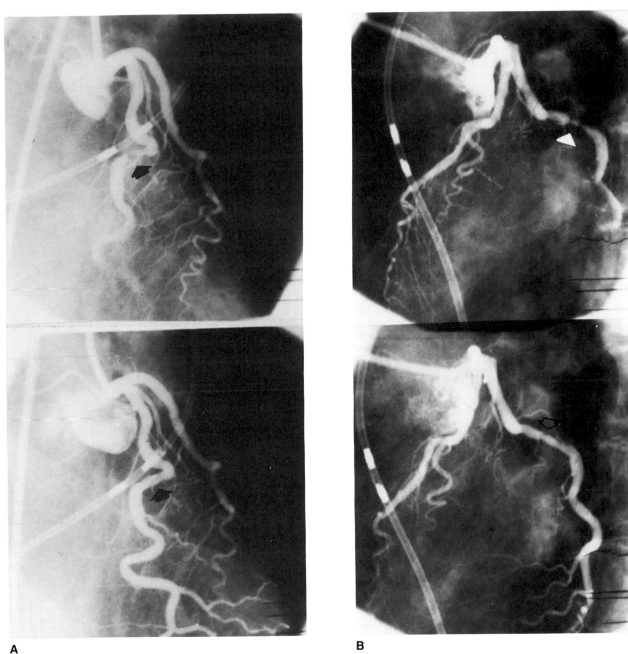

A   **B**

Figure 29. **A.** Tortuous left circumflex (LCFX) coronary artery in RAO-15° position with 15° caudal angulation exposing lesion (arrow) near right heart catheter, before (above) and after (below) angioplasty. **B.** LCFX in LAO-60° projection of same patient, showing lesion (arrow) before (above) and after (below) angioplasty. Note: Length of left main stem, angulation of LCFX from left main, and curve of LCFX proximal to stenosis are not well appreciated in this view (compared to RAO projection).

Swan Ganz multipurpose pacemaker catheter, the proximal lesions in the left anterior descending coronary artery can often be juxtaposed to the right heart catheter in the right anterior oblique projection (Fig. 30). With midzonal left anterior descending coronary lesions, the left lateral projection places the right heart catheter in proximity to the lesion (Fig. 31). Stenosis in the left circumflex coronary artery can be placed in a spatial juxtaposition to the right heart catheter in the right anterior oblique projection (Fig. 29A); lesions in the midzone of the right coro-

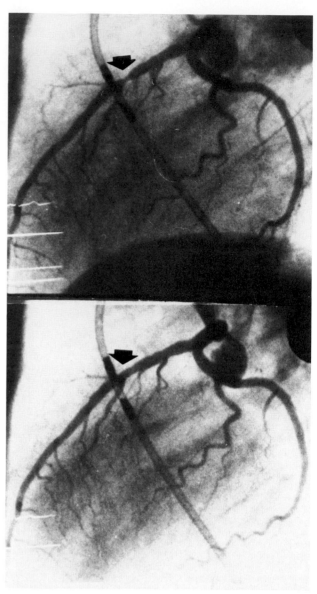

Figure 31. LAD in left lateral position. Note the juxtaposition of lesion (arrow) to multipupose right heart catheter before (above) and after (below) angioplasty. (Reprinted from Ref. 17.)

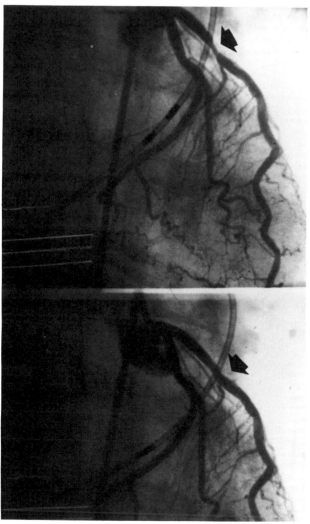

Figure 30. Left anterior descending (LAD) coronary artery in RAO-30° projection. Note the juxtaposition of lesion (arrow) and multipurpose pacemaker right heart catheter (Myler;USCI) aiding in spatial coordination of lesion before (above) and after (below) angioplasty. (Reprinted from Ref 17.)

nary artery can also be localized near the right heart catheter in the right anterior oblique (Fig. 32), and proximal lesions in the left lateral (Fig. 33) projections. These recommendations may help the physician to position the dilation balloon across the lesion during angioplasty.

High-resolution fluoroscopy and cine arteriography are absolute necessities for the performance

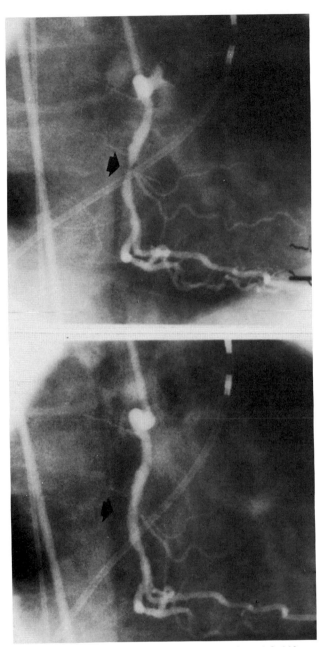

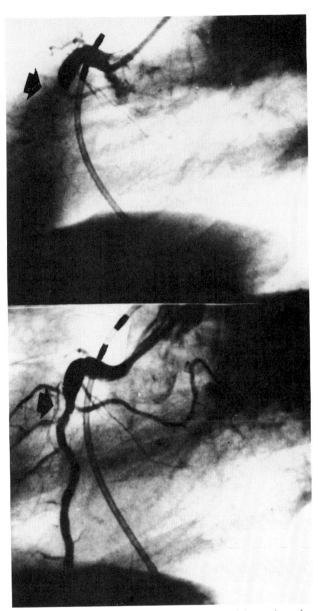

Figure 33. RCA in left lateral position with total occlusion (arrow) just below "crossing" of right heart catheter in this plane, before (above) and after (below) angioplasty.

Figure 32. Right coronary artery (RCA) in RAO-30° position with lesion (arrow) overlapping right heart catheter before (above) and after (below) angioplasty.

of coronary angioplasty. Bimodal imaging systems and biplane cinefluoroscopy may be of great help for laboratories performing coronary angioplasty. A high-resolution videotape or videodisc system with "freeze-frame" capability is recommended for coronary angioplasty procedures.

The patient who has been prepared with salicylates, nitrates, and calcium channel blockers before admission to the cardiac catheterization laboratory receives nifedipine, either orally or sublingually, at the laboratory. If an intravenous line has not been put into place in the ward, it should be inserted in the cardiac catheterization laboratory before proceeding with the control arteriogram that precedes the angioplasty procedure.

TRANSFEMORAL PTCA TECHNIQUE

Selection and Use of Guiding Catheters

Guiding catheter selection is based on careful analysis of the preangioplasty coronary arteriogram (Table 7). Particular reference should be made to the size of the aortic root, the origin of the target coronary artery from the aorta, and angulations of the specific branch to be entered by the dilation catheter. Large aortic roots may require a Judkins guiding catheter with a number 5 or 6 curve or an Amplatz II curve when entering the left coronary artery. A superior takeoff of the left main stem coronary artery may require the use of a Judkins 3.5 curve for coaxial entrance and a stable position in the left coronary orifice.

Selection of a guiding catheter for the right coronary artery may also be determined by the size of the aortic root and the proximal segment of the right coronary artery. If the proximal right coronary has a "shepherd's crook" configuration (Fig. 34), it may require a right Amplatz I or II curve or even a left Amplatz I or II curve to achieve coaxial entrance and a stable position for introducing the dilation catheter into the vertical portion of the right coronary artery. When attempting severely stenotic lesions, it may be necessary to have a stable "backup" position of the guiding catheter to cross the stenotic lesion with the balloon catheter. This may require a number 9 French or a high-flow 8F Judkins guiding catheter, which gives more support than a standard number 8 French guiding catheter. With lesions that are very severe, it is conceptually important to place the guiding catheter in line with the segment of artery that is stenotic, therefore achieving maximum coaxial direct force when attempting to cross the lesion with a dilation catheter. In sharply angulated left circumflex coronary arteries (as in "shepherd's crook" right coronary arteries) the Amplatz catheters have proved to be quite helpful in this regard.

Saphenous vein grafts are placed in a variety of ways by different cardiovascular surgeons. It, therefore, is necessary to have an excellent diagnostic coronary arteriogram prior to angioplasty to evaluate

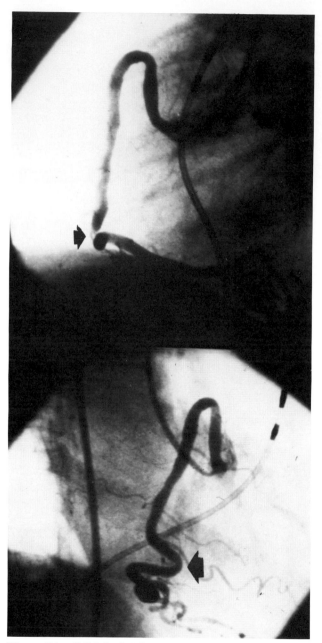

Figure 34. RCA in LAO-60° (above) and RAO-30° (below) projections, showing proximal shepherd's crook configuration and distal lesion (arrow).

the manner in which the saphenous vein graft originates from the ascending aorta. The saphenous vein graft to the right coronary artery is best entered by a straight or multipurpose guiding catheter (Fig. 35). The saphenous vein graft to the left anterior descending coronary artery is usually entered with the

TABLE 7   Selection of a Guiding Catheter

Fit of diagnostic catheter

Anatomy of proximal vessel
   Length of left main stem
   Angulation of target artery
   Tortuosity of vessel

Anatomy of aortic root and sinuses of Valsalva

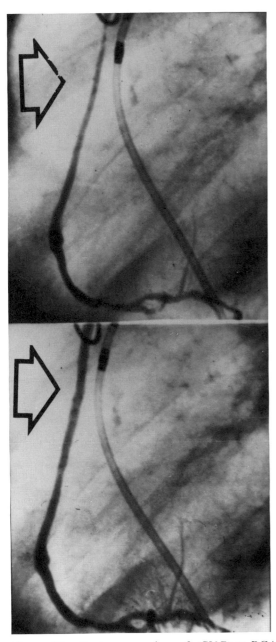

Figure 35. Saphenous vein graft (SVG) to RCA in LAO-60° position, showing (relatively) vertical position of graft vis-à-vis aorta. Graft cannulation achieved with straight (unformed) guiding catheter. Long proximal segmental lesion demonstrated (arrow) before (above) and after (below) angioplasty.

Judkins 4 curve right coronary guiding catheter, but occasionally a Judkins 5 curve, an Amplatz, or even a multipurpose or El Gamal catheter may be necessary (Fig. 36). The saphenous vein graft to the left circumflex coronary artery is often best cannulated by either a Judkins 4 curve right guiding catheter, an Amplatz, a multipurpose, an El Gamal, or a specially modified catheter to enter the origin of the saphenous vein graft, which may be in a superior and sometimes eccentric position (Fig. 37).

Since the guiding catheter is nontapered, it is introduced through a specially designed three-piece introducer sheath system. The placement of this introducer system demands great care. An introduction angle of less than 45° allows safe entrance to the femoral artery (Fig. 22). More vertical entry angles may cause kinking of the sheath, compromising smooth introduction of the guiding catheter through it. The introduction of this three-part system is made over a 0.035 or 0.038 inch safety guidewire, using the standard Seldinger technique (47). Should marked tortuosity or obstruction be encountered in the ileofemoral system, then alternative entry sites, perhaps the contralateral femoral artery or the brachial approach, should be utilized (Table 8).

When the sheath system is in position, the two Teflon introducer catheters that are part of the introducer system are removed from the single outer sheath. Then with the sheath in place, its side arm is flushed and attached to a high-pressure continuous infusion system. The sheath has a diaphragm at the proximal end to prevent back bleeding. It is often preferable to leave the safety guidewire in place during these manipulations to retain the position in the aorta (particularly in cases with tortuous ileofemoral systems); then the guiding catheter can readily be advanced over the safety guidewire into the ascending aorta. Once the guiding catheter is in position in the ascending aorta, the safety wire guide can be removed. The guidewire is a 0.038 or 0.063 inch Teflon-coated guidewire with a floppy or J-shaped tip. The 0.063 inch guidewire with J curve has been recommended for introduction of the guiding catheter into the ascending aorta. However, we have had excellent results with the smaller (0.038 inch) guidewires and have used these without complication since 1980.

The selection of the guiding catheters depends on a number of factors, including the width of the ascending aorta, the location and spatial position of the vessel to be cannulated, and the anatomic configuration of the vessel itself. It is vitally important to understand that the guiding catheter must cannulate the orifice of the target coronary artery coaxially to allow easy access of the dilation catheter into the distal segment of this vessel. An eccentri-

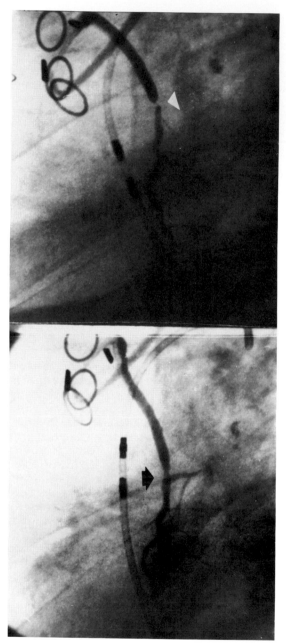

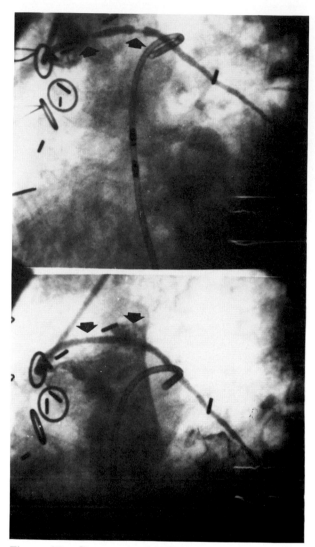

Figure 37. Seven-year-old SVG to LCFX in LAO-60° position, showing graft cannulation with Judkins right guiding catheter. Note two arrows denoting lesions before (above) and after (below) angioplasty.

Figure 36. SVG to LAD in LAO-60° position, showing cannulation of graft with multipurpose (King) guiding catheter. Lesion (arrow) at insertion site of SVG and LAD, before (above) and after (below) angioplasty.

cally placed diagnostic coronary catheter can deliver an adequate infusion volume of contrast material to achieve satisfactory angiography. However, a guiding catheter must be properly seated in a spatial relation coaxial to the target coronary artery to allow direct access of the dilation catheter and wire into the target vessel via the guiding catheter and to afford necessary backup support for the guiding catheter to deliver the dilation catheter across a severe stenosis. The importance of guiding catheter selection and positioning cannot be overemphasized

TABLE 8 Femoral versus Brachial Approach

Access to ascending aorta

Sinuses of Valsalva and aortic root anatomy

Length of left main

Angulation of LAD, LCFX, and RCA*

*LAD = left anterior descending; LCFX = left circumflex; RCA = right coronary artery.

since this relation of the guiding catheter and the native coronary artery is essential to any successful coronary angioplasty. Experience will help select the appropriate guiding catheter in an individual case; one can use the initial diagnostic coronary angiogram as a guide. For example, if a Judkins coronary catheter has been utilized during the diagnostic coronary arteriogram, the knowledge of the size and form utilized and its ease of cannulation of the target coronary artery orifice may give a preliminary clue to the selection of a guiding catheter.

The guiding catheter differs from the diagnostic arteriographic catheter in several important ways: (1) the guiding catheter is not tapered, (2) the tip of the guiding catheter is shorter, (3) the guiding catheter is composed of three different layers, and (4) the guiding catheter is thin-walled. Therefore, the guiding catheter must be handled with some care and dexterity because it will readily "twist" and become useless. Although there are a variety of shapes and curves available for the femoral guiding catheter system (Table 2), not all the possible variations of coronary artery anatomy can be met by the available preformed guiding catheters. For these reasons, a heat gun that can deliver more than 650°F (the melting point of the Teflon innerlining of the guiding catheter) can be employed for adapting the guiding catheter to unusual coronary anatomy (Fig. 8).

Once the appropriate guiding catheter position is achieved in the coronary ostium in a stable and coaxial fashion and confirmed in several fluoroscopic projections, the preangioplasty diagnostic arteriogram can be obtained, recording on 35 mm cine film as well as on a videotape or a videodisc system, the latter to achieve a freeze-frame analysis of the specific views and to record the angioplasty procedure.

The guiding catheter is then attached to a two- (or three-) port manifold via the standard Y connector (clear) with a pressure and contrast line available (Figs. 18–20, 24). The tip of the guiding catheter can be turned anteriorly or posteriorly to direct the dilation catheter and guidewire as necessary. For example, a left Judkins style guiding catheter tip can be positioned more anteriorly (once the tip is en-gaged in the left coronary ostium) by counterclockwise torque. This maneuver will position the "elbow" of the catheter posteriorly in the ascending aorta and, by the fulcrum action on the catheter tip in the left main stem, turn the tip anteriorly in the direction of the anterior descending artery. Contrarily, clockwise rotation of the guiding catheter will turn the elbow anteriorly so the tip is "fulcrumed" posteriorly in the direction of the circumflex branch. Often, to direct the dilation catheter into the left anterior descending artery (and to avoid the intermediate branch), in addition to counterclockwise rotation, slight advancement of the guiding catheter will permit anterior direction. To engage the circumflex branch, in addition to clockwise torque, slight withdrawal of the guiding catheter may achieve the desired cannulation of the posterior branch.

Selection and Use of Balloon Catheters

Whereas the guiding catheter selection is determined to a great extent by the coronary anatomy, dilation catheter selection is determined in great measure by the pathology of the coronary artery lesion (Table 9). The more severe the stenosis, in general, the more likely will be the choice of a "low profile" balloon catheter (Figs. 12 and 15). The selection of balloon size is determined by the diameter of the coronary artery in the region of the stenosis. The result of angioplasty should be a near-normal "patency," vis-à-vis the adjacent coronary artery diameter. There is a tendency to "oversize" balloons slightly for saphenous vein graft stenoses, in vessels with restenosis following an initially successful angioplasty and perhaps in patients with associated coronary spasm. In contrast, "undersizing" balloons is advised in very eccentric lesions on bends or curves in the artery, in females, elderly patients, and those with associated metabolic abnormalities (e.g., diabetes, renal insufficiency) to avoid the somewhat increased risk of dissection in these subsets (Table 10).

The latest USCI steerable wire shaft is 0.014 inch (0.356 mm) outer diameter and will probably replace the 0.016 inch steerable wire presently in use. The

TABLE 9 Factors in Selection of a Dilation Catheter

Profile of catheter

Shaft quality (stiffness)

Distal lumen (pressure)

Balloon compliance

Balloon size (inflated)

TABLE 10   Dilation Balloon Sizing

Equal to nondiseased segment of vessel to be dilated

"Oversize" balloons: saphenous vein grafts, recurrences, spasm

"Undersize" balloons: lesions that are eccentric on "bends" in arteries; females; associated metabolic abnormalities (e.g., diabetes, renal insufficiency); elderly

Compliance; reproducible size of balloons at high pressures (PVC versus PE*)

*PVC = polyvinylchloride; PE = polyethylene. (Reprinted from Refs. 89 and 90.)

selection of the wire, whether a "steerable" or a "flexible" tip, is an important consideration (Table 11). The steerable tip has a tapered mandrel distally, which allows the tip to be shaped and has an ability to retain memory for this curve (Figs. 16 and 17). The major advantage of this type of steerable wire is that it allows directional torque control through the length of the target coronary artery proximal and distal to the stenosis. In contrast, the flexible or very flexible (floppy) tip wire has less torque control than the steerable design. An example of coronary anatomic "preference" for the steerable wire would be a very tortuous vessel with a severe eccentric stenosis that required more control of the direction of the wire (Figs. 38–40). In contrast, the floppy-tipped wire may be preferable in a long, irregular severe stenosis with small branches within the length of the stenosis in which the flexible wire can be "floated" along the bloodstream to reach a position distal to the stenosis in the coronary artery (Figs. 41 and 42).

Once the appropriate steerable wire and dilation balloon catheter have been prepared with the balloon filled with a contrast/saline mixture, the central lumen is flushed with contrast material and attached to a three- (or four-) port manifold via the steerable Y connector (yellow). This connector permits the manipulation of the steerable wire and allows pressure measurement and contrast injection through the dilation catheter. The dilation catheter, whose inflation and deflation characteristics have been

TABLE 11   Factors in Guidewire Selection

Steerability
Flexibility
Formability (tip memory)
Trackability
Visibility (fluoroscopic)

checked again, is then introduced through a standard Y connector (clear) that will be attached to the guiding catheter. With the steerable wire advanced several centimeters beyond the tip of the dilation catheter (and with back bleeding observed through the guiding catheter), the whole dilation catheter system is introduced into the guiding catheter slowly and steadily, permitting back bleeding to continue. Then with an antegrade flush of saline or contrast through the standard Y connector, it is secured to the Lurlock hub of the guiding catheter to seal the system. As the dilation catheter is further advanced slowly through the guiding catheter, attention should be paid to the possibility of small bubbles within the standard Y connector (Figs. 19–21).

If bubbles should appear, the system should be disconnected and antegrade flush through the standard Y connector and back bleed repeated through the guiding catheter to achieve a bubble-free seal. As the dilation catheter led by the steerable wire advances through the guiding catheter toward its distal tip, which has been engaged in the coronary ostium, slight clockwise or counterclockwise rotation of the guiding catheter may permit a coaxial introduction of the steerable wire into the appropriate target branch with greater ease. The selective cannulation by the steerable wire therefore is facilitated by manipulating the guiding catheter, as well as by preforming or shaping the distal tip of the steerable wire. The steerable wire can then be advanced into the target artery, traverse the stenosis, and achieve a position in the distal coronary artery well ahead of the dilation catheter. This distal position of the steerable wire permits a more stable introduction and more facile trackability of the dilation catheter over the steerable wire. This steerable wire is indeed a "safety wire" since it achieves coaxial luminal cannulation of the target vessel and the stenosis and permits correct directional movement of the dilation catheter over it. As the dilation catheter tracks over the steerable wire, its position should be checked fluoroscopically with selective contrast injections through both the dilation and guiding catheters. In addition, continuous pressure monitoring through the two catheters will indicate the entering and crossing of the stenosis by the dilation catheter. If the stenosis is very severe and the atheromatous lesion is "hard," then it may be necessary transiently to advance the guiding catheter into a more stable position in the target artery to permit the back-up power to "push" the balloon catheter across the lesion. When the stenosis has been crossed by the balloon segment of the dilation catheter, the guiding catheter then can be withdrawn slightly to relieve the obstruction of the artery; then the pressure gradient across the stenosis can be ob-

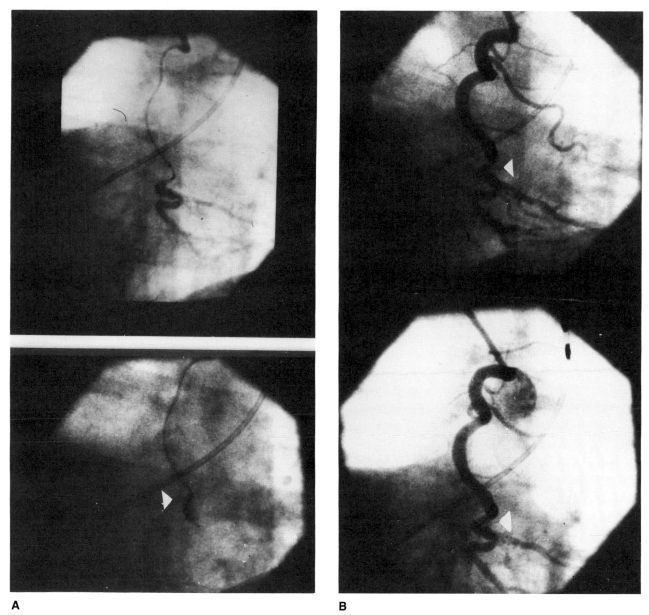

A   **B**

Figure 38.   A. Dilation catheter advancing into tortuous RCA in RAO-30° projection. Contrast material injected distally via catheter (above) and balloon (arrow) inflated on lesion (below). **B.** Tortuous RCA in RAO-30° projection, showing lesion (arrow) before (above) and after (below) angioplasty.

tained. Contrast injection via the guiding catheter can aid in evaluating the proportional relation between the location of the lesion and the balloon length. It is often helpful to have a right heart pacemaker catheter in place in the pulmonary artery for marking the position of the coronary stenosis (in at least one angiographic plane) in juxtaposition to this catheter for spatial reference (Figs. 29–33).

Only when the balloon is properly positioned across the stenosis and confirmed fluoroscopically and by pressure measurements is it time to inflate the balloon by an inflation device. Balloon inflation, always performed under fluoroscopic control, is done slowly. The end point is full balloon inflation ("sausage-shaped") with no indentation on the normal inflated silhouette. We generally hold this initial balloon in-

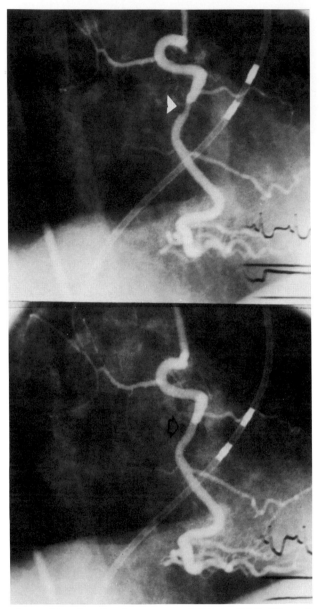

Figure 39. Proximally tortuous RCA in RAO-30° position with lesion (arrow) just distal to S-shaped curve before (above) and after (below) angioplasty.

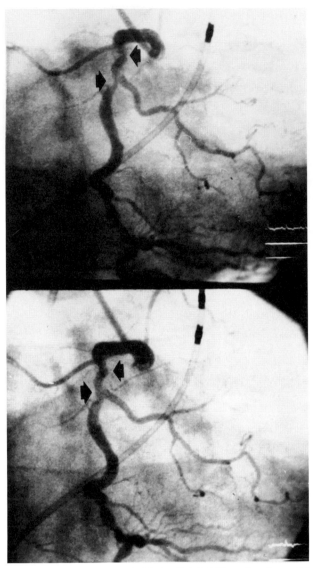

Figure 40. Tortuous RCA in RAO-30° projection with two lesions (arrows), the distal one involving large acute branch before (above) and after (below) angioplasty.

flation for 10–20 s to achieve appropriate "cracking" of the atheromatous plaque. The balloon is then deflated rapidly, and a recording of the postdilation pressure gradient is obtained. With subsequent inflations, atmospheric pressure necessary to achieve full balloon inflation is also recorded. If, with the initial inflation, for example, the balloon reaches full inflation at 8 atmospheres, subsequent full balloon

inflations may be achieved at lower pressures. Conceptually, all subsequent dilations are intended to "re-form or remodel" the arterial lumen by compacting the atheroma and perhaps expressing some liquid material from the plaque. These latter inflations require time, and therefore longer duration (40–90 s) of inflations is necessary. Preliminary evidence from our laboratory and others (48) suggests that these prolonged inflations not only yield a more satisfactory initial postangiographic appearance, but also may play a role in decreasing the previously

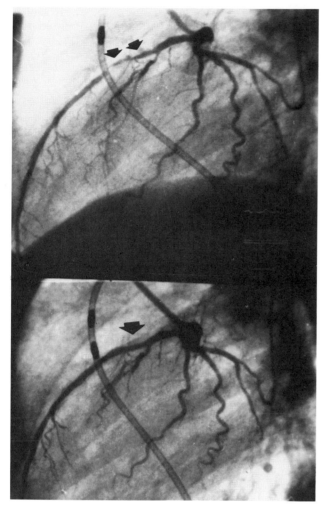

Figure 41. LAD in left lateral projection, showing long irregular midzonal stenosis (arrows) before (above) and after (below) angioplasty in diabetic woman.

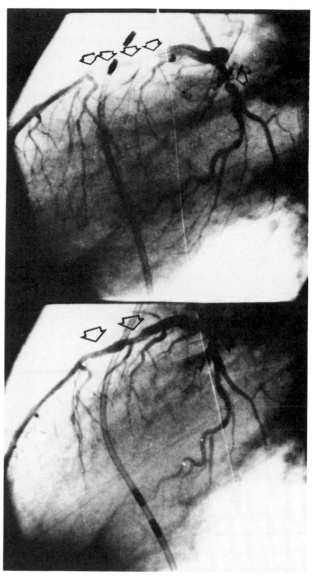

Figure 42. Left lateral projection of left coronary artery, showing very long irregular subtotal stenoses (arrows) of LAD and discrete lesion (arrow) of LCFX before (above) and after (below) angioplasty.

reported incidence of restenosis (49,50), which also may be affected by antiplatelet therapy (51).

The end-points for successful angioplasty are the following:

1. Full balloon inflation ("sausage-shaped")
2. Significant angiographic improvement in luminal diameter (which can be assessed by intermittent guiding catheter contrast injection during the procedure)
3. Marked decrease in the pressure gradient across the lesion that has been dilated
4. Improvement in coronary flow

Full balloon inflation conceptually indicates the cracking of the plaque. At the present time, the USCI balloons are tested to 10 atmospheres, but somewhat higher pressures can be achieved, particularly with the smaller balloon catheters if intraballoon pressure is increased very slowly. There are certainly lesions that will not yield to full balloon inflation even at 14 atmospheres of pressure because of "hardness" of the atheroma (due to a larger percentage of fibrous and calcific material within the plaque). In the near future, balloon material will be of such strength that pressures of 20 or even 30 atmospheres can be readily achieved without the balloon's failing. Prototypes of these balloons are being tested. The current USCI balloons made of PVC have moderate compliance at higher pressures, i.e., a noted modest increase in the inflated outer diameter of the balloon at pressures of 6 atmospheres and above. (The ACS balloons are polyethylene and have better compliance characteristics.) Incidentally, when these balloons fail, they "tear" and the pressure falls to zero. When this occurs, contrast material from the balloon, expressed distally in the coronary artery, can be observed fluoroscopically. In case of a ruptured balloon, the dilation catheter, wire, and guiding catheter should be removed in toto from the coronary artery and withdrawn into the descending aorta. It is not advisable to withdraw a "torn balloon" into the guiding catheter when it is still in the coronary ostium because part of the balloon material may possibly embolize into the coronary artery.

The second indicator for a successful completion of angioplasty relates to angiographic appearance following dilation. These contrast injections are made either through the guiding catheter or the guiding and dilation catheters, the latter having been withdrawn proximal to the stenosis while the steerable wire remained in the distal artery segment. On-line fluoroscopic and videotape analysis may permit a fairly accurate qualitative assessment of luminal diameter. (More quantitative angiographic evaluation, of course, will be obtained by analyzing the post-angioplasty cinearteriogram. If the angiographic appearance of the stenosis is not satisfactory, the dilation catheter can be advanced over the steerable wire and across the stenosis and repeat the dilation procedure.

If the balloon has been "undersized" in relation to the adjacent coronary artery diameter, it may then be desirable to use a larger balloon catheter. In this situation, the steerable wire can be withdrawn from the dilation catheter positioned distal to the stenosis, and a 0.014 inch (0.356 mm) diameter, 300 cm long exchange wire can be introduced through the dilation catheter into the distal coronary artery. With the guiding catheter in a stable position in the coronary ostium, the dilation catheter can then be withdrawn slowly over the exchange wire under constant fluoroscopic monitoring. Once the initial dilation catheter is removed, a previously prepared larger dilation catheter can then be advanced over the exchange wire with appropriate attachments made with both Y connectors. This exchange of dilation catheters must be slow and careful, with continuous fluoroscopic evaluation of the distal position of the exchange wire and of the guiding catheter positions. Once the larger dilation catheter has entered the target artery with continuous pressure recording and intermittent contrast injection, the larger balloon can be positioned across the lesion, and a repeat angioplasty can be performed.

The steerable wire not only permits an exchange of balloon catheters but also allows repositioning of a dilation catheter throughout the angioplasty procedure. It is during the 30-minute period following angioplasty that the highest incidence of abrupt reclosure of the "dilated segment" occurs, resulting from the development of a minidissection (invasion of the media), the creation of an intimal flap, or possibly intense localized coronary spasm.

After angioplasty, one should not abruptly remove the steerable wire or dilation and guiding catheters, particularly if preliminary angiographic evidence suggests suboptimal dilation or a significant, persistent pressure gradient. If abrupt reclosure occurs during this half-hour following angioplasty (with resultant chest pain and ECG ST segment abnormalities), then the dilation catheter can easily and safely be tracked again over the steerable wire, which has remained far distal to the stenosis. If, on the other hand, the steerable wire has been removed and abrupt reclosure occurs (with resultant clinical manifestations of severe angina and ECG changes of current injury that cannot be reversed by intracoronary nitroglycerin or sublingual nifedipine), recrossing the lesion may be more difficult and hazardous. Approximately half of the abrupt reclosures that occur during the first 30 minutes following angioplasty can be reopened and the need for emergency coronary artery bypass surgery averted.

The third major determinant of coronary angioplasty is the measurement of pressure gradients. Abolishment of a large pressure gradient across the stenosis suggests a successful hemodynamic result and may indicate whether or not the situation following angioplasty is stable, i.e., a changing pressure gradient. With regard to this latter point, if the pressure gradient continues to vary, this may indicate an anatomically or physiologically unstable situation and may necessitate leaving the steerable wire in a distal position and waiting a longer period of time before removing the dilation and guiding catheters. There is some evidence that the lower the pressure gradient following angioplasty, the less likely is a recurrence (52,53). That is, a significant persistent pressure gradient may be an early indicator of a tendency to later recurrence and, thus, might suggest that one should try to achieve as low a residual gradient as possible (less than 15 mmHg) at the termination of an angioplasty.

The measurement of pressure gradient may in itself be an indicator as to whether or not a moderate stenosis should be treated by the angioplasty technique. A lesion of 50 percent or less may not be associated with a pressure gradient and may show no functional impairment in coronary flow dynamics (Pitt, personal communication; 54). If invasive and noninvasive testing provides no objective evidence that the lesion is responsible for myocardial ischemia, it may be advisable to leave this lesion alone, particularly in patients undergoing multivessel angioplasty. The reason for this recommendation is that abrupt reclosure may follow angioplasty even

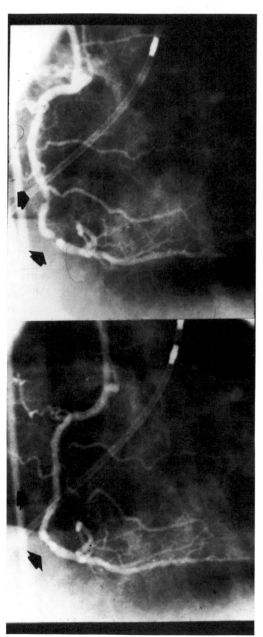

Figure 43. RCA in RAO-30° projection, revealing tandem lesions (arrows) before (above) and after (below) angioplasty.

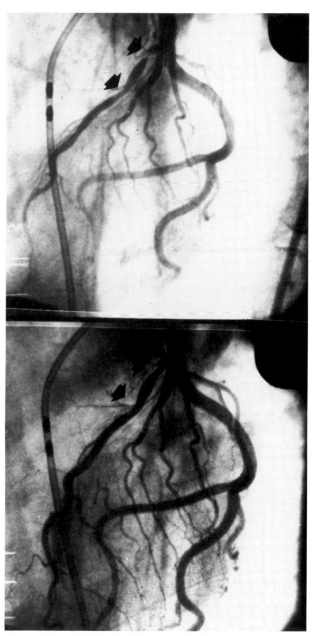

Figure 44. LAD in LAO-60° projection, showing proximal and midzonal lesions (arrows) before (above) and after (below) angioplasty.

in lesions of moderate severity; in addition, late restenosis may occur to a more severe degree than in the initial untreated lesion (55). Thus, the patient who has angioplasty of a moderate stenosis that is hemodynamically, functionally, and clinically insignificant may be exposed to the risk of abrupt reclosure and the possibility of restenosis at a higher degree of severity. These moderate lesions can be reevaluated readily by noninvasive and invasive studies at periodic intervals, and, if they progress spontaneously and are responsible for myocardial ischemia, angioplasty can be applied to them at this later occasion.

With tandem lesions, we generally dilate a proximal lesion before the distal one, allowing more accurate assessment of pressure gradients at each stenosis (Figs. 43 and 44). With regard to multivessel angioplasty, the most hemodynamically significant vessel stenosis is performed initially, with less severe vessel lesions attempted subsequently, assuming a satisfactory result by angiographic and pressure gradient determination in the vessel just completed. Since coronary flow is directly proportional to the radius of the coronary artery stenosis and to the pressure difference across the stenosis (56), the hemodynamic significance of an atherosclerotic lesion can be better appreciated by measurement of the pressure gradient, as well as the luminal diameter of the lesion (Table 12).

Finally, the measurement of coronary flow is in an exciting state of evolution. By the use of hyperemic flow studies using digital subtraction angiography, current investigations indicate that vasodilator coronary flow reserve does improve after successful coronary angioplasty (54). Coronary flow (57,58) (Fig. 45) and left ventricular contractility (59,60) (Figs. 46 and 47) have been measured directly, as well as by nuclear scintigraphic techniques (61,62), in the days and months following angioplasty to quantify improved coronary flow.

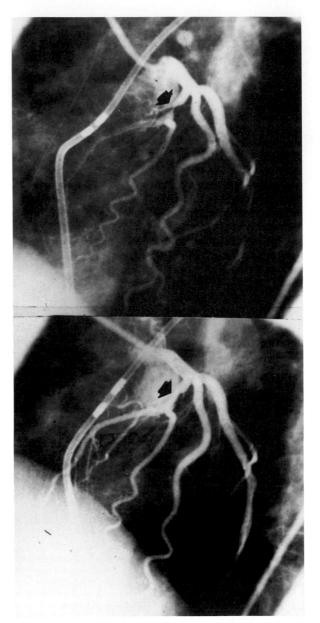

Figure 45. LAD in LAO-60° cranial-25° projection, showing lesion (solid arrow) before (above) and after (below) angioplasty. In these "timed" frames, note (qualitative) improvement in appearance of distal LAD flow (open arrow) after angioplasty.

POSTDILATION ARTERIOGRAM

It is only when the parameters of angioplasty success (increased luminal diameter, reduction of pressure gradient, and improved coronary flow) are satisfied (and a suitable waiting period following the

TABLE 12   Coronary Angioplasty: Importance of Pressure Measurements

Coronary flow directly proportional to pressure gradient (Poiseuille's law)

Pressure gradient related to functional significance (coronary flow reserve)

Pressure gradient indicates hemodynamic significance and whether lesion should be treated (PTCA)

Persistent significant pressure gradient following PTCA may indicate
   1. Unstable lesion
   2. Need for further angioplasty
   3. Likelihood of recurrence

then even the same point in the cardiographic cycle, preferably on the Q wave (end diastole), can be compared for pre- and postangioplasty cine arteriogram frames.

In most postangioplasty arteriograms, intraluminal linear filling defects, "haziness," and, occasionally, minute extraluminal contrast are noted. In the early days of coronary angioplasty these were considered to be complications of the procedure. However, it soon became apparent that these angiographic findings were to be expected and in no way were necessarily complications of angioplasty but, rather, along with the previously noted indicators of success, were angiographic features of the

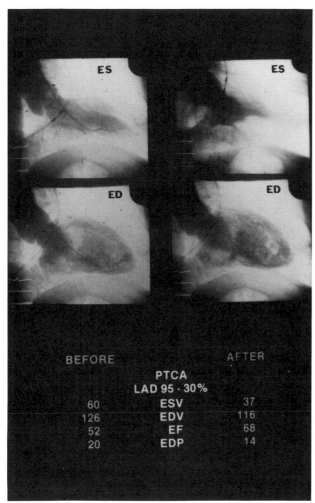

Figure 46. Left ventriculogram at end-systole (ES) and end-diastole (ED) performed just before (left) and immediately after (right) LAD angioplasty, showing a decrease in end-systolic volume (ESV), end-diastolic volume (EDV), and end-diastolic pressure (EDP) and an increase in ejection fraction (EF).

procedure has passed) that the dilation catheter and steerable wire are withdrawn. Then, utilizing the guiding catheter as an angiographic tool (or, if this catheter is too large for the orifice of the target artery, it is exchanged for a diagnostic angiographic catheter), a postangioplasty cine arteriogram is obtained in precisely the same projections and obliquities as in the preangioplasty control arteriograms. The quantitative measurements of percentage of diameter stenosis and the calculated cross-sectional area can be determined subsequently on evaluating the cine arteriogram in the same views (63,64). If an ECG tracing is available on the cine frame,

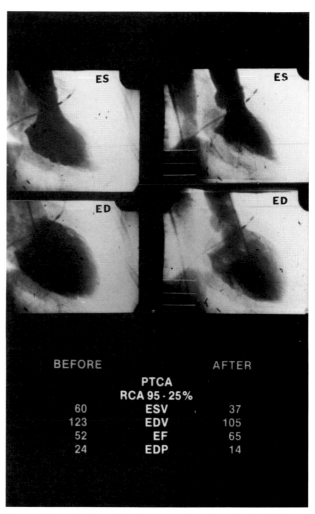

Figure 47. Left ventriculogram at end-systole (ES) and end-diastole (ED) performed just before (left) and immediately after (right) RCA angioplasty, showing decrease in ESV, EDV, and EDP, and increase in EF.

procedure itself. These angiographic features have been euphemistically called "controlled injuries" (Figs. 48 and 49). There appears to be little correlation between the presence or absence of these angiographic manifestations of the controlled injury and the presence or absence of late recurrence. In the NHLBI Registry for PTCA, the mean diameter improvement following angioplasty was 54 percent (17). With newer balloon catheters of higher pressure tolerance and the dilation of even more severe stenotic lesions, the future improvement of diameter stenosis may be even greater than that indicated by the past NHLBI Registry data.

Ultimately, the success of angioplasty in a given patient is determined not only by the improvement of anatomic pathology, but by normalization of the preangioplasty clinical abnormalities defined by angina pectoris, exercise treadmill test, and nuclear scintigraphic studies.

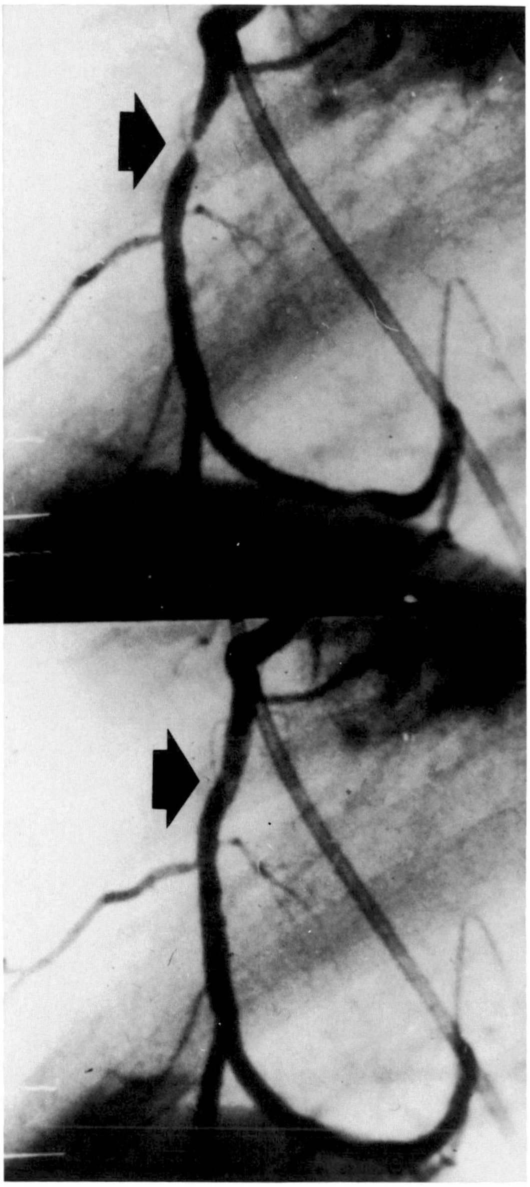

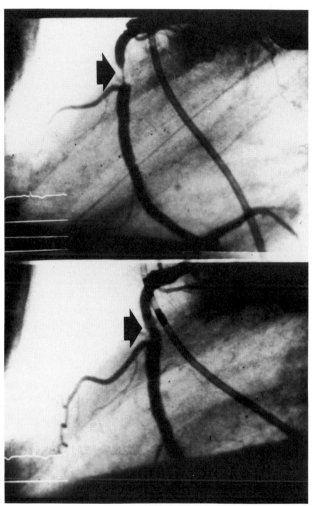

Figure 48. RCA in left lateral position, showing lesion (arrow) before (above) and after (below) angioplasty. Note in the postangioplasty angiogram (below), the small intraluminal filling defect (arrow).

Figure 49. RCA in left lateral position, showing lesion (arrow) before (above) and after (below) angioplasty. Note in the postangioplasty angiogram (below) slight irregularity and haziness in outline of vessel segment at the site of the lesion (arrow).

POSTANGIOPLASTY PATIENT CARE

In the early postangioplasty period following transfer from the cardiac catheterization laboratory, it is mandatory that patients be in a unit where they can be closely monitored for clinical, electrocardiographic, and hemodynamic progress. An intensive or semiintensive care unit facility is recommended for the initial 24 hours following angioplasty. Femoral venous and arterial sheaths should remain in place for approximately 2 or 3 hours following angioplasty to allow the effects of heparin to be attenuated. In addition, the arterial line may allow the direct monitoring of arterial pressure, particularly if intravenous nitroglycerin (used frequently in unstable patients or those suspected or known to have associated coronary spasm) is administered. Protamine is usually not used to correct the effects of heparin unless persistent bleeding is noted after removal of the sheaths. However, if protamine is utilized, it should be administered in low dosage (e.g., 10–20 mg) and over at least a 10-minute period. Protamine should be avoided in neutral protamine Hagedorn (NPH)-insulin-dependent diabetics (65).

DISCHARGE MEDICATIONS AND FOLLOW-UP SCHEDULE

Aspirin or dipyridamole initiated a day or two prior to angioplasty is continued for a period of 6 months after angioplasty. The precise dosage of these medications has not been determined, but at present we use enteric-coated aspirin, 325 mg administered twice daily prior to, and then once daily for 6 months following, angioplasty. Dipyridamole at a dose of 75 mg tid may also be utilized on the day before and for 6 months following angioplasty. These, of course, are used to decrease the platelet response to the angioplasty procedure (51). (However, as yet no controlled randomized trial of low- or high-dose aspirin or dipyridamole has been done in humans.) In addition, because of the tendency to coronary spasm, nitrates and calcium channel blockers are utilized before, during, and for 1–3 months after angioplasty. In patients who have a suspected or known tendency to coronary spasm (in addition to the "fixed" coronary stenosis), these medications may be used for longer duration and at higher dosages (46).

It is recommended that the patient have a repeat exercise study within the first week following angioplasty and again 3 months, 6 months, and 2 years later. This "noninvasive" method may give the objective evidence of myocardial perfusion status during the follow-up period (61,62). Should the patient's symptoms recur, or exercise or radioisotope perfusion studies again become abnormal, then the patient should undergo repeat coronary arteriography. If restenosis has occurred, a repeat transluminal angioplasty can be performed at the time of the arteriogram. The window of recurrence in coronary angioplasty is almost universally noted to be within 6 months from the time of the procedure. Extremely rare examples of later recurrence have been noted. Indeed, if a patient has a documentation of a continued patency 6 months after coronary angioplasty, and some months or years later the anginal pattern recurs, it is considerably more likely that a new lesion is responsible for the angina than a very late recurrence of the "original" stenosis (66).

ANATOMIC AND ANGIOGRAPHIC CRITERIA FOR PTCA

Indications

At present, coronary angioplasty is indicated in patients with angina pectoris relatively refractory to medical therapy who have objective evidence of myocardial ischemia documented by exercise testing (treadmill or bicycle ergometry) or nuclear scintigraphic studies (thallium 201 or multiple-gated wall motion: MUGA), plus angiographic evidence of flow-limiting coronary artery lesion(s) (Table 13). Initially, patients with discrete (short) lesions in single arteries were considered ideal candidates for coronary angioplasty (Figs. 30–32; 50A, B; and 51). The reason for this rather restricted indication centered

TABLE 13 Indications for Coronary Angioplasty

Clinical

   Angina pectoris (relatively) refractory to medical treatment

   Objective evidence of myocardial ischemia

Anatomic/pathologic

   Relatively discrete (short) stenoses

   Subtotal or "recent" total occlusions

   Single vessel disease

   Multivessel disease

   Saphenous vein graft stenosis (discrete)

   Tandem lesions

   Major branch bifurcation lesions (kissing balloon technique)

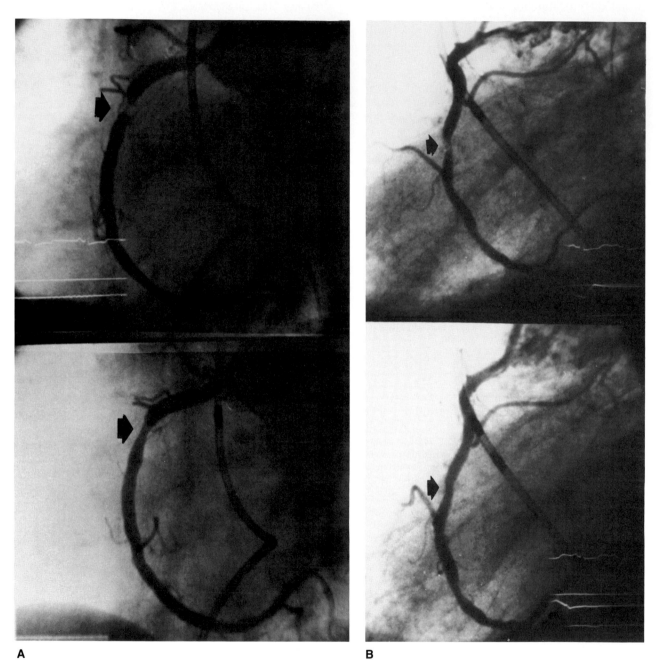

Figure 50. **A.** RCA in LAO-60° projection, showing isolated severe discrete stenosis (arrow) before (above) and after (below) angioplasty. The configuration of this vessel (C-shaped) is ideal for angioplasty. (Contrast Figs. 34, 40, 41, 52, and 53.) **B.** RCA in left lateral projection, showing lesion (arrow) before (above) and after (below) angioplasty. This anatomic configuation of RCA is also quite favorable for the angioplasty technique.

A

B

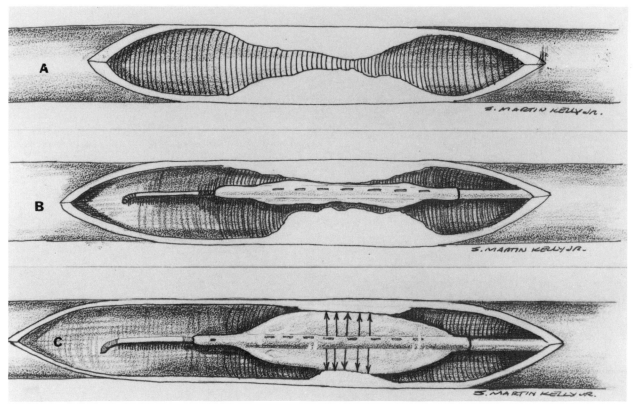

Figure 51. Schematic showing concentric stenosis (**A**), dilation balloon within stenosis (**B**), and balloon inflated, resulting in "compaction" of atheroma and enlargement of lumen (**C**). (Reprinted from Refs. 15 and 17.)

around three factors: (1) technologic limitations of the catheter systems available in the early years of coronary angioplasty, (2) relative inexperience of the operators, and (3) single vessel lesions being the best models to demonstrate "normalization" of subjective and objective parameters of ischemia.

However, with increased experience and improvement in technology (e.g., guiding and dilation catheters), case selection has widened considerably. Unusual or aberrant coronary ostia are presently more readily cannulated by the currently available forms of guiding catheter or by reshaping the catheter with a heat gun. Tortuous coronary arteries, eccentric stenoses, and branching vessels can be traversed with facility by the steerable dilation catheter system (Figs. 29A, B; 34; 38–42; 52, 53). Severe stenotic lesions can be crossed with the low profile steerable dilation catheters. So-called hard atheromatous lesions can be more successfully cracked with balloons that have higher pressure tolerance.

In several major coronary angioplasty centers (27,30), it has become apparent that patients with complex patterns of coronary anatomy or pathology can be treated by coronary angioplasty with similar success and complication rates to those noted previously in patients with single vessel disease. Experience at Seton Medical Center shows the improvement in the success rate during the coronary angioplasty era to be due in part to the introduction of the new guiding catheters in 1980 and of the steerable dilation catheters in 1983. In addition, in our center, there has been a shift in the type of cases undergoing coronary angioplasty. From 1978 to 1982, 5 percent of patients who had coronary angioplasty were of the "complex" type; however, during the 1983–1984 period, 54 percent were in this category (67). Complex angioplasty procedures performed at our center include multivessel disease (39 percent) (Fig. 54 A, B); tandem lesions (22 percent) (Figs. 41, 43, 44); recent total occlusions (68) without (Figs. 33, 55–59) and with (Fig. 60) infarction (16 percent); major branch bifurcation lesions requiring the "kiss-

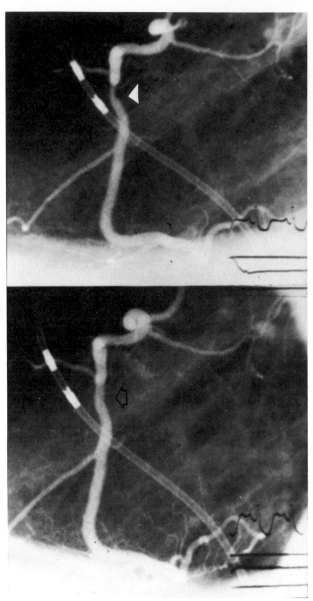

Figure 52. RCA in LAO-60° projection, showing tortuous proximal segment with stenosis (arrow) before (above) and after (below) angioplasty.

ing balloon'' technique (69) (6 percent) (Figs. 61–66); vein graft angioplasty (12 percent) (Figs. 35–37); and left main stem angioplasty (70) (5 percent) (Fig. 67). Thus, despite a shift to more complex cases, there has been an increase in success rate to 92 percent with no significant increase in emergency surgery (presently 3 percent), as a result of increased experience and improved technology (67). Whether restenosis rates in complex angioplasty will prove to be significantly higher than in single lesion disease

remains to be seen, although preliminary evidence would suggest otherwise (27).

Contraindications

With improvement in technology and widened experience, many former ''contraindications'' (Table

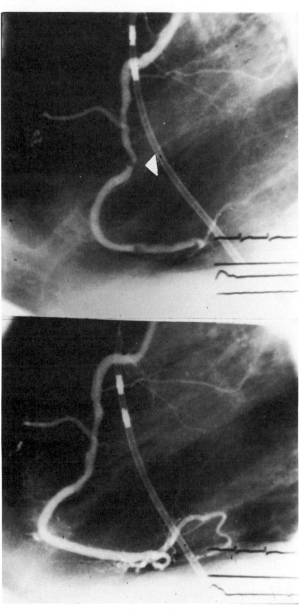

Figure 53. RCA in left lateral projection, showing eccentric stenosis (arrow) within tortuous midzonal segment before (above) and after (below) angioplasty. This configuration of vessel and lesion is more likely to result in dissection.

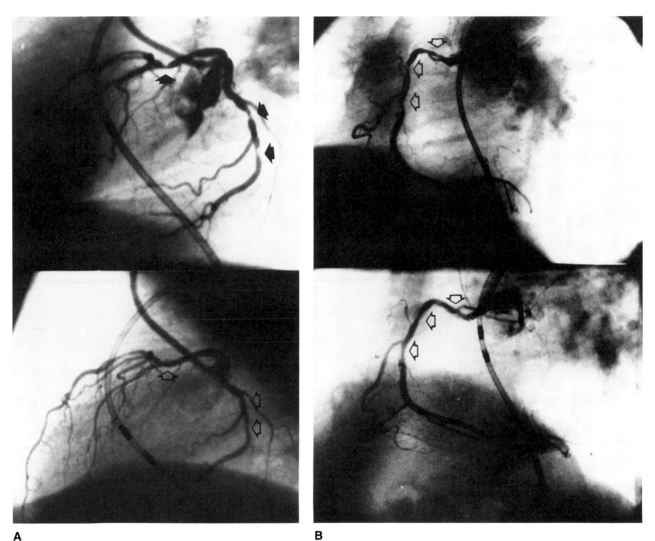

Figure 54. A. Left lateral projection of left coronary artery, showing stenoses (arrows) in LAD and LCFX before (above) and after (below) coronary angioplasty. **B.** LAO-60° projection of RCA, showing stenoses (arrows) before (above) and after (below) coronary angioplasty (in same 72-year-old male).

14) have fallen away. Eccentric lesions, recent total occlusions, calcified lesions, tandem lesions, and multivessel disease, all of which were considered relative contraindications in the early years of angioplasty, are now routinely treated by this technique. Yet, there are still cases in which coronary angioplasty is not recommended. Vessels that have been known to be totally occluded for more than 6 months (and perhaps an even shorter period) cannot be readily traversed with coronary dilation catheters and steerable wires presently in use (71) because of the organization and fibrosis of the thrombus. Ul-

cerative lesions (Fig. 68) and long (greater than 2 cm) segmental areas of stenosis (Figs. 41, 42, 69), although often successfully "dilated," are certainly not ideal for coronary angioplasty (72,73), and it would appear from our experience that the recurrence rate in these lesions is higher than in those that are discrete (short). Areas of major branch bifurcation in which the possibility of sacrificing a large side branch is approximately 14 percent (74,75) probably should be avoided unless the kissing balloon technique is utilized. This approach requires the placement of two dilation catheters via two guid-

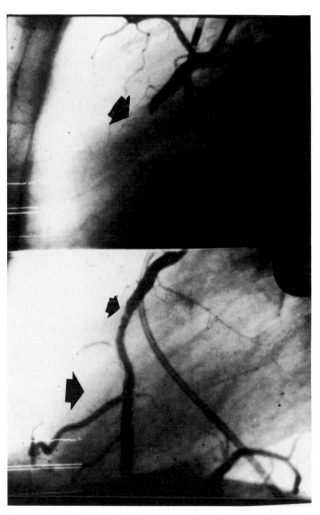

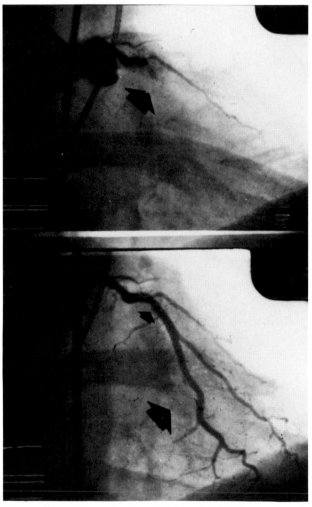

Figure 55. RCA in left lateral projection, showing total occlusion (large arrow) at site of previously noted severe stenosis before (above) and after (below) angioplasty. Note small intraluminal filling defect (small arrow) in postangioplasty frame. There was no evidence of embolization in distal vessel (large arrow).

Figure 56. "Selective" LCFX in RAO-30° caudal 15° projection, showing "recent" total occlusion (arrow) at site of severe stenosis before (above) and after (below) angioplasty. There was an excellent collateral from the RCA to the LCFX noted before, and absent immediately after, angioplasty (see also Figs. 34A,B in Ref. 17.)

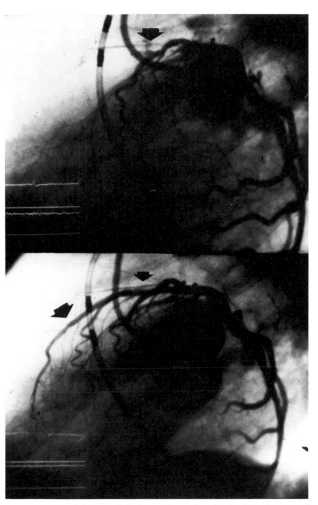

Figure 57. LAD in left lateral projection, showing total occlusion (arrow) at site of previously noted severe stenosis just distal to first major diagonal branch before (above) and after (below) angioplasty. Note faint collaterals to LAD from LCFX and diagonal branch before angioplasty (RCA also supplied collateral filling of distal LAD prior to angioplasty.)

ing catheters in the branches of the target artery (Figs. 61–66). "Old" saphenous vein grafts in which the angiographic appearance would suggest diffuse disease should certainly be avoided since the occlusive process is characterized by diffuse and friable atheromatous and thrombotic material that can be readily embolized into the distal native circulation (76–84)(Fig. 70A–C).

Conceptually, coronary angioplasty should be applied to patients in whom "complete revascularization" of diseased vessels responsible for myocardial ischemia can be achieved. With rare exception, if coronary bypass surgery can effect more complete revascularization in a given patient than coronary angioplasty can, then the former should be recommended. For example, if a patient has a very old total occlusion of a dominant right coronary artery and a severe stenosis of a left anterior descending coronary artery with retrograde collateral filling of the distal right coronary artery via the left coronary system and has a viable inferior left ventricular wall, then (because coronary angioplasty could successfully treat the left anterior descending artery but could not reopen the old total occluded right coronary artery) we would recommend double aortocoronary bypass surgery. On the other hand, with a similar example of coronary pathology, if the left ventricular inferior wall showed no evidence of viability (by thallium 201 scintigraphy), it might be reasonable to treat only the left anterior descending coronary artery by either angioplasty or coronary bypass surgery. There may be situations in which the risk of surgery is great because of noncardiac pathology (e.g., pulmonary, renal, or metabolic dis-

TABLE 14　Contraindications for Coronary Angioplasty

Clinical

　Very poor cardiac surgical risk (except in special circumstances)

Anatomic/pathologic

　Orificial stenoses

　Old total occlusions

　Long segmental stenoses (?)

　Ectatic vessels (with stenoses)

　Heavily calcified lesions (?)

　Ulcerative lesions (?)

　Left main stem ("unprotected")

　Old diffusely diseased saphenous vein grafts

　"Incomplete" revascularization (versus coronary artery bypass graft) except in special circumstances

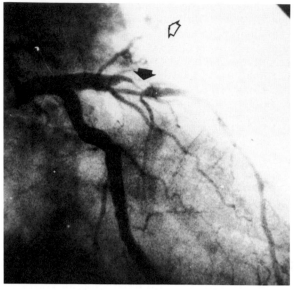

A

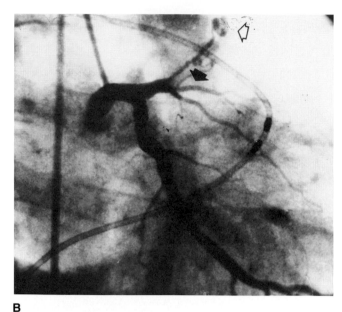

B

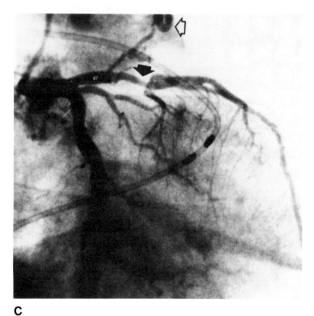

C

Figure 58. **A.** RAO-30° projection in 166-kg male, showing severe discrete stenosis (arrow) of LAD during initial arteriogram. **B.** Total occlusion (arrow) 1 day later. No infarction in interim, presumably because of "effective" collaterals (from RCA). Also visible is coronary-pulmonary fistual (open arrow). **C.** After successful angioplasty.

eases), and angioplasty is more reasonable. On the other hand, certain patients, especially those with very poor left ventricular function, may be put at greater risk by angioplasty (especially if complicated) than by elective surgery.

One must select patients, to some degree, defensively, always asking, What will happen to the patient if there is abrupt reclosure of the target vessel during coronary angioplasty? In this manner, a more rational approach to the selection of patients for coronary angioplasty can be realized. Similarly, one must relate stenotic lesions considered for coronary angioplasty to the vessel distribution of myocardial blood flow, left ventricular myocardial segment ischemia, and, particularly, the benefit of coronary angioplasty to the patient. Simply dilating lesions

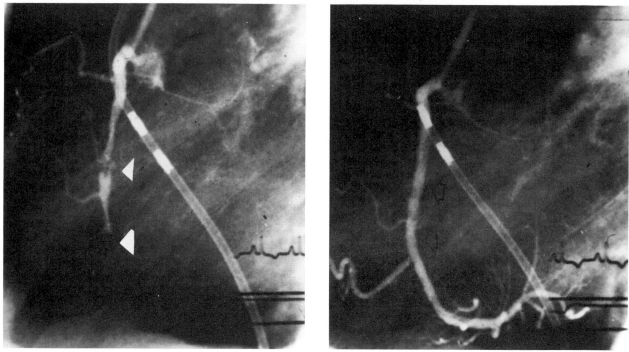

Figure 59. RCA in left lateral projection, showing two midzonal stenoses (arrows), the proximal one with a filling defect (presumably thrombus) and the distal one totally occluded at site of a recently noted severe stenotic lesion before (left) and after (right) angioplasty. (Low-dosage intracoronary streptokinase was used after angioplasty to clear minor distal intraluminal filling defects.)

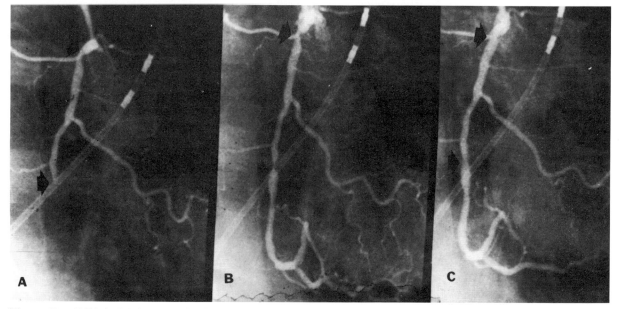

Figure 60. RCA in RAO-30° projection, showing severe proximal stenosis and total midzonal occlusion (arrows) in patient with acute evolving inferior myocardial infarction before (**A**) and after (**B**) intracoronary infusion of streptokinase. Note in (**B**) severe residual stenosis at site of total occlusion. **C.** Patient was subsequently treated with angioplasty with excellent results at both proximal and midzonal sites.

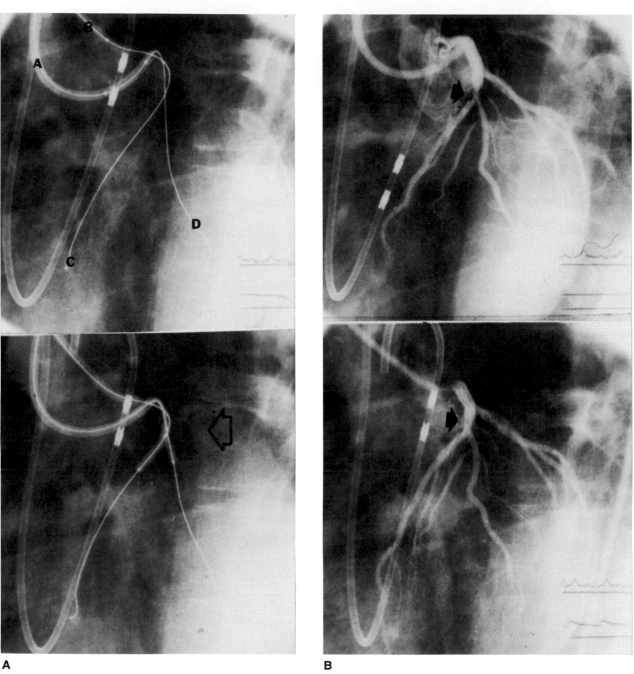

A

B

Figure 61.   **A.** LAO-45° cranial 25° projection of left coronary artery. Upper view shows brachial (A) and femoral (B) guiding catheters and steerable wires in LAD (C) and first major diagonal (LADD) (D). Lower view shows inflated dilation balloon catheters at sites of ''bifurcation'' lesions (arrow). **B.** LAO-45° cranial 25° projection. Arrow points to bifurcation lesions before (above) and after (below) kissing-balloon angioplasty.

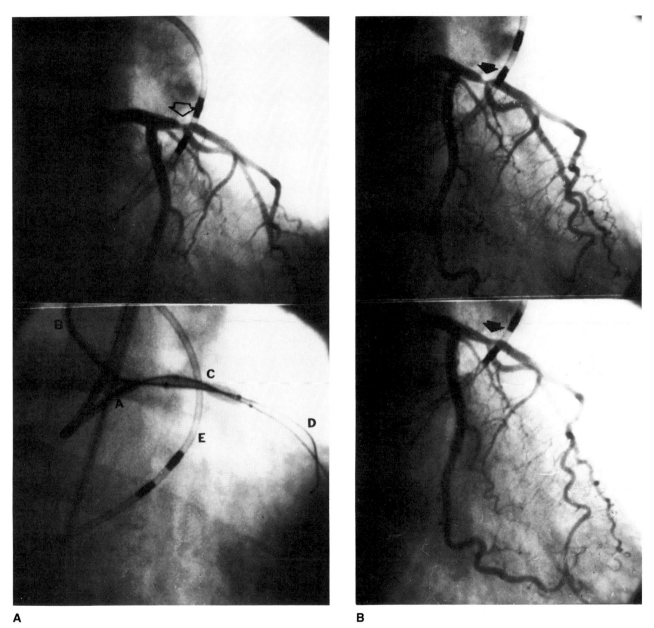

Figure 62.   A. RAO-30° projection, showing LAD/LADD bifucation lesion (arrow) before (above) and during (below) kissing-balloon angioplasty. Note brachial (A) and femoral (B) guiding catheters and both dilation balloon catheters (C) and wires (D). Multipurpose right heart catheter (E) aids in spatial coordination of lesion. **B.** RAO-30° projection, showing bifurcation lesion (arrow) before (above) and after (below) kissing-balloon angioplasty.

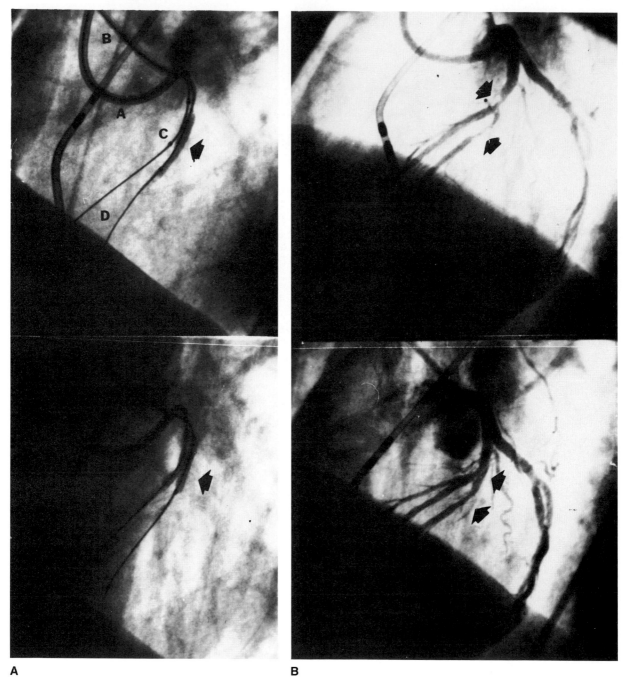

A   **B**

Figure 63. **A.** LAO-60° cranial 30° projection, showing brachial (A) and femoral (B) guiding catheters and dilation catheters (C) and steerable wires (D) in LAD and LADD. In the upper view, balloon is inflated in LADD (arrow) and below in both LAD and LADD (arrow). **B.** LAO-60° cranial 30° projection. Arrow indicates bifurcation lesion in LAD and proximal LADD before (above) and after (below) angioplasty.

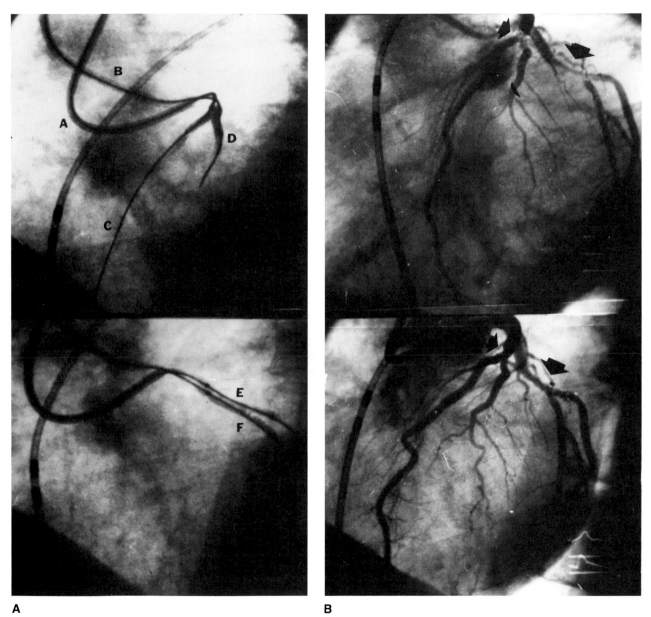

Figure 64. **A.** LAO-60° cranial 25° projection. Upper view discloses brachial (A) and femoral (B) guiding catheters and dilation catheters and steerable wires in LAD (C) and LADD (D). Lower view shows same in LCFX (E) and obtuse marginal (F) branches. **B.** LAO-60° cranial 25° projection. Arrows indicate bifurcation lesions involving the LAD, LADD, and LCFX and obtuse marginal branches before (above) and after (below) kissing-balloon angioplasty.

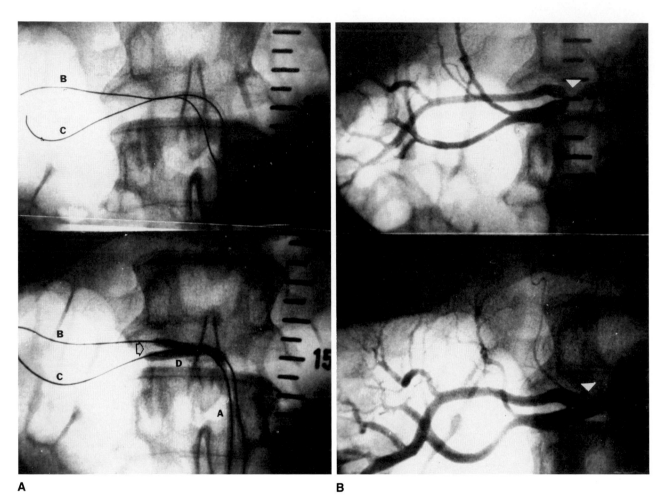

A   **B**

Figure 65. **A.** Two renal guiding catheters (A) in aorta with steerable wires in upper (B) and lower (C) branches of right renal artery. In lower view dilation balloons (D) are simultaneously inflated (arrow) on "bifurcation" lesion. **B.** Bifurcation lesion (arrow) right renal artery stenosis before (above) and after (below) kissing-balloon renal angioplasty Although the patient was on therapy, severe renovascular hypertension (blood pressure: 190/120) was documented before angioplasty. Afterward, blood pressure was 135/80 with no antihypertensive therapy.

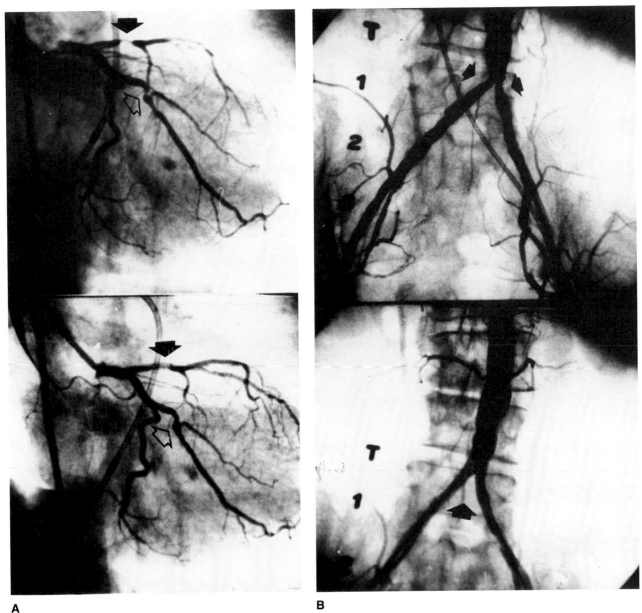

Figure 66. **A.** RAO-30° projection, showing severe stenosis (arrow) involving bifurcation of LAD, and LADD, and LCFX (open arrow) before (above) and after (below) angioplasty. **B.** Anterior-posterior projection of aortic bifurcation. Arrows show stenotic lesions of proximal common iliac arteries before (above) and after (below) angioplasty in same patient.

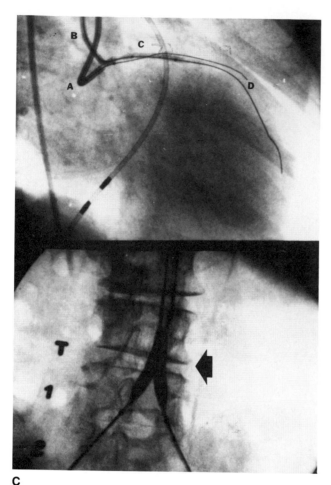

C

Figure 66 (*Continued*).   C. Upper view shows RAO-30°
projection of brachial (A) and femoral (B) guiding catheters, dilation catheters (C), and wires (D) in LAD and
LADD. Lower view is anterior-posterior projection of
aortoiliac bifurcation, showing iliac angiioplasty catheter
balloons (arrow) inflated simultaneously on iliac lesions.
These kissing-balloon coronary and iliac angioplasties (plus
the LCFX PTCA) prevented a triple coronary artery bypass graft (CABG) and a double aortofemoral bypass operation in this 66-year-old female.

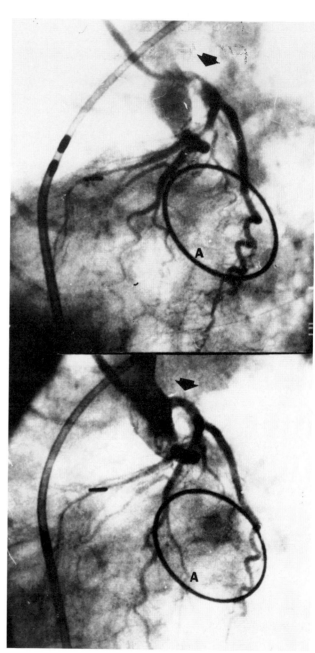

Figure 67.   LAO-60° cranial 25°, showing severe left main
stem stenosis (arrow) before (above) and after (below)
angioplasty. Patient had prior cardiac surgery to insert a
mitral valve prosthesis (A) and saphenous vein bypass
grafts to the LAD (patent) and LCFX (occluded), the
latter resulting in recurrent symptoms and signs of ischemia.

without regard to pathophysiology of myocardial ischemia is an inappropriate use of the coronary angioplasty technique.

Yet, although left main stem coronary stenosis is almost invariably associated with involvement of other coronary arteries and is generally considered a contraindication to coronary angioplasty, patients who have "protected" left coronary systems (that is, at least one patent saphenous vein graft to a left coronary arterial branch) may be considered reasonable candidates for angioplasty to achieve improvement in blood flow to the left coronary circulation in which the saphenous vein graft has occluded, particularly if nuclear scintigraphy has shown this branch to be the cause of reversible myocardial ischemia (70) (Fig. 67).

It should be emphasized that there is a somewhat greater (two to three times) risk of mortality in performing coronary angioplasty in patients who have undergone previous coronary bypass surgery, in particular those in whom multivessel or complex angioplasty is contemplated (27,85) (Figs. 35, 36, 67, 69, 71A, B, 72A, B). This risk is no doubt at least in part due to the increased difficulty and prolonged operative time necessary to perform a repeat median sternotomy and to expose the heart. Thus, again, one should select these patients "defensively" and always have immediate operating room availability. Often an intra-aortic balloon assist device is recommended in performing coronary angioplasty in patients with depressed left ventricular function, and may also be necessary to stabilize the patient, on the way to surgery, who has had an abrupt reclosure of a "dilated" coronary arterial segment (86).

The pattern of case selection for coronary angioplasty has progressed slowly and carefully, evaluating and reevaluating types of cases suitable for this technique. This evolutionary process continues at present, influenced greatly by technologic advances and the shared experience of the international angioplasty community.

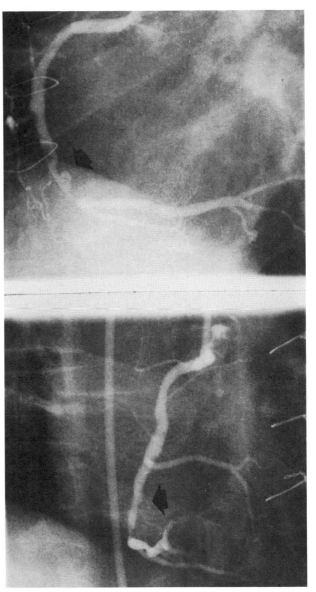

Figure 68. RCA in LAO-60° projection, showing ulcerative lesion (arrow) above. Note that "ulcer" is not evident in RAO projection (below) in this postoperative patient.

ASSESSMENT OF CURRENT PTCA TECHNIQUES

Although enormous progress has been made in the technologic advances of the coronary angioplasty systems, further improvement can be anticipated. The guiding catheter can be made smaller (number 7 or 8 French) with a stiffer shaft to achieve greater power and torque control and a softer tip for safety. The inner diameters of the guiding catheter are now larger to improve pressure monitoring and contrast injection. The configurations of the guiding catheters will continue to evolve to allow facile cannulation of even the most unusual coronary anatomic patterns. The dilation catheter will also continue to improve, with shafts that are "stiffer," tips that are softer, profiles that are lower, and balloons that can achieve pressures of 20–30 atmospheres. Steerable

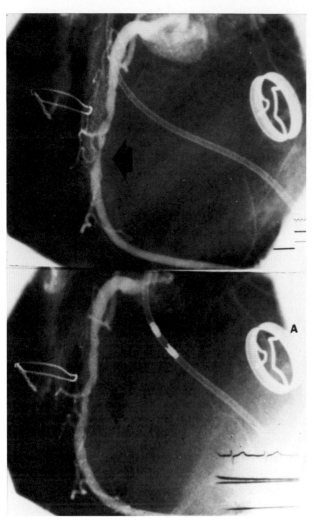

Figure 69. RCA in left lateral projection, showing long segmental stenosis (arrow) before (above) and after (below) angioplasty. Patient previously had mitral valve replacement (A).

TABLE 15   Coronary Angioplasty Future Directions

Guiding catheters
  Smaller OD
  Larger ID
  Stiffer shafts
  Softer tips
  More configurations
  More torque and memory
Dilation catheters
  Lower profile
  Stronger balloons
  Stiffer shafts
  Softer tips
Steerable wires
  More torque control
  Better tip memory
  Better trackability

more sophisticated, and one could speculate that in the future a discrete, old total occlusion (that cannot be crossed by a dilation catheter) might be opened (if only slightly) with a laser system and the residual stenosis further treated with balloon angioplasty.

The NHLBI of the National Institutes of Health established a registry for coronary angioplasty in 1979. These data, although obtained from the initial experience (1977–1981) with coronary angioplasty and with more "primitive" equipment, have been invaluable in demonstrating that coronary angioplasty can be performed successfully with a relatively low risk (18). A future prospective trial (88), focused on complex angioplasty, should reveal the present incidence of success, complications, and recurrence in patient subsets, showing directions for this remarkable new nonoperative treatment for coronary artery disease.

wires can improve with greater distal tip memory, torque control, flexibility, and trackability. At present, all manufacturers of angioplasty equipment (Table 15) are focusing considerable effort on meeting these needs.

Whether laser angioplasty will play a role in the treatment of coronary arterial occlusive disease remains speculative (87). There is no doubt that argon, Nd:YAG, and excimer (ultraviolet) laser systems can recannulate occluded coronary arteries, but the safety of this technology is, at present, suspect. However, laser angioplasty will certainly become

Acknowledgments: To Drs. David Clark, Colman Ryan, Robert Dunlap, my associates and friends, thank you for your help in the development of this chapter. To Miss Mary Murphy and Mrs. Jodi Rosen, thanks for the resource material from SFHI/SMC data registry for PTCA. To Mr. John McMorrow, Mr. Benito Hidalgo, Miss Suzanne Jones, and staff I am indebted for their participation in the development of PTCA in the United States. To Dr. Eric Topol, gratitude for his help in preparing several tables included in this chapter. And finally, acknowledgment to my dear friends and colleagues, Drs. Andreas Gruentzig and Simon Stertzer, without whom there would not be the procedure of coronary angioplasty as we know it today.

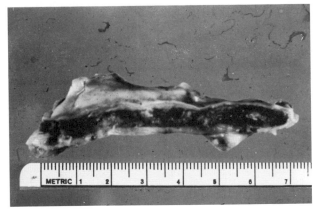

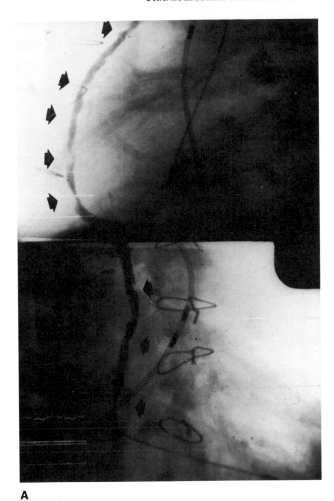

Figure 70. **A.** Five-year-old SVF to RCA in LAO-60° (above) and RAO-30° (below) projections, showing diffuse intraluminal filling defects (arrows). **B.** Resected SVG (at time of reoperation), showing gross findings of diffuse atheromatous and thrombotic disease (which was very friable). **C.** Microscopic findings of SVG, showing thrombus (A) and underlying atheroma (B).

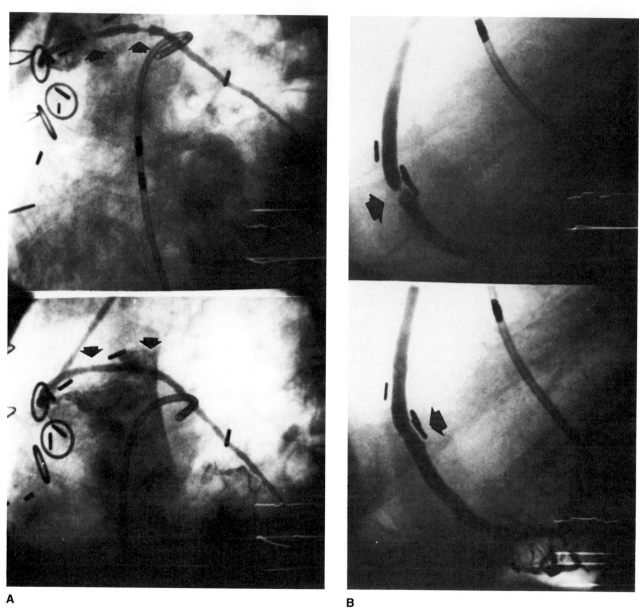

A   **B**

Figure 71. **A.** Seven-year-old SVG to LCFX in LAO-60° position, showing two lesions (arrows) before (above) and after (below) angioplasty (see Fig. 37). **B.** Seven-year-old SVG to RCA in LAO-60° projection, showing severe discrete stenosis (arrow) at site of clips before (above) and after (below) angioplasty (patient had SVG-LCFX angioplasty at the time).

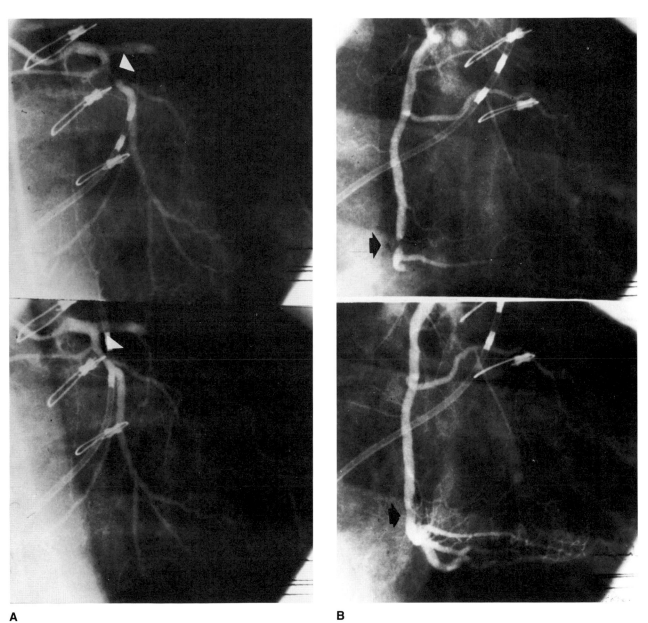

Figure 72.   A. LCFX in RAO-15° caudal 15° projection, showing severe stenosis (arrow) in proximal segment before (above) and after (below) angioplasty. Patient had prior SVG LAD (patent) and SVG LCFX (occluded). **B.** RCA in RAO-30° projection of the same patient, showing a "new" lesion (arrow) in distal vertical segment before (above) and after (below) angioplasty, performed at the same time as LCFX angioplasty.

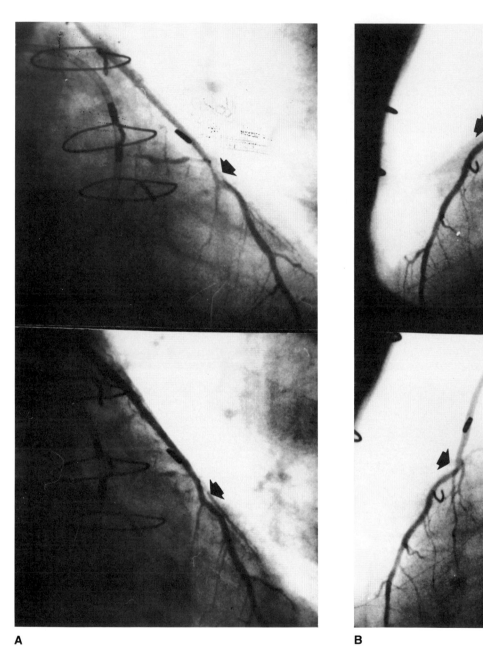

A   B

Figure 73. Left internal mammary artery to LAD in RAO-30° (**A**) and left later (**B**) projections, showing insertion site stenoses (arrows) before (above) and after (below) angioplasty.

REFERENCES

1. Dotter CT, Judkins MP: Transluminal treatment of arteriosclerotic obstruction: Description of a new technique and preliminary report of its application. Circulation 30:654, 1964.

2. Zeitler E, Schoop W, Zahnow W: The treatment of occlusive arterial disease by transluminal catheter angioplasty. Radiology 99:19, 1971.

3. Gruentzig AR: Die perkutane transluminale Rekanalisation chronischer arterieller Verschulusse (Dotter-Prinzip) mit einem doppellumigen Dilatations-Katheter. Fortschr Roentgenstr 124:80, 1976.

4. Judkins MP: Selective coronary arteriography: I. A percutaneous transfemoral technique. Radiology 89:815, 1967.

5. Amplatz K, Formanck G, Strangler P, Wilson W: Mechanics of selective coronary artery catheterization via femoral approach. Radiology 89:1040, 1967.

6. Gruentzig AR: Transluminal dilatation of coronary artery stenosis. Lancet 1:263, 1978.

7. Stertzer SH, Myler RK, Bruno JP, Wallsh E: Transluminal coronary artery dilatation. Pract Cardiol 5:25, 1979.

8. Gruentzig AR, Senning A, Siegenthaler WE: Nonoperative dilatation of coronary artery stenosis: Percutaneous transluminal coronary angioplasty. N Engl J Med 301:61, 1979.

9. Gruentzig AR, Myler RK, Stertzer S: Coronary angioplasty (PTCA)—present state of the art. (Abstract) Circulation 59–60(suppl II):264, 1979.

10. Stertzer SH, Myler RK, Wallsh E, Bruno MS, DePasquale NP: Dilatation of obstructed coronary arteries by percutaneous transluminal angioplasty. J Cardiovasc Med 5:1059, 1980.

11. Proceedings of the Workshop on Percutaneous Transluminal Coronary Angioplasty (June 15–16, 1979), publication 80-2030, US Dept of Health, Education, and Welfare, March 1980.

12. Myler RK: Percutaneous transluminal angioplasty. Arch Inst Cardiol Mex 50:401, 1980.

13. Myler RK (ed): Angiography and Angioplasty. Pro Clinica, New York, 1980.

14. Dotter CT, Gruentzig AR, Schoop W, Zeitler E (eds): Percutaneous Transluminal Angioplasty. New York, Springer-Verlag, 1983.

15. Myler RK, Gruentzig AR, Stertzer SH: Technique and clinical indications for percutaneous transluminal coronary angioplasty, in Mason DT, Collins JJ (eds): Myocardial Revascularization: Medical and Surgical Advances in Coronary Disease. New York, Yorke Medical, 1981, p 431.

16. Myler RK, Gruentzig AR, Stertzer SH, Kent K, Dorros G, Spring D, Kaltenbach M: Transluminal Coronary Angioplasty, in Bruschke AVG, Van Herpen G, Vermeulen FEE (eds): Coronary Artery Disease Today. Amsterdam, Excerpta Medica, 1982, p 253.

17. Myler RK, Gruentzig AR, Stertzer SH: Coronary angioplasty in Rapaport E (ed): Cardiology Update III, New York, Elsevier North-Holland, 1983, chap 1.

18. Proceedings of the Workshop on Percutaneous Transluminal Coronary Angioplasty (June 7–8, 1983). Am J Cardiol 53(no. 12):1984 (suppl).

19. Detre KM, Myler RK, Kelsey SF, Van Raden MV, Mitchell H: Baseline characteristics of patients in National Heart, Lung and Blood Institute Percutaneous Transluminal Coronary Angioplasty Registry. Am J Cardiol 53:7C, 1984.

20. Kelsey SF, Mullin SM, Detre K, Mitchell H, Cowley MJ, Gruentzig AR, Kent KM: Effect of investigator experience on percutaneous transluminal coronary angioplasty. Am J Cardiol 53:56C, 1984.

21. Meier B, Gruentzig AR: Learning curve for percutaneous transluminal coronary angioplasty: Skill, technology or patient selection. Am J Cardiol 53:65C, 1984.

22. Gruentzig AR, Meier B: Current status of dilatation catheters and guiding systems. Am J Cardiol 53:92C, 1984.

23. McAuley BJ, Simpson JB: Advances in guidewire technology. Am J Cardiol 53:94C, 1984.

24. Fogarty TJ, Kinney TB, Finn JC: Current status of dilatation catheters and guiding systems. Am J Cardiol 53:97C, 1984.

25. Hartzler GO, Rutherford BD, McConahay DR: Percutaneous transluminal coronary angioplasty for acute myocardial infarction. Am J Cardiol 53:117C, 1984.

26. Gold HK, Cowley MJ, Palacios JF, Vetrovec GW, Atkins CW, Block PC, Leinbach RC: Combined intracoronary streptokinase infusion and coronary angioplasty during the myocardial infarction. Am J Cardiol 53:122C, 1984.

27. Dorros G, Stertzer SH, Cowley MJ, Myler RK: Complex coronary angioplasty: Multiple coronary dilatations. Am J Cardiol 53:126C, 1984.

28. Faxon DP, Detre KM, McCabe CH, Fisher L, Holmes DR, Cowley MJ, Bourassa MG, Van Raden M, Ryan TJ: Role of percutaneous transluminal coronary angioplasty in the treatment of unstable angina: Report from the National Heart, Lung and Blood Institute Percutaneous Transluminal Coronary Angioplasty and Coronary Artery Surgery Study Registries. Am J Cardiol 53:131C, 1984.

29. Rapport E: Current clinical topics: Potential future uses for percutaneous transluminal coronary angioplasty. Am J Cardiol 53:136C, 1984.

30. Hartzler GO: Percutaneous transluminal coronary angioplasty in multivessel disease. Cathet Cardiovasc Diagn 9:537, 1983.

31. Przybojewski JZ, Weich HFH: Percutaneous transluminal coronary angioplasty. S Afr Med J (special issue) January 25, 1984.

32. Williams DO, Gruentzig AR, Kent KM, Myler RK, Stertzer SH, Bentivogli L, Bourassa M, Block PC, Cowley MJ, Detre K, Dorros G, Gosselin A, Simpson J, Passamani E, Mullin S: Guidelines for the performance of percutaneous transluminal coronary angioplasty. Circulation 66:693, 1982.

33. Weaver WF, Myler RK, Sheldon WC, Huston JT, Judkins MP: Guidelines for physician performance of percutaneous transluminal coronary angioplasty. Cathet Cardiovasc Diagn 11:109, 1985.

34. Gruentzig AR: Perkutane Dilatation von Koronarstenosen-Beschreibung eines neuen Kathetersystems. Klin Wochenschr 54:543, 1976.

35. Gruentzig AR, Schneider HJ: Die perkutane dilatation chronischer Koronarstenosen-Experiment und Morphologic. Schweiz Med Wochenschr 107:1588, 1977.

36. Gruentzig AR, Riedhammer HH, Turina M: Eine neue Methode zur Perkutanen. Verh Dtsch Ges Kreislaufforsch 42:282, 1976.

37. Gruentzig AR, Turina MI, Schneider JA: Experimental percutaneous dilatation of coronary artery stenosis. Circulation 54:81, 1976.

38. Gruentzig AR, Myler RK, Hanna EH, Turina MI: Coronary transluminal angioplasty, abstracted. Circulation 55–56(suppl III):84, 1977.

39. Kimbiris D, Iskandrian A, Saras H, Goel I, Bemis CE, Segal BL, Mundth E: Rapid progression of coronary stenosis in patients with unstable angina pectoris selected for coronary angioplasty. Cathet Cardiovasc Diagn 10:101, 1984.

40. Pasternak RC, Baughman KL, Fallon JT, Block PC: Scanning electron microscopy after coronary transluminal angioplasty of normal canine coronary arteries. Am J Cardiol 45:581, 1980.

41. Block PC, Myler RK, Stertzer SH, Fallon JT: Morphology after transluminal angioplasty in human beings. N Engl J Med 305:382, 1982.

42. Block PC, Baughman K, Pasternak RC, Fallon JT: Transluminal angioplasty: Correlation of morphologic and angiographic findings in an experimental model. Circulation 61:778, 1980.

43. Baughman KL, Pasternak RC, Fallon JT, Block PC: Transluminal angioplasty of post mortem human hearts. Am J Cardiol 48:1044, 1981.

44. O'Gara PT, Guerrero JL, Feldman B, Fallon JT, Block PC: Effect of dextran and aspirin on platelet adherence after transluminal angioplasty of normal canine coronary arteries. Am J Cardiol 53:1695, 1984.

45. Swanson KT, Vlietstra RE, Holmes D, Smith HC, Reeder GS, Bresnahan JF, Bove AA: Efficacy of adjunctive dextran during percutaneous transluminal coronary angioplasty. Am J Cardiol 54:447, 1984.

46. Corcos T, David PR, Val PG, Robert J, Mata LA, Waters DD, Bourassa MG: Percutaneous transluminal coronary angioplasty (PTCA) in patients with variant angina: A three-year experience. (Abstract) J Am Coll Cardiol 5:446, 1985.

47. Seldinger SI: Catheter replacement of the needle in percutaneous arteriography: A new technique. Acta Radiol [Diag.] (Stockh) 39:368, 1953.

48. Kaltenbach M, Beyer J, Walter S, Klepzig H, Schmidts L: Prolonged application of pressure in transluminal coronary angioplasty. Cathet Cardiovasc Diagn 10:213, 1984.

49. Holmes DR, Vlietstra RE, Smith HC, Vetrovec GW, Kent KM, Cowley MJ, Faxon DP, Gruentzig AR, Kelsey SF, Detre KM, Van Raden MJ, Mock MB: Restenosis after percutaneous transluminal coronary angioplasty (PTCA): A report from the PTCA registry of the National Heart, Lung and Blood Institute. Am J Cardiol 53:77C, 1984.

50. Meier B, King SB, Gruentzig AR, Douglas JS, Hollman J, Ischinger T, Gallan K, Tankersley R: Repeat coronary angioplasty. J Am Coll Cardiol 4:463, 1984.

51. Faxon DP, Sanborn TA, Haudenschild CC, Ryan TJ: Effect of antiplatelet therapy on restenosis after experimental angioplasty. Am J Cardiol 53:72C, 1984.

52. Marantz T, Williams DO, Reinert S, Gewirtz H, Most AS: Predictors of restenosis after successful coronary angioplasty. (Abstract) Circulation 70(suppl II):176, 1984.

53. Leimgruber PP, Roubin GS, Rice CR, Tate JM, Gruentzig AR: Influence of intimal dissection after coronary angioplasty (PTCA) on restenosis rate. (Abstract) Circulation 70(suppl II):175, 1984.

54. O'Neill WW, Vogel RA, Walton JA, Bates ER, Colfer HT, Aueron FM, Lefree T, Pitt B: Criteria for successful coronary angioplasty as assessed by alterations in coronary vasodilatory reserve. J Am Coll Cardiol 3:1382, 1984.

55. Ischinger T, Gruentzig AR, Hollman J, King SB, Douglas JS, Meier B, Bradford J, Tankersley R: Should coronary arteries with less than 60% diameter stenosis be treated by angioplasty? Circulation 68:148, 1983.

56. Grossman W, McLaurin LP: Clinical measurement of vascular resistance and assessment of vasodilator drugs, in Grossman W (ed): *Cardiac Catheterization and Angiography*. Philadelphia, Lea and Febiger, 1980, p 116.

57. Williams DO, Riley RS, Singh AK, Most AS: Restoration of normal coronary hemodynamics and myocardial metabolism after percutaneous transluminal coronary angioplasty. Circulation 62:653, 1980.

58. Williams DO, Riley RS, Singh AK, Gewirtz H, Most AS: Evaluation of the role of coronary angioplasty in patients with unstable angina pectoris. Am Heart J 102:1, 1981.

59. Carlson EB, Myler RK, Cowley MJ, Vetrovec GW: Improved left ventricular function immediately after successful PTCA in patients with depressed ejection fraction. Personal information.

60. Serruys PW, Wijns W, van den Brand M, Meij S, Slager C, Schuurbiers JCH, Hugenholtz PG: Left ventricular performance, regional blood flow, wall motion, and lactate metabolism during transluminal angioplasty. Circulation 70:25, 1984.

61. Kent KM, Bonow RO, Rosing DR, Ewels CJ, Lipson LC, MacIntosh CL, Bacharach SL, Green MV, Epstein SE: Improved myocardial function during exercise after successful percutaneous transluminal coronary angioplasty. N Engl J Med 306:441, 1982.

62. DePuey EG, Leatherman LL, Leachman RD, Dear WE, Massin EK, Mathur VS, Burdine JA: Restenosis after transluminal coronary angioplasty detected with exercise-gated radionuclide ventriculography. J Am Coll Cardiol 4:1103, 1984.

63. Meier B, Gruentzig AR, Goebel N, Pyle R, von Gosslar W, Schlumpf M: Assessment of stenoses in coro-

nary angioplasty: Inter- and intraobserver variability. Int J Cardiol 3:159, 1983.

64. Serruys PW, Reiber JHC, Wijns W, van den Brand M, Kooijman CJ, ten Katen HJ, Hugenholtz PG: Assessment of percutaneous transluminal coronary angioplasty by quantitative coronary angiography: Diameter versus densitometric area measurements. Am J Cardiol 54:564, 1984.

65. Stewart WJ, McSweeney SM, Kellett MA, Faxon DP, Ryan TJ: Increased risk of severe protamine reactions in NPH insulin-dependent diabetics undergoing cardiac catheterization. Circulation 70:788, 1984.

66. Williams DO: Presented at Coronary Angiography— 1984, Harvard Medical School Symposium, Boston, June 11–14, 1984.

67. Myler R: Coronary angioplasty: Widening indications/current state of the art. Proceedings of Conventional and New Approaches to the Prevention and Treatment of Acute and Chronic Myocardial Ischemia. Bethesda, Heart House Learning Center (Am. Coll. Cardiol.), 1984.

68. Wexman M, Stertzer SH, Myler RK, Cieszkowski J, Clark D, George B, Murphy M, Rosen J: Coronary angioplasty: Recent total occlusions. (Submitted for publication)

69. George B, Myler RK, Stertzer SH, Clark D, Cieszkowski J, Millhouse F, Topel E, Murphy M, Rosen J: Coronary angioplasty: Bifurcation lesions—the ''kissing balloon'' technique. (Submitted for publication)

70. Stertzer SH, Myler RK, Insel H, Wallsh E, Rossi P: Transluminal coronary angioplasty in left main stem coronary stenosis. Int J Cardiol (in press).

71. Holmes DR, Vlietstra RE, Reeder GS, Bresnaham JF, Smith HC, Bove AA: Angioplasty in total coronary artery occlusion. J Am Coll Cardiol 3:845, 1984.

72. Meier B, Gruentzig AR, Hollman J, Ischinger T, Bradford J: Does length or eccentricity of coronary stenoses influence the outcome of transluminal dilatation? Circulation 67:497, 1983.

73. Hall DP, Gruentzig AR: Influence of lesion length on initial success and recurrence rates in coronary angioplasty abstracted. Circulation 70(suppl II):176, 1984.

74. Meier B, Gruentzig AR, King SB, Douglas JS, Hollman J, Ischinger T, Aueron F, Galan K: Risk of side branch occlusion during coronary angioplasty. Am J Cardiol 53:10, 1983.

75. Leimgruber PP, Moldenhauer RT, Libow MA, Douglas JS, Gruentzig AR: Fate of occluded side branches after coronary angioplasty, abstracted. Circulation 70 (suppl II):296, 1984.

76. Block PC, Cowley MH, Kaltenbach M, Kent KM, Simpson J: Percutaneous angioplasty of stenoses of bypass grafts or of bypass graft anastomotic sites. Am J Cardiol 53:666, 1984.

77. Clark D, Stertzer SH, Myler RK, George B, Cieszkowski J, Murphy M, Rosen J: Coronary angioplasty: Saphenous vein graft and internal mammary artery angioplasty. (Submitted for publication)

78. Dorros G, Johnson W, Tector AJ, Schmahl TM, Kalush SL, Janke L: Percutaneous transluminal coro-

nary angioplasty in patients with prior coronary artery bypass grafting. J Thorac Cardiovasc Surg 87:17, 1984.

79. Jones EL, Douglas JS, Gruentzig AR, Craver JM, King SB, Guyton RG, Hatcher CR: Percutaneous saphenous vein angioplasty to avoid reoperative bypass surgery. Ann Thorac Surg 36:389, 1983.

80. Douglas JS, Gruentzig AR, King SB, Hollman J, Ischinger T, Meier B, Craver JM, Jones E, Waller JL, Bone DK, Guyton R: Percutaneous transluminal coronary angioplasty in patients with prior coronary bypass surgery. J Am Coll Cardiol 2:745, 1983.

81. Bourassa MG, Enjalbert M, Campeau L, Lesperance J: Progression of atherosclerosis in coronary arteries and bypass grafts: Ten years later. Am J Cardiol 53:102C, 1984.

82. Campeau L, Lesperance J, Bourassa MG: Natural history of saphenous vein aortocoronary bypass grafts. Mod Concepts Cardiovasc Dis 53:59, 1984.

83. Campeau L, Enjalbert M, Lesperance J, Bourassa MG, Kwiterovich P, Wacholder S, Sniderman A: The relationship of risk factors to the development of atherosclerosis in saphenous vein grafts and the progression of disease in the native circulation: A study 10 years after aortocoronary bypass surgery. N Engl J Med 311:1329, 1984.

84. Aueron F, Gruentzig AR: Distal embolization of a coronary artery bypass graft atheroma during percutaneous transluminal coronary angioplasty. Am J Cardiol 53:953, 1984.

85. Dorros G, Cowley MJ, Janke L, Kelsey SF, Mullin SM, Van Raden M: In-hospital mortality rate in National Heart, Lung and Blood Institute Percutaneous Transluminal Coronary Angioplasty Registry. Am J Cardiol 53:17C, 1984.

86. Alcan KE, Stertzer SH, Wallsh E, De Pasquale NP, Bruno MS: The role of intra-aortic balloon counterpulsation in patients undergoing percutaneous transluminal coronary angioplasty. Am Heart J 105:527, 1983.

87. Choy DS, Stertzer SH, Myler RK, Marco J, Fournial G: Human coronary laser recanalization. Clin Cardiol 7:337, 1984.

88. Fisher LD, Holmes DR, Mock MB, Pettinger M, Vlietstra RE, Smith HC, Ryan TJ, Judkins MP, Gosselin AJ, Faxon DP: Design of comparative clinical studies of percutaneous transluminal coronary angioplasty using estimates from the coronary artery surgery study. Am J Cardiol 53:138C, 1984.

89. Duprat G, David PR, Lesperance J, Val PG, Fines P, Robert P, Bourassa MG: An optimal size of balloon catheter is critical to angiographic success early after PTCA, abstracted. Circulation 70(suppl II):295, 1984.

90. Schmitz HJ, von Essen R, Meyer J, Effert S: The role of balloon size for acute and late angiographic results in coronary angioplasty, abstracted. Circulation 70(suppl II):295, 1984.

91. Busch UW, Sebening H, Beeretz R, Heinze R: Reliability of pressure recordings via catheters used for transluminal coronary angioplasty. Tex Heart Inst J 11:110, 1984.

14

Brachial Approach to Transluminal Coronary Angioplasty

SIMON H. STERTZER, M.D.

INTRODUCTION

The brachial approach to percutaneous transluminal coronary angioplasty is an interventional modification of the Sones (1,2) technique of coronary arteriography. It arose from the impelling developments in selective angioplasty conceived by Andreas Gruentzig (3,4). The brachial approach was predicated on the remarkable ease with which a single diagnostic catheter could be used by a trained operator to cannulate both the left and right coronary systems. Martin Kaltenbach and I were the first to adapt Gruentzig's technique of transluminal coronary angioplasty to a brachial guiding system (5,6). Dorros and I subsequently published the first report detailing the nature of this approach to coronary angioplasty in April of 1978 (6).

The original brachial guiding catheter system was indeed primitive, sharing with the early femoral guiding catheters the difficult and unwieldly drawbacks of an all-Teflon catheter. The first Gruentzig system from the femoral approach was modified by reducing the all-Teflon guiding catheter to a more complex extruded catheter having only a lining of Teflon. In conjunction with USCI, I then devised a specialized modification of the woven Dacron Sones catheter (Fig. 1), which was first fitted with a Teflon lining to conform to the required coefficient of friction of the polyvinylchloride dilation balloon. This woven Dacron catheter was rendered flexible in its terminal 3 inches by attenuating the thickness of its inner Teflon lining. The woven Dacron brachial type of catheter is manufactured in a number 8.3 or 8.0 French outer diameter and is entirely Teflon-lined despite the thinned-out last 2 or 3 inches. By attenuating the Teflon of this otherwise nontapered 8.3 catheter, the operator is then able to direct the guiding catheter up from the sinus of Valsalva into the left main, left anterior descending, circumflex, or right coronary arteries, "superselectively," in the majority of instances. More recently, the flexible tip guiding catheter has been customized with a longer flexible tip, as well as bent tips, according to the particular demands of an individual angioplasty case. Four-, five-, and six-inch tips have been specifically designed with straight and bent stylets for selective applications in coronary bypass grafts. This catheter is now being adapted to serve as a multipurpose guiding catheter for femoral entry, the length being extended to 110 cm.

Whereas the Sones taper is generally 1 or 1.5 inches, experience has shown that a 3-inch flexible guiding tip is unquestionably the most useful in dealing with the vicissitudes of anatomy of the aortic root and

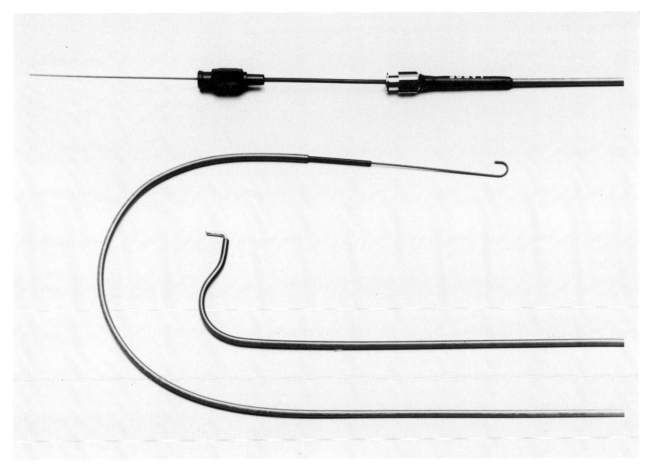

Figure 1. **Top**. Hub of guiding catheter with introducer and guidewire. **Middle**. Straight tip of brachial guiding catheter (Stertzer), showing introducer and J wire. **Bottom**. Bent-tip brachial guide with forming stylet in place.

coronary arteries in middle-aged and elderly patients. This is particularly true in hypertensive loss of elasticity of the innominate artery in the aortic arch.

The author has found that in addition to the attenuated Teflon lining of the guiding catheter tip, the introduction of a small, medium, or large stylet is frequently of some value in adjusting the guiding catheter to the size of the aortic root in an individual case. The size of the sinus of Valsalva cannot generally be determined from the plain chest radiograph but must instead be determined from the actual cine angiogram of the patient, or from fluoroscopy at the time of the angioplasty procedure itself. Whereas side holes were initially felt to be important to preserve flow in those arteries that appeared to be momentarily obstructed by the guiding catheter, they have not proved necessary in actual practice. The

technique of preloading and dilating the right coronary artery to avoid obstruction of the right main branch is described later. This maneuver obviates side holes in these brachial guiding catheters.

Therefore, the guiding catheter side holes, although still commercially available, have been eliminated from the standard USCI product. On the other hand, the introduction of a bent tipped (Castillo-Amplatz) stylet to some of the guiding catheters has clearly facilitated selective entry into the circumflex coronary artery, as well as into certain types of high and posterior takeoffs of the left main stem.

Despite the customized modifications, 85 percent of brachial approach angioplasty procedures can be performed with a medium stylet, 3 inch flexible, or straight tip Stertzer guiding catheter from the right brachial artery. Approximately 6 percent of the cases will require a bent tip stylet, and 2 or 3 percent may require either a large or small size stylet to accommodate a specific aortic root problem. One or two percent will require longer or shorter (i.e., 4–6 inch, or 1½–2 inch) flexible tips. This leaves approximately 2–3 percent of the cases in whom, for reasons illustrated by Figure 2, the brachial angioplasty cannot be performed from the right arm at all. In those 2 or 3 percent the operator should elect either to use the left arm or to go directly to the femoral approach. In those instances in which the left arm is employed, the brachial cutdown is performed from the left side, after which the operator returns to the right side of the table for routine guiding catheter manipulation.

Two specific instances preclude the use of the brachial guiding catheter in the right arm. First, there are those cases in whom the guiding catheter is totally immovable in the innominate artery (Fig. 2). This is indeed a rare event and is sometimes avoided by moving the right arm away from the patient's side to a 30 or 45° angulation. General tortuosity of the innominate artery is quite common and is usually very easily handled by careful catheter manipulation adjusted by critically timed deep inspiration. Second, there are cases in which aortic uncoiling is so severe that the guiding catheter moves posteriorly in a loop before turning in a hairpin to reenter the ascending aorta. These situations are diagrammed in Figure 3 and indicate that the femoral approach or the left arm approach is preferable. It should

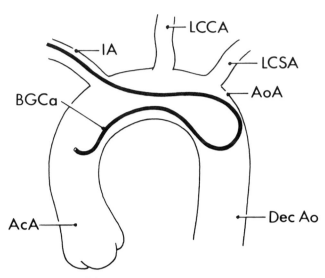

Figure 3. LCCA: left common carotid artery; BGCa: brachial guiding catheter; DecAo: descending aorta; AoA: aortic arch.

nonetheless be emphasized that brachial guiding catheter problems that preclude an angioplasty procedure should not be encountered more than 2 or 3 percent of the time in experienced hands.

GUIDING CATHETER MANIPULATION

The brachial guiding catheters differ significantly from the Sones catheter not only in the distribution of the Teflon lining but in the absence of taper in the terminal portion of the flexible tip as well. As a result, the brachial guiding catheters are prefitted with an introducer catheter (Fig. 1) that facilitates entry into the arteriotomy site. The introducer protects the artery from guiding tip trauma. This system has been so successful that no proximal entry damage (i.e., proximal dissection) to the brachial artery has ever been observed in 1008 usages of this guiding catheter in one center. Indeed the incidence of pulse loss in the right radial artery is lower than in general catheterization since the patients are fully heparinized during the angioplasty procedure. Local trauma to the brachial artery during excessive manipulation of a number 8.3 guiding catheter in a small vessel may sometimes occur but is significantly reduced by the operator's experience and ability to repair the vessel. However, no more than 1 percent of brachial catheterization cases should fall into the hands of a vascular surgeon for repair if the operator is sufficiently trained in the use of this system.

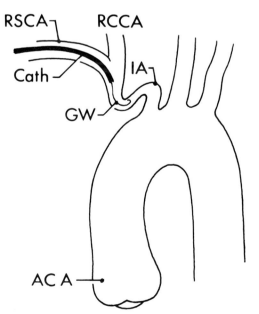

Figure 2. RSCA: right subclavian artery; RCCA: right common carotid artery; IA: innominate artery; AcA: ascending aorta.

The introduction of the brachial guiding catheter (Fig. 4), which is described later, is facilitated by the introducer as well as by the use of a flexible J guidewire that is fitted with a movable core so that it can become floppy in its last 3 or 4 inches. In the early days of this procedure, a Sones diagnostic catheter was used to perform the initial study, after which it was replaced by a guiding system over a 0.035 inch exchange wire. Experience has shown that this is no longer necessary, and all guiding catheter introductions are now made with a preloaded system. This means that the guiding catheter, with its introducer, is fitted with the flexible tip J wire, which is then advanced into the brachial artery according to the technique described next and shown in Figure 4. Note that the first catheter to enter through the arteriotomy is the introducer catheter, over the J wire which is already placed into the artery.

After entry into the brachial system, the guiding catheter is advanced to the root of the aorta over the guidewire. All of the inner hardware is then removed from the guiding catheter, and diagnostic, as well as therapeutic, procedures are performed. Should a guiding catheter need to be changed during the course of a procedure, the same technique is carried out without an exchange wire. In those cases in which entry into the ascending aorta was extremely difficult from either arm, it is then recommended to exchange the guiding catheter over a proper exchange wire.

Two general areas of comment regarding the brachial approach from the right arm should be stressed before going into further detail about the use of these tools. The first area relates to the so-called kinking of the juncture of the guiding catheter shaft with the flexible portion where the Teflon becomes attenuated. Because this "fishing rod" type of kinking is almost certainly due to operator inexperience and improper catheter handling, the angiographer should seek more expert advice before proceeding. This kinking is not an inherent quality of the number 8.3 catheter, and it is now a rare event. However, the emergence of a number 8.0 French, woven Dacron USCI-Stertzer catheter, for patients with smaller vessels, may indeed tend to kink more readily than the number 8.3 catheter with a firmer shaft. Repeated complaints, however, about kinking of the catheter at the junction of Teflon attenuation are almost certainly prima facie evidence of its misuse.

The second area related to the inability to control this guiding catheter involves the type of aorta that is so uncoiled in the arch that the catheter proceeds with a loop into the arch posteriorly and then returns anteriorly into the ascending aorta (Fig. 3). In these cases, the catheter advancement results in an increasingly large and posteriorly directed loop. When this cannot be easily corrected by deep breaths, it is highly likely that the brachial approach from the right arm is not feasible. As stated previously, the left arm or femoral approach should be immediately elected. On the other hand, the inability to hold a femoral guiding catheter in the right, circumflex, or left main coronary caused by anatomic configuration problems encountered from the groin should encourage a femoral angiographer to abandon that technique and proceed to the brachial method if he is experienced in it. This is preferable to inordinate femoral catheter exchanges or reshaping of the femoral catheter, which is not recommended by the manufacturer.

CATHETER DESIGN CHARACTERISTICS

The flexible tip brachial USCI-Stertzer guiding catheter is a woven Dacron number 8.3 or 8 French catheter with a 3 inch flexible tip and a medium "curve." Packaged sets may render small or large size curve for an appropriate match with the sinus of Valsalva, but the workhorse of brachial angioplasty is the 3 inch medium flexible tip. This is an untapered catheter and is particularly interesting in that the tip length is precisely twice that of the stan-

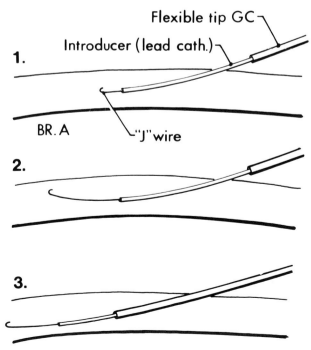

Figure 4. Brachial artery cutdown.

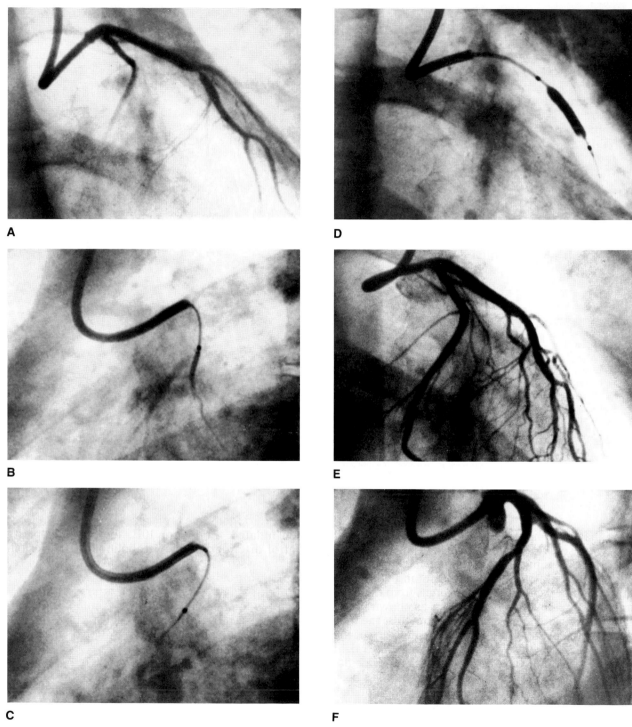

Figure 10. **A.** Anterior descending artery lesion in right anterior oblique view. Stenosis seen at the level of diagonal branch. **B.** Brachial guiding catheter with standard balloon in the diagonal branch. **C.** Brachial guiding catheter—left anterior oblique view. Dilation catheter redirected into the left anterior descending stenosis. **D.** Balloon inflation in right anterior oblique view. **E.** After dilation, right anterior oblique view. **F.** After dilation, left anterior oblique view.

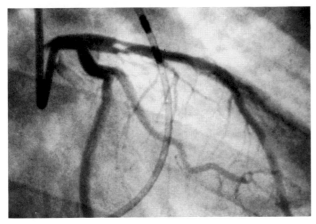

A

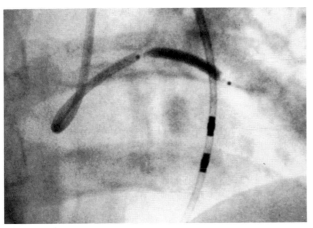

B

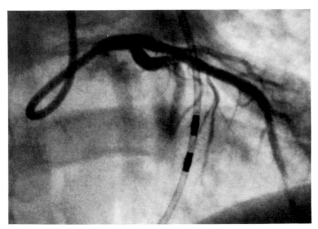

C

Figure 11. **A**. Anterior descending lesion in right anterior oblique view, upper vessel. **B**. Balloon inflation. **C**. After angioplasty.

to 5 mmHg. The use of the large 3 inch flexible tip guiding catheter is demonstrated also in Figure 12.

Flexible guiding catheters with 4, 5, and 6 inch tips are frequently used in the straight and bent tip varieties for direct entry into bypass vein grafts. The large loop 3 inch flexible tip guiding catheter is illustrated quite clearly in the high takeoff anterior descending coronary artery in Figure 12A. Dilation of a proximal anterior descending lesion is demonstrated in Figure 12B. Observe the large loop bridging this wide sinus of Valsalva and aortic root. Figure 12C shows the postangioplasty result. Another example of the very large loop 3 inch flexible guiding catheter is seen in Figure 13. Figure 13A shows the very large loop bridging the wide aorta into the high takeoff anterior descending coronary artery and also shows a high-grade proximal lesion of the anterior descending artery. Balloon inflation with a 3.7 mm nonsteerable USCI catheter is shown in Figure 13B. Figure 13C shows the postangioplasty arteriogram.

At the time of this writing, percutaneous transluminal angioplasty via the brachial or femoral approach has been successful in recanalizing the total occlusion (10) of the anterior descending artery in over 98 patients at the San Francisco Heart Institute. Figure 14 presents an example of this in a patient who did not have transmural infarction. In Figure 14A, the stump of the anterior descending coronary artery is in the right anterior oblique projection just to the left of the Myler catheter. The brachial guiding catheter is injecting most of the contrast material into the circumflex system. Balloon inflation is seen in Figure 14B, and reconstitution of the anterior descending artery lumen is seen in Figure 14C. Total occlusion without associated transmural infarction usually occurs in the presence of collateral circulation and is observed during the interim between the original diagnostic angiogram and the proposed transluminal angioplasty.

The ideal time to dilate the nonthrombotic total obstruction is during the few days immediately after it has taken place. Sometimes, this is associated with a clinical event of prolonged anginal pain or change in symptoms but is not related to the type of thrombus that occurs in full-blown acute myocardial infarction (11). Transluminal angioplasty in patients with thrombotic obstruction (12) associated with acute myocardial infarction is discussed in Chapter 21. Indeed, it is our feeling that a total occlusion that silently occurs between the diagnostic arteriogram and an electively scheduled angioplasty procedure is generally of such a nature that its composition is largely platelet material with only small amounts of fibrin. This generally encroaches upon a lumen that is already compromised in more than

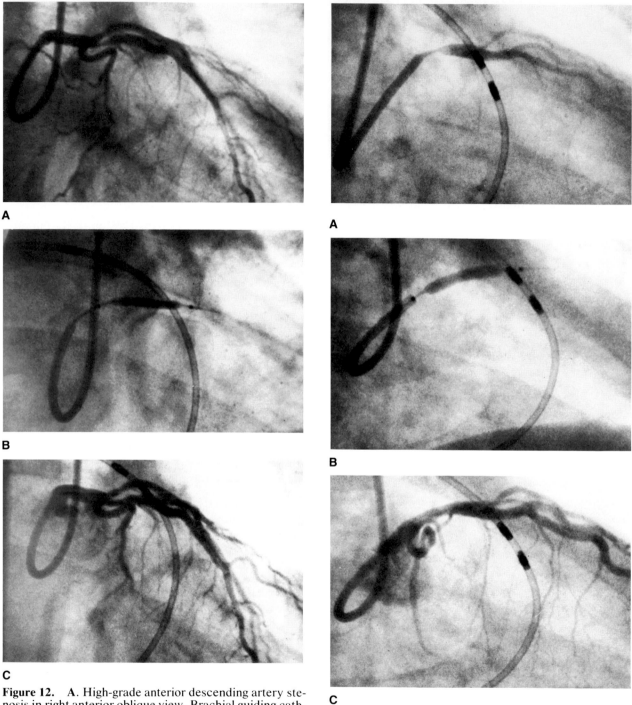

Figure 12.   **A**. High-grade anterior descending artery stenosis in right anterior oblique view. Brachial guiding catheter; stenosis seen between diagonal and circumflex vessels. **B**. Balloon inflation. **C**. After angioplasty.

Figure 13.   **A**. Large loop brachial guiding catheter—right anterior oblique projection. High-grade anterior descending artery stenosis, superselective, at orifice of the vessel. **B**. Balloon inflation. **C**. After angioplasty, right anterior oblique position.

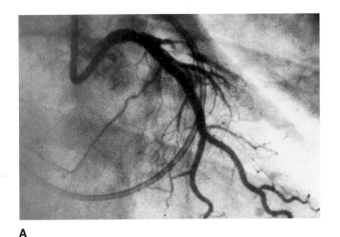

A

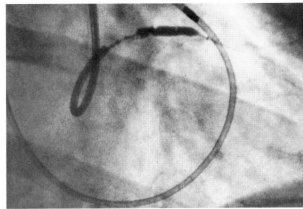

B

Figure 14.   A. Total obstruction of left anterior descending coronary artery just beyond obtuse marginal take-off. **B.** Balloon inflation. **C.** Reconstitution of anterior descending coronary artery in right anterior oblique position.

C

90 percent of its diameter. The platelet-fibrin matrix is easily remedied by angioplasty if it is not totally organized. Indeed, excellent results have been obtained in these categories of patients with total occlusion at the San Franciso Heart Institute.

One of the great advantages of brachial transluminal angioplasty in all three coronary vessels is the superselective quality of the flexible tip guiding catheter. This means that angioplasty can be facilitated by the ability to manipulate the guiding catheter tip well into the anterior descending, circumflex, or right coronary artery. Figure 15B illustrates how selective entry into the anterior descending system permits enough back pressure to force the dilation catheter through the extremely tight lesion seen in Figure 15A. Superselective guiding catheter advancement, when properly used, is helpful not only in developing forceful guiding catheter support for tight lesions but also in directing the balloon catheter at certain points of vessel tortuosity where inordinate bending of the steerable or nonstandard balloon

catheters may otherwise be required. The postangioplasty result in this case is shown in Figure 15C.

Superselective placement of the guiding catheter into the left anterior descending artery facilitating the balloon catheter entry into a rigid lesion is seen most graphically in Figure 16. In Figure 16A and B, an extremely high grade anterior descending lesion can be appreciated. In Figure 16C, the balloon is being inflated. In Figure 16D and E, the postangioplasty arteriograms show improvement and demonstrate the extent to which the anterior descending artery is entered with the guiding catheter to permit the dilation catheter to be forced through the tight lesion. This also permits excellent selective visualization of the vessel at the termination of the procedure. In Figure 17A, one can appreciate a high-grade tight anterior descending lesion immediately to the right of the pacing catheter; a Sones catheter opacifies the vessel. In Figure 17B, the guiding catheter is advanced up to the lesion to assist in the forcing of the dilation catheter through the lesion

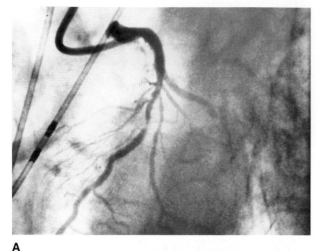

A

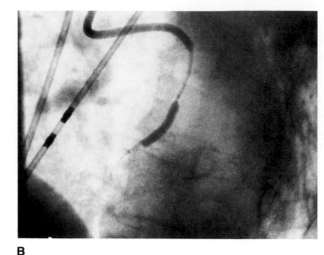

B

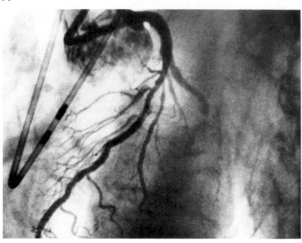

C

Figure 15. **A**. High-grade midanterior descending artery stenosis in left anterior oblique position. **B**. Superselective tracking of dilation catheter by brachial guide in left anterior oblique position; balloon inflation. **C**. After angioplasty. Guiding catheter retracted to proximal portion of left main stem.

that was otherwise impassable. The balloon is inflated in Figure 17C; note that the guiding catheter was retracted during the dilation. Figure 17D demonstrates the dilated segment. This maneuver corresponds to the counterclockwise rotation and advancement of the femoral guiding catheter into the anterior descending coronary artery under circumstances of severe lesion inflexibility.

An ideal example of superselective brachial guiding catheter entry into a vessel facilitating the "force-through" of an extremely tight lesion in the circumflex system is illustrated in Figure 18. In this case, an obtuse marginal vessel was the single vessel whose lesion caused a highly significant degree of angina pectoris in a 53-year-old patient. Objective evidence of ischemia was obtained by treadmill thallium. Figure 18A shows the guiding catheter, which was advanced right up to the circumflex lesion. Such force

was necessary to cross the lesion with the balloon catheter that the dilation catheter remained partly inside the guiding catheter during inflation. The hourglass configuration of the expanding balloon appears immediately to the left of the pacing catheter in Figure 18B. In Figure 18C, at 12 atmospheres of pressure, the lesion finally yielded. In 18D, the extensive, excellent opening of the lesion in the obtuse marginal artery is visible. The patient was relieved of his symptoms, and objective evidence of ischemia was reversed.

The role of transluminal coronary angioplasty in the treatment of proximal left anterior descending disease and in left main coronary stenosis is not appreciably different from that of the femoral or brachial approach. The treatment of left main disease is discussed in Chapter 13, on the transfemoral approach to angioplasty. Proximal left anterior de-

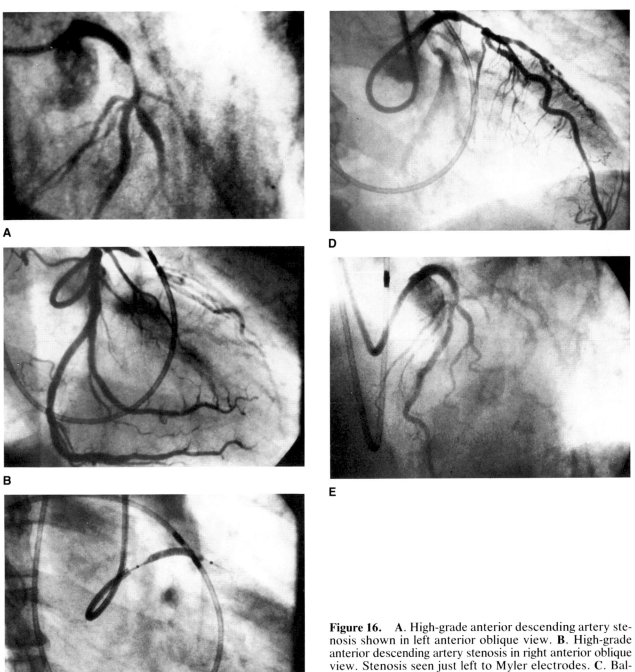

Figure 16. **A**. High-grade anterior descending artery stenosis shown in left anterior oblique view. **B**. High-grade anterior descending artery stenosis in right anterior oblique view. Stenosis seen just left to Myler electrodes. **C**. Balloon inflation, right anterior oblique view. **D**. Result of angioplasty. There are mild residual stenosis and spasm in right anterior oblique view. **E**. Marked improvement of anterior descending artery stenosis after angioplasty; left anterior oblique view.

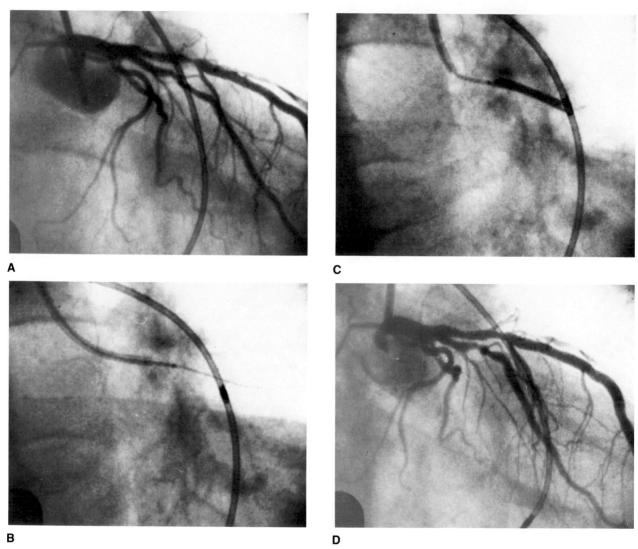

A

B

C

D

Figure 17. **A.** Stenosis in the midanterior descending artery just beyond pacing catheter. **B.** Superselective guide entry up to the lesion with balloon force-through. **C.** Guiding catheter retracted during balloon inflation. **D.** Result of angioplasty is seen directly under pacing catheter.

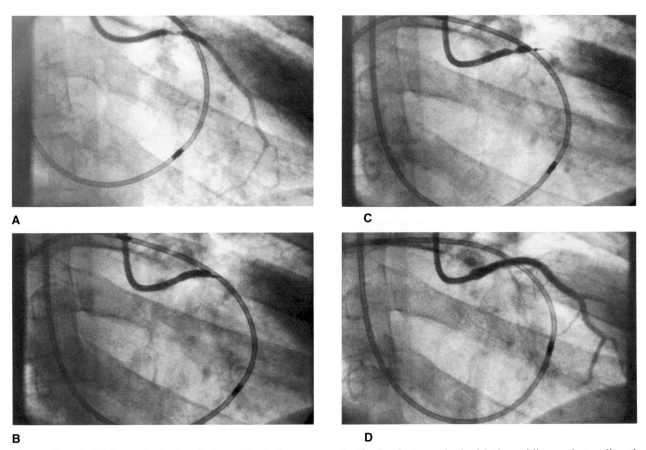

Figure 18.   **A**. High-grade, isolated circumflex lesion, superselectively photographed with the guiding catheter directly at the lesion. **B**. Balloon forced through with proximal portion of dilation catheter still within guide. **C**. Full balloon inflation. **D**. Result of angioplasty.

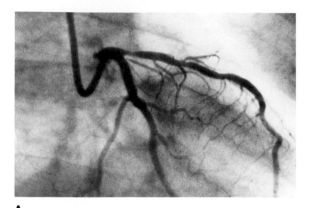

A

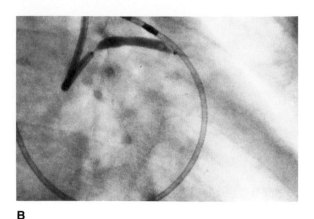

B

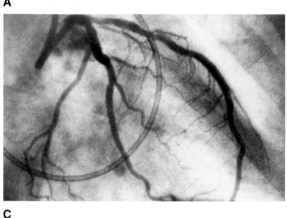

C

Figure 19. **A.** Origin stenosis of left anterior descending artery photographed with brachial guiding catheter in right anterior oblique view. **B.** Inflation of a 3.7 mm balloon. **C.** Result of angioplasty.

scending disease, however, may be divided into the so-called ideal, or hourglass, lesion beyond the ostium of the anterior descending artery and those lesions that originate almost at the origin, where it bifurcates with the circumflex artery from the left main stem. This origin lesion may have a higher recurrence rate and is particularly delicate since frequently the inflated balloon must lie across the left main stem trunk or circumflex orifice to induce a proper dilatation. Figure 19 demonstrates a proximal anterior descending lesion whose proper treatment involved the placement of a 3.7 mm balloon across the orifice of the circumflex and left main stem. This tight lesion required 9 atmospheres to reduce the gradient and to improve the midballoon constriction. The hazy area just beyond the origin of the left main stem in Figure 19C shows the post-angioplasty results. The prognosis of this type of lesion is not clear. It is the feeling of this author that a higher recurrence rate may accompany this type of lesion. The brachial guiding catheter is seen in all three views in the right anterior obliquity.

A similar situation is illustrated in Figure 20A, where a Sones catheter identifies a very high grade,

extremely proximal anterior descending artery lesion in an extremely short left main stem coronary artery. Indeed, by either the femoral or the brachial approach, this type of anterior descending artery may be entered selectively by almost any guiding catheter chosen. The brachial guiding catheter in Figure 20B has just been withdrawn from the left main coronary orifice as the inflated number 3.7 balloon straddles the short main stem and the high-grade anterior descending lesion. (In some patients, especially young women, when this lesion occurs as an isolated event, it has been speculated that it may represent a pathologic process different from coronary atherosclerosis. Indeed it may be the only lesion present, and it may be seen occasionally in premenopausal females. Its recurrence rate is generally higher than that of the angioplasty series in the older age group, and it appears to have an elasticity that does not predispose to optimal dilatation.) In Figure 20C, the effect of a successful dilation in this type of lesion can nonetheless be observed. In Figure 20D, the collateral circulation is visible transseptally from the right, into the anterior descending coronary system before dilation. In Figure 20E, the collateral circulation is no longer observable immediately after dilation. Figure 21, on the other hand, demonstrates the very proximal anterior descending lesion, again in the setting of the very short left main stem coronary artery. In this case, however, the process is indeed atherosclerosis, and the guiding catheter needed to be pointed directly into the vessel for the passage of the guidewire and 3.0 balloon. Because of obstruction that is caused, the guiding catheter was withdrawn once the balloon catheter was properly positioned in the lesion by that maneuver, as seen in Figure 21B. The difference in luminal caliber and distal flow between 21A and 21C can easily be appreciated.

The angulation of the guiding catheter superiorly into the dorsal surface of the left anterior descending coronary artery is an undesirable way to begin the passage of the guidewire and balloon catheter into the left main coronary artery. It is adjusted for by having the patient take a deep breath and rotating the guiding catheter clockwise so that it faces the anterior descending lumen directly. This rotation will tend (in vessels such as the one in Fig. 21) to totally obstruct the vessel momentarily. However, as soon as the balloon has been directed in the proper coaxial position, the guiding catheter can be rotated counterclockwise, withdrawing it out of the vessel altogether, as shown in Figure 21B.

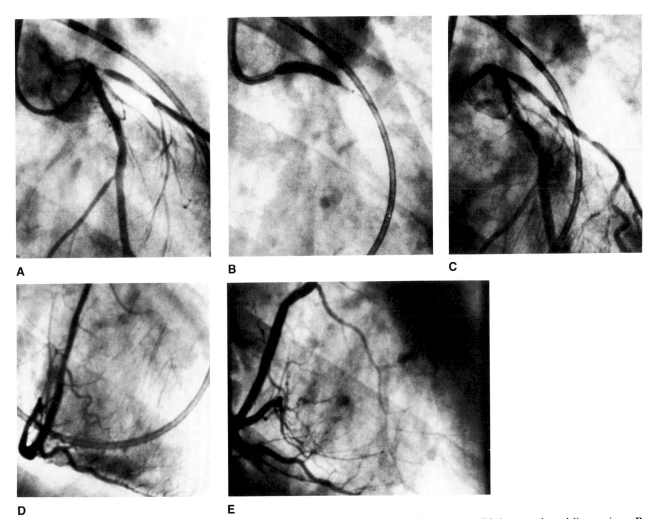

A   B   C

D   E

Figure 20.   **A**. High-grade stenosis at the origin of the left anterior descending artery. Right anterior oblique view. **B**. A 3.7 mm balloon has been inflated in the stenosis. **C**. Origin of left anterior descending artery after angioplasty, filmed through the guiding catheter. **D**. Right-to-left trans-septal collateral seen prior to angioplasty. **E**. Disappearance of collateral circulation after successful angioplasty.

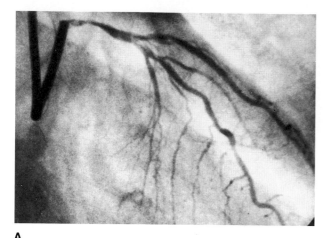

A

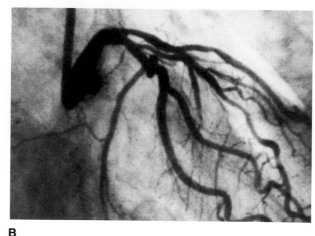

B

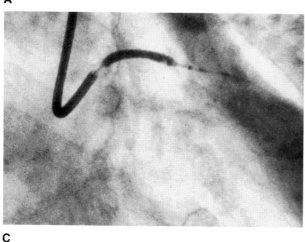

C

Figure 21. **A**. Long segment, high-grade stenosis of anterior descending artery filmed through guiding catheter, selectively injecting anterior descending ramus. **B**. Balloon inflation—right anterior oblique view. **C**. Reconstitution of anterior descending artery.

Selective entry of the brachial guiding catheter into an anterior descending, circumflex, or even a right coronary artery in an effort to gain otherwise unnegotiable access to the central lumen of a coronary vessel is not a feature of the brachial approach only. In Figure 22, an Amplatz guiding catheter from the leg was the only guiding catheter that permitted seating of a femoral guiding catheter into the left circumflex system so that enough power could be used to cross the 3.0 balloon through this tight lesion. In Figure 22B, the balloon is inflated; Figure 22C shows the postangioplasty result. The form of the Amplatz catheter is identical to the form assumed by the brachial guiding catheter with the bent tip (Fig. 1C) that is sometimes used for posteriorly directed left main stem coronary vessels, as well as for difficult circumflex lesions of this type.

As in femoral angioplasty, the choice of an incorrect brachial guiding catheter can present difficulties that tend to frustrate the operator and mis-

guide one's understanding of the technique. A high-grade midanterior descending lesion is seen to the right of the pacing catheter and opacified with a diagnostic Sones catheter, as shown in Figure 23A. A 3 inch brachial guiding catheter was used by the author to treat this lesion, as is seen in Figure 23B, but considerable manipulation of the guiding catheter was necessary because 3 inches of flexible tip was superfluous. As can be seen by the excessive slack of the flexible tip observed in Figure 23C, a 2 inch guiding catheter would have been preferable. The case illustrated in Figure 24, however, reinforces the importance of having 3–4 inches of flexible loop on this extremely rigid lesion, which required upward of 14 atmospheres of pressure to yield. This exceeded the limits of the manufacturer's specified balloon tolerance, and ultimately the balloon ruptured as the procedure ended. Nonetheless, considerable increase in intraluminal diameter can be appreciated in Figure 24A versus 24C. In Figure

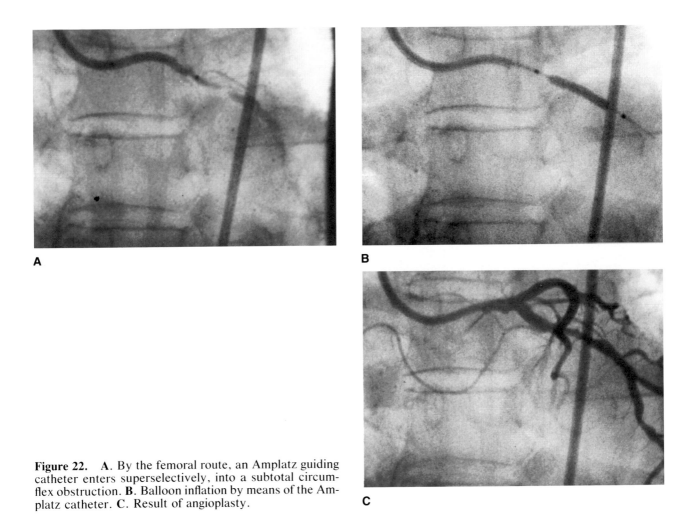

Figure 22.   **A.** By the femoral route, an Amplatz guiding catheter enters superselectively, into a subtotal circumflex obstruction. **B.** Balloon inflation by means of the Amplatz catheter. **C.** Result of angioplasty.

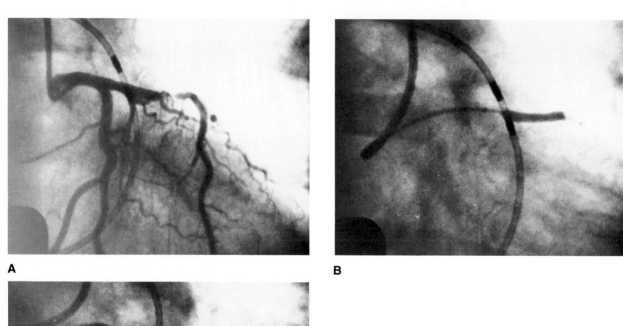

A

B

C

Figure 23. **A**. High-grade stenosis of anterior descending artery at level of pulmonary conus, filmed with Sones catheter. **B**. Guiding catheter withdrawal and balloon inflation of left anterior descending artery stenosis in right anterior oblique view. **C**. Postangioplasty guiding catheter injection in right anterior oblique view.

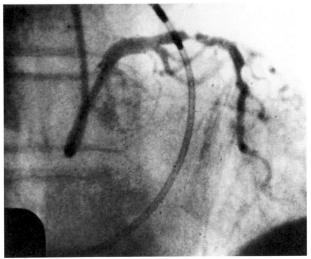

A

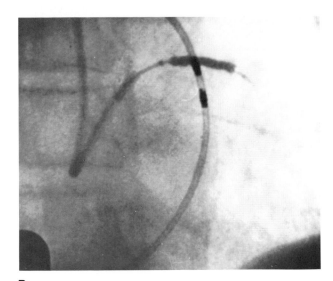

B

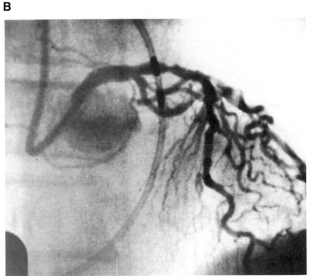

C

Figure 24. **A**. High-grade anterior descending artery stenosis filmed using large loop brachial guiding catheter. **B**. Large loop brachial guiding catheter placed selectively in anterior desending artery for balloon inflation. **C**. Result of angioplasty.

24A, the 4 inches of flexible brachial tip was used to gain power in the left coronary sinus to force the balloon catheter through the lesion. This large loop is sometimes used to gain similar entry into left anterior descending and circumflex saphenous vein grafts.

Right Coronary Artery

Perhaps the greatest advantage of the selective seeking tip of the flexible guiding catheter in the brachial method (13,14) is its use in the right coronary artery. Figure 25A shows the guiding catheter injection into

a severely stenosed proximal right coronary artery depicted in the left anterior oblique projection. Figure 25B shows that the actual luminal diameter of this stenosis was stringlike and extremely inflexible. Tiny branches around the stenosis had to be avoided. This maneuver involved the passage of the brachial guiding catheter right down to the lesion with injection of tiny amounts of contrast media through the dilation catheter to facilitate passage of the guidewire through the true lumen. It also permitted the great backup force needed to slide the number 3.7 balloon through a very tight lesion. The effect of dilation is seen in the postdilation arteriograms of Figure 25D and 25E.

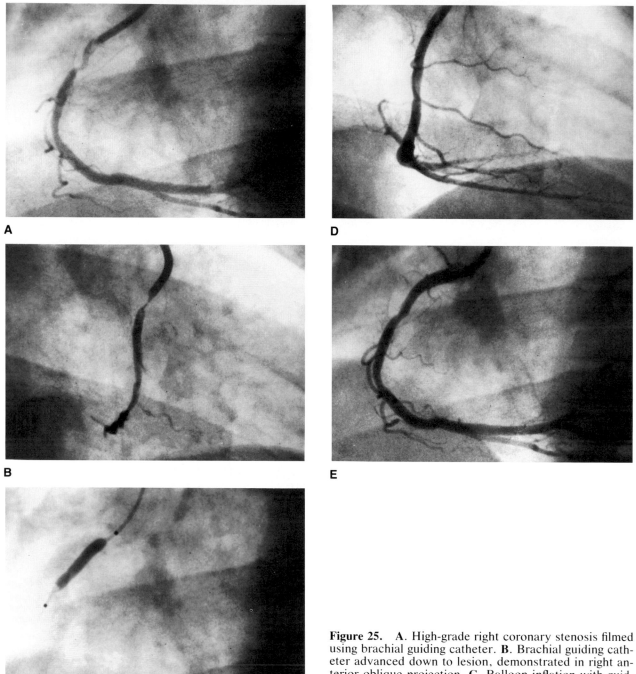

Figure 25. **A.** High-grade right coronary stenosis filmed using brachial guiding catheter. **B.** Brachial guiding catheter advanced down to lesion, demonstrated in right anterior oblique projection. **C.** Balloon inflation with guiding catheter totally withdrawn during inflation; left anterior oblique view. **D.** Result of angioplasty in right anterior oblique view. **E.** Result of angioplasty in left anterior oblique view.

The more classic use of the brachial method in dealing with difficult right coronary problems is seen in Figure 26. In 26A, the 2 inch brachial guiding catheter is observed identifying a high-grade stenosis in the vertical portion of the right coronary C loop. This stenosis was so tight that only a 2 mm balloon (Fig. 26B) could be passed through. This created an improved intraluminal channel (Fig. 26C). The only fashion in which the 2 mm balloon could be negotiated through the stenosis was by clockwise rotation of the guiding catheter down to the lesion itself, after requesting the patient to take a deep breath, and simply forcing the balloon system through. In our experienced hands, this maneuver alone has never dissected the right coronary artery. The channel created in Figure 26C permitted a dilation with a 3.7 balloon, as is seen in Figure 26D, with the result obtained in Figure 26E. This can usually be done at the same sitting, and with steerable equipment, the use of an exchange wire will safeguard against dangers of recrossing.

One of the areas of right coronary angioplasty that occasionally present serious clinical difficulty is the "shepherd's crook" type of lesion. This lesion is seen in Figure 27, where the proximal initial (usually upward) curve of the right coronary is affected by a high-grade lesion (Figure 27B). A low profile 20-20 USCI catheter appears within the lesion in Figure 27A. It will be noticed that to avoid obstruction of the vessel the guiding catheter has been withdrawn to the root of the aorta. This maneuver has replaced the necessity for side holes in the brachial method. A number 8 French guiding catheter, as well as the number 8.3 French guiding catheter with side holes, is nonetheless available for the rare instance in which such a device may be required. In Figure 27C the postangioplasty result is observed. This type of lesion not only is a difficult challenge at times but may have a higher predisposition to plaque splitting and emergency surgery. Such a propensity has been particularly noted in diabetic females.

A more classic right coronary stenosis is shown in Figure 28A in a patient with severely unstable angina pectoris. In Figure 28B this lesion is in the process of dilation. In order to cross the lesion in Figure 28B, the guiding catheter had to be advanced temporarily all the way down to the lesion. Once the lesion was crossed, the guiding catheter was withdrawn to the proximal segment of the artery before proceeding with balloon inflation. Figure 28C shows post angioplasty results. In Figure 29, 11 atmospheres of pressure was necessary to break this extremely hard midright coronary lesion. Figure 29C shows an excellent angiographic result in the left anterior oblique projection. In Figure 29A, the guid-

ing catheter is in a typical control angiographic position. In Figure 29B, the guiding catheter has been removed to the aorta so that it does not obstruct the vessel during the angioplasty procedure.

The advantage of using a brachial catheter in the right coronary artery to penetrate the vessel deeply relates to the force that may be judiciously applied to cross a difficult lesion and to directing the catheter through an eccentric stenosis. In Figure 30A, a high-grade stenosis can be seen in the left anterior oblique position in a proximal right coronary artery. The filling defect that is apparent below the guiding catheter tip is due to layering of contrast. In this nonsteerable system, crossing the lesion was difficult because of its basic eccentricity. In Figure 30B, the balloon has already crossed the lesion, but the guiding catheter has been brought down very close to the lesion in an effort to facilitate the direction of the wire tip of the balloon catheter through the eccentric stenosis itself. Indeed, with deep breathing by the patient, varying position of the guiding catheter, and use of steerable or preformed nonsteerable balloon catheter, passage through an eccentric stenosis is greatly facilitated. The result in Figure 30C is excellent despite the irregular indentation of this highly nonuniform stenotic lesion.

Bypass Vein Grafts

The advantages of transluminal coronary angioplasty by the brachial method are best appreciated in the right coronary artery. Anterior descending coronary angioplasty is frequently identical in efficiency by either the femoral or brachial technique. Certain circumflex lesions are generally better approached by one method or the other, depending on the specific problems encountered when the less desirable technique is employed. The technique of choice is frequently impossible to determine before actually entering the aorta with one guiding catheter or the other.

Saphenous vein graft dilation, on the other hand, is a particularly difficult subject about which to generalize. Dilation of right vein grafts frequently entails a vertical descent into the graft itself. This is almost always extremely easy with the use of the right brachial artery technique. The left brachial technique and a straight or El Gamal catheter from the groin are generally acceptable alternative approaches. In circumflex grafts, experiences have found the femoral approach to be totally satisfactory, although the brachial method is clearly feasible. The anterior descending grafts are quite variable and sometimes best approached by the brachial

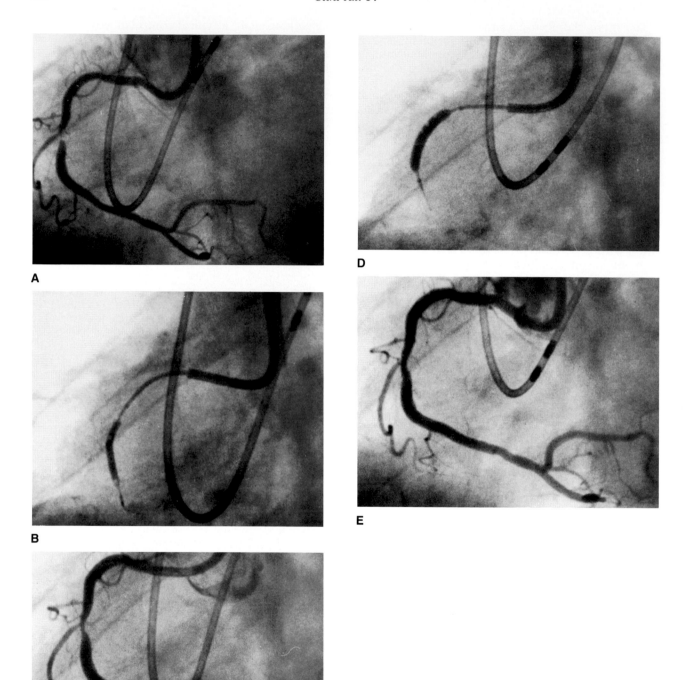

A

B

C

D

E

Figure 26.   **A.** Midright coronary stenosis with guiding catheter at orifice of vessel, Myler catheter in pulmonary artery; left anterior oblique view. **B.** Guiding catheter used to facilitate entry of 2.0 mm balloon into high-grade stenosis. **C.** Improvement in lumen of the artery after angioplasty. **D.** Nonsteerable, larger balloon replaced through coronary stenosis to effect full dilatation. **E.** Final result; left anterior oblique view.

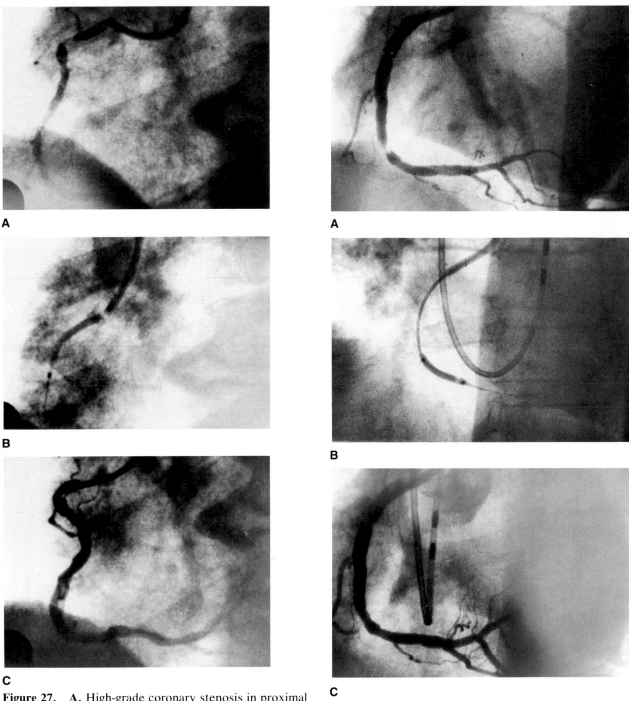

Figure 27.   A. High-grade coronary stenosis in proximal portion of right coronary as it turns into acute marginal branch. **B.** Balloon inflation of the stenosis. **C.** Shepherd's crook anatomy after dilation; left anterior oblique view.

Figure 28.   A. High-grade right coronary stenosis; left anterior oblique view. **B.** Distal balloon inflation. **C.** Result of angioplasty in left anterior oblique view.

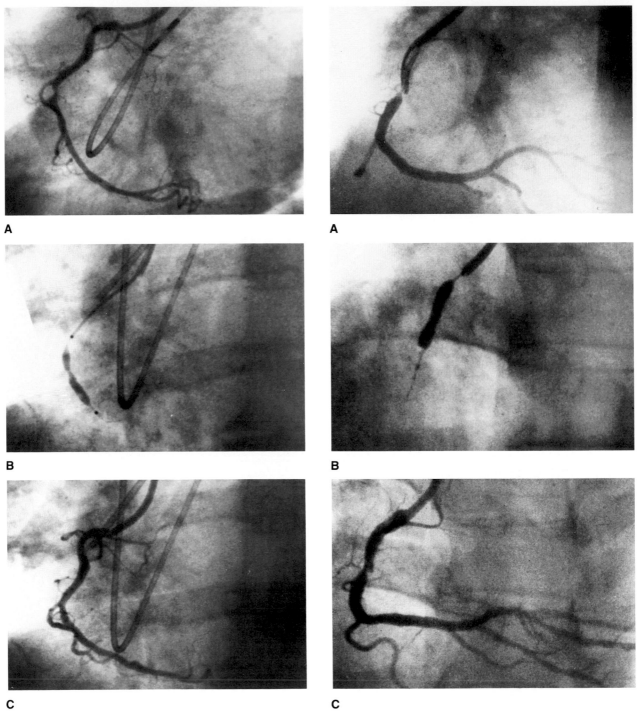

Figure 29. **A**. Midright coronary stenosis; left anterior oblique view. **B**. Note indentation during initial balloon inflation. **C**. Result of angioplasty; left anterior oblique view.

Figure 30. **A**. High-grade right coronary stenosis; proximal right coronary view. **B**. Inflation with a 3.7 mm balloon. **C**. Result of angioplasty; left anterior oblique view.

method. This is clearly an individual anatomic decision and is a subject that cannot be clarified until the procedure is actually begun.

In Figure 31, for example, a high-grade stenosis is seen in the midsegment of a vein graft to the anterior descending artery. By the brachial method, in Figure 31B, inflation of the graft stenosis can be observed. In Figure 31C, flow has been reestablished into the graft with excellent distal perfusion. Right graft stenosis at the distal anastomosis site is illustrated by Figure 32. This is characteristically the most successful area for a graft to be dilated and is best diagnosed and treated within the first year of the graft operation. In Figure 32B, the balloon is inflated, and in Figure 32C the postdilation result is

clearly identified. Figure 33 demonstrates a technique facilitating deep penetration into a complex graft.

In Figure 33A the brachial guiding catheter has entered an anterior descending graft, but the dilation catheter cannot be extended around the curve at the tip of the guidewire because of inability to develop back pressure. The so-called Carr maneuver, named after Dr. Matthew Carr of Florida Medical Center, is employed to deepen the graft penetration by the guiding catheter. The balloon is inflated at its point of farthest entry and gently tugged upon, causing simultaneous advance of the guiding catheter over the balloon catheter. The guiding catheter then advances into the graft, providing further back pres-

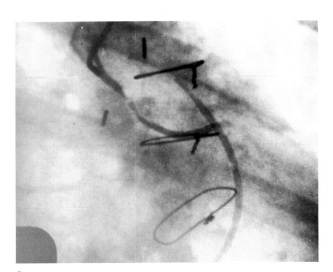

A

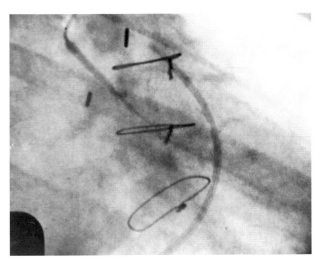

B

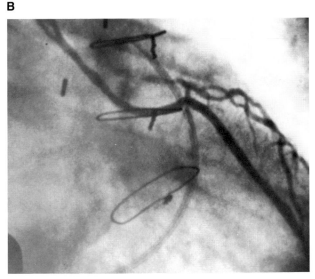

C

Figure 31. **A.** Midanterior descending graft 3 months after surgery; right anterior oblique view. **B.** Balloon inflation. **C.** Result of angioplasty shows marked increase in distal filling of the graft and perfusion of the anterior descending artery.

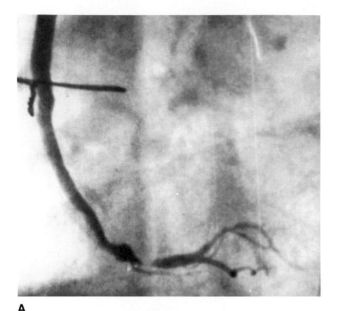

A

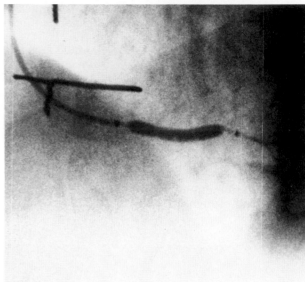

B

C

Figure 32. **A**. High-grade insertion stenosis of right coronary artery saphenous vein graft; left anterior oblique view. **B**. Close-up view of 3.7 mm balloon inflation in distal anastomosis. **C**. Result of angioplasty; left anterior oblique view.

sure for balloon advancement, as seen in Figure 33B. At the termination of this maneuver, as depicted in Figure 33C, one observes the excellent flow into an anterior descending coronary vessel whose distal insertion site stenosis has been dilated by combination of the brachial guiding catheter approach and the Carr maneuver.

CONCLUSION

In summary, the brachial approach to transluminal coronary angioplasty is a very useful adjunct to the femoral procedure. Experts in brachial catheterization can perform angioplasty from the right or left arm with a success rate clearly equal to that of the femoral approach. Nonetheless, these techniques

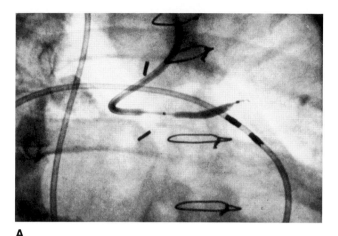

A

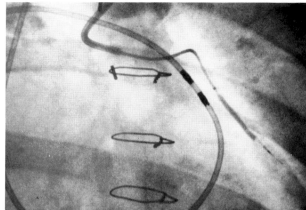

B

Figure 33. **A**. Brachial guiding catheter entry into the origin of left anterior descending artery graft. Balloon inflation facilitating Carr maneuver. **B**. Dilation catheter advancement into distal left anterior descending coronary stenosis. **C**. Postdilation view of anterior descending artery stenosis with marked antegrade and moderate retrograde filling of the anterior descending system.

C

should not be considered competitive. In 85 percent of all cases, either approach will entail the same degree of ease or difficulty. It is reiterated, however, that 10 percent of cases will be markedly facilitated in time and degree of difficulty by selecting the superior approach. Sometimes the selective cannulation of a circumflex or anterior descending vessel from the arm is the only way to avoid multiple guiding catheter exchanges and other unwieldy maneuvers from the groin. Contrariwise, inordinate tortuosity of the innominate system may render entry into the left main coronary orifice from the right brachial approach an extremely arduous task. Under those circumstances the left arm or the Judkins femoral approach may quickly alter the degree of difficulty represented by a given case. Frequently, however, the ideal approach cannot be predicted before the case is in progress.

The ability to go from one approach to the other

so greatly widens the armamentarium of the skilled operator that many more successful results can be achieved in a shorter time. Most strikingly, the ability to perform both the diagnostic arteriogram and the dilation procedure with the brachial guiding catheter has proved to be unquestionably easier than using the shorter tipped Sones catheter. This knowledge will clearly permit wider familiarity with the brachial approach of angioplasty. Meticulous care to the artery in terms of cutdown and repair are of course essential to ensure the proper utilization of this technique.

Finally, certain lesions, especially those located in the right coronary artery and in grafts, may be dilatable only by a brachial approach. It is certainly prudent for all angioplasty operators to glean some knowledge of the potential of the brachial approach whether or not they intend to incorporate it into their own laboratory procedures.

REFERENCES

1. Sones FM, Shirey EK, Proudfit WL, Wescott RN: Cine coronary aortography. Circulation 20:773, 1959.
2. Sones FM, Shirey EK: Cine coronary arteriography. Mod Concepts Cardiovasc Dis 31:735, 1962.
3. Gruentzig AR: Transluminal dilatation of coronary artery stenosis—experimental report, in Zeitler E, Gruentzig A, Schoop W (eds): *Percutaneous Vascular Recanalization*. Berlin, Springer-Verlag, 1978, pp 57–65.
4. Gruentzig AR, Senning A, Siegenthaler WE: Nonoperative dilatation of coronary-artery stenosis: Percutaneous transluminal coronary angioplasty. N Engl J Med 301:61–68, 1979.
5. Dorros G, Stertzer SH, Bruno MS, Kaltenbach M, Myler RK, Spring DA: The brachial artery method to transluminal coronary angioplasty. Cathet Cardiovasc Diagn 8:233–242, 1982.
6. Dorros G, Stertzer S, Kaltenbach M, Bruno MS: The brachial artery approach to (percutaneous) transluminal coronary angioplasty, abstracted. Circulation 62(suppl III): III–161, 1980.
7. Dorros G, Stertzer S, Kaltenbach M, Myler R, Spring S (Lutheran Hospital of Wilwaukee, Inc., and Medical College of Wisconsin, Milwaukee, Wisconsin): Coronary angioplasty: Comparison of brachial and femoral artery approaches. Presented at American College of Cardiology, San Francisco, March 15–19, 1981.
8. Dorros G, Stertzer S, et al: Coronary angioplasty: A comparison of brachial and femoral artery methods. Presented at Fourth Symposium on Coronary Heart Disease, Frankfurt/Main, May 18–19, 1981.
9. Dorros G, Stertzer SH, Myler RK, Kaltenbach M, Spring D: Transluminal coronary angioplasty: Comparison of the brachial and femoral methods. Cathet Cardiovasc Diagn 9:547–553, 1983.
10. Stertzer SH, Myler RK, Wallsh E, Bruno MS, DePasquale NP: Dilatation of obstructed coronaries by percutaneous transluminal angioplasty. J Cardiovasc Med 5:1059–1064, 1980.
11. Kimbiris D, Iskandrian A, Saras H, et al: Rapid progression of coronary artery stenosis in patients with unstable angina pectoris as seen in patients scheduled for coronary angioplasty, abstracted. J Am Coll Cardiol 1:724, 1983.
12. Rentrop P, Blanke H, Karsch KR, Kaiser H, Kostering H, Leitz K: Selective intracoronary thrombolysis in acute myocardial infarction and unstable angina pectoris. Circulation 63:307–317, 1981.
13. Krajcer Z, Boskovic D, Angelini P, Leatherman LL, Springer A, Leachman RD: Transluminal angioplasty of right coronary artery: Brachial cutdown approach. Cathet Cardiovasc Diagn 8:553–564, 1982.
14. Krajcer Z, Boskovic D, Angelini P, Springer A, Leatherman LL, Leachman RD: Transluminal angioplasty of right coronary artery: Improved success rate with brachial approach, flexible-tip guiding catheter, abstracted. Circulation 64(suppl IV):IV–252, 1981.

15

Independently Movable Guidewire Techniques for Percutaneous Coronary Angioplasty

STEPHEN N. OESTERLE, M.D.
JOHN B. SIMPSON, M.D.

INTRODUCTION

The treatment of atherosclerotic peripheral vascular disease by transvascular intervention was originally introduced by Dotter and Judkins over 20 years ago (1). The dilation catheter system they used was eventually modified for use in coronary arteries by Andreas Gruentzig. The initial clinical publications appeared in the United States literature in 1979 when Gruentzig described his first coronary interventions in humans (2).

The early experience with angioplasty in the United States was acquired with first-generation Gruentzig catheters, which were designed with a guidewire fixed to the tip of the balloon catheter. As angioplasty techniques evolved with this catheter system, it became clear that the fixed guidewire had limitations in dealing with varying anatomic differences in the length and curvature of coronary arteries. These limitations often precluded angioplasty attempts on branching or tortuous vessels.

At Stanford University in 1979, John Simpson began development of an alternative catheter system for percutaneous transluminal coronary angioplasty (PTCA). The fundamental difference from the system introduced by Gruentzig was the incorporation of a removable guidewire in the central lumen of the dilation catheter, similar to the concept long used in angiography (radiology) for selecting vessels in peripheral vascular system. This Simpson system permitted the independent movement of a guidewire into and out of the balloon catheter, allowing for repetitive shaping of the wire as the dilation catheter was advanced along the coronary vessel. The guidewire could be advanced, withdrawn, reshaped, and reinserted into the dilation catheter so that serial advancement of the guidewire and the dilation catheter could proceed smoothly toward the target lesion. This design also offered the theoretical advantage of enabling the operator to withdraw the dilation catheter from an area of dilatation (if symptomatic myocardial ischemia became problematic) while leaving the guidewire well beyond the stenotic segment in the artery. This guidewire position would maintain access across the area of previous dilatation. Once coronary blood flow was restored and the oxygen debt to the distal muscle repaid, the dilation catheter could again be safely advanced across the area of dilatation over the guidewire without fear of dissecting beneath the "split" atheroma into the intima. Leaving the guidewire distal to the area of stenosis also obviates the necessity of renegotiating branch vessels, both proximal and distal to the area of stenosis. Figure 1 illustrates the fundamentals of this concept.

The selection of movable guidewires for coronary angioplasty was initially quite limited. Until 1982 we exclusively used a Cook standard 0.018 inch guidewire that could be preshaped to facilitate advancement of the dilation catheter into proximal coronary segments. This was an extremely stiff wire, and we found that coronary dissections and plaque disruptions were not uncommon during attempts to cross lesions. Figure 2 is a schematic diagram of the Cook standard wire, illustrating the stiff nature of the safety wire. Figure 3 illustrates the potential complications

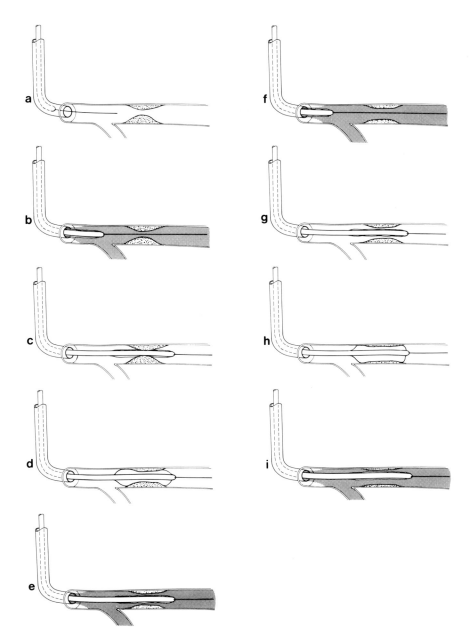

Figure 1. Fundamentals of a movable guidewire system. **a.** Guiding catheter, dilation catheter, and movable guide in the proximal coronary artery. **b.** Contrast injected through guiding catheter, demonstrating wire position. **c.** Dilation catheter advanced over guidewire into the area of stenosis. **d.** Inflation of balloon. **e.** Balloon deflated and contrast injected through guiding catheter, documenting persistent flow in the distal vessel. **f.** Dilation catheter withdrawn into guiding catheter with guidewire left across the area of stenosis to maintain position. Guiding catheter injection demonstrates angiographic improvement in the area of previous dilation. **g.** Dilation catheter readvanced into the area of stenosis over guidewire. **h.** Repeat inflations. **i.** Balloon deflated and contrast injections through guiding catheter, demonstrating angiographic improvement and persistence of flow in the distal vessel.

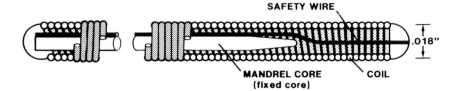

Figure 2.   The Cook standard wire.

encountered with this wire in attempts to cross the irregular lumen of a high-grade stenosis.

The first significant evolution in angioplasty guidewire technology occurred when we began to "fracture" this standard wire. This fracture was accomplished by exceeding the tensile strength and actually breaking the safety wire. When properly performed, this maneuver left the Cook wire with a floppy distal tip that significantly improved both the safety and success rate for crossing high-grade stenoses. To some degree this floppy tip was flow-directed and had the potential to be "streamed" through an area of stenosis. Alternatively, it had the ability to prolapse across areas of narrowing with minimum disruption of the plaque. The advent of this first prototype "floppy" guidewire significantly increased our success in coronary angioplasty. Nevertheless, this modified wire had several major drawbacks. First, it required a certain degree of operator understanding and experience to fracture the wire successfully in a useful way. Second, the maneuver of fracturing the safety wire defeated its essential purpose which was to prevent the mandrel from penetrating the coils of the wire. Figure 4 demonstrates one of the potential complications encountered with this fractured wire.

Another major limitation of this modified standard 0.018 inch guidewire was the external diameter. The 0.018 inch shaft precluded both measurement of the distal pressure gradient and injection of constant medium through the tip of the dilation catheter as the guidewire rested in the central lumen of the balloon catheter.

Despite the multiple limitations of this early floppy guidewire, we were able to use a movable guidewire successfully to perform coronary angioplasty (3).

Since that report, not only the guidewires but also the dilation and guiding catheters have been extensively modified. There now is an array of movable guidewires that allow for various degrees of shapability, flexibility, and steerability. In addition there has been marked reduction in the external shaft diameter of these wires, allowing for continuous measurement of pressure and distal injection of contrast while the wire rests within the central lumen of the dilation catheter. All of the catheter manufacturers have converted to this movable guidewire system, which utilizes an independently movable and steerable guidewire.

In this chapter we will discuss the details of the Simpson angioplasty system, including specifics of guidewire construction and use, dilation catheter construction, and various guiding catheter options. With a well-based understanding of the technical features of the angioplasty catheter system, the various techniques and strategies applied to a particular clinical problem become more predictable. The selection of an appropriate guiding catheter for various anatomies will be discussed. The criteria for selecting dilation catheters will be reviewed, along with the advantages of distal contrast injections and the measurement of trans-stenotic pressure gradients. Most importantly, the strategy for guidewire selection will be extensively discussed. Following these sections on equipment and strategy will be a brief summary of our clinical results over a 5-year period, using a movable guidewire system. The discussion concludes with a general guide and formula for patient selection based upon this experience.

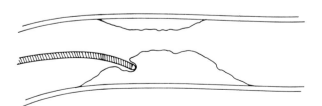

Figure 3.   "Stiff" guidewire disrupting intimal plaque.

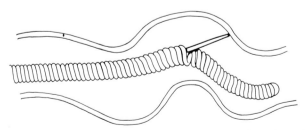

Figure 4.   "Fractured" safety wire protruding through coils of guidewire and injuring intima.

FUNDAMENTALS OF EQUIPMENT

Dilation Catheters

The Simpson-Robert (ACS) dilation catheter is manufactured from two concentric pieces of polyolefin tubing. The outer member is contiguous with the balloon wall and provides for a *deflated* outer diameter varying from 0.045 to 0.065 inches, depending upon the catheter selected. The central lumen accepts a spectrum of guidewires up to 0.018 inches in diameter. Figure 5 is a diagram of the various components of this catheter system. Clearance between the inner and outer members of the dilation catheter allows for infusion of contrast down the shaft of the catheter to inflate the distal balloon segment. While the balloon is initially filled with contrast, a stainless steel hypodermic tubing (vent tube) is advanced to the tip of the balloon catheter to purge the air bubble. When the balloon is completely filled with contrast, the hypodermic tubing is pulled back from the tip of the catheter and sealed. There are two gold markers on the shaft of the dilation catheter that facilitate fluoroscopic demarcation of the balloon and one additional gold marker at the tip.

The Simpson-Robert catheter has five inflated diameter sizes: 2.0 mm, 2.5 mm, 3.0 mm, 3.5 mm, and 4.0 mm. The length of the balloon segment is standardized for all catheters at 20 mm. The total length of the catheters ranges from 120 to 140 cm. We have found this selection of catheters sufficient for all but the extremely unusual case. Figure 6 is a graphic illustration of the Simpson-Robert balloon catheters in profile with the standard 0.038-inch guidewire as a scale reference. As can be seen, the 2.0-mm dilation catheter has a deflated external diameter similar to that of the standard 0.038-inch guidewire. As is evident in this diagram, the substantial differences in the deflated external diameters of the various catheters may be quite important in providing for a successful crossing of a high-grade stenosis.

The component of flexibility in the dilation catheter is of importance when selecting the most appropriate catheter. The dilation catheter should have sufficient flexibility to allow the balloon and shaft to negotiate the abrupt bends commonly found at the takeoff of the left circumflex artery or the branching points of obtuse marginal segments. Similarly, the proximal right coronary artery often has a steep upward takeoff (the so-called shepherd's hook), and catheter flexibility is the key to successful advancement of the dilation catheter into the distal vessel. Although flexibility is frequently desirable, the dilation catheter must also have sufficient stiffness to allow advancement of the catheter without buckling and to provide sufficient support while crossing areas of high-grade narrowing. The tip of the Simpson-Robert dilation catheter is tapered. To some degree, there is a bougenage effect as the catheter first enters the stenosis. Insufficient catheter stiffness, however, is rarely a reason for

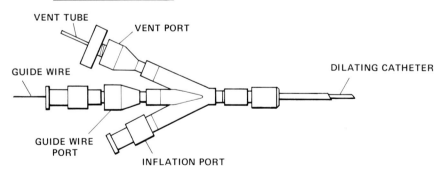

TRIPLE-ARMED ADAPTER

VENT TUBE

VENT PORT

GUIDE WIRE

DILATING CATHETER

GUIDE WIRE
PORT

INFLATION PORT

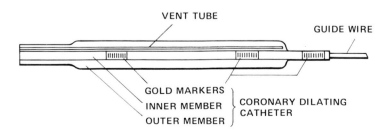

VENT TUBE

GUIDE WIRE

GOLD MARKERS
INNER MEMBER   CORONARY DILATING
OUTER MEMBER   CATHETER

Figure 5. The ACS Simpson-Robert dilation catheter.

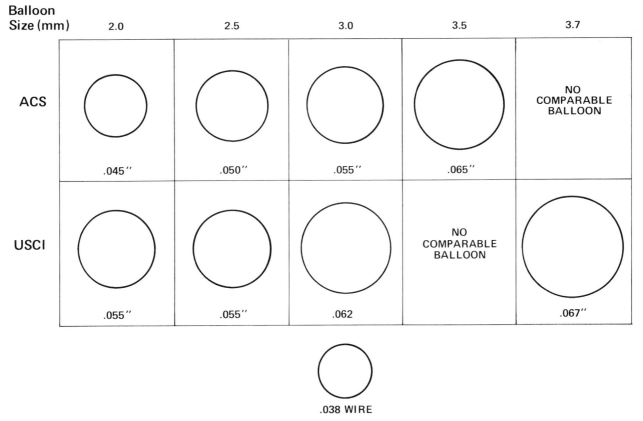

Figure 6. Relative deflated balloon profiles.

primary failure to cross an area of stenosis. We are more frequently foiled by the catheter's inability to negotiate tortuosities.

We have found *stiffness* of the catheter to be an asset with straightforward left anterior descending coronary artery lesions and *flexibility* to be of greater importance in approaching stenoses in both the left circumflex and right coronary arteries. The strategy for selection of various dilation catheters will be discussed in a later section.

Guiding Catheters

The small diameter and marked flexibility, combined with relatively long length, make the dilation catheter essentially a "noodle." In addition it has little torque control; hence, it is necessary to "load" it into a guiding catheter that can be moved and torqued in the aortic root. The guiding catheter is engaged in the coronary ostium and directs the dilation catheter and guidewire as they enter the proximal coronary artery. The guiding catheter also sup-

ports the proximal shaft of the dilation catheter as it is advanced across stenoses. Without the support of a guiding catheter, the dilation catheter will buckle either in the aorta or proximal coronary artery, precluding forward movement in the coronary artery. The fixed curves in the guiding catheter tip (similar to those in the standard angiographic catheters) are also used to advantage with various guidewire applications.

The Simpson system utilizes several different sizes and configurations of guiding catheters depending on the coronary anatomy encountered. The standard ACS guiding catheter is a single lumen catheter with number 8.8 French (we often call it *9 French* for convenience) outer diameter. These guiding catheters have a laminar construction with an outer layer of polyolefin and an inner layer of teflon.

First-generation left guiding catheters were manufactured with out-of-plane tips to facilitate subselective positioning in either the left anterior descending or left circumflex coronary arteries. The newer guiding catheters have had a substantial improvement in torque. We have found that one can now achieve the same subselective positioning of the left guiding catheters with simple clockwise or counter-clockwise torque.

Guiding catheters are also manufactured with a number 8 French outer diameter (USCI–Myler guiding catheter). Although contrast injections are more difficult, in certain situations the smaller guiding catheter is very useful.

Guidewires

Movable guidewires are the key feature of the Simpson angioplasty system. With an understanding of these guidewire constructions, one can predict both their uses and limitations in a given clinical case. Before the details of each wire are described, the features of an ideal guidewire will be outlined. Figure 7 lists the cardinal virtues of a movable guidewire for coronary angioplasty. *Shapability* is the ability to modify the shape of the wire to facilitate subselection of coronary segments at abrupt turns and bifurcations. *Steerability* is a feature that allows for changes in the direction of the wire as it is advanced into more distal portions of the coronary vessel by torque control.

In order for a steerable wire to be useful, at least two requirements must be satisfied. First, it cannot be straight; it must have a curve or J shape on its tip in order to select a direction by torque control. Second, one must be capable of mechanically transmitting torque along the entire shaft of the wire, so that the shaped tip of the wire is steered toward a desired direction.

Flexibility is of key importance in attempting to cross stenoses. The distal tip of the wire should be sufficiently flexible to allow passage through the irregular or tortuous lumen of a coronary lesion without "lifting" the plaque or dissecting the coronary artery. *Visibility* is of varying importance, depending on the imaging equipment used. The diameters of guidewires are small enough that fluoroscopic resolution becomes a limiting factor, especially in larger patients. The movement of the guidewires is a delicate maneuver, and a clear image is obviously of great importance. Various metal alloys, including gold, platinum, and tantalum, have been incorporated into the recent generation of guidewires to enhance fluoroscopic imaging. The standard 0.018-inch guidewires have, until recently, lacked enhanced fluoroscopic visibility, and use of these metals in guidewires has dramatically improved the ease of guidewire manipulation under fluoroscopy.

The "ideal" guidewire has yet to be introduced. Invariably, a compromise in one of the desirable features is necessary to improve a specific feature of the guidewire; for example, increasing torque usually decreases flexibility. Increasing flexibility will generally diminish the shapability of a wire. Because of these practical compromises, we have found it necessary to develop a "family" of guidewires, and we frequently utilize all of them in the course of a single angioplasty procedure. Before addressing the strategy of guidewire selection, we will discuss the engineering design of the various guidewires currently in use. One may be able to use the guidewires effectively and safely by understanding their fundamental features.

The standard 0.018-inch guidewire is illustrated in Figure 2. As mentioned, this wire was used as the first removable guidewire for coronary angioplasty in 1979. It is essentially a coil, not unlike a tightly wound "slinky", wrapped around a mandrel, with a safety wire welded between the mandrel and the end of the coil. This wire has a uniform outer diameter of 0.018 inches along the entire shaft. The safety wire inside the coil shell is malleable and allows for shaping of the wire. Although the coil is quite flexible, the safety wire inside buttresses it to such an extent that true "floppiness" is lost. As previously mentioned, this wire can be made into a floppy wire by applying an excessive pull at the distal tip and actually fracturing the welded junction between the coil shell and the safety wire.

Figure 8 is a schematic diagram of the ACS 0.018 inch standard wire. This wire is similar in construction to the standard guidewire but has several important modifications. First, the distal tip is significantly more flexible than that of the Cook standard wire. Instead of a stiff safety wire, a flexible shaping ribbon extends from the mandrel to the tip of the coil. While preventing coil detachment, this shaping ribbon also provides for modifications in the configuration of the distal tip of this wire. As in the Cook standard guidewire, there is a uniform outer diameter of 0.018 inches, precluding distal pressure measurements and contrast injections when engaged with the balloon catheter. Although the distal tip is significantly more flexible than that of the Cook standard guidewire, we still do not think this is an ideal crossing wire.

Figure 9 is a schematic representation of a second-

GUIDE WIRE FEATURES

- SHAPABILITY
- STEERABILITY
- FLEXIBILITY
- VISIBILITY

Figure 7.

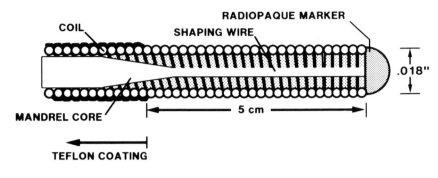

Figure 8. ACS standard guidewire.

8/83

generation wire, the ACS pressure-dye-torque (PDT) wire, introduced in 1982. The wire combines the flexibility features of a coreless wire at the distal end with a stiff main shaft, facilitating torque control. The coil is limited to the distal end of the mandrel and includes a short safety ribbon that also allows for shaping of the tip. The addition of a J configuration at the tip of the PDT wire provides for transmitted torque to direct the tip. The outer diameter of the distal coil is 0.018 inches, which is connected to a 0.009 inch proximal steel shaft. The lower diameter of the shaft allows a margin of space within the central lumen of the balloon catheter and permits distal pressure measurements and contrast injections when the distal coil segment of the wire is advanced beyond the tip of balloon catheter. The PDT wire also has gold markers in the distal coil for an improved fluoroscopic visibility.

The ACS floppy wire is illustrated in Figure 10. This wire has an identical construction to that of the ACS standard wire with one exception: The distal coil has no shaping ribbon, providing maximum flex-

ibility in the tip. As previously mentioned, the floppy tip can be either flow-directed or prolapsed across most of the lesions with maximum success and minimal complications. Although torque is not a key feature of this wire, there is potential for changing the tip direction of the floppy wire with torque along the shaft. The ACS 0.018 inch exchange wire (300 cm in length) has the same floppy distal tip construction.

The *USCI steerable wire* is depicted in Figure 11. This wire was also introduced in 1982 and is conceptually similar to the ACS PDT wire. The shaft has an attenuated outer diameter of 0.016 inches and consists of an outer hypotube with a solid, stainless steel inner core. The hypotube ends 25 cm proximal to the top of this wire. The inner core is tapered, with a platinum coil wound over the distal 25 cm. This wire has a thin safety wire, which provides support for the distal coil and gives it shaping potential. By shaping the distal tip into a J configuration, one can apply torque along the shaft of this wire, changing the direction of the tip. Its low profile

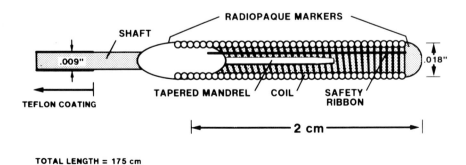

Figure 9. ACS PDT straight guidewire.

8/83

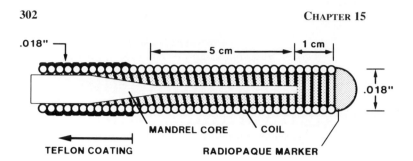

TOTAL LENGTH = 175 cm

8/83

Figure 10. ACS floppy guidewire.

shaft also allows for distal contrast flushes and pressure measurements; however, the PDT wire, which has both of these features, is more effective. The tip of the USCI wire is slightly stiffer than that of the ACS PDT wire. What the USCI wire loses in flexibility it gains in enhanced torque.

PROCEDURE STRATEGY

Guiding Catheter Placement

A critical review of the predilation coronary arteriogram is essential before selecting the most appropriate guiding catheter. Not only is the vessel involved important but so is the location of the lesion within the artery. Figure 12 illustrates three common anatomical situations encountered when approaching a proximal left anterior descending (LAD) stenosis. In Figure 12-1 there is a short left main artery that bifurcates into the LAD and left circumflex arteries. Although a standard Judkins left

number 4.0 angiographic catheter would usually be appropriate for angiographic study, a Judkins left number 3.5 guiding catheter will seat in the left main coronary ostium with an upward direction of the catheter tip. This superior and anterior position provides a favorable direction for advancing the guidewire into the anterior descending artery. As the guidewire is extruded out of the guiding catheter, it will usually seek the superior aspect of the left main artery toward the proximal left descending coronary artery. In this anatomical situation, the Judkins left number 4 guiding catheter will seat in a more horizontal position, providing equal access to the left circumflex artery. A similar strategy (Fig. 12-2) for a short guiding catheter is employed when one wants to introduce a wire selectively into the LAD coronary artery in a situation in which the LAD and left circumflex arteries have a common origin. When the left main coronary segment is long (Fig. 12-3), the upward seating of the guiding catheter is less advantageous, and a Judkins left number 4 guiding catheter will provide greater support for advancement of the dilation catheter across the stenosis.

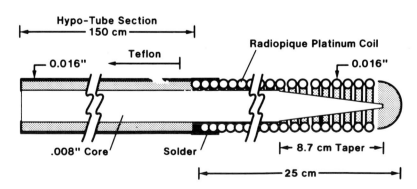

8/83

Figure 11. USCI steerable guidewire.

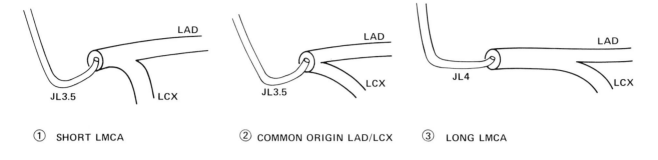

① SHORT LMCA   ② COMMON ORIGIN LAD/LCX   ③ LONG LMCA

Suggest: 9F, Judkins Configuration, JL3.5 or JL4
8F, Amplatz—rarely needed

Figure 12. LAD guiding catheter selection.

With long left main segments, the number 3.5 guiding catheter is not advantageous because the superior orientation will be lost by the time the wire has reached the bifurcation of the LAD and left circumflex arteries.

We generally use a number 9 French guiding catheter when working in the left coronary system because damping is rarely a problem. The larger guiding catheters have better torque and allow for better injection of contrast around the balloon catheter.

Angioplasty in the right coronary artery is frequently more problematic than in the LAD artery primarily because of difficulties associated with the guiding catheter. Seating of the guiding catheter within the right coronary ostium can often be difficult, and advancement of the balloon catheter through a tortuous right coronary artery can be precluded by poor guide support. A common right coronary artery configuration is schematically illustrated (Fig. 13A) as

the so-called shepherd's hook. The standard Judkins right number 4 catheter configuration will usually seat well in this vessel; however, the marked upward initial takeoff and tortuosity of the artery diminish guiding catheter support as the dilation catheter is advanced into the vessel and across the high-grade stenosis. The right coronary configuration illustrated in Figure 13B is also frequently encountered. This anatomy tends to draw the guiding catheter deep into the artery. Damping is often a problem in the right coronary artery regardless of the guiding catheter selected; when it is a major problem, we frequently change to a number 8 French guiding catheter. By going to a smaller guiding catheter one compromises both torque and ability to inject contrast around the dilation catheter. If damping continues to be a problem with the number 8 French catheter, we reattempt engagement with a guiding catheter with Amplatz configuration. ACS has re-

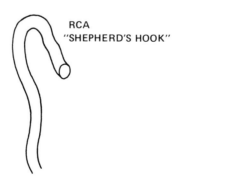

RCA
"SHEPHERD'S HOOK"   RCA

Suggest: Judkins R4 (9F/8F)   Suggest: Judkins R4 (9F/8F)
         Amplatz rarely useful   Amplatz R1/R2 (9F/8F)

Problems: Guideline catheter <u>SUPPORT</u>   Problems: Good support, but
          for dilitation catheter is <u>POOR</u>   damping is common

Figure 13. Right coronary artery guiding catheter selection (see text for details).

A   B

Suggest: 9F Judkins R4
Amplatz R1

Problems: TORQUE, STABLE POSITION WITHIN OSTIUM, DILITATION CATHETER SUPPORT can all be problematic.

Figure 14. Bypass graft guiding catheter selection.

cently introduced a new generation of right coronary guiding catheters with side holes. While compromising proximal dye injections, these side holes effectively "vent" the proximal coronary and allow continued flow through the vessel. The right guiding catheters are also utilized for selective engagement of saphenous vein grafts (Fig. 14).

Guiding catheter problems in the left circumflex artery are similar to those encountered in the right coronary artery. Providing adequate support for advancement of the dilation catheter can be difficult. Figure 15 illustrates several anatomical variations of the circumflex artery. In 15-1 there is an abrupt takeoff of the circumflex coronary artery in a long left main segment. Selection of a guiding catheter plays little role in the successful advancement of a guidewire into the left circumflex. Likewise the guiding catheter provides minimal support for advancement of the dilation catheter in the left circumflex after the shaft has made the first abrupt bend out of the left main segment. Figure 15-2 shows an oblique takeoff of the left circumflex coronary artery. Guiding catheter tip configuration plays a minimal part in selecting the circumflex, although this anatomy is more favorable for supporting advancement of the dilation catheter across a stenosis. Figure 15-3 illustrates a short left main where torquing of the guiding catheter posteriorly can help select the left circumflex coronary artery. This is also an anatomy favorable for supporting the dilation catheter as it is advanced into the left circumflex. A number 9 French Judkins left number 4 guiding catheter is generally selected for all three of these commonly encountered anatomies.

The role of guiding catheters has a critical impact on the success rate of coronary angioplasty, especially for the left circumflex and right coronary arteries. Recent development efforts have been directed toward a more flexible guiding catheter that could be introduced deeper into the coronary arteries. At the present time, we are limited to the standard number 8 and 9 French guiding catheter systems with their attendant problems; Figure 16 summarizes the relative merits and disadvantages of both systems.

The guiding catheters presently used are stiff catheters and *cannot* be manipulated as conventional angiographic catheters are. If they are used improperly, the potential for proximal coronary dissection is high. With careful attention to technique, the guiding catheter can be used safely and imaginatively to enhance success during angioplasty.

Dilatation Catheter Use

The strategy for selecting dilation catheters centers around several factors: (1) native vessel size, (2) morphology of the lesion, and (3) presence of multiple stenoses.

The size of the native vessel is the most important determinant for selecting a dilation catheter. We choose the appropriate dilation catheter after esti-

Guiding Catheter Potential			
Selecting LCX	—	+	++
Support for Dilitation Catheter	—	+++	++

Suggest: 9F Judkins L4

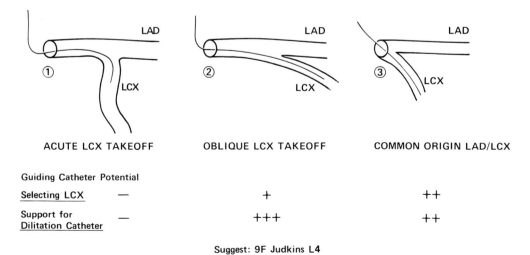

Figure 15. Left circumflex coronary artery guiding catheter selection (see text for details).

		↓DAMPING	↓SHEATH SIZE	↑TORQUE	↑SUPPORT	DISTAL PRESSURE/ CONTRAST
Figure 16. Relative merits of guiding catheters.	9F	—	—	++	++	+++
	8F	++	++	+	+	—

mating the diameter of the native vessel at both the proximal and distal sides of the stenosis to be dilated. The larger the native vessel, the larger the dilation catheter selected. Although some feel the degree of stenosis should dictate the selection, we have found that most dilation catheters can usually cross an area of stenosis if a wire can be successfully positioned through and beyond the lesion. Occasionally an exchange for a smaller dilation catheter is necessary; however, the result is compromised unless a repeat dilation with a larger balloon is performed. A stenosis may be so critically narrowed that initial dilation requires a catheter with the lowest deflated profile. As mentioned previously, a partial dilation employs a low profile catheter, and the catheter is then exchanged over a 300 cm exchange wire for a larger dilation catheter.

Morphology of the lesion to be dilated plays a lesser role in the selection of a dilation catheter: whether the lesion is tubular, focal, or diffuse has little impact. If the lesion is tubular and quite lengthy, one may select a customized dilation catheter with a longer balloon length. Alternatively, one can per-

form sequential inflations as one moves the dilation catheter over the guidewire placed through and beyond the length of the stenosis.

When there are multiple stenoses to be dilated, the issue of selecting dilation catheters becomes more critical. Frequently, one will encounter sequential stenoses within a single arterial segment (Fig. 17). This is a variation on the diffuse atherosclerotic lesion already noted. The selected dilation catheter may be appropriate for a proximal stenosis, but the more distal stenosis may be in the tapering portion of the native vessel. The dilation catheter selected for the proximal lesion may be oversized for the distal lesions; conversely, a dilation catheter sized for the most distal lesion may be inappropriately small for the proximal lesions. One needs to plan for possible exchange over a long guidewire in situations such as these.

Multivessel angioplasty presents another situation requiring careful dilation catheter selection. Frequently one encounters high-grade stenoses in both the major vessel and several side branches. The dilation catheter most appropriate for the major

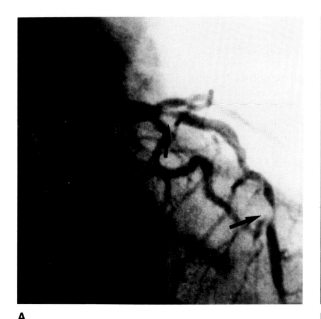

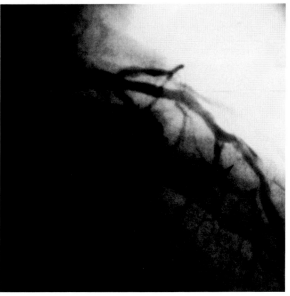

A   **B**

Figure 17. **A.** Sequential stenoses in a single arterial segment. **B.** Catheter was oversized for distal lesion and dissection occurred.

vessel is oftentimes oversized for the branch vessels. The initial selection obviously will depend on which lesions can be addressed primarily and which ones secondarily.

Once a balloon catheter has been selected and the lesion crossed and dilated, the issues of inflation pressures and duration arise. There are many different opinions and practices in this regard. We tend to utilize maximum inflation pressures in the 100- to 120-psi range. The average duration of inflations is 30 seconds. We occasionally go to higher pressures and longer inflations when performing a repeat angioplasty in the setting of restenosis.

Furthermore, we usually attempt angioplasty with larger inflated balloon sizes when dealing with restenoses. Since we have little understanding of the mechanisms of restenoses, these strategies are, at present, based on supposition.

Guidewire Application

We have covered some strategy concerning the selection of guidewires while discussing the distinctions among various guidewire constructions. Again, a complete understanding of the technical features of each guidewire will permit more imaginative and efficacious use.

The ACS 0.018 inch standard guidewire is very useful for introducing the dilation catheter into the proximal aspect of the LAD or left circumflex coronary arteries. The proximal portion of the LAD artery has numerous septal and diagonal branches. A straight guidewire will commonly "dive" into one of these side branches or, alternatively, enter the circumflex artery. As illustrated in Figure 18, one can preshape the ACS standard guidewire with a Judkins left configuration. As the guidewire is advanced out through the guiding catheter, the tip of the secondary curve will seat in the tip of the guiding catheter, directing the wire in a superior direction. This preformed (anterior descending artery) curve will generally keep the guidewire out of both the

circumflex artery and the other proximal side branches of the anterior descending artery.

Alternatively, if the left circumflex artery is to be selected, the wire can be preshaped with a "circumflexcurve" to favor subselective cannulation of the circumflex artery. As seen in Figure 19, the secondary curve is 180° opposite to the primary curve. The secondary curve will again seat in the guiding catheter, now directing the tip of the guidewire inferiorly. This shaping of the standard guidewire is particularly helpful when the left circumflex coronary artery has an abrupt takeoff from a long left main artery.

Both left anterior descending artery and circumflex artery modifications to the standard wire are useful for subselecting the proximal segment of respective arteries. With the guidewire placed into the proximal artery, the dilation catheter can be advanced over the wire toward the target lesion. This wire is relatively stiff and is not intended to be used to cross the lesion. When the dilation catheter is advanced into the proximal segment of the intended artery, the preshaped standard (0.018 inch) stiff wire is replaced by a floppy-tipped guidewire for crossing the target lesion.

The floppy guidewire has several strategic advantages. We have found it to be the safest wire for crossing stenoses because it usually does not disrupt an eccentric plaque or dissect the coronary artery. The floppy distal tip of the guidewire will "seek" the channel of a high-grade stenosis because it is relatively flow-directed. Once positioned beyond the stenosis, the stiff shaft of the floppy guidewire provides an excellent support for the dilation catheter to follow. This wire is particularly useful in negotiating the proximal right coronary artery when it is exceptionally tortuous. The 0.018 inch floppy guidewire's main drawback is also characteristic of the standard wire: its caliber precludes both pressure measurements and distal contrast injections.

The PDT wire combines both a flexible tip for crossing and a shapable tip for steerability. In addition, its low profile shaft allows for pressure mea-

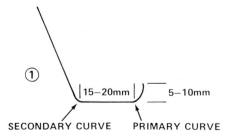 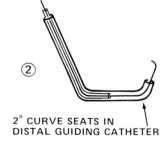

Figure 18. LAD modification of the ACS standard guidewire.

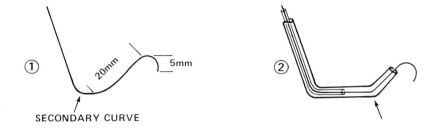

Figure 19. Left circumflex coronary artery. Modification of the ACS standard guidewire.

surements and satisfactory contrast injections at the balloon catheter tip.

The ability to measure pressure at the tip of the dilation catheter is an important feature of an angioplasty system. Not only does the trans-stenotic pressure gradient indicate when the balloon catheter has crossed the lesion, but it also provides an objective assessment of hemodynamic improvement by balloon dilation. One of our desired end points for dilation has been the reduction in the trans-stenotic pressure gradient to 10 mmHg or less. If the tip of the dilation catheter wedges into a small side branch distal to the dilated lesion, it can cause an artificially elevated postdilation gradient. This can be prevented by confirming the position of the catheter tip by contrast injections through the balloon catheter. The PDT wire is sufficiently flexible for safe crossing of the coronary lesions. We prefer either this PDT wire or the floppy wire for crossing most lesions, once the balloon catheter has been positioned in the proximal segment of the target artery.

The J tipped PDT wire can be torqued to change the direction of the guidewire advancement. The PDT J wire is strategically important in negotiating the course of the artery when there are multiple side branches. By advancing and withdrawing while applying torque, one can usually avoid the side branches encountered. Alternatively, this PDT J wire can subselect the side branches of the left anterior descending and circumflex arteries. We have also found that the PDT J wire is useful when attempting to cross eccentric lesions. Guidewires can hang up on shelves

or within the ulcer created in an eccentric lesion. One can usually torque a PDT J wire through such meandering channels with a significantly better success rate than without.

The USCI steerable wire has similar application capabilities to those of PDT guidewire. The USCI wire is quite shapable and is an excellent wire for selecting the proximal LAD or left circumflex arteries. The USCI wire has exceptional fluoroscopic visibility.

The ACS HI-torque floppy (Fig. 20) is a prototype guidewire that is a second-generation PDT wire. It is as flexible and shapable as the PDT wire, while improving torque and visibility. The 0.016 inch shaft of this catheter reduces the potential for contrast injections and high-fidelity pressure measurements.

The ACS SOF-T guidewire is a second generation standard wire. It is schematically diagrammed in Figure 21. The tip has enhanced flexibility and visibility. Although shapability is diminished, its uses are similar to those of the ACS standard wire. This wire is suitable for crossing stenoses.

Table 1 lists the family of removable guidewires currently used in coronary angioplasty, contrasting the relative merits of these various guidewires. All of these wires can be left distal to the lesion in an artery, thus providing a safe and easy access for recrossing the dilated lesion as necessary. Over any of these guidewires, one can advance the dilation catheter into an area of stenosis, perform multiple inflations, and withdraw the dilation catheter to assess flow while leaving the wire straddling the area

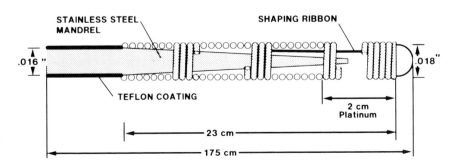

Figure 20. ACS high-torque floppy guidewire.

tempted. Again, no controlled comparison with surgery has been conducted, and the potential benefit of angioplasty for each of these patients is unclear.

Although left main coronary artery disease could be considered two-vessel disease, we have eschewed angioplasty on isolated left main coronary artery lesions. These lesions are technically quite straightforward, but the 5 percent potential for coronary occlusion seems unacceptable in the face of the anticipated infarction and possible death.

Angiographically documented total occlusions had been thought to be inaccessible to coronary angioplasty. Recently, there have been multiple reports of successful dilation of vessels with total occlusions (6,7). We have occasionally approached total occlusions with dilation catheters in situations in which the distal vessel has been well defined through collateral flow during injection of the contralateral artery. If the occlusion has been present for more than 3 months, angioplasty is rarely successful in reopening the vessel.

Experience with movable guidewires has proved to be of value as an adjunct during the treatment of acute myocardial infarction with streptokinase. Operators in some centers proceed directly to angioplasty in the setting of an acute myocardial infarction, without antecedent thrombolytic therapy. The decision to intervene in acute myocardial infarction with streptokinase alone, angioplasty alone, or the combination of thrombolysis and dilation varies from institution to institution. There has been no controlled comparison of these approaches. At Stanford University, streptokinase is given intravenously in the emergency room via the femoral vein conduit. Patients are then taken to the cardiac catheterization laboratory, where coronary angiography is performed. If the acutely occluded vessel has not recanalized after streptokinase, we will frequently probe the area of thrombosis with a guidewire. The wire is advanced through the angiographic catheter via a rotating Y connector with an O ring. Utilizing the same techniques for guidewire advancement in angioplasty, one can advance the wire to the area of occlusion. Gentle probing of the thrombosis is possible, and we have frequently been able to pass a guidewire into the distal vessel. The guidewire can then be removed, and the residual channel provides additional surface area for circulating streptokinase to interact with the thrombus. We have found the USCI steerable wire to be ideal in this situation. Its relatively stiff distal coil provides some support for traversing the thrombus, however, it is not so stiff as to dissect the coronary artery easily. A patient left with a residual high-grade stenosis following streptokinase is frequently taken back to the catheterization laboratory within 48 hours. Abrupt re-

closure following streptokinase generally occurs in the 48- to 72-hour postinfusion "window," when the effects of streptokinase are attenuated and coagulation parameters return to normal (8–10). We have found this to be the most appropriate time to perform a semielective angioplasty in patients who have undergone a streptokinase infusion.

Each physician must devise an individual patient selection algorithm. Initially criteria should be extremely conservative. We would recommend that one begin angioplasty on straightforward left anterior descending coronary lesions that are proximal and generally not greater than 95 percent stenotic. As one develops facility in handling the guidewires and guiding and dilation catheters, more complex angioplasty can be undertaken.

Patient selection should also be guided by the quality of surgical backup in the institution where one performs angioplasty. Without exception, angioplasty should be performed only in an institution where cardiopulmonary bypass capabilities are available. Angioplasty on patients who have significant myocardium at jeopardy should be performed after sufficient consultation with the surgical staff in an institution where there is a high level of confidence in the standby cardiac surgery team.

CLINICAL RESULTS

Since Simpson began to use the movable guidewire system for coronary angioplasty in 1978, this technique has been applied to over 400 coronary lesions. During that time there has been a continual evolution of the angioplasty system, with technological advances not only in guidewires but in guiding catheters and dilation catheters. We have reviewed our cumulative experience with this system over the years 1979–1983, spanning Simpson's investigational work at Stanford and his subsequent clinical experience at Sequoia Hospital in Redwood City, California. Preliminary results are presented here to provide a standard of reference for effectiveness of this angioplasty system.

An extensive data base has been coded and compiled for computer analysis. It includes demographic information, clinical histories, and carefully detailed angiographic and hemodynamic parameters for the lesion dilated. Independent observers estimated the percentage of lesion stenosis in both pre- and postdilations. The morphology of the lesion (predilation) and the incidence of dissection were carefully documented.

Table 2 provides a summary of the Simpson PTCA data base spanning the years 1979–1983. There were 335 isolated lesion dilations performed on 291 pa-

TABLE 2   Simpson PTCA Data Base

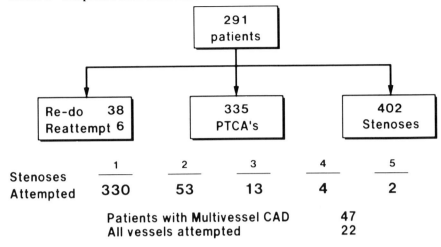

Stenoses Attempted	1	2	3	4	5
	330	53	13	4	2

Patients with Multivessel CAD   47
All vessels attempted   22

Stanford & Sequoia
1979-1983

tients. Attempted dilations numbered 402. As this table indicates, the majority of attempts involved isolated single lesions. When more than one lesion was attempted in a patient, this frequently involved sequential lesions within the same coronary vessel. The actual incidence of dilations involving multivessels, in which all vessels were diseased, was 22 cases.

When analyzed for individual lesions attempted, the overall success rate was 83 percent (Table 3) and 78 percent success without complication. Within the PTCA failure group there was a 6.8 percent incidence of overall complications requiring either emergency or same-day "elective" coronary bypass surgery. Fifty-three percent (36/68) of the failed cases were not associated with emergency bypass grafting. [Author's remark: *Same-day elective surgery* meant that the patient underwent bypass surgery the same day after an unsuccessful angioplasty without complication, when the surgical team was at hand and the operating room was available. These patients neither had clinical symptoms for ischemia

TABLE 3   PTCA Outcome

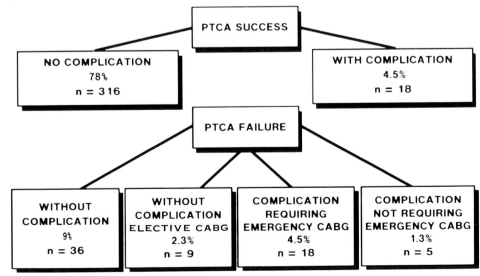

From Stanford plus Sequoia experience, 1979–1983, 402 stenoses.

plasty with and without thrombolytic therapy for treatment of acute myocardial infarction. Am Heart J 106:965–973, 1983.

9. Goldberg S, Urban PL, Greenspan A, Lebentahl M, Walinsky P, Moroko P: Combination therapy for evolving myocardial infarction: Intracoronary thrombolysis and percutaneous transluminal angioplasty. Am J Med 72:994–997, 1982.

10. Gold HK, Leinbach RC, Palacios IF, Yasuda T, Block PC, Buckley MJ, Atkins CW, Daggett WM, Austen WG: Coronary reocclusion after selective administration of streptokinase. Circulation 68:150–154, 1983.

11. Hall D, Gruentzig AR: Percutaneous transluminal coronary angioplasty: Current procedure and future direction. AJR 142:13–16, 1984.

16

Percutaneous Coronary Angioplasty in Patients with Multivessel Disease

GEOFFREY O. HARTZLER, M.D.

INTRODUCTION

Balloonists have an unsurpassed view of the scenery, but there is always the possibility it may collide with them (H. L. Mencken: *The Bend in the Tube*).

Percutaneous transluminal coronary angioplasty (PTCA) is established as an effective alternative therapy to coronary artery bypass surgery for the management of selected patients with symptomatic single vessel disease (1–6). The success and safety of the procedure, and its morbidity and mortality as performed in the near-ideal circumstance have been well characterized by the National Heart, Lung and Blood Institutes PTCA registry (5–8). These data have documented an acceptable low incidence of complications including death, procedure-related infarction, and need for emergency coronary bypass surgery, while achieving modest success for percutaneous coronary angioplasty performed by inexperienced and beginning operators.

Restrictive criteria for patient selection in the early stage of angioplasty development limited the procedure to the uncommon patient with recent onset of refractory angina resulting from a single, proximal, concentric, noncalcified, subtotal obstructive lesion in the absence of compromised left ventricular function (10). Recommended by the developers of the technique, these guidelines for patient selec-

tion were appropriate to minimize patient risk while improving the likelihood of a successful procedure at a time when technical skills were immature, catheters primitive, and outcomes uncertain. However, with time, experience, and many technical advances, angioplasty has evolved to a procedure with potential benefit for far more than the initially predicted 5 percent of patients undergoing coronary angiography (3,11–13).

Application of percutaneous angioplasty to "higher risk" subgroups of patients has been controversial. However, considered as a group, patients "at risk," including those with poor left ventricular function, prior coronary artery bypass surgery, acute myocardial infarction, advanced age, and multivessel disease, constitute a majority of patients with symptomatic coronary artery disease. Consequently, it is not unexpected that percutaneous angioplasty would be performed in these settings (14–23), particularly as increasing experiences and technical skills combined with improved catheter designs have facilitated the procedure.

DEFINITION OF MULTIVESSEL PTCA

Here the terms "multivessel dilation" or "multivessel angioplasty" will refer to dilation of two or more coronary lesions during the same angioplasty procedure. This includes dilation of two ("tandem") or more sites of disease within a single major coronary artery (e.g., left anterior descending, right, circumflex), two or more lesions including a major artery and its branches (e.g., left anterior descending and its diagonal branch) or in two or more separate coronary arteries and their branches. Additionally, the term will include dilation of diffuse arteriosclerotic segments of coronary arteries extending greater than 2 cm in length and into two or more coronary segments as determined by the CASS (Coronary Artery Surgical Study) coding system (24).

CONCEPTUAL AND PRAGMATIC CONSIDERATIONS

Clearly, coronary artery bypass surgery is established as a definitive therapy for patients with single or multiple vessel coronary artery disease. Risks and outcome of the procedure have been characterized in numerous reports and serve as the basis for comparison with any alternative therapy. Still, is the comparison fair? Is it appropriate to equate the reported experiences of a "mature" surgical procedure from high-volume centers with the results of the NHLBI Registry which was collected in the early stages of a new and evolving technique? These latter data can only be considered immature and preliminary in that they represent a combined experience from over 120 contributing centers, the majority having performed fewer than 50 procedures utilizing relatively primitive catheter designs (5,25–27). Because of limited experience and appropriately restraining guidelines for patient selection, only a very small percentage of patients underwent multiple vessel or multiple lesion dilation. However, preliminary reports from experienced centers suggest that multiple vessel angioplasty can be performed with early results and risks not dissimilar from those of single vessel angioplasty or substantially different from those associated with multiple vessel bypass graft surgery (22–23). At the present time a definitive body of data that precisely and accurately characterizes contemporary coronary angioplasty does not exist, further compromising a comparison of clinical results from these two separate procedures.

Just as coronary artery bypass surgery evolved from a single vessel procedure to the placement of multiple bypass grafts, it is rational and logical that coronary angioplasty would ultimately be applied to patients with multivessel disease. If dilation of a single, proximal lesion can be accomplished safely

and effectively, why should a patient with two (or even more) proximal lesions be denied the procedure? If improved technical skills and advanced design technologies allow dilation of a distal or branch artery lesion, why should the procedure be denied to a patient with two (or even more) distal or branch stenoses? Similarly, what logic dictates that separate coronary stenoses should be electively dilated on separate days? This approach may be analogous to performing two separate coronary bypass operations for treatment of double vessel coronary disease.

Assuming that coronary angioplasty can be performed in the setting of multiple vessel disease with an acceptable benefit and risk ratio, several additional factors support a decision for the procedure. Cost containment is a major consideration. The shortened hospital stay, lesser procedure-related costs, and reduced professional fees relative to coronary artery bypass graft surgery result in a base cost of angioplasty for single vessel dilatation of about one-third the cost of bypass surgery (28). There is every reason to believe that these reduced costs will exist for multiple vessel dilation procedures. This has been the experience within our own institution.

Lesser patient morbidity compared with that of coronary bypass surgery is another major consideration favoring the decision for elective angioplasty of single *and* multiple coronary obstructions. This feature needs little emphasis as physicians and patients alike are aware of the dramatic differences, especially if they have had a previous bypass operation. The avoidance of general anesthesia and thoracotomy are compelling reasons to consider coronary angioplasty particularly for the elderly, those with compromised left ventricular function, and those patients with prior bypass surgery potentially facing a second or third operation.

TECHNICAL CONSIDERATIONS OF MULTIPLE VESSEL ANGIOPLASTY

Obviously, it would be impossible to learn the process of auscultation, coronary angiography, or multiple vessel angioplasty through the reading of a manuscript. Consequently, this author will not attempt to create a "definitive" treatise conveying all subtleties of patient selection, catheter choice and manipulation, and patient management before and after coronary angioplasty. Rather, I will describe in general terms the approaches utilized by the author and his colleagues in the performance of complex angioplasty, while providing more detailed discussion of selected procedural aspects. These comments and guidelines result from a large experience concentrated within a small number of invasive cardiologists, and cannot be considered "definitive" or universally applicable. Opinions and practice are subject to change and our approaches to coronary angioplasty continue to evolve coincident with increasing experience and new technical innovations.

RISK CONSIDERATIONS OF MULTIVESSEL PTCA

Does simultaneous multiple lesion dilation have increased risk relative to single lesion angioplasty? Or relative to multivessel coronary bypass surgery? Not infrequently, multivessel angioplasty may be less hazardous than dilation of a single lesion. For example, angioplasty of two or three isolated, relatively proximal and concentric, discrete lesions in a single patient may be far less challenging and risky than dilation of a single, distally located obstruction in a tortuous arterial segment with immediately adjacent branches.

Patients undergoing multiple lesion coronary angioplasty at our institution are older than the subgroup of patients with single vessel disease, and have had a greater association of "high-risk" features including poor left ventricular function and prior coronary bypass surgery. The presence of extremely poor left ventricular function is clearly associated with increased procedural mortality, both for patients undergoing single or multiple lesion dilation. Still, coronary angioplasty in these settings may be the best management solution when compared with the also "high-risk" therapies of continued medical management or coronary bypass surgery. For each patient, the presence of these additional risk factors needs to be considered relative to the experience of an individual dilator and the anticipated outcome of alternative therapy available within the institution.

CONSIDERATIONS OF CORONARY ANATOMY

In general, the angiographic information allowing a decision for bypass surgery is the simple demonstration of a relatively proximal high-grade lesion with adequate distal vessel. However, more elaborate anatomic information is required to support a decision for coronary angioplasty. The segment of coronary artery proximal to the obstruction needs clear delineation with determination of its precise origin, the presence of tortuosity, and its relationship to branch vessels. The target lesion itself within the artery requires critical analysis in terms of its length, relationship to branch vessels, the presence or absence of tortuosity, eccentricity, ulceration, and calcification. The coronary segment immediately distal to the stenosis also requires analysis with particular reference to its length and tortuosity as these factors influence the type of balloon catheter and the guidewire selected for the procedure. Consequently, goal-directed, high-resolution, quality cine coronary arteriograms including caudally and cranially angled views are essential for proper patient selection. This author believes that every coronary arteriogram should be performed and reviewed with the intent to offer the patient coronary angioplasty, if clinically indicated and anatomically appropriate.

In general, a branch vessel arising within or adjacent to a stenotic segment will remain uncompromised by dilation of the primary lesion. However, the dilating physician does need to consider the relative importance of the branch and possible result should it become occluded. Many variations upon this theme exist, such that each patient and his unique coronary anatomy require individual consideration. When the branch vessel is of only moderate or small size and supplies "limited" myocardium, it may be a very reasonable decision to dilate the parent vessel while accepting the risk of side branch occlusion without a commitment for an urgent bypass operation. Frequently, an acutely occluded branch vessel can be reopened utilizing directionally changeable (steerable) guidewires. Should the vessel remain patent, which is the rule rather than the exception, any existing stenosis in the branch itself may then be electively dilated.

A unique "two-balloon" technique for the protection of an adjacent or bifurcation vessel while its companion branch undergoes dilation has been reported. However, this technique is technically rigorous and impractical to apply to the majority of circumstances. This author believes that with experience, the occurrence of branch occlusion can be reliably predicted based upon a careful assessment of the lesion morphology, geometry, and relationship to the branching point, by careful study of the arteriogram prior to the procedure.

HEMODYNAMIC CONSIDERATIONS

As discussed earlier, the presence of extremely poor left ventricular function increases the risk of multivessel angioplasty just as it increases the risk of coronary bypass surgery. In our own experience, when patients are properly selected and carefully managed throughout the procedure, the risk is minimized and outcome acceptable, particularly in relation to alternative therapies. In this circumstance, even transient balloon occlusion of a major artery such as the left main coronary or left anterior descending coronary may not be well tolerated. This may produce acute left ventricular dysfunction leading to systemic hypotension and reduced coronary perfusion which persists following deflation of the intracoronary balloon. Elective placement of an intra-aortic balloon pump prior to the coronary angioplasty procedure can facilitate dilation in this circumstance by "unloading" the left ventricle and supporting coronary perfusion immediately following balloon deflation. We use elective intra-aortic balloon pumping and shorter balloon inflation in patients with extremely poor left ventricular function, and in selected patients undergoing left main coronary angioplasty.

The uncommon occurrence of transient bradycardia during manipulation of a coronary artery may be poorly tolerated in patients with poor left ventricular function. Although we do not use a temporary pacemaker during routine coronary angioplasty, a transvenous electrode catheter is placed prior to angioplasty in this unique group of patients. If an intra-aortic balloon pump has been placed, it can be synchronized to a ventricular-paced rhythm should profound hemodynamic compromise and bradycardia occur.

Low molecular weight dextran is administered routinely to patients undergoing coronary angioplasty in our institution. However, this agent should be empirically reduced or avoided entirely in those patients undergoing single or multivessel angioplasty in the setting of extremely poor left ventricular function.

PREVIOUS CORONARY BYPASS SURGERY

In our experience, prior cardiac surgery has not emerged as a significant predictor of increased risk or compromised outcome. However, it must be acknowledged that a previous cardiac operation clearly will delay surgical reentry into the mediastinum should urgent bypass surgery be required because of a angioplasty-related complication. Additionally, suitable veins for grafting may be slow to acquire, further delaying an urgent operation. In this setting, the decision to perform balloon dilation may be correct and carry less risk than a second or third operation but requires an experienced angioplasty team.

MANAGEMENT OF COMPLICATIONS

The most common procedure-related complication is that of acute coronary occlusion. In our experience, in patients who have been pretreated with aspirin and dipyridamole, and who have received 10,000 units of heparin intravenously in addition to 500 cc of intravenous dextran, acute occlusion resulting from thrombosis is extremely rare. And it uncommonly results from coronary spasm. In the majority of circumstances, acute occlusion results from extreme plaque and arterial wall disruption with or without formation of a subintimal or perhaps medial hematoma. Based upon an extensive experience, we believe that the first and most appropriate maneuver in the catheterization laboratory following an occurrence of the procedure-related acute occlusion is to attempt to recross the site of obstruction. Often, this can be accomplished with a floppy guidewire or a relatively floppy but directionally steerable guidewire followed by the passage of a balloon for repeat dilation. On most occasions, this maneuver will restore patency and stabilize the lesion although repeated balloon inflations frequently for prolonged periods and with higher pressures may be required. We have chosen to continue intravenous heparin for at least 24 hours in this circumstance, assuming that the patient no longer has evidence of continued myocardial ischemia. With this approach, a majority of lesions remain unobstructed and improved without the need for urgent coronary bypass surgery.

On one occasion during a multivessel angioplasty, a left anterior descending coronary lesion would not remain patent following what would otherwise be adequate balloon expansion utilizing a guidewire system. Because of immediate vessel closure associated with chest pain and ST segment elevation, we ultimately elected to readvance the balloon catheter well through the occluded segment, withdraw the guidewire and then manually infuse oxygenated blood obtained from the femoral artery through the central lumen of the balloon catheter. This allowed a 30 to 40 cc/minute infusion and immediately stabilized the patient with relief of chest pain and return of ST segments to baseline. He was then much more leisurely prepared for an operation and underwent successful single vessel bypass grafting without enzymatic or electrocardiographic evidence of myocardial infarction.

When performing balloon angioplasty in high-risk patients, it must be acknowledged that the consequences of an acute occlusion may preclude an urgent operation. Consequently, the procedure should be performed by the most experienced dilators with expertise in the management of critically ill patients.

SELECTION OF CATHETERS AND GUIDEWIRES

Guiding Catheters

The choice of a proper guiding catheter is as important to the success of the dilation procedure as is the choice of the balloon catheter itself. Without adequate proximal support, critical lesions cannot be crossed. A guiding catheter which is too rigid or inappropriately matched to the coronary size may inflict an ostial injury which can cause potentially disastrous consequences. The guiding catheter may obstruct the coronary ostium producing myocardial ischemia with angina, arrhythmia, or hemodynamic collapse limiting the dilation attempt. In other circumstances, it may be necessary to advance the guiding catheter well into the left main, left anterior descending, or right coronary artery in order to gain enough support to advance the balloon catheter through a lesion. In general, we commence the dilation procedure utilizing 8 French guiding catheters having the technical advantage of small size and increased flexibility. However, if more rigid proximal support is required, the catheter can be exchanged for a 9 French size.

Unique coronary anatomy also determines the most appropriate guiding catheter. A superiorly oriented right coronary artery referred to as a "shepherd's crook" is often best dealt with utilizing a left Amplatz guide catheter. On occasion, a standard Judkin's type right coronary guide may prove optimal, particularly when "knuckled" within the aortic root to gain support against the opposite aortic wall. Alternatively, a straight-tipped catheter such as a "graft guide" may effectively cannulate unusually oriented or anomalously arising arteries.

It must be remembered that the guiding catheter is potentially a "lethal weapon" and that its characteristics of stiffness, torqueability, and flexibility not only differ from those of more routine angiographic catheters but undergo fairly rapid loss of shape during the procedure. One should not hesitate to exchange or modify a guiding catheter as required, particularly for multiple lesion dilation cases. Pointless maneuvers with an inadequate guiding catheter not only prolong the procedure but create risk for the patient.

Balloon Catheters

The dramatic technical evolution of balloon angioplasty catheters has clearly facilitated the perform-ance of coronary angioplasty in patients with multiple vessel disease. We now approach the majority of lesions with movable wire and directionally changeable balloon catheter systems. Still, at the present time, some lesions are best approached with fixed-wire catheters and can only be crossed and dilated with instruments having low profile and extreme tip rigidity. It is not the intent of this manuscript to critique and compare balloon catheters produced by the different manufacturers. However, substantial physical differences exist which, when appreciated, influence the balloon catheter choice relative to the lesion being dilated. There are inherent differences in catheter tip flexibility, deflated profile, balloon response to high pressures, and to the ease or guidewire manipulation resulting from material and design differences.

A major cause of plaque and vessel disruption is due to dilation with oversize balloons. Consequently, the choice of proper balloon diameter is critical to the success of the procedure. We do not routinely make formal measurements of the angiographic coronary arterial dimension prior to balloon catheter selection. With experience, however, it is possible to accurately select a 2.0, 2.5, 3.0 mm, or larger balloon based upon "eyeball" comparisons with the routine angiographic catheter dimensions. To err on the side of underdilation is preferable to overdilation of the artery. The lesion can be recrossed and dilated with sequentially larger balloons if required. On the other hand, an oversized balloon may cause dissection and sudden occlusion of the target artery.

We do not hesitate to use high inflation pressures (10 to 15 atmospheres) if required for full expansion of the stenotic lesion. However, routine use of high pressures greatly exceeding that pressure required to open the lesion should be avoided. In general, we limit the pressure applied to 1 to 2 atmospheres greater than that required to achieve full balloon expansion. However, we generally apply four to six balloon inflations at each site for up to 30 to 60 seconds, or as tolerated by the patient. Frequently, longer inflation times do result in opening of a lesion at lower pressures such that the use of extremely high inflation pressures with risk of balloon rupture or arterial wall injury may be minimized.

Guidewires

In the author's opinion, the most significant evolutionary development in angioplasty technology has been the creation of functional guidewire systems.

However, "steerable" or less euphemistically "directionally changeable" balloon catheters and movable guidewires have increased the technical complexity of the angioplasty procedure. These "three-handed" systems often require simultaneous manipulation of guiding catheter, balloon catheter, and guidewire while being attentive to proximal and distal pressures, proximal and distal contrast injections, and the patient's state of well-being. Fortunately, guidewire systems have increased the inherent procedural safety while allowing access to more distal and complex lesions. These guidewire systems will further enhance the future of coronary angioplasty.

A single, perfect, guidewire does not and probably will not come to exist. This statement is not an indictment of engineering capabilities but is a recognition that patients and coronary artery disease are highly variable. We have learned through experience that multiple guiding catheters, balloon catheters, and guidewires with different physical properties may be required to perform not only single but multiple lesion angioplasty in a given patient.

Highly flexible or floppy guidewires are potentially the least traumatic and frequently prove best-suited for approaching distal lesions, crossing diffuse segments of disease, and negotiating irregular or eccentric obstructions.

Stiffer wires may be more or less flexible depending upon numerous design features. Although the property of increased stiffness creates inherently greater risk of mechanical trauma to the arterial wall, the resultant increased torque control is highly desirable and allows directional change. Stiffer wires rarely have the desired tip configuration upon withdrawal from the package. Added curvature is usually required with the degree and arc of curve determined by the anatomy of the artery and lesion being negotiated.

Regardless of the guidewire being utilized, it should be well-visualized fluoroscopically. Its position must be readily determined even while using angled views in the obese patient. Second, the guidewire must move freely when advanced and withdrawn through the balloon catheter. It cannot be bound or restricted by Y-connectors or the balloon catheter itself as these adverse factors reduce the ability to feel resistance occurring within the coronary artery. Effective use of multiple guidewires is not a "chance" event. Rather, it results from logical, stepwise selection and modification of guidewires, each with differing physical properties, as required to safely negotiate the artery and lesion undergoing dilation.

MANAGEMENT PROTOCOL BEFORE AND AFTER DILATION

For 24 to 48 hours prior to elective angioplasty, patients receive an "antiplatelet" regimen consisting of aspirin, one tablet three times daily, and dipyridamole, 75 mg three times daily. Beta adrenergic blockers are discontinued to further reduce the likelihood of coronary spasm complicating the procedure and a low-dose calcium antagonist is commenced. Routine premedication consists of meperidine, 50 to 75 mg, promethazine hydrochloride, 25 mg, and atropine, 0.5 mg by intramuscular injection.

In the laboratory prior to coronary angioplasty, low molecular weight dextran is rapidly infused in an attempt to further prevent acute platelet aggregation and acute thrombosis. Patients receive verapamil, 5 mg intravenously, lidocaine, 75 mg intravenously, and isosorbide dinitrate, 5 mg sublingually. Heparin, 10,000 units intravenously, is administered after arterial puncture and before cannulating the target coronary artery.

Following balloon dilation, an intravenous heparin infusion is continued for from 12 to 24 hours to maintain a partial thromboplastin time of two to four times control values. Aspirin, dipyridamole, and a calcium antagonist are continued in addition to a cutaneous nitroglycerin preparation. If angina or equivocal ECG changes occur while the patient is in the hospital following a procedure, repeat coronary angiography is performed.

Routine follow-up consists of a baseline treadmill exercise test at 24 to 36 hours following angioplasty, at 2 months, and at 1 year. Repeat coronary angiography is recommended if angina recurs or the exercise study suggests recurrent stenosis.

DILATION PROCEDURE

Here the use of guidewire-directed balloon catheters has been emphasized. However, on occasion, standard fixed-wire balloon catheters of earlier design are preferable depending upon the character of the lesion being dilated. For example, we frequently choose to enter sites of total occlusion with a 20–30G or 20–20DG balloon catheter (United States Catheter and Instrument, Billerica, MA).

These catheters have a centrally fixed guidewire with relatively rigid tip which increases the risk of vessel injury but does allow application of greater force to sites of rigid stenosis or complete occlusion while guiding the catheter with subselective injections of contrast material. Extensive discussion or catheter designs and comparative indications for use of individual products is not within the scope of this text.

Frequently, multilesion angioplasty is little more than single lesion dilation at multiple sites. Discrete lesions are crossed as described in previous sections, followed by initial balloon inflation to 3 to 4 atmospheres pressure. It is vitally important to observe the balloon fluoroscopically during initial inflations as valuable information guiding the remainder of the procedure is obtained. Balloon indentation during inflation by the lesion confirms proper positioning. The balloon dimension can be compared with that of the native coronary artery and any apparent mismatch of size be considered for an exchange to a smaller or a larger balloon as indicated. The pressure at which the lesion opens can be observed influencing the choice of subsequent inflation pressures.

The approach to dilating diffuse lesions exceeding 2 cm in length is not dissimilar. However, overlapping balloon inflations are required. When utilizing fixed-wire balloon catheters, it was most prudent to manipulate the balloon completely through the lesion followed by sequential, overlapping inflations as the balloon was withdrawn. With the guidewire-directed balloons, the most effective approach is to perform overlapping inflations progressing from proximal to distal over the guidewire positioned as distally as possible. Each dilated segment is treated as a discrete lesion, receiving multiple balloon inflations with incrementally increased pressures until full expansion of the balloon is achieved.

In general, we inflate the balloon for 30 to 60 seconds at each site, gradually increasing the pressure of subsequent inflations until the lesion opens fully as judged fluoroscopically. Most lesions receive one or two dilations using pressures of 1 to 2 atmospheres greater than that required to initially produce full balloon expansion. The true end-point of a successful dilation is the demonstration of a wide vessel patency with no more than 20 to 30 percent residual stenosis upon immediate postdilation angiography. If this goal is not achieved, the lesion may be recrossed and dilated with higher pressures or larger balloons depending upon the clinical circumstance and the judgment of the dilating physician.

With increasing experience, we have found the routine measurement of crossing and postdilation pressure gradients to be of limited value. Rarely does this information influence the catheter choice, number of balloon inflations, or pressures utilized. Precise distal pressure measurements are difficult to obtain, subject to much artifact, and frequently inaccurate despite attention to technical detail. The mere introduction of the catheter into a small coronary artery creates a relative obstruction and inaccurate measurements both before and after dilation. The obligate small lumens produce damped and meaned pressure wave forms of limited utility. The introduction of movable, central guidewire systems has further complicated and prolonged accurate assessment of distal pressures. Consequently, we have abandoned their routine measurement, particularly during multiple lesion dilations where duration of the procedure may be prolonged.

CLINICAL CASE EXAMPLES

Two cases will be presented, each illustrating the author's approach to multiple vessel dilation, including patient selection, catheter and guidewire choices, and technical manipulations.

CASE EXAMPLE 1: The first case is a 22-year-old woman with a familial hyperlipidemia, premature coronary artery disease, and accelerating angina pectoris (Figure 1). Four months previously, she presented with unstable angina and electrocardiographic abnormalities at rest. A coronary arteriogram revealed small and diffusely diseased coronary arteries with high grade or subtotal occlusions of the proximal and mid segments of the anterior descending coronary artery, proximal and mid segments of the diagonal branch, distal right coronary artery, posterolateral branch, and posterior descending artery. Clinical and anatomic factors favoring coronary angioplasty at that time rather than coronary bypass surgery included her young age, the likelihood of rapidly progressive disease, small stature (approximately 100 lb, 5 ft. 1 in.), small coronary arteries, and the presence of diffuse and distal atheromatous disease. All sites of obstruction were successfully dilated using small balloon catheters (2.0 and 2.5 mm in diameter) with relief of angina and reversion of her treadmill exercise test to "negative." The postdilation management program consisted of aspirin, one tablet twice daily, dipyridamole, 75 mg three times daily, dilatiazem, 60 mg three times daily, and continuation of a complex lipid lowering regimen.

Three months later she again presented with rapidly progressive angina. Repeat coronary angioplasty was performed at the time of diagnostic coronary angiography which demonstrated restenosis at some but not all previously dilated sites and progression of disease within the circumflex artery, which now contained a discrete, high-grade obstruction.

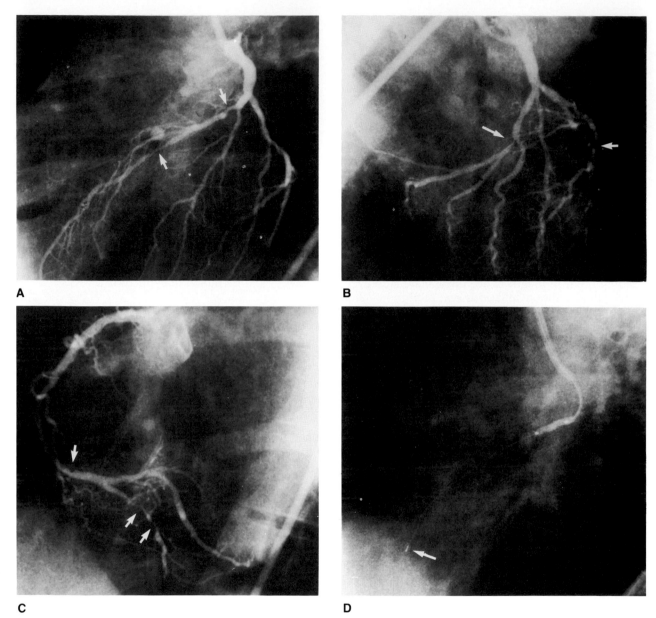

Figure 1. **A.** LAO cranial projection. Arrows indicate two sites of high-grade left anterior descending artery lesions. All arteries are small and diffusely diseased. **B.** RAO caudal projection. Long arrow indicates site of subtotal mid-circumflex artery obstruction. Short arrow indicates site of mid-left anterior descending artery lesions. **C.** LAO cranial projection. Arrows indicate sites of distal right coronary and mid-posterior descending artery stenoses. **D.** LAO cranial projection. A relatively small (2.5 mm diameter) balloon catheter is inflated within the proximal left anterior descending coronary lesions. Note (arrow) the very distal tip location of guidewire, appropriate to create balloon and guiding catheter stability. A small, flexible, guiding catheter (USCI) was combined with an Advanced Catheter Systems' "gold floppy" guidewire and 2.5 mm balloon catheter.

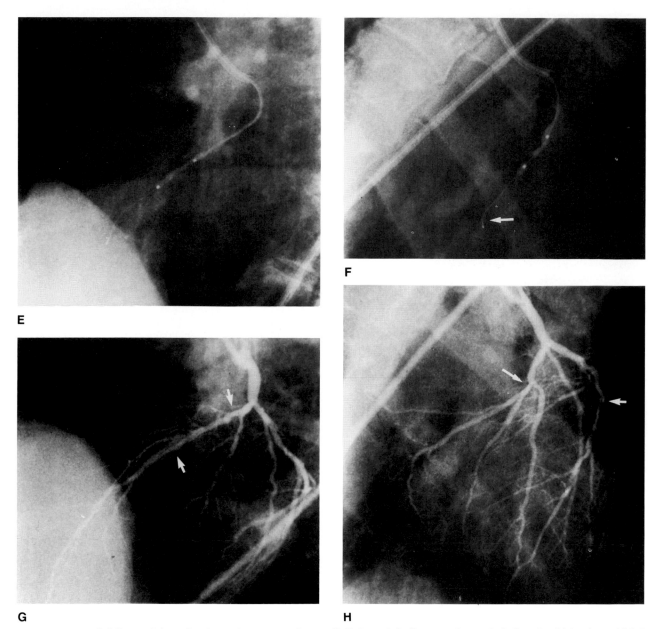

Figure 1. **E.** LAO cranial projection. An extremely small (2.0 mm) balloon catheter is inflated within the mid-left anterior descending coronary stenosis. **F.** RAO caudal projection. A 2.5 mm balloon catheter is inflated within the distal circumflex obstruction. A directionally changeable "gold standard" guidewire with tip shaped to facilitate steering (arrow) was initially directed through the lesion and positioned distally. **G.** LAO cranial projection. Immediate post-PTCA results with arrows indicating sties of previous high-grade obstruction. **H.** RAO caudal projection. Immediate post-PTCA result. Arrows indicate sites of previous high-grade obstructions.

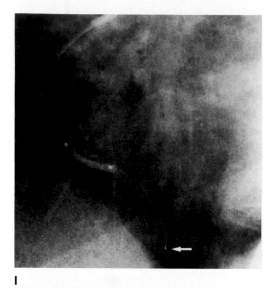

I

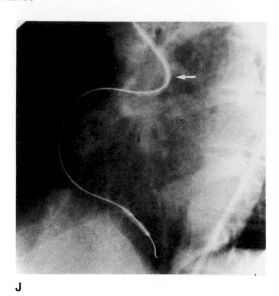

J

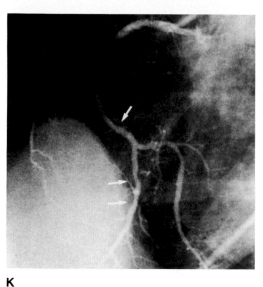

K

Figure 1. **I.** LAO projection. A 3.0 mm balloon is inflated within the distal right coronary artery. Note distal tip location of floppy guide-wire (arrow) passed into and across the posterior descending coronary artery stenoses. A relatively small, flexible, 8-French guiding catheter has been utilized. **J.** LAO projection. 2.5 mm balloon inflated within the posterior descending stenoses. The guidewire was exchanged for one with more body and stiffness (USCI platinum) allowing greater "push" of the balloon into and through the distal lesion. Note intentional buckling (arrow) of the guiding catheter within the aortic root to gain firmer ostial seating. **K.** LAO cranial projection. Immediate post-PTCA result. Arrows indicate sites of previous high-grade stenoses.

Factors favoring percutaneous angioplasty as opposed to coronary bypass surgery included the patient's preference and the presence of distal disease within the right and circumflex coronary arteries. Additionally, operative risks were increased because of prior thoracotomy, advanced age, peripheral artery insufficiency, and the need for continued anticoagulation created by the valve prosthesis.

Prior to angioplasty, he was begun on aspirin and dipyridamole and his prothrombin time was allowed to drift downward to 16 seconds. The procedure was performed from the right femoral artery utilizing standard percutaneous technique without complication.

CLINICAL RESULTS OF MULTIVESSEL PTCA

In the first 1400 consecutive dilation procedures performed by a group of three cardiologists at the Mid-America Heart Institute, 1971 or 88 percent of 2243 lesions were successfully dilated (Table 1). There

CASE EXAMPLE 2:   The second case is a 70-year-old physician with prior aortic valve replacement who presented with progressive angina. Diagnostic catheterization and coronary arteriography demonstrated a normally functioning valve prosthesis but mildly dilated left ventricle with reduced global contractility. High-grade obstructive lesions were present within the distal right coronary artery, posterolateral branch, and posterior descending artery. The distal circumflex artery was subtotally occluded (Figure 2).

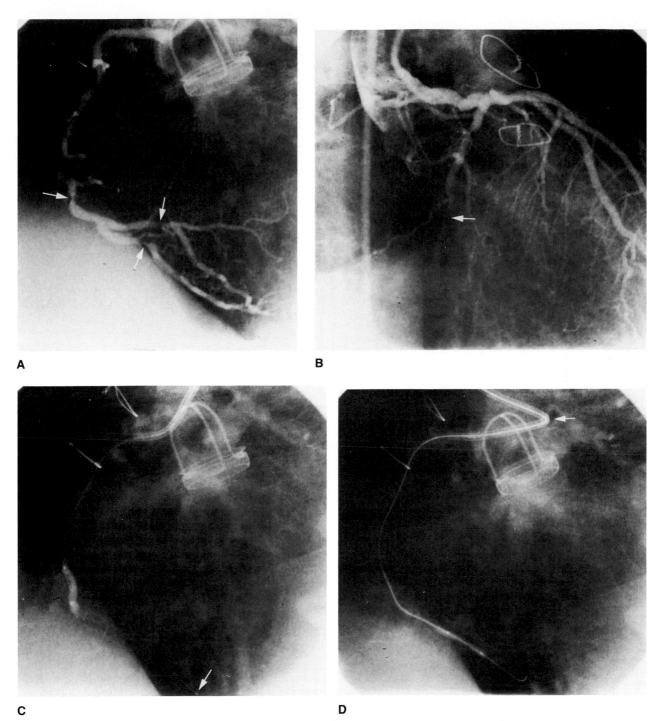

Figure 2. **A.** LAO projection. Arrows indicate sites of high-grade stenoses within the distal right coronary artery, posterolateral branch and posterior descending branch. Note tortuous coronary arteries. A sternal wire and the aortic ball-valve prosthesis are apparent. **B.** RAO projection. Arrow indicates subtotal obstruction of the mid-circumflex artery with slow distal run-off. **C.** LAO projection. 3.0 mm (diameter) balloon inflated within the distal right coronary stenosis. Note distal location of "floppy" guidewire tip within the posterior descending artery. **D.** LAO projection. 2.5 mm balloon forced across the posterior descending branch stenosis utilizing a directionally changeable platinum guidewire providing greater body and support. Also note knuckling of the guiding catheter within the aortic root (arrow) to obtain greater ostial support. Following dilatation of this lesion, the balloon catheter was directed into the postero-lateral segment utilizing the same guidewire and similar maneuvers.

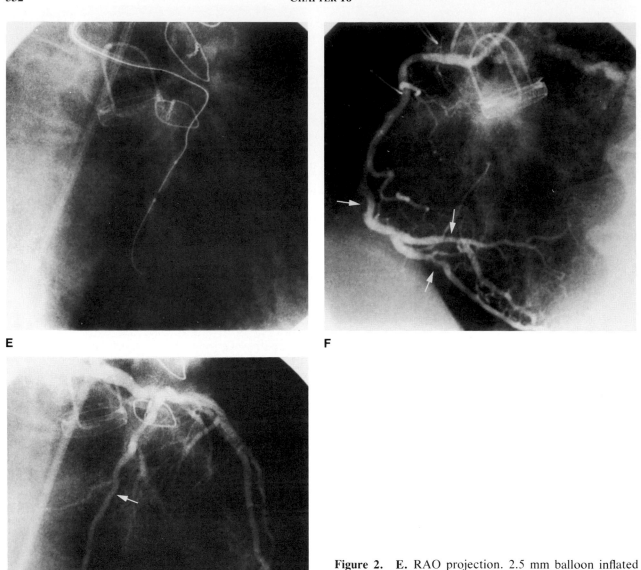

E

F

G

Figure 2. **E.** RAO projection. 2.5 mm balloon inflated within the mid-circumflex artery obstruction. Note the near right-angle origin of the circumflex artery from the left main coronary artery requiring use of a relatively stiff, directionally changeable platinum guidewire combined with deep-seating of the guiding catheter. **F.** LAO projection. Immediate post-PTCA result. Arrows indicate sites of previous high-grade stenoses. **G.** RAO projection. Immediate post-PTCA result. Arrow indicates site of previous subtotal occlusion.

were 176 patients 70 years of age or older, 143 patients with prior coronary bypass surgery, 161 patients with poor left ventricular function defined as ejection fraction less than or equal to 45 percent, and 98 patients who underwent angioplasty as primary therapy for acute myocardial infarction.

Five hundred seven procedures (36 percent) included dilation attempts of two to eight stenoses (Tables 1 and 2). Of 1350 lesions attempted, 1241 or 92 percent were successfully crossed and dilated. The anatomic distribution of lesions attempted varied widely and is summarized in Table 2. Four

TABLE 1 Multiple Lesion PTCA

Lesions	No. of Patients		No. of Attempts	Success Rate	
Single	893	(64%)	893	730	(82%)
Double	301	(21%)	602	541	(90%)
Triple	131	(9.4%)	393	367	(93%)
Quadruple	38	(2.7%)	152	143	(97%)
Quintuple	24	(1.7%)	120	112	(93%)
Sextuple	9	(0.6%)	54	50	(93%)
Septuple	3	(0.2%)	21	20	(95%)
Octuple	1	(0.07%)	8	8	(100%)
Total	1400	(100%)	2243	1971	(88%)

TABLE 2 Multiple Lesion PTCA (_n_ = 507 Patients)

Vessel Category	No. of Patients	
Double lesion dilation		
Single artery (sequential)	93	(31%)
Artery and branch	39	(13%)
Two arteries	169	(56%)
Total	301	(59%)
Triple lesion dilation		
Single artery (sequential)	19	(15%)
Two arteries	80	(61%)
Three arteries	32	(24%)
Total	131	(28%)
Four, five, six lesion dilation		
Single artery and branches	2	(3%)
Two arteries and branches	29	(39%)
Three arteries and branches	44	(58%)
Total	75	(15%)

hundred ten patients or 81 percent had successful dilation of all lesions attempted. "Incomplete revascularization" (e.g., two of three lesions attempted) occurred in 17 percent of patients with the majority being rendered asymptomatic or improved by the procedure. Consequently, 466 patients or 92 percent of 507 multilesion angioplasty cases were clinically successful as judged by interview, improved functional class and treadmill exercise testing.

Major procedure-related complications (Table 3) included transmural myocardial infarction in ten patients (2 percent). Urgent coronary bypass graft surgery was required in seven patients (1.4 percent). There were six deaths resulting in 1.2 percent mortality.

Meaningful long-term follow-up for the entire multivessel angioplasty series does not exist. The marked increase in frequency of multivessel procedures performed within our institution is relatively recent, creating only a small population of patients available for analysis at greater than 1 year after dilation. Elective repeat coronary arteriography at 1 year has been difficult to obtain because of long-distance referral patterns and because many patients who are asymptomatic decline the follow-up procedures.

The first 90 consecutive patients in our series undergoing multiple vessel angioplasty were followed from 7 to 29 months (Table 4). At a mean follow-up of 1 year, 77 patients or 86 percent were asymptomatic, without clinical evidence of restenosis. Eleven patients or 12 percent had recurrent angina or changing treadmill exercise test results leading to repeat angiography and a confirmation of restenosis of one or more previously dilated lesions. Six of the 11 patients underwent repeat dilation with all attempts successful. The remaining five patients underwent elective coronary artery bypass surgery. Two deaths occurred during the follow-up period. Both patients had prior myocardial infarctions and were receiving antidysrhythmic drugs for control of chronic ventricular ectopy. Consequently at 1 year follow-up, 83 patients or 92 percent of those undergoing multiple vessel angioplasty were asymptomatic or improved, although 7 percent required a repeat dilation procedure.

TABLE 3 Major Complications (1350 Lesions/507 Patients)

Category	No. of Patients	
Urgent CABG	7	(1.4%)
Myocardial infarction	10	(2.0%)
Death	6	(1.2%)
Total	23	(4.6%)

TABLE 4 Follow-up Results (90 Consecutive Multiple PTCA Cases)*

Conditions	Patients	
Asymptomatic	77	(86%)
Restenosis	11	(12%)
PTCA Repeat	6	(7%)
CABG	5	(6%)
Death	2	(2%)

*Follow-up period ranged from 7 to 27 months with a mean of 12 months.

MULTIVESSEL PTCA
IN PERSPECTIVE

At the present time, no multivessel or multiple lesion angioplasty experience can be strictly compared with the results of coronary bypass surgery. However, several observations, conclusions, and recommendations can be made, supported by the experience from our institution.

Multiple vessel dilation can be performed safely and effectively in a majority of well-selected patients. Multiple vessel dilation can be performed with mortality equivalent to that reported for coronary bypass surgery in similar patient populations. The advantages of angioplasty are obvious—lesser patient morbidity, more rapid recovery with an early return to full activity, and reduced overall cost. In most circumstances, failed multivessel angioplasty will not preclude subsequent elective coronary bypass surgery.

In approaching a patient with multiple vessel disease, the first lesion dilated should be the most significant obstruction. Although this determination may be uncertain, in general, the lesion will be a proximal high-grade obstruction within the major circulation. Following successful dilation of this lesion, functionally less significant lesions of the distal circulation and peripheral branches should be approached. These judgments will differ somewhat from patient to patient dependent upon many factors including anatomic variation, prior myocardial infarction, the presence or absence of collateral circulation, and the experience of the operating physician. Clearly, the optimal approach for each patient will require individualization; but by dilating the most significant stenosis first, clinical improvement can be anticipated should angioplasty of lesser obstructions prove impossible.

At the present time, complex angioplasty including multiple lesion dilation should be performed cautiously in well-selected patients by the more experienced operator with demonstrated ability and success in single lesion angioplasty. The use of directionally changeable guidewire systems increases the likelihood of success, the ability to cross and dilate more complex lesions, and the inherent safety of the procedure.

Many patients successfully managed by initial multilesion dilations will undoubtedly require repeat dilation and coronary bypass surgery as their disease progresses, becoming less suitable for balloon dilation. However, the time of elective coronary bypass surgery can be and will have been deferred, allowing more years of symptom-free life prior to a first, second, or perhaps third bypass operation.

The major question of long-term patency remains unanswered. The incidence of restenosis for predominantly single vessel angioplasty has approached 30 percent but varies greatly in reports from different centers (29–31). However, the pattern of restenosis in patients with multiple vessel disease may not be strictly analogous to those with single vessel obstruction. The dilated lesions in predominately older patients with diffuse coronary disease may respond differently from those of younger patients with single coronary stenosis. Our preliminary experience with multiple vessel dilation suggests that the incidence of clinically recognized restenosis on a per patient basis is no higher than that of patients undergoing single lesion angioplasty. However, some patients clearly have progression of disease at all previously dilated sites while others may develop a single, discrete restenosis. Factors contributing to the variable pattern of restenosis including its absence are unknown. Future investigation will be directed toward identifying optimal dilatation protocols and treatment regimens capable of minimizing restenosis in patients undergoing single and multiple lesion dilation.

The medical and surgical communities may ultimately insist upon a randomized trial comparing multiple vessel angioplasty with multiple vessel coronary bypass surgery. Obviously, such a study will be fraught with difficulty in its performance and interpretation of results. Sufficient experience with multivessel dilations does not yet exist to allow a meaningful multiple center study or comparison with bypass surgery. Some physicians have been unable to recognize or at least acknowledge the value of any therapy proving less effective than coronary bypass surgery. However, one cannot minimize or ignore the high cost, or the physical and psychological trauma associated with bypass surgery, which must be weighted negatively in any valid comparison.

The future of coronary angioplasty lies within the dissemination and teaching of technical skills, further refinement of catheter systems, further characterization and limitation of restenosis, and in the development of new innovations allowing application of the procedure to larger groups of patients with coronary artery disease.

REFERENCES

1. Greuntzig AR, Meier B: Percutaneous transluminal coronary angioplasty. The first five years in the future. Int J Cardiol 2:319, 1983.

2. Stertzer SH: Transluminal coronary angioplasty—1981. Investigative technique or established procedure? Arch Intern Med 142:679, 1982.

3. Hamby RJ, Katz S: Percutaneous transluminal coronary angioplasty: Its potential impact on surgery for coronary artery disease. Am J Cardiol 45:1161, 1980.

4. Kober G, Scherer D, Koch M, Dowinsky S, Kaltenbach M: Transluminal coronary angioplasty. Early and long-term results in 250 procedures. Herz Kreislauf 6:309, 1982.

5. Kent KM, Bentivoglio LG, Block PC, Cowley MJ, Dorros G, Goselin AJ, Greuntzig A, Myler RK, Simpson J, Stertzer SH, Williams DO, Bourassa MG, Kelcy SF, Detre KM, Mullin S, Passamani E: NHLBI Percutaneous transluminal coronary angioplasty (PTCA) Registry: four years' experience (Abstr). Am J Cardiol 49:904–1982.

6. Kent KM, Bentivoglio LG, Block PC, Cowley MJ, Dorros G, Goselin AJ, Greuntzig A, Myler RK, Simpson J, Stertzer SH, Williams DO, Fisher L, Gillespie MJ, Detre K, Kelcy S, Mullin SM, Mock MB: Percutaneous transluminal coronary angioplasty: A report from the registry of the national heart, lung and blood institute. Am J Cardiol. 49:2011, 1982.

7. Faxon DP, Ryan TJ, McCabe CH, Kelcy SF, Detre K, Members of the NHLBI PTCA Registry: Determinants of a successful percutaneous transluminal coronary angioplasty (NHLBI—PTCA Registry) (Abstr). Am J Cardiol. 49:905, 1982.

8. Cowley M, Bentivoglio L, Block P, Bourassa M, Detre K, Dorros G, Gosselin A, Greuntzig A, Kent K, Myler R, Simpson J, Stertzer S, Williams D, Mullin S, Mock M, participating centers: Emergency coronary artery bypass surgery for complications of coronary angioplasty: NHLBI PTCA registry experience (Abstr). Circulation 54 (Suppl IV):IV-193, 1981.

9. Doros G, Bentivoglio L, Block PC, Bourassa M, Cowley M, Detre K, Gosselin A, Greuntzig A, Kent K, Myler R, Simpson J, Stertzer S, Williams D, Mullin S, Mock M, participating centers, PTCA Registry, NHLBI: Fatal complications of percutaneous transluminal coronary angioplasty (PTCA) (Abstr). Circulation 65(Suppl IV):IV-254, 1981.

10. Myler RK, Greuntzig AR, Stertzer SH: Technique and clinical indictions for percutaneous transluminal coronary angioplasty, in *Myocardial Revascularization*, Mason DT, Collins JJ Jr, eds. New York, Yorke Medical Books, 1981, pp 431–444.

11. Rapaport E: Editorial: Percutaneous transluminal coronary angioplasty. Circulation 60:969, 1979.

12. Editorial: Percutaneous transluminal coronary angioplasty. Lancet 1:235, 1979.

13. Berger SM, Gorfinkel HJ: Candidates for transluminal coronary angioplasty. Am J. Cardiol 48:810, 1981.

14. Alcan KE, Stertzer SH, Wallsh E, De Pasqualle MP, Bruno MS: The role of intra-aortic balloon counterpulsation in patients undergoing percutaneous transluminal coronary angioplasty. Am H J 105:527, 1983.

15. Douglas JS, Greuntzig AR, King SB III, Hollman J: Long-term results of percutaneous transluminal angioplasty for aortocoronary saphenous vein graft stenosis (Abstr). Circulation 66(Suppl II):11–124, 1982.

16. Hartzler GO, Rutherford BD, McConahay DR: Percutaneous coronary angioplasty with and without prior streptokinase infusion for treatment of acute myocardial infarction (Abstr). Am J Cardiol 49:1033, 1982.

17. Meltzer RS, Van den Brand M, Serruys PW, Fioretti P, Hugenholtz PG: Sequential intracoronary streptokinase and transluminal angioplasty in unstable angina with evolving myocardial infarction. Am H J 104:1109, 1982.

18. Meyer J, Merx W, Schmitz H, Erbel R, Kiesslich T, Doerr R, Lambertz H, Bethge C, Krebs W, Bardos P, Minale C, Messmer BJ, Effert S: Percutaneous transluminal coronary angioplasty immediately after intra-coronary streptolysis of transluminal myocardial infarction. Circulation 66:905, 1982.

19. McCallister BD, Hartzler GO, Rutherford BD, McConahay DR: Palliative percutaneous transluminal angioplasty or unstable angina in patients over 70 years of age (Abstr). Circulation 64(Suppl IV):IV-255, 1981.

20. McCallister BD, Hartzler GO, Reed WA, Johnson TW: Percutaneous transluminal coronary angioplasty in elderly patients: A comparison to coronary artery bypass surgery (Abstr). J Am Coll Cardiol. 1:656, 1983.

21. Mock M, Holmes D Jr, Vlietstra R, Detre K, Gersh B, Orszulak T, Schaff H, Piehler J, Van Raden M, Passamani E, Kent K, Kelcy S, Greuntzig A, and participants in NHLBI Registry: Percutaneous transluminal coronary angioplasty (PTCA) in patients $\geq$ 60 years of age registered in the NHLBI registry (Abstr). Circulation 66(Suppl II):II-329, 1982.

22. Dorros G, Stertzer SH, Cowley M, Kent K, Williams D: Complex transluminal coronary angioplasty: Multivessel disease and multiple dilatations (Abstr). Circulation 66(Suppl II):II-329, 1982.

23. Hartzler GO, Rutherford BD, McConahay DR, McCallister SH: Simultaneous multiple lesion coronary angioplasty. A preferred therapy for patients with multiple vessel disease (Abstr). Circulation 66(Suppl II):II-5, 1982.

24. Killip T (ed): The National Heart, Lung and Blood Institute Coronary Artery Surgical Study (CASS). Circulation 63(Suppl I):I-1, 1981.

25. Levy RI, Mock MB, Willman VL, Passamani ER, Frommer PL: Percutaneous transluminal coronary angioplasty: A status report. N Engl J Med 305–399, 1981.

26. Holmes DR Jr, Vlietstra RE, Mock MB, Smith HC, Cowley MJ, Kent KM, Detre KM: Follow-up of patients undergoing percutaneous transluminal coronary angioplasty (PTCA): A report from the NHLBI PTCA registry (Abstr). Am J Cardiol 49:916, 1982.

27. Dorros G, Bentivoglio LG, Block PC, Bourassa M, Cowley M, Detre K, Goselin A, Greuntzig A, Kent K, Myler R, Simpson J, Stertzer S, Williams D, Mullin S, Mock M, participating centers, PTCA Registry,

NHLBI: Fatal complications of percutaneous trans-luminal coronary angioplasty (PTCA) (Abstr). Circulation 64(Suppl IV):IV-254, 1981.

28. Jang GC, Block PC, Cowley MJ, Greuntzig AR, Dorros G, Holmes DR, Kent KM, Leatherman LL, Myler RK, Stertzer SH, Sjolander M, Willis WH, Vetrovec GW, Williams DO: Comparative cost analysis of coronary angioplasty and coronary bypass surgery: Results from a national cooperative study (Abstr). Circulation 66(Suppl II):II-124, 1982.

29. Dangoisse V, Val PG, David PR, Lesperance J, Crepauj, Dyrda I, Bourassa MG: Recurrence of stenosis after successful percutaneous transluminal coronary angioplasty (PTCA) (Abstr). Circulation 66(Suppl II):II-331, 1982.

30. Holmes DR, Vlietstra RE, Smith HC, Vetrovec GW, Cowley MJ, Kent KM, Detre KM, Myler R: Restenosis following percutaneous transluminal coronary angioplasty (PTCA): A report from the NHLBI PTCA Registry (Abstr). Am J Cardiol. 49:905, 1982.

31. Jutzy KR, Berte LE, Alderman EL, Ratts J, Simpson JB: Coronary restenosis rates in a consecutive patient series one year post-successful angioplasty (Abstr). Circulation 66(Suppl II):11–331, 1982.

17

Intraoperative Balloon Angioplasty of the Coronary Artery

NOEL L. MILLS, M.D.

INTRODUCTION

A logical step in the treatment of occlusive coronary artery disease has been the use of balloon-tipped catheters in the operating room after initial reports for coronary dilation by way of the percutaneous route (1). Diffuse distal disease often occurs in the coronary arteries at locations that are not amenable to coronary bypass surgery and where runoff of proximally placed bypass grafts is impeded. The objective of dilating coronary arteries at the time of bypass surgery, therefore, is primarily to offer an increased runoff for the bypass graft and secondly to offer collateral circulation through the dilated segment to other areas of left ventricular myocardium.

Intraoperative transluminal angioplasty of the coronary arteries has been carried out since 1979 (2). The technique is still evolving, and its long term outlook will depend upon late clinical results, which are difficult to obtain, mainly because angiographic studies for a late follow-up present a difficult task. Although the most widespread interest in percutaneous coronary angioplasty induced advances in the percutaneous balloon catheter design, the catheter development for the intraoperative application of balloon angioplasty has lagged. Intraoperative dilations of potentially correctable lesions are often unsuccessful because of poor design of balloon catheters.

The rationale for intraoperative coronary angioplasty centers on the theme that artery segments to be dilated are those that are not amenable to coronary bypass surgery or endarterectomy. Since complete revascularization improves the results of coronary bypass surgery, adjunctive balloon dilation during bypass surgery may be considered an effort to revascularize more completely the coronary artery segments that otherwise would not be revascularized. Many such lesions have high grades of stenoses and the myocardium may suffer ill effect on a later closure of such artery remote from the hospital environment. Surgeons have dilated such lesions with rigid graduated probes since the inception of coronary bypass surgery. That technique, however, generated little enthusiasm because of the necessity of forcefully pushing on the lesions, which has a tendency to disrupt the external elastic membrane at the endarterectomy site. Compared to dilating a lesion with a rigid anatomic probe, balloon dilation is more rational, although the design of the intraoperative balloon catheters needs improvement.

Intraoperative transluminal angioplasty to date has been performed, in conjunction with vein bypass surgery, by introducing the balloon catheter through the coronary arteriostomy at the site of the proposed graft anastomosis. Intraoperative angioplasty has not been used in arteries that are not bypassed by vein grafts although they may have distal obstructing disease.

The single most obvious difference between intraoperative angioplasty and percutaneous angioplasty is the fact that during intraoperative angioplasty, the target artery is exposed to the direct vision of the operating surgeon. The surgeon can place the balloon catheter appropriately across the lesion by using direct visual control, palpation by hands, and measurements of catheter length between arteriotomy and the location of the target lesion.

PATHOPHYSIOLOGIC CONSIDERATIONS

Just as certain types of coronary lesions are more amenable to endarterectomy, so are certain coronary lesions more amenable to operative angioplasty. It is generally noted that coronary arteries are more atherosclerotic under pathologic examination than can be seen on an arteriogram of the same area. However, in studying the effects of balloon dilation in cadaver coronary arteries, rather sobering observations have been made (3). Compaction of the atherosclerotic plaque by balloon dilation in coronary arteries contributes to only a minimal increase in lumen size. The dominant mechanism by which the atherosclerotic plaques are dilated in coronary arteries is separation of the junction between the arterial wall and the atherosclerotic plaque with radial displacement of the separated plaque. Along with this mechanism, there are distention of the vessel wall and longitudinal fracture in one or more areas along the length of the atherosclerotic lesion. A small percentage (10–12 percent) of dilatation is effected by extrusion of fluid content of the plaque (4). The relatively low incidence of embolization or thrombosis after dilation may be due to the laminar flow across the dilatated area. A "remodeling" is felt to occur in the artery during the postdilation period, but the degree to which any endothelium resembling a normal lining repairs the dilatated area is not known.

Dilation at the point of arterial bifurcations often presents undesirable results because of linear tears, avulsion of the branch, occlusion by plaque material or embolization. Therefore, balloon angioplasty by either percutaneous or operative route has generally been avoided in the region of a major bifurcation.

The atherosclerotic disease in the coronary artery may range from (1) soft, semisolid plaque material, to (2) areas of fibrosis with a paucity of cholesterol, or (3) lesions with varying degrees of calcification and solidification. Often the surgeon may get a "feel" for the type of disease at hand and that intuition

helps in making a reasonable decision whether or not to use operative angioplasty. For instance, if at operation the atherosclerotic lesion appears to comprise much liquid cholesterol and debris, there may be an increased chance of embolization of cholesterol debris into the distal coronary branches beyond the anastomosis site, thereby jeopardizing runoff vessels (5). Therefore, in such situations, operative angioplasty may be wisely avoided.

PATIENT SELECTION

Patients selected for operative angioplasty generally fall into that subgroup with extensive coronary disease usually involving all three coronary artery systems. Heredity and diabetes mellitus have been frequent concomitant features of operative transluminal angioplasty (OTA) patients yet offered no contraindication for its use. Since November, 1980, OTA has been used in approximately 8 percent of patients undergoing coronary artery bypass. In approximately 10 percent an angioplasty was not planned from data on the preoperative coronary angiogram yet was performed because of operative findings. Coronary arteries that are ideal for operative angioplasty are those in which a coronary artery bypass is being placed at a point where the coronary vessel is as free of disease as possible and 1.5 mm or larger in diameter. The lesion to be dilated is remote from the arteriotomy site, and the coronary artery beyond the lesion is in an area not generally feasible for coronary bypass. The prime example is the left anterior descending (LAD) coronary artery that has a second significant lesion at the apex with a vessel thereafter relatively free of disease (Fig. 1A). Second, lesions that obstruct flow to a branch, e.g., diagonal lesions, but are not large enough for a bypass graft may be dilated (Fig. 1B). Proximal tandem lesions may allow increased runoff between the lesions by dilating the more distal lesion through the coronary arteriotomy when otherwise two grafts would not be indicated (Fig. 1C). Operative angioplasty is contraindicated in several settings:

1. It is not used to accomplish dilation of a diseased coronary artery in order to enlarge it to accept a bypass graft; in this pathologic entity an endarterectomy is indicated.
2. Angioplasty is not used to dilate an anastomosis through the graft that has been narrowed because of faulty technique. A repetition of the anastomosis is indicated.
3. Angioplasty is not used to open a proximal lesion that may offer competition from the native circulation with the bypass graft.
4. Angioplasty of the coronary artery is not used to

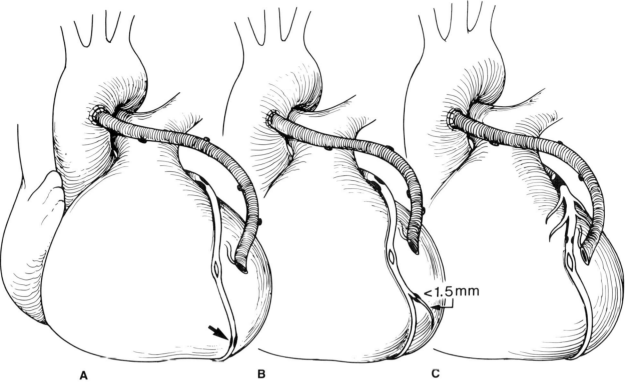

Figure 1. Indications for operative transluminal angioplasty. **A.** LAD coronary artery with proximal lesion and a second significant lesion at the apex. Operative dilation has been shown to offer collateral circulation via the distal vessel. **B.** Lesion at origin of branch too small to accept a graft that is accessible via the coronary arteriotomy. **C.** Tandem lesions may be dilated to offer flow to the intervening myocardial supply.

open a proximal lesion that could embolize debris into the distal coronary bed (Fig. 2).

ANGIOPLASTY DEVICES

Catheters

A number of catheters have become available for OTA since the initial experience with long percutaneous catheters. No single catheter to date offers ideal technology; each has advantages and disadvantages. Balloon sizes range from 1½ to 2½ mm in diameter and are generally 20 mm in length. The balloon diameter may be chosen with the help of the cardiac catheter used during angiography when the exact diameter of the tip of the catheter is known. The coronary artery size may be judged by using the catheter tip as a reference point. Rarely is a 3 mm diameter balloon indicated other than during the course of dilating a proximal tandem LAD lesion. During the course of an operative angioplasty, two different balloon systems may be used to complement each other to achieve a successful result.

Coaxial System

The catheter most commonly used for operative angioplasty has been a 1 mm diameter catheter with a 2 mm diameter balloon that is 20 mm in length.* A guidewire 8 mm long on the end of the catheter helps direct it through the lesion. (Balloon catheters without the guidewire are found to be totally unsatisfactory because of "hang-up" of catheter tip on ledges of the lesions.) Guidewire-tipped catheters have the advantage of (1) multiple use when more than one lesion is to be dilated and (2) slow control of dilating pressure. The size of the catheter, however, is a disadvantage because occasionally lesions can be traversed with the guidewire, although the catheter with the deflated balloon will not cross the area to be dilated. A significant disadvantage of the catheter is that the balloon is placed too far from the catheter tip. Often the catheter tip will traverse the lesion, yet because of a bifurcation and hence smaller coronary artery size distally it can be passed no farther. The balloon therefore cannot be centered across the

*USCI, Billerica, Masschusetts 01821.

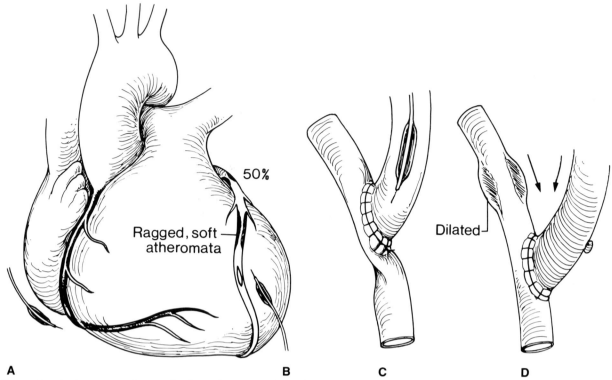

A B C D

Figure 2. **A.** Intraoperative angioplasty is not used to dilate a diseased artery to allow it to accept a graft; endarterectomy is preferred. **B.** Angioplasty should be avoided when there is a potential to create embolic debris that may jeopardize the distal coronary bed and bypass graft. **C.** An anastomosis narrowed by faulty technique should be redone rather than angioplastied prior to proximal anastomosis suturing. **D.** Angioplasty is avoided when the opened vessel may compete with a bypass graft.

lesion to be dilated (Fig. 3). Lack of a taper from the guidewire to the tip of the catheter also causes hang-up on firm coronary lesions, preventing catheter passage. Rarely, if ever, does the deflated balloon itself offer resistance to catheter passage across the lesion. On occasion the guidewire tip of the catheter will bend at the guidewire catheter interface, resulting in a failure of the catheter and incurring the expense of another catheter.

A coaxial system, in which the catheter is advanced over a guidewire after the guidewire has traversed the lesion, was used and found to have many of the same technical limitations as the previously mentioned device.

Linear Extrusion (Fogarty-Chin) Catheter

An ingenious catheter system successfully used in larger peripheral arteries has been applied to the coronary arterial tree.† The balloon is housed within the catheter and with pressure the balloon is extruded from the tip of the catheter to traverse the lesion as it unfolds. Lateral force is applied to the lesion with very little forward force in larger arteries. Compliance of the lesion is mandatory for the

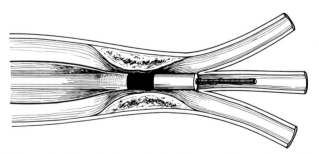

Figure 3. Length from the balloon to the catheter tip prevents the balloon from reaching the lesion when the catheter advance is prevented by a smaller coronary size or branches.

†Edwards Laboratories, Santa Ana, California 92711.

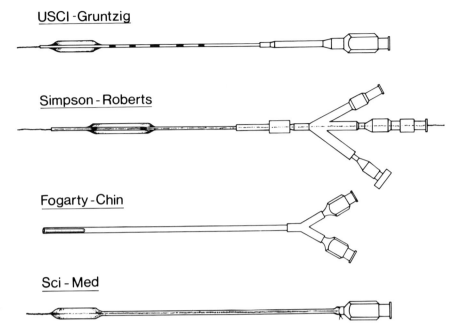

USCI-Gruntzig

Simpson-Roberts

Fogarty-Chin

Sci-Med

Figure 4. Catheter types for operative transluminal angioplasty.

success of this catheter. The device is limited for coronary use, however, by the uncontrolled, sudden extrusion of the balloon from within the catheter. Also, the catheter tip having no taper has a tendency to hang up on noncritical lesions as it is advanced down the coronary artery to the site of proposed dilation.

All catheter systems have the disadvantage of leaks at bonding sites or of the irradiated balloons. The linear extrusion system has the disadvantage of only a single use. Reinsertion of the balloon within the catheter is difficult and is discouraged. An ideal device would be one with no catheter, in which the balloon tapers from the guidewire and continues up to the syringe to enclose the guidewire only‡ (Fig. 4).

DILATION TECHNIQUE

After coronary arteriotomy, the artery is carefully calibrated at a point just prior to the lesion. The epicardium over the lesion is open whenever possible and the lesion visualized and palpated. If cardiac fat or depth of artery prevents direct visualization, the locus of the lesion is identified by measuring and comparing the length of the calibrated probe from the lesion to the arteriotomy. After calibration of the coronary artery, balloon size is selected, i.e.,

‡SCIMED, Minneapolis, Minnesota 55441.

1.0 mm calibrated probe = 1.5 mm balloon. If a 1 mm calibrated probe can be passed across the lesion, the lesion has a 95 percent chance of being successfully dilated. With the coaxial system, the catheter is passed down to the lesion and with gentle pressure it is passed across the lesion to a point where the lesion is in the center of the balloon. A 2.5 mL saline-filled syringe is connected to a manometer, and 7.5–10 atmospheres of pressure are applied for a 30-second interval. This is carried out for three successive dilations and deflations of the balloon, and the device is withdrawn. The site of the lesion is marked with a silver clip for future angiographic reference, and the appropriate-size calibrated probe is passed across the dilated lesion.

If a 1 mm probe could be passed before dilation as well as after dilation, yet a 1.5 mm probe could not be passed after dilation, a large-diameter balloon catheter may be tried with caution. Care is taken to dilate only the area of the lesion and not to dilate normal coronary artery. Although operative angiography is available, it has not been used because of the considerations of crossclamp time, increased chance of infection, and the fact that failure would not warrant any further manipulation of the area (6). On two occasions a 0.5 mm Teflon pressure catheter was passed through a side branch of the saphenous vein graft across the dilated lesion to confirm absence of a pressure gradient. Because of fear of disrupting a successfully dilated plaque this technique has not been used more frequently (Fig. 5).

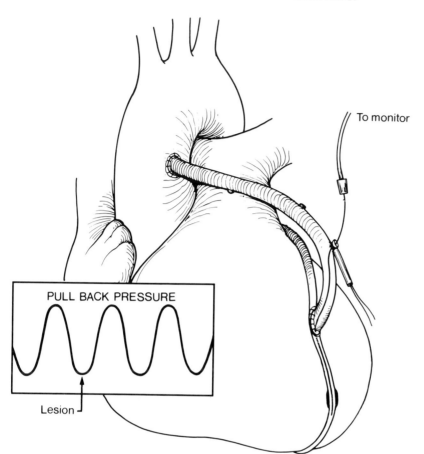

To monitor

PULL BACK PRESSURE

Lesion

Figure 5. A tiny catheter withdrawn across a dilated lesion reveals no gradient across a previous 95 percent stenosis.

Dilations with the linear extrusion catheter are carried out likewise through the arteriotomy used for coronary bypass. The catheter is passed to a point 1 cm proximal to the lesion, and manometric pressure measured by a 2.5 mL saline-filled syringe is used to extrude the balloon from the catheter. After successful traversal of the lesion, full dilating pressure is applied to 10 atmospheres and held for 30 s. Three separate inflations are used prior to withdrawing the catheter. Some resistance to withdrawal of the catheter has been found with all devices used yet has not been associated with any untoward effects.

CLINICAL RESULTS

Operative transluminal angioplasty was first performed in November 1980, and to date the author has dilated 92 lesions in 79 patients, 8 of whom were females. Ages ranged from 34 to 73 years. Fifty-five patients had internal mammary artery bypass and 20 had concomitant endarterectomy. One patient

had repair of ventricular aneurysm, and one had aortic valve replacement. The average number of bypass grafts in the group was 4.1. Of the 92 lesions that had dilation attempted, 48 were in the distal LAD location, 12 were in the proximal LAD, and 21 were in the right coronary artery (RCA) system. Nine lesions were in the branches of the circumflex coronary artery, and in two patients diagonal lesions were dilated (Table 1). The overall success rate for attempted dilation of 92 lesions was 79 percent.

TABLE 1 Operative Transluminal Angioplasty (79 Patients and 92 Lesions)

Locus	Success	Failure	Total
Distal LAD	43	6	48
Proximal LAD	9	3	12
RCA system	13	8	21
Circumflex system	8	1	9
Diagonal	1	1	2
Total	73 (79%)	19 (21%)	92 (100%)

A search for perioperative myocardial infarction was carried out in all patients by using serial electrocardiograms (ECGs) and creatine phosphokinase)(CPK-MB) isoenzymes. An isotope scan of the myocardium was performed if the CPK or ECG indicated perioperative myocardial infarction. Two patients had overt myocardial infarcts, one being in the area of operative dilation. Congestive heart failure, prolonged cardiac drug support, or intra-aortic balloon pump was not present in the series. Five other patients had isoenzyme changes suggestive of myocardial infarction but without ECG or radionuclide scan changes. All patients had been administered aspirin, 600 mg per day, and Persantine, (dipyridamole) 75 mg per day, according to the Mayo Clinic protocol (7).

There have been no early or late deaths in the group. The most common cause of an "unsuccessful dilation" at operation was inability to cross the plaque with the catheter device. In two instances, the lesion would not dilate because of circumferential calcification that could not be cracked even though the lesion was traversed by the balloon (8). In the latter part of the study, the linear extrusion catheter (Fogarty-Chin) was used in 10 patients. In five cases there was successful dilation with a good operative result. In one patient the lesion could not be dilated with any type of device. In another, the balloon pushed the plaque distally, rather than traversing it. There were two instances in which the balloon extruded from the catheter rapidly and perforated the coronary artery at the proximal origin of the lesion. This necessitated an extensive endarterectomy with closure of the coronary in one patient (Table 2).

Calcification and length of lesion have directly affected results of operative angioplasty (5). The success rate of dilating calcified lesions has been approximately 60 percent. However, when a long lesion, i.e., 2 cm or greater, is associated with calcification, the success rate drops accordingly, as expected (5).

Postoperative angiographic studies of patients who have had coronary artery bypass surgery and who are clinically enjoying a good result continue to be difficult to obtain. In studying the coronary arteries angiographically for up to 1 year in 27 patients who had 29 lesions with attempted angioplasty, a number of important findings have emerged. Of 25 lesions that had a "successful" angioplasty at the time of operation, a good result was obtained in 18 or 72 percent. There were three lesions in which the stenosis was unchanged, and one lesion had improvement of approximately 50 percent. Two lesions angioplastied had coronary artery occlusion of the area dilated. A graft was completely occluded to a coronary artery that had extensive dilation of noncritically stenotic areas in addition to the critically stenotic lesion. The 72 percent success rate of OTA found on postoperative angiography demonstrates the need for pursuing studies of the operative angioplasty technique. The three lesions that had stenoses that were not significantly improved by operative angioplasty occurred early in the study period, when lower inflation pressures of the operative balloon were used. With manometric pressures during dilation of 7–10 atmospheres, this problem has not recurred.

Four lesions that were "failures" at operation because of inability of the catheter to pass a lesion were studied angiographically. In one instance, a coronary artery perforation at a bifurcation by the linear extrusion catheter system revealed a healing of the coronary artery perforation with no false aneurysm or other untoward effect. In two instances, the coronary artery at the lesion and distal to it remained patent. In one lesion in which the catheter

TABLE 2   Fogarty-Chin Angioplasty

Patient	Stenosis, Percentage	Operative Result	Postoperative Angioplasty	Locus
1	70	Unable to dilate calcified lesion		Circumflex
2	65	Balloon disrupted plaque		LAD
3	50	Successful	Good result	LAD
4	95	Successful	Incomplete relief of stenosis	LAD
5	40	Successful		
6	95	Perforation of coronary	Perforation sealed	RCA
7	80	Perforation of coronary		LAD
8	80	Successful		LAD
9	65	Successful		LAD
10	70	Successful		LAD

would not be passed, closure of the coronary artery at that point was found. No myocardial infarctions could be identified in the patients with these four lesions.

By careful study of the postoperative angiograms, a supply to other coronary arterial systems, i.e., LAD to distal right, has been identified distal to successfully angioplastied lesion in two patients.

COMPLICATIONS

By far, the most common complication associated with operative angioplasty was balloon or catheter leak. Until recently, it appeared that a balloon rupture or catheter leak offered no more than a "nuisance factor" for the operating surgeon. However, recently a 2 mm guidewire-tipped catheter with a 20 mm long balloon was being used to dilate a lesion in the distal one-third of the LAD coronary artery. Dilation was carried out in routine fashion. During the second inflation at 7–10 atmospheres of pressure with the balloon snugly situated across the lesion, a linear rupture of the dilating balloon occurred. A perforation of the coronary artery at that point followed, with extrusion of fluid in the subepicardial area adjacent to the dilated vessel (Fig. 6). The problem was repaired by performing an extensive manual core endarterectomy and suturing the tear with 8-0 proline over a 1.0 mm catheter. No arrhythmias, myocardial infarctions, or other untoward sequelae followed. The patient was discharged on the seventh postoperative day. Other than the perforations associated with use of the linear extrusion system mentioned, known complications have been nil. Spasm, arrhythmia, or late hemorrhage has not been identified.

Angioplasty of long segments is probably unwise, especially if normal intima is disrupted along the way. With low flow and smaller arteries the tendency toward thrombosis is great in spite of administration of antiplatelet aggregate drugs. In certain instances, distal 3-cm- to 4-cm-long lesions have been dilated with the use of two catheter systems. The coaxial and the linear extrusion systems are not mutually exclusive; in fact, in this setting, they may complement one another. One example is a distal LAD lesion approximately 2½ cm long around the apex, in which the Fogarty-Chin catheter cannot be advanced to the lesion because of minor degrees of stenosis and the USCI catheter cannot be made to advance across the lesion. The USCI catheter is therefore used to dilate minor stenoses down to the severe stenosis in order that the Fogarty-Chin can be advanced down to that point. The Fogarty-Chin catheter is then used to dilate the long distal tight lesion.

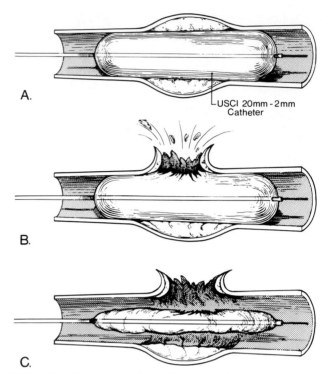

A.

USCI 20mm-2mm Catheter

B.

C.

Figure 6. Rupture of the midportion of the angioplasty catheter resulted in laceration of the coronary artery in one patient.

In repeated failure of operative angioplasty, inability to traverse the lesion with the catheter stands out as the most common cause. Circumferential calcification may prevent dilation because catheter balloon pressure is insufficient to disrupt the concentric ring of calcium. A ridge on the catheter-guidewire interface often causes the catheter to hang up on firm atheromatous material, preventing passage across the lesion. The distance from the catheter tip to the balloon in the coaxial systems prevents passage of the catheter far enough across a lesion to effect dilation when immediate branching follows the lesion. Liberated debris from proximal lesions presents a hazard for failure not only of the angioplasty but of the bypass graft. Damage to normal intima by long areas of dilatation up and down the coronary artery may result in coronary thrombosis.

CONCLUSION

Operative angioplasty is safe and offers a chance for more complete myocardial revascularization in patients with severe diseased distal coronary atherosclerosis. The catheter technology has lagged behind that of the percutaneous techniques, and design improvements are clearly warranted. Steerable systems will probably not offer technical advantages because generally the lesions are identified and palpated during the technique. Care must be taken to avoid situations in which embolic debris may cause a hazard to the bypass graft distally. Plaque disruption, dissection, and coronary artery damage should be negligible with appropriate care. Most importantly, coronary arteries that can be bypassed or endarterectomized should be treated with those techniques and angioplasty reserved for coronary arteries generally not amenable to the usual bypass techniques. Clinically, in following patients postoperatively, there is no discernible difference between those undergoing standard coronary artery bypass surgical techniques and those in whom the technique has been complemented by operative angioplasty.

REFERENCES

1. Greuntzig AR: Transluminal dilatation of coronary artery stenosis. Lancet 1:263, 1978.
2. Walsh E, Franzoni AJ, Closs RH, Bruno MS, Stichen F, Stertzer SH: Transluminal coronary angioplasty during saphenous coronary bypass surgery. A preliminary report. Ann Surg 191:234, 1980.
3. Mills NL, Doyle DP: Does operative transluminal angioplasty extend the limits of coronary artery bypass surgery? A preliminary report. Circulation 66 (Suppl I):I–26, 1982.
4. Fogarty TJ, Chin A, Shoor PM, Blair GL, Zimmerman JJ: Adjunctive intraoperative arterial dilatation. Simplified instrumentation technique. Arch Surg 116:1391, 1981.
5. Mills NL, Ochsner NL, Doyle DP, Kalchoff WP: Technique and results of operative transluminal angioplasty in 81 consecutive patients. J Thorac Cardiovasc Surg 86(5):689, 1983.
6. Jones EL, Spencer BK: Intraoperative angioplasty in the treatment of coronary artery disease. J Am Coll Cardiol 3:970, 1983.
7. Cheesbro JH, Clements IP, Foster V: A platelet inhibiting drug trial and coronary artery bypass operations. N Engl J Med 307:73, 1982.
8. Mills NL, Doyle DP: Experience with operative angioplasty of the coronary arteries. Am Heart J 107(4):383, 1984.

18

Percutaneous Transluminal Angioplasty in Patients with Failed Coronary Bypass Graft

JAY HOLLMAN, M.D.

INTRODUCTION

Coronary artery bypass graft (CABG) surgery is currently performed on an estimated 159,000 patients annually (1). Although the vast majority of patients will be angina-free in the first year after bypass surgery, about 30 percent will have angina at 5 years and 38 percent at 10 years (2). Patients with recurrent angina following bypass surgery, who have known the complete relief of angina from revascularization, are often dissatisfied with the partial relief of angina afforded by medical therapy. When coronary angioplasty can restore them to an angina-free state, they are among the most grateful patients. The reasons for recurrent angina after bypass surgery are three: 1) incomplete revascularization, 2) progression of disease in ungrafted or grafted vessels, and 3) graft failure.

As bypass vein graft surgery has improved technically, smaller vessels are grafted, and more grafts are placed. This has reduced the number of patients with recurrent angina due to incomplete revascularization and to progression of the disease (2). Increased numbers of cases surviving a decade following their bypass surgery have increased the number of late graft failures due to atherosclerosis in the graft (3). The consequences of incomplete revascularization following bypass surgery are not trivial. Even though mortality is improved in most subsets of patients after bypass surgery, it remains at about 2 percent per year with one-half of the mortality due to cardiac causes (4). The rate of myocardial infarction is about 2.4 percent per year after bypass surgery. This infarction rate is not significantly reduced over the control population (5). It is speculative but not illogical to hypothesize that close follow-up and early recatheterization of the symptomatic patients may result in reduction of morbidity and mortality of the individuals who have previously had bypass surgery.

It is important to realize that repeat bypass surgery is a technically more difficult procedure associated with a higher morbidity and mortality even in the most experienced centers. The mortality rate in repeat coronary bypass surgery was four times that of initial surgery. Blood loss requiring transfusions is more frequent in repeat operations, and the perioperative infarction rate is three times as likely that of first-time bypass surgery (6,7).

In summary, after coronary bypass surgery pa-

346

tients have a 1 percent per year incidence of cardiac-related death and a 2.4 percent incidence of myocardial infarction, although they enjoy improved survival over most of the medically treated patients. Moreover, the symptomatic relief of bypass surgery diminishes with time. Early experience from the NHLBI PTCA Registry suggested that percutaneous coronary angioplasty in the patients who had a previous bypass might be associated with a higher mortality (8,9). (There was 8.1 percent mortality in patients with prior bypass surgery versus 0.7 percent in patients without prior bypass surgery.) However, transluminal angioplasty experiences in 122 postbypass surgery patients at Emory University suggest that balloon angioplasty can be performed with a primary success rate and complication rate comparable to those of patients who had never had bypass surgery (10). Thus, percutaneous angioplasty is a prudent alternative to repeat bypass surgery in patients with recurrence of severe angina following bypass surgery or in those with failed bypass vein grafts.

PATIENT SELECTION

The percutaneous balloon angioplasty of the patient with prior bypass surgery encounters some specific anatomic and clinical features. These factors influence the outcome of angioplasty and thus affect the strategy of patient selection.

Most of the anatomic factors associated with percutaneous angioplasty of the patient with prior bypass surgery can be defined by diagnostic arteriogram. Because the bypass vein graft is a circulation conduit between the base of the aorta and the segment of a major epicardial coronary artery, the failed bypass grafts can be grouped into four specific anatomic types (10). The number one consideration is the graft failure at the site of proximal anastomosis, which is a surgically sutured junction between the aortic root and the vein graft. The second area is the distal anastomosis junction, where the vein graft is connected to the epicardial coronary artery. The third area of graft failure is the vein graft body itself, encompassing the entire length of the vein between the two suture points. The fourth area of concern is in the native target vessel distal to the distal anastomosis junction.

These anatomic factors affect decision making about angioplasty in two ways in these patients. The first is the longer-term results after a successful angioplasty. As will be discussed in the clinical result section, the short-term and medium-term outcomes of the graft angioplasty vary among the different anatomic types of graft failures: the distal anasto-mosis junction, the vein body itself, or the proximal anastomosis junction. Technically, all three anatomic types are commonly approached through the graft channel. The technical difficulty may vary from patient to patient, but technical success of dilation may not vary significantly by the type of anatomic location of the lesion. A flow-limiting lesion in the native vessel distal to the anastomosis junction is not a direct result of bypass surgery, but such a lesion technically is associated with graft because most often these distal lesions have to be approached through the vein graft itself because the proximal segment of the native vessel may have been already occluded, denying access of the balloon catheter to the distal lesion via the main channel of the native artery.

The patients who have had bypass surgery in the past present certain specific clinical problems in regard to performing a percutaneous transluminal angioplasty of the previously bypassed arteries. Postoperative patients are a high-risk group should an emergency bypass surgery become necessary during percutaneous angioplasty. For this reason, even though we perform four to six angioplasty procedures daily at the Cleveland Clinic, each postbypass angioplasty patient is reviewed and discussed with the cardiac surgery staff before any such procedure. Specifically, we discuss which therapy is really in the best interest of the patient, transluminal angioplasty versus repeat bypass surgery. What strategy should be employed should the target artery have sudden closure during angioplasty? Is the amount of myocardium at risk large enough to justify an emergency bypass surgery should the balloon dilation fail?

In general terms, if left ventricular function is severely compromised, elective repeat bypass surgery is a better option. The number, extent, and type of previous surgical procedures is also a consideration. Some surgeons do not consider patients with two prior failures of bypass surgery candidates for a third bypass surgery. In this situation, percutaneous coronary angioplasty may have to be performed without surgical backup, or no procedure may be performed at all.

The presence of a patent internal mammary graft slows the rate at which the patient can be safely placed on bypass pump. Extensive pericardial and mediastinal scarring (as in a patient with previous left ventricular aneurysm resection, pericardial plication, or postpericardial poudrage) makes emergency bypass surgery difficult to perform, thus increasing the risk of percutaneous coronary angioplasty. Patients who had prior vein strippings of the lower extremities are also a difficult category. Important considerations are also the probability of primary success and recurrence after the procedure. As primary success improves and emergency surgery rate decreases, the angioplasty alternative improves over repeat bypass surgery. Honest discussions, based on the angioplasty and surgery statistics at one's particular center, may help to formulate the best decision for the patients. In some circumstances angioplasty may be able to revascularize a patient when bypass surgery cannot (Fig. 1); in such circumstances, angioplasty has an advantage.

The most effective means of treating a patient with a failed prior bypass surgery is to dilate the native vessel. This is not always possible since the native vessel is frequently occluded. If one has a choice between dilating a failed vein graft versus a native artery, dilating the native vessel is usually preferred. Patency of one graft makes dilation of the native vessel more safe and in some cases feasible (Fig. 2). Left main dilation can be performed if the left anterior descending artery or the circumflex artery has a patent graft. The advantage of dilating native vessels is that recurrence rates are known, and the chance of embolism is negligible when compared to a higher incidence of embolization in patients with balloon angioplasty of the vein graft.

PATHOPHYSIOLOGY OF FAILED VEIN GRAFTS

When dilation of the native vessel is not possible, what can be done for the stenotic graft itself? What happens to the vein graft when the narrowed segment of a vein graft is dilated with a balloon? Data at present would clearly suggest that many graft lesions can be successfully dilated with good initial and short-term results (6 months) (9). Pathologic study of grafts from surgical and postmortem specimens have shown that the pathophysiology of the graft stenosis and graft occlusion may not have a single etiologic process (15). Selection of a patient for balloon dilation of the failed bypass grafts should ideally consider the underlying pathology of the lesion to be dilated. Early graft closure in less than 1 month is commonly due to graft thrombosis and may occur within hours of a surgery (11). Reopening of these thrombosed grafts using thrombolytic agents such as streptokinase has been described (12); however, a high incidence of hemorrhage and pericardial tamponade can be expected in the patients who had recent bypass surgery (13). Angioplasty is unlikely to be successful in early graft closure even if suturing technique with overlying thrombus is the underlying etiology. Thrombus does not compress, and dilation of a fresh suture site might well result in rupture of the anastomosis site and massive hemorrhage.

After 1 month all vein grafts (Fig. 3) show intimal hyperplasia (11). Whether it represents organization of fibrin thrombus that lines the vein graft shortly after implantation of a response of the vein graft to arterial pressures (arterialization) is unknown (14). This intimal hyperplasia or intimal fibrous plaque is morphologically indistinguishable from the fibrous plaque of atherosclerosis. Intimal hyperplasia alone may be responsible for an isolated graft stenosis at the site of anastomosis or in the graft body.

As early as 11 months after bypass grafting (in hyperlipidemia patients) (15) or more usually after about 3 years (16) in normolipidemic patients, atherosclerotic changes occur in the grafted bypass vein. These changes differ from atherosclerosis in the native coronaries in that foam cells frequently erode irregularly through the intima of the grafted vein (16) (Fig. 4). Surgeons and pathologists handling these atherosclerotic grafts are impressed by the friability of an old vein graft.

DILATION TECHNIQUE OF BYPASS GRAFTS

The key to performing angioplasty of the bypass vein graft is obtaining a stable guiding catheter position. The technical problems, such as side branches encountered in the native coronary artery dilation, are not present with angioplasty of the bypass vein grafts, but the tortuosity of the bypass grafts can make a passage of the balloon catheter quite difficult unless the guiding catheter is firmly seated at the ostium of the target vein graft.

Careful examination of the diagnostic angiography prior to the angioplasty procedure is the first step in performing bypass graft dilation. Although surgeons may place bypass grafts anywhere along the ascending aorta, bypass grafts are usually made at certain sites. Left anterior descending grafts are usually placed low on the left side of the ascending aorta; circumflex grafts are usually higher on the left anterior ascending aorta. Right coronary grafts are usually lower on the right side of the ascending aorta. The catheter of choice for the lower left side of the aorta is a right Judkins type or a left Amplatz type. Higher on the left side a bypass graft type is useful; if it is not effective, then the type of catheter used lower on the aorta is useful. The right-sided grafts are best reached by a bypass graft type guide or a right Judkins type.

Once the guiding catheter has engaged the vein graft ostium, the balloon catheter is cautiously advanced down the bypass vein graft. If more force is required, the guiding catheter is advanced over the balloon catheter, deeper into the vein graft. Such a maneuver is usually safe once the balloon catheter

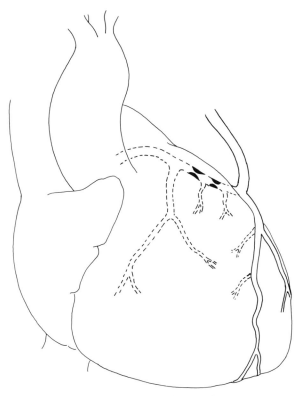

FIRST SEPTAL PERFORATOR ISOLATION

A

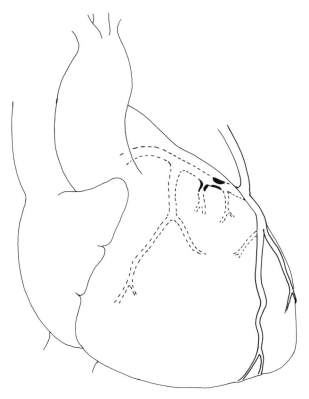

DISEASE IN A LARGE FIRST SEPTAL PERFORATOR

B

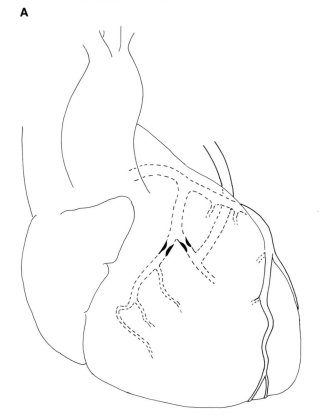

DISEASE OF CIRCUMFLEX IN A.V. GROOVE

C

Figure 1. **A.** Often, a first large septal perforator is isolated by disease both proximally and distally. The vessel often lies under the pulmonary artery and is difficult to bypass. In this case, proximal angioplasty is performed percutaneously, or intraoperative angioplasty is performed retrogradely. **B.** Bypass of a large septal perforator is a difficult surgical procedure. PTCA of these vessels can be done. **C.** The circumflex in the atrioventricular (AV) groove is a difficult and risky vessel to bypass. Often the vessels proceeding from the circumflex in the AV groove are small, but the amount of myocardium supplied by the vessels together is fairly large. In such cases, PTCA of the circumflex may be the treatment of choice.

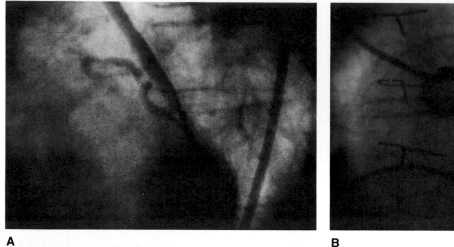

A

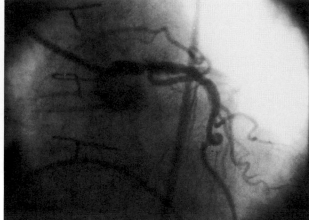

B

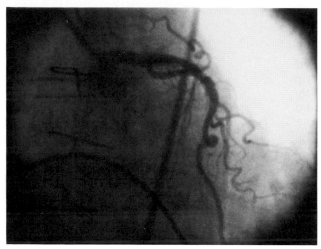

C

Figure 2. Left main dilation in a postoperative patient. Saphenous vein bypass graft to circumflex was patent. **A.** Left main stenosis before (**B**) and after (**C**) PTCA. Dilation was not imprudent in this case since the presence of a patent bypass graft would have allowed emergency bypass surgery with a reasonable chance of success. Fortunately it was not required.

is placed deep enough. Stenoses occurring at the distal insertion site within the first year after bypass surgery are usually soft and generally easy to cross and dilate despite their apparently severe narrowing, provided that the guiding catheter is firmly anchored. The choice of balloon catheters is less im-

portant than the choice of guiding catheters in dilating the graft cases. If the guiding catheter is properly engaged and provides the support needed, almost any balloon catheter will work. The advantage to an over-the-wire system is that the wire can be safely advanced through the vein graft and the stenotic

TABLE 1 Comparison of Early and Late Morphologic Changes* of the Vein Graft

Category	Early	Late
Pathology	Intimal hyperplasia	Atherosclerosis
Age of graft	6 weeks–3 years	3 years or older
Site	Anastomosis	Graft body
Recurrence	Low	High
Embolic episode	None (?)	Present

*Adapted from Ref 16.

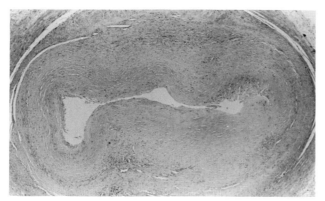

Figure 3.   Intimal hyperplasia in a bypass graft, resulting in occlusive disease.

lesion with gentle pressure. When more force is required to advance the balloon through the lesion, one can be confident, with a wire in place, that the power being applied will safely advance it through the site of the obstruction.

The diameter and length of the balloon should match the target artery or graft size. The lesions at the distal anastomosis site occur usually in the relatively small vessels. Not infrequently, a 2.5 mm balloon is sufficient for dilation of distal insertion sites. Occasionally in a small caliber distal vessel, only a 2.0 mm balloon is required. In general, a small amount of oversizing is desirable to achieve optimal dilatation and hemodynamic results. The proximal anastomosis site requires using 3.5, 3.7, 4.0, and 4.2 mm balloons. Oversizing of balloon catheters in the proximal anastomosis sites may help improve hemodynamic results and may decrease recurrence rates.

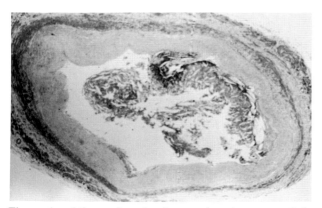

Figure 4.   Atherosclerotic changes in a bypass graft in place for years. Note looseness and friability of material in the lumen, contrasting with the firm organized material in Figure 3.

In dilating the body of an old vein graft, caution is justified because of the embolic risk. Use of an oversized balloon is not encouraged in dilating the body segment of an old vein graft. Occasionally, in dilation of distal anastomosis lesions, a steerable system will be required to direct the balloon catheter into the main lumen of the native artery and away from the side branches. In very severe lesions, a steerable wire is needed to find the lumen; however, most lesions at the anastomotic junctions are concentric and can be crossed easily with a relatively atraumatic floppy wire.

During coronary artery or graft angioplasty, measurements of the pressure gradient across the lesion are used as guides to the selection of inflation pressures and number of inflations to obtain desirable results. Pressure gradients, however, are not always completely accurate, because of the wedging of the catheter tip and the relative size of the balloon catheter in a distal artery or in a stenotic lesion.

One particular difficulty is dilation of Y-shaped vein grafts. This situation is solved by the use of guidewire. A relatively sharp curve is shaped on the tip of a steerable wire, which is advanced around the angle associated with the Y graft. In some cases, however, a sharp curve at the tip of the guidewire may prevent its advance down the vein graft once the Y curve is negotiated and passed. If the balloon catheter is advanced over the wire, it can be moved down beyond the Y junction. The curved wire is then exchanged for a more straight wire for an easy advancement through the straight course of the vein until the wire crosses the lesion and dilation is achieved. When an exchange of guidewire is required, an ACS balloon or a Meditech balloon is preferred, since these balloons allow wire to pass more freely (Fig. 5). Teflon-coated guidewires, however, make exchanges possible with any catheter system.

CLINICAL RESULTS

Data on bypass graft dilation are limited despite the large number of patients who have undergone bypass surgery. The NHLBI PTCA Registry report by Kent et al. (9) presented 25 cases of bypass vein graft angioplasty with a primary success rate of 56 percent. Douglas et al. (10) at Emory University reported 62 cases of percutaneous angioplasty on patients with failed bypass surgery. Their primary success rate in these patients with failed bypass grafts was 88 percent (30/34) for dilation of distal anastomosis junction, 80 percent (4/6) for dilation near the proximal anastomosis, and 96 percent (22/23) for

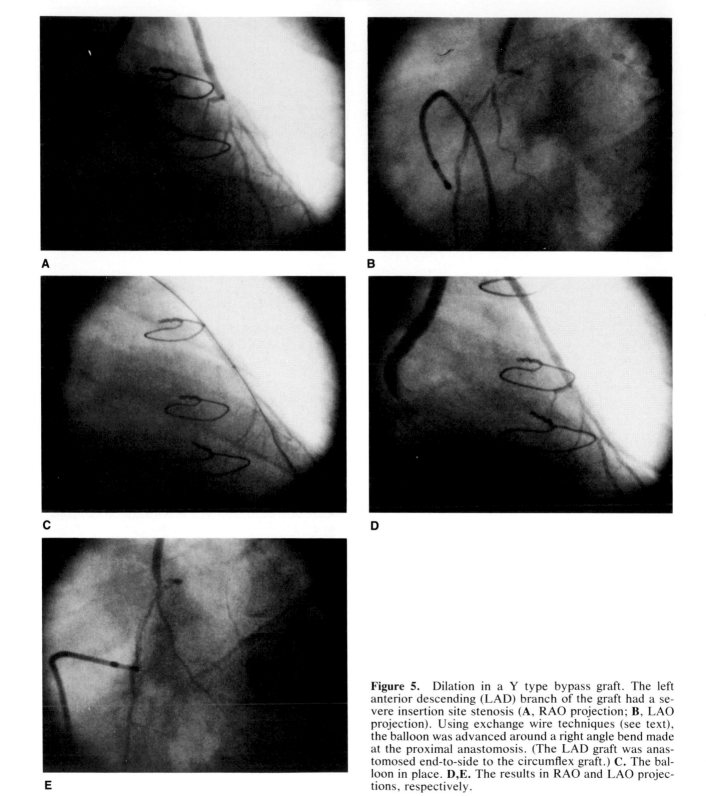

Figure 5. Dilation in a Y type bypass graft. The left anterior descending (LAD) branch of the graft had a severe insertion site stenosis (**A**, RAO projection; **B**, LAO projection). Using exchange wire techniques (see text), the balloon was advanced around a right angle bend made at the proximal anastomosis. (The LAD graft was anastomosed end-to-side to the circumflex graft.) **C.** The balloon in place. **D,E.** The results in RAO and LAO projections, respectively.

TABLE 2 Angioplasty of CABG Patients

Category	Cases	Percentage
Graft body	18	58
Proximal anastomosis	8	26
Distal anastomosis	5	16
Total	31	100

dilation of vein graft body. Recurrence rate was very high (61 percent) in patients who had angioplasty of the vein graft body; however, the patients who had undergone dilation of the distal anastomosis junction had a far lower and more reasonable recurrence rate, 21 percent, on follow-up angiograms.

The clinical experience with graft angioplasty at Cleveland Clinic is summarized in Table 2. Although we had a similar number of patients with stenotic lesions in the proximal anastomosis and graft body, the number with stenosis of the distal anastomosis site is much greater in the Emory series. This difference may be in part due to the difference in distal anastomosis construction. All distal anastomoses at the Cleveland Clinic are performed with interrupted suture technique; the Emory surgeons use a continuous suture method.

The lower primary success in dilating graft body stenosis may reflect case selection and an early reliance on Stertzer guiding catheters, which did not supply the power for crossing lesions. Two patient outcomes were not successful because of the inability to force the balloon catheter through the lesion

TABLE 3 Graft Body Dilation

Category	Cases	Percentage
Successful	13/18	72
Continued success	6/13	46
Recurrence	7/13	54
Unsuccessful	5/18	28
Emergency surgery	2/5	40
No surgery	3/5	60

TABLE 4 Angioplasty of Anastomosis Sites

Category	Success Rate Percentage (Ratio)	Recurrence Rate Percentage (Ratio)
Distal anastomosis	100 (5/5)	20 (1/5)
Proximal anastomosis	88 (7/8)	43 (3/7)

(Fig. 6). One patient, only 7 months after surgery, had diffuse intimal hyperplasia not responding to multiple inflations; with a new balloon and the use of a larger catheter this might have been successfully dilated (Fig. 7). Two patients with a graft body stenosis (10 years after surgery) showed sudden occlusion of their grafts prior to crossing the lesion. In one case, extensive guiding catheter manipulation occurred during the procedure. In the later case, the graft closed with the first guiding catheter injection

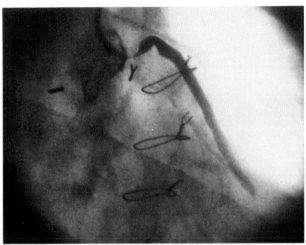

A

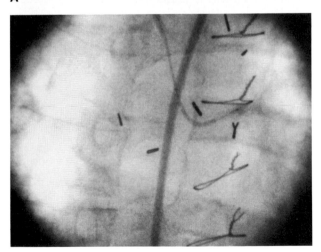

B

Figure 6. An unsuccessful proximal anastomosis dilation attempt. **A.** The severe proximal stenosis. **B.** The USCI DG 20-20 is partially across the obstruction (note the guiding catheter backed up into the aorta). Such a lesion could likely be crossed by a guidewire system and a guiding catheter with more power (e.g., a left Amplatz catheter).

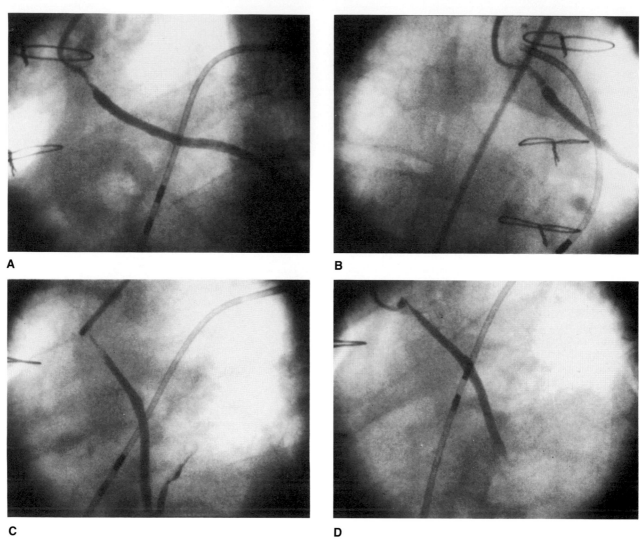

Figure 7. This young patient, 7 months after bypass surgery, illustrates intimal hyperplasia unresponsive to balloon pressure of 8–10 atmospheres applied over 15 times in each graft. Circumflex graft before (**A**) and after (**B**) PTCA in LAO projection. LAD graft before (**C**) and after (**D**) angioplasty. Although a slight measurable increase in the diameter narrowing was present in the circumflex graft, the patient was not clinically improved, later undergoing elective bypass surgery. The graft removal at surgery showed intimal hyperplasia.

(Fig. 8); pathologic examination of the graft in that case showed hemorrhage into the atherosclerotic plaque where angioplasty was to be attempted. There was no evidence of embolism or arterial dissection.

CLINICAL PERSPECTIVE

There are still a number of unanswered questions regarding angioplasty of bypass via graft, but a number of principles have emerged. Early graft occlu-

sion of less than 1 month due to thrombus will not favorably respond to balloon angioplasty. Graft occlusion after the second month may be due to intimal hyperplasia or atherosclerosis. The lesions due to intimal hyperplasia can likely be dilated with an acceptable complication rate. Atherosclerotic lesions may cause embolization and may be associated with a high recurrence rate.

Distinguishing between intimal hyperplasia and atherosclerotic lesions is not always possible by clinical and angiographic criteria. In the absence of

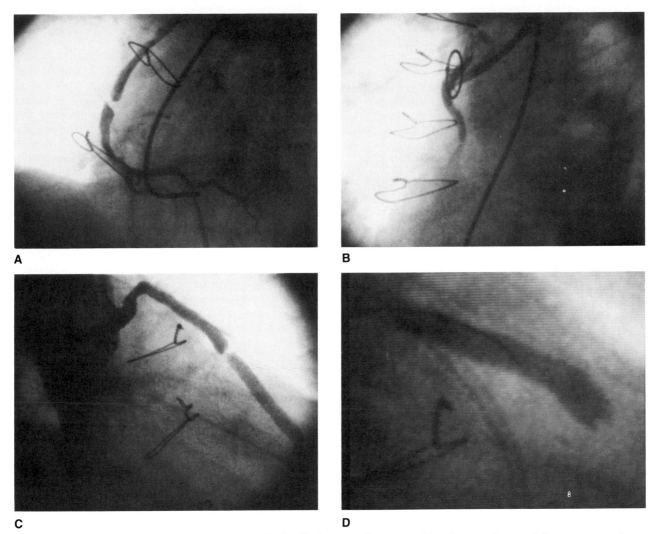

Figure 8. Graft closure occurring in late graft body dilations. **A.** A severe midgraft stenosis in a right coronary artery. **B.** Graft closure after PTCA attempt (**A,B**: LAP projection). Fortunately, the patient was stable and surgery was not required. **C,D.** A patient with a graft body stenosis in a left anterior descending graft. PTCA was successful in this graft 2 months earlier. This lesion represents a late graft body recurrence. With the first guiding catheter injection, the graft occluded completely, necessitating emergency bypass surgery (**C**, before PTCA; **D**, after PTCA attempt in the RAO projection, showing total abrupt occlusion).

hyperlipidemia, most graft lesions in the first 2 or 3 years following bypass surgery are probably due to intimal hyperplasia. Three years after bypass surgery, atherosclerotic changes are more likely the cause of stenosis. In general, if a graft appears smooth except for an isolated stenosis, graft angioplasty may be worthwhile. Diffusely diseased grafts after 8–12 years of bypass surgery probably should not be dilated. Stenosis of the anastomosis site, both proximal and distal, appears to be safe, although the number of proximal dilations reported is small. Re-

currence after dilation of the anastomosis site appears to be acceptable.

AREAS OF FUTURE RESEARCH

Late vein graft stenosis is best treated by prevention. Technical advances improved the early patency (3), but further improvement by technical modifications is unlikely. The association between graft atherosclerosis and hyperlipidemia makes it reasonable to recommend risk factor modification to patients after bypass surgery. Whether that will improve the results after graft angioplasty remains to be proved, but it certainly is worthy of further investigation.

Chesebro and colleagues at the Mayo Clinic reported the results of a carefully planned study of the use of antiplatelet drugs during the postoperative period. The use of antiplatelet drugs (aspirin and dipyridamole) appears to improve the patency of bypass grafts at 1 month and 1 year (17,18). Ideally the antiplatelet drugs would improve the long-term patency of the bypass vein grafts. Sufficient data on the effects of antiplatelet and anticoagulant drugs in patients undergoing balloon dilations of native coronary arteries or coronary vein bypass grafts are lacking.

Investigations to correlate angiographic appearance of bypass grafts with pathologic examination would be useful. Is it possible to identify a graft with atherosclerotic changes that may be susceptible to embolism if percutaneous transluminal coronary angioplasty is attempted? What is the frequency of embolism in dilation atherosclerotic grafts? Is this a theoretical or real danger? Could a laser be safely used in old graft body lesions? The answers to these and other questions await careful investigations.

REFERENCES

1. Coronary Artery Surgery Study (CASS): A randomized trial of coronary artery bypass surgery. Circulation 68:939–950, 1983.
2. Loop FE, Cosgrove DM, Taylor PC, Golding LAR, Lytle BW, Gill CC, Stewart RW: Coronary bypass surgery—the total experience at the Cleveland Clinic Foundation, in Hurst JW (ed): *Clinical Essays on the Heart*. New York, McGraw-Hill, 1984, vol 2, pp 131–138.
3. Campeau L, Enjalbert M, Lesperance J, Vaislic C, Grondin CM, Bourassa MG: Atherosclerosis and late closure of aortocoronary saphenous vein grafts: Sequential angiographic studies at two weeks, one year, five to seven years, and 10 to 12 years after surgery. Circulation 68(Suppl II):II-1–II-7, 1983.
4. Khonsari S, Staff A: Coronary artery bypass surgery: The total experience of the Oregon Health Science University, in Hurst JW (ed): *Clinical Essays on the Heart*. New York, McGraw-Hill, 1984, vol 2, pp 173–179.
5. Miller DG, Stinson EB, Shumway NE: Evolution of coronary bypass surgery at Stanford University, in Hurst JW (ed): *Clinical Essays on the Heart*, New York, McGraw-Hill, 1984, vol 2, pp 187–206.
6. Loop FD, Cosgrove DM, Kramer JR, et al: Late clinical and arteriographic results in 500 coronary artery reoperations. J Thorac Cardiovasc Surg 81:675–685, 1981.
7. Loop FD, Lytle BW, Gill CC, Golding LAR, Cosgrove DM, Taylor PC: Trends in selection and results of coronary artery reoperations. Ann Thorac Surg 36:380–388, 1983.
8. Dorros G, Cowley MJ, Simpson J, et al: Percutaneous transluminal coronary angioplasty: Report of complications from the National Heart, Lung and Blood Institute PTCA Registry. Circulation 67:723–730, 1983.
9. Kent KM, Bentivoglio LG, Block PC, et al: Percutaneous transluminal coronary angioplasty: Report from the Registry of the National Heart, Lung and Blood Institute. Am J Cardiol 49:2011, 1982.
10. Douglas JS Jr, Gruentzig AR, King SB III, Hollman J, Ischinger T, Meier B, Craver JM, Jones EL, Waller JL, Bone DK, Guyton R: Percutaneous transluminal coronary angioplasty in patients with prior coronary bypass surgery. J Am Coll Cardiol 2:745–754, 1983.
11. Bulkley BH, Hutchin GM: "Accelerated atherosclerosis": A morphologic study of 97 saphenous vein coronary artery bypass grafts. Circulation 55:163–169, 1977.
12. Rentrop P, Blanke H, Karsch KR, et al: Recannalization of an acutely occluded aortocoronary bypass by intragraft fibrinolysis. Circulation 62:1123–1126, 1980.
13. Holmes DR, Chesebro JH, Vlietstra RE, Orszulak TA: Steptokinase for vein graft thrombosis—a caveat. Circulation 63:729, 1981.
14. Brody WR, Kosek JC, Angell WN: Changes in vein grafts following aortocoronary bypass induced by pressure and ischemia. J Thorac Cardiovasc Surg 64:847–854, 1972.
15. Lie JT, Laurie GM, Morris GC Jr: Aortocoronary bypass saphenous vein graft atherosclerosis: Anatomic study of 99 vein grafts from normal and hyperlipoproteinemic patients up to 75 months postoperatively. Am J Cardiol 40:906–914, 1977.
16. Smith SH, Geer JC: Morphology of saphenous vein-coronary artery bypass grafts. Arch Pathol Lab Med 107:13–18, 1983.
17. Chesebro JH, Clements IP, Fuster V, et al: A platelet inhibitor drug trial in coronary artery bypass operations: Benefit of perioperative dipyridamole and aspirin therapy on early postoperative vein graft patency. N Engl J Med 307:73–78, 1982.
18. Chesebro JH, Fuster V, Elveback LR, et al: Effect of dipyridamole and aspirin on late vein graft patency after coronary bypass operations. N Engl J Med 310:209–214, 1984.

19

Emergency Bypass Surgery of Patients Undergoing Percutaneous Coronary Angioplasty*

DOUGLAS A. MURPHY, M.D.
JOSEPH M. CRAVER, M.D.

INTRODUCTION

During the past few years, percutaneous transluminal coronary angioplasty (PTCA) has undergone a remarkable growth and application as a new revascularization modality in the care of patients with coronary artery disease (1). With the increasing utilization of coronary angioplasty and the growing number of institutions around the country that offer this service, an increasing number of cardiac surgeons will need to be involved in the surgical support of coronary angioplasty patients. Since 4 to 7 percent of coronary angioplasty procedures resulting in acute ischemic complications (2) necessitating emergency bypass revascularization, all cardiac surgeons must be familiar with the medical and surgical problems associated with this group of patients in their often difficult and urgent clinical circumstances. To be successful and effective in managing these acute emergency patients with acute ischemia resulting from coronary angioplasty, it is essential to have established a cooperative and effective team approach between the physician who performs the angioplasty, the cardiac anesthesiologist, and the cardiac surgeon who will perform the emergency bypass surgery. In this chapter, we will discuss our experiences with patients requiring emergency bypass surgery as a result of attempted but failed coronary angioplasty at Emory University Hospital.

Coronary bypass surgery matured at Emory in the mid-1970s and is now being performed in over 2000 patients per year. Percutaneous coronary angioplasty was initiated in September 1980 at Emory University Hospital, 383 cases of coronary angioplasty were performed in the first year while angioplasty volume has increased to 110 cases per month in 1983. Fifty-eight patients have had angioplasty complications requiring emergency bypass surgery. The initial surgical support required for our angioplasty program and the results when emergency bypass surgery was required have been presented in earlier publications (4,6). Our experience supporting angioplasty will be updated and amplified in this chapter along with discussions regarding our management practices for these patients in a high risk state.

*This report was supported in part by a grant from The Wasie Foundation.

357

SPECTRUM AND ONSET OF CLINICAL SYMPTOMS

Most of the acute coronary injuries arising from angioplasty procedures manifest themselves immediately during the procedure in the cardiac catheterization laboratory, but also can present as a delayed sudden closure of the coronary artery after leaving the cardiac catheterization laboratory having had an initially successful procedure. The spectrum of acute myocardial ischemia from complicated angioplasty has ranged from acute chest pain without electrocardiographic changes to chest pain with electrocardiographic changes of transmural injury with cardiogenic shock.

This broad clinical spectrum of acute ischemia is summarized in Table 1 on 58 patients who underwent emergency bypass revascularization at Emory University Hospital, as a result of coronary angioplasty complications. Although the onset of an acute ischemic injury usually manifests itself immediately in the cardiac catheterization laboratory, the acute ischemic manifestations from arterial injury were delayed for 12 to 24 hours after the procedure in 12 percent of our patients. This potential for the delayed occlusion of the coronary artery, which was initially dilated successfully, requires an availability of surgical intervention for a period of 24 hours following any coronary angioplasty procedure.

PATHOPHYSIOLOGY OF INJURY

The morphologic changes observed at the site of balloon dilatation in an atherosclerotic coronary lumen suggest a longitudinal splitting of the intimal layer of an artery (3). Such an intimal disruption or a "controlled injury" to the arterial wall tended to remain localized in the immediate vicinity of the balloon dilatation site, possibly due to fibrotic nature of the tissues around the atherosclerotic plaque (Figure 1). This intimal injury sets the stage for a possible dissection of the intima leading to a complete

TABLE 1 Clinical Manifestations

Clinical Findings	Patient Number	%
Chest pain	58	100
ST elevation	41	73
Hypotension*	12	21
Cardiac arrest†	8	14

*Systolic pressure less than 90 mmHg.
†Episodes of ventricular fibrillation or asystole occurring prior to the induction of anesthesia.

luminal occlusion and blockage of the oxygenated blood flow through the artery. Such dissections account for the acute ischemia in approximately two-thirds of the patients (Table 2) undergoing emergency revascularization following angioplasty at Emory (4,5). A less common form of intimal injury can be induced in the artery distal to the balloon dilated segment by the tip of the balloon catheter or by the tip of the guidewire (6). This type of intimal injury in the less diseased distal segment may lead to a dissection that will propagate downstream to the distal coronary artery (Figure 2).

Although less frequent, a coronary artery occlusion can occur from the formation of an intraluminal clot due to the exposure of thrombogenic surfaces at the site of balloon induced intimal injury. This form of coronary occlusion during angioplasty procedure is limited by the administration of antiplatelet agents prior to angioplasty procedure and intra-arterial administration of heparin during the angioplasty procedure. Severe myocardial ischemia and injury in the absence of dissection or complete occlusion (2,7) has been observed in patients who had only coronary spasm or in patients who had spasm superimposed on any of the above mechanisms of coronary occlusion due to angioplasty complications.

PROFILE OF ANGIOPLASTY PATIENTS

Patients who are selected for percutaneous coronary angioplasty usually have angina of recent onset, fewer number of coronary arteries diseased, and good left ventricular function, although multivessel angioplasties are being proposed and investigated at present. A less common group of patients are those with previous bypass surgery who are undergoing coronary angioplasty of native coronary arteries or saphenous vein bypass grafts.

Nearly all patients who are candidates for coronary angioplasty should be considered candidates for bypass surgery should the angioplasty procedure be unsuccessful or complications arise during the procedure. Occasionally, however, patients who are not considered candidates for elective bypass surgery because of associated medical conditions, poor left ventricular function, or severely diseased distal coronary vasculature could still be considered candidates for angioplasty. In these patients who are not considered candidates for elective bypass surgery, a firm decision should be made before the angioplasty procedure is undertaken whether an emergency bypass surgery should be performed or

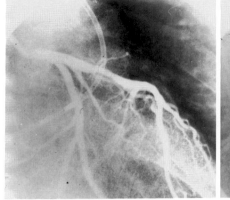
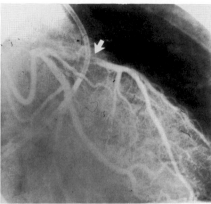

Figure 1. A. Coronary arteriogram revealing a stenosis of the left anterior descending artery. **B.** Following transluminal angioplasty a localized dissection can be identified (arrow) at the site of balloon dilatation.

A   B

not, should the angioplasty fail and complication occurs.

STRATEGY FOR PATIENT MANAGEMENT

Successful management of patients with acute myocardial ischemia induced by angioplasty complications depends on well-coordinated and cooperative efforts between the angioplasty team and the surgical team. The approaches to the care of these emergency patients include a diagnostic arteriogram to determine the cause of ischemia and to assess the extent of jeopardized myocardium. Furthermore, the steps in caring for these patients should include an attempt to reverse the ischemia using nonsurgical methods, as well as efforts to minimize the severity of ischemia and to stabilize the hemodynamics, and then, if deemed appropriate, the safe and speedy conduct of anesthesia and bypass surgery.

It is our policy to carry out emergency bypass surgery on these patients, if the patient following

angioplasty procedure has a preinfarction myocardial ischemia which is refractory to nonsurgical treatments. The decision to proceed with emergency bypass surgery is reserved for those patients who are in preinfarction state and when medical and interventional techniques have failed to reverse the ischemic process. Furthermore, the decision for emergency bypass surgery is reserved for those additional patients with ischemic chest pain when there is a significant amount of myocardium in jeopardy as determined by diagnostic angiography regardless of whether ischemic electrocardiographic changes are present or not.

CORONARY ARTERIOGRAPHY

When acute myocardial ischemia develops as a result of angioplasty complications, a prompt coronary arteriography will allow for recognition of the mechanisms of ischemia (5). The selection for a specific methodology from among many therapeutic modalities of restoring circulation such as intracor-

TABLE 2   **Angiographic Profile of Pathology**

Coronary Artery	Dissection		Occlusion		Total	
	n = 35	(%)	n = 23	(%)	n = 58	(%)
Left main	2	(3)	0	(0)	2	(3)
Left anterior descending	16	(28)	19	(33)	35	(60)
Right coronary	17	(30)	3	(5)	20	(35)
Saphenous vein graft	0	(0)	1	(1)	1	(1)
Total	35	(60)	23	(40)	58	(100)

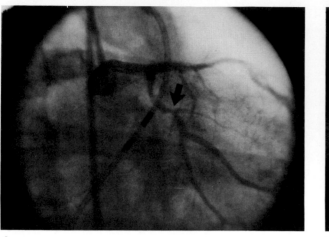
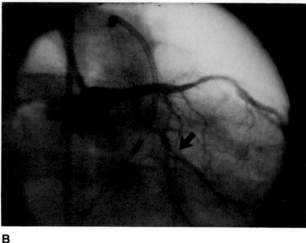

A   B

Figure 2. **A.** Coronary arteriogram demonstrating a proximal circumflex stenosis (arrow). **B.** Following transluminal angioplasty, a dissection can be identified originating in a large marginal branch to the site of the successful balloon dilatation (arrow).

onary administration of nitroglycerin for spasm, intracoronary administration of streptokinase for thrombus, or operative bypass revascularization will depend on the findings of a diagnostic arteriogram (8). Coronary arteriography will also allow for an assessment of the amount of myocardium in jeopardy and will aid in determination for the need of an emergency surgery. If the cause of chest pain is due to occlusion of a small artery such as a small diagonal branch after a successful dilatation of a large left anterior descending artery, the patient deserves medical management. Whereas dissection or occlusion of the left anterior descending artery in proximity to the left main bifurcation or near a large diagonal artery requires an expeditious bypass revascularization.

If the problem occurs while the patient is still in the cardiac catheterization laboratory where the angiography facilities are available, a diagnostic coronary injection to define the true status of the complication will take a minimum of time. However, in a patient who has already left the cardiac catheterization laboratory and develops acute signs of ischemia on the ward, a diagnostic arteriogram will undoubtedly cause a loss of additional time between the onset of ischemia and surgical intervention. In these patients who develop myocardial ischemia after they have left the cardiac catheterization laboratory, the need to perform diagnostic coronary arteriography depends on the outcome of the earlier angioplasty procedure. If the patient is being observed for an angiographically recognized but clinically silent intimal dissection of a major coronary artery,

then emergency revascularization should be promptly carried out without the benefit of further diagnostic arteriography. Alternatively, if the earlier angioplasty was uncomplicated, regardless whether it was successful or not, the patient should have the immediate diagnostic arteriography to determine if the patient has severe coronary spasm, intraluminal clot formation, or luminal occlusion directly caused by the arterial injury. Diagnostic arteriography in patients who develop sudden chest pain after leaving the cardiac catheterization laboratory has resulted in an emergency bypass rate of less than 4 percent at Emory Hospital by identifying those patients who are best suited for nonsurgical therapy (5).

INTRA-AORTIC BALLOON PUMP

Employing intra-aortic balloon pumping in patients who develop acute myocardial ischemia as a complication of balloon angioplasty has several theoretical advantages (5). Counterpulsation with intra-aortic balloon may reduce myocardial oxygen demand during systole and enhance the coronary flow to the ischemic myocardial zone during diastole by augmentation of diastolic pressure. In patients who develop acute ischemia during angioplasty, it is frequently observed that the electrocardiographic signs of myocardial ischemia improves with initiation of intra-aortic balloon pumping (5,9). Furthermore, the intra-aortic counterpulsation may also be beneficial by raising mean arterial pressure and cardiac output in patients with hemodynamic deterioration due to acute myocardial ischemia. Stabilization and improvement in hemodynamic parameters in patients with acute myocardial ischemia may prevent extension of the ischemic zone, thus avoiding cardiac arrest and end-organ injury, and provide an orderly transfer of the patient to the operating room. When a complication occurs during an angioplasty and acute myocardial ischemia becomes refractory to medical management, the intra-aortic balloon pump should be employed while the patient is still in the cardiac catheterization laboratory (5). If a patient develops a lesser degree of myocardial ischemia on the ward after leaving the cardiac catheterization laboratory, a balloon counterpulsation should be seriously considered, especially if the surgical standby has been disbanded and a delay in prompt surgical revascularization is anticipated. Intra-aortic balloon pump should also be considered in such situations as when the patient has had previous cardiac surgery and epicardial scarring which will require prolonged dissection before bypass grafting can be achieved.

A rapid implementation of counterpulsation in patients who develop acute injury during angioplasty can be facilitated by a number of prearranged preparations: (a) placement of balloon pump console and intra-aortic balloon catheters in the angioplasty suite, (b) immediate availability of skilled balloon pump technicians, and (c) prepping both groin areas prior to the angioplasty procedure. Intra-aortic balloon insertion can be achieved very rapidly using percutaneous Seldinger puncture and guidewire technique. The potential complications arising from the use of intra-aortic balloon pump can be minimized by limiting the insertion of the balloon catheter to experienced personnel, use of fluoroscopy, and removal of the balloon catheter during the postoperative period as soon as the patient's condition allows such a removal.

COORDINATION OF SURGICAL SUPPORT

When the decision is made to proceed with emergency bypass surgery, time is of the essence. Surgical support for an angioplasty procedure performed during the initial two months at Emory Hospital involved a cardiac surgical operating room standing empty with a full complement of personnel available. In the third and fourth months, angioplasty procedures were performed at times when a cardiac operating room was empty between elective operations or at the end of the day. From that time forward, there had been no formal coordination between angioplasty procedures and the cardiac sur-

gical schedule. Patients who require emergency myocardial revascularization are accommodated in the first operating room available. There are four cardiac surgical operating rooms at Emory Hospital. In smaller institutions, more strict coordination of the angioplasty and elective cardiac surgical schedules may be necessary to allow for immediate availability of cardiac surgical personnel and facilities.

When a complication of acute ischemia occurs during angioplasty procedure and it appears that emergency revascularization may be required, the surgery, anesthesia, and operating room teams are alerted immediately. A member of the cardiac anesthesia team is dispatched to the angioplasty suite to assess the patient's status and to assist in transporting the patient to the operating room. Simultaneously, an operating room is prepared; coordination of this effort is directed by the cardiac anesthesiologist in charge. The practice of early notification of a possible angioplasty problem has expedited many later critical situations and has resulted in only minor inconvenience to the elective schedule when the potential emergency surgery proved to be unnecessary.

ANESTHESIA FOR FAILED ANGIOPLASTY PATIENTS

While the emergency patient is still in the angioplasty suite, the cardiac anesthesia team may insert arterial pressure monitoring catheters, administer pharmacologic agents, and institute ventilation assistance as necessary to maintain adequate hemodynamics during transport to the operating room. Insertion of comprehensive monitoring catheters prior to induction of anesthesia has not significantly delayed surgery in our experience. Over 95 percent of our patients have received pulmonary artery catheters prior to anesthesia induction. The majority of patients were anesthetized with diazepam and fentanyl or morphine.

Vasopressor agents were required in over 75 percent of the patients prior to and during cardiopulmonary bypass to counteract the vasodilatory effects of nifedipine and nitroglycerin given during and immediately following the angioplasty (12). Inotropic agents were needed in approximately 30 percent of patients after cardiopulmonary bypass to maintain satisfactory hemodynamics. Pulmonary artery catheters inserted prior to induction permitted the rational selection of vasoactive and cardiotonic drugs on the basis of objective hemodynamic data and facilitated the salvage of these acutely ischemic patients.

SURGICAL TECHNIQUE

As soon as the induction of anesthesia is achieved, a median sternotomy is performed. Intraoperative findings frequently include discoloration and dysfunction (Table 3) of left ventricle in the distribution of the injured coronary artery (4,5). Less commonly, there may be areas of periarterial hemorrhage proximally or distally along the injured segment of the artery. Cannulation of the heart is rapidly carried out and cardiopulmonary bypass is instituted as soon as possible. In the case of patients with previous cardiac surgery who are hemodynamically unstable despite intra-aortic balloon pumping, institution of cardiopulmonary bypass is applied via the femoral arteries while the heart is being exposed. Following institution of cardiopulmonary bypass, no extended time should be wasted for systemic cooling prior to aortic crossclamping and administration of cardioplegia. Persistent ST-segment elevation on electrocardiogram has been frequently seen in this group of patients despite an empty beating heart on cardiopulmonary bypass. Myocardial preservation is achieved with intra-aortic administration of cold crystalloid cardioplegia and iced saline solution in the pericardial well. If more than one vein graft is planned, the injured artery(ies) should be bypassed first and 200 ml of cold blood cardioplegia administered through the vein graft perfusing the area of ischemic myocardium immediately following distal graft anastomosis. Cardioplegia administered via the aorta often cannot reach this ischemic area due to the acute arterial occlusion and lack of collateral vessels. Additional cold blood cardioplegia is administered via the vein graft to injured areas every 5 minutes if multiple distal grafts are required until aortic flow is restored on completion of the proximal anastomosis. We have also used the internal mammary artery for the bypass graft in this clinical setting, when only one graft was needed. However, the additional time required to harvest and prepare it for use and the inability to infuse cold solutions distally through it to injured myocardium if multiple grafts are required limit its desirability for most emergency patients.

Following completion of vein graft anastomoses and reestablishment of flow to the distribution of the injured artery, the heart is rested on cardiopulmonary bypass in a warm, empty, beating state for 30 minutes. It has been during this period of reperfusion that improvement in color and function of the previously ischemic muscle has frequently been observed (4,5).

The location of the coronary arteriotomy for distal vein graft anastomosis is based on the known presence of atherosclerotic lesions and is not influenced by the mechanism of coronary artery injury by the angioplasty procedure. Techniques of vein graft anastomosis are those used in routine revascularization except in the case of distal coronary dissection (6). Distal dissection should be suspected when the coronary artery at the point of anticipated arteriotomy has a bluish discoloration which persists even after administration of crystalloid cardioplegia. The diagnosis is confirmed by arteriotomy revealing both false and true arterial lumens (Figure 3). Technical management of this lesion involves reapproximation of the dissected vessel wall layers with anastomosis of the vein graft to the true lumen. Temporary stay sutures are useful in realigning the layers during the procedure of a routine running anastomosis (Figure 4).

Although proximal coronary artery ligation has been advocated in patients undergoing revascularization for coronary artery dissection (10), this has been shown to be unnecessary in our experience (6). In the case of extended distal coronary dissection, reconstruction of the distal vessel and reestablishment of flow in the true arterial lumen via vein graft has been demonstrated on late restudy to obliterate the false lumen and maintain distal vessel patency despite restored patency of the proximal dissected native coronary artery segment (6).

In patients who undergo balloon angioplasty of a prior vein graft itself, special technical considerations may be necessary in cases of acute myocardial ischemia resulting from the balloon dilatation of an old saphenous vein graft. In this particular setting, angioplasty of atherosclerotic vein grafts can result in embolization of atherosclerotic debris to the distal coronary artery. If such a complication can be recognized before or during the operation, balloon embolectomy of the involved coronary artery can be performed or a new saphenous vein graft can be placed beyond the site of embolic debris.

In our experience of 58 angioplasty patients undergoing emergency bypass revascularization, approximately 30 percent of the patients will require

TABLE 3   Operative Findings

Description of Findings	Patient Number	%
LV dysfunction, discoloration	21	36
Periarterial hemorrhage	10	17
No abnormal finding	27	47
Total	58	100

from 6[
the pro
cardiog
chest le
sion of
balloon
tiated t
above .
duced,
The pa
cardiop
grafts v
and rig
comple
pain. T
was hy
weanec
sure wa
and no
tient de
of his e
remaine
sure be
which (
diac fil
chanica
piratory
oxygen
tient d
monia
day wit

CASE E)
ferior ir
mal left
posteric
akinetic
nary sy
tractility
normal.
tery cau
angioca
only te
severel
rested r
laborate
other su
where t
was "c
revascu
formed,
lowing a
after the
anterior
minutes
was we

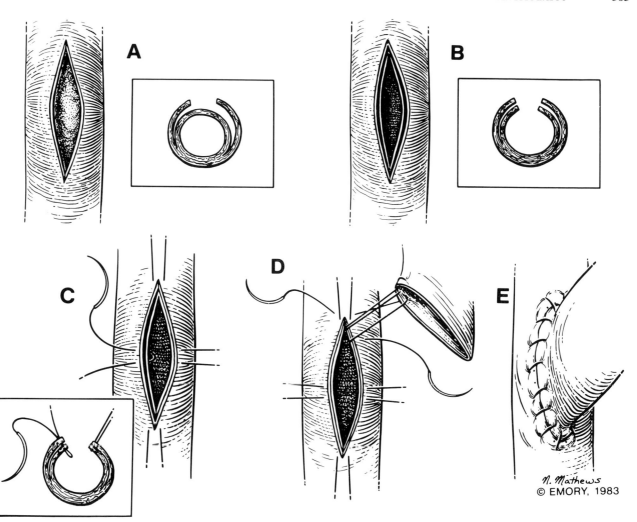

Figure 3. Distal coronary artery dissection and repair. **A.** The initial coronary arteriotomy opens only the false lumen exposing the convex surface of the true lumen. **B.** A second arteriotomy opens the true arterial lumen. **C.** The arterial wall layers are reapproximated by temporary stay sutures. **D,E.** Vein graft anastomosis using running technique.

inotropic drug support to be successfully separated from the cardiopulmonary bypass (12). It is also observed that by applying intra-aortic balloon pumping to all patients with persistent ST-segment elevation preoperatively, it was rarely necessary to employ balloon pumping in additional patients (5).

POSTOPERATIVE PATIENT CARE

The most common postoperative problem encountered in the patients who had undergone bypass surgery following angioplasty complication is medias-

tinal hemorrhage despite adequate reversal of heparin (4,5). Routine administration of aspirin for a period of time to these patients prior to balloon angioplasty often result in significant platelet dysfunction which may not be readily reversed by administration of platelet concentrates during surgery. Systemic hypotension in the immediate postoperative period in these patients is also common, even though the cardiac output is adequate. This phenomenon is largely due to decrease in peripheral vascular resistance secondary to administration of nitroglycerin and Nifedipine given routinely before, during, and after angioplasty to reduce the possibility of a coronary

spa
qui
age
sist

PF
M

An
sur
slij
of
nin
the
car
the
co
isc
mu
sei
my
isc
no
car
lik
im
ne
ex
ica
in
on
cr
les
co
the
ab
ris
sis
ce
th
in
ha
an
gr
tir
ac

th
ge
In
sij
tic
el

and return of global perfusion. In contrast, patients with critical lesions in only one vessel but only moderate disease in other systems and good ventricular function have closely paralleled the experience of single-vessel disease patients.

So long as the restenosis rate following angioplasty for a single vessel remains approximately 33 percent (1,17), the surgical alternative for treatment of multivessel disease seems preferable. Furthermore, neither the long-term results nor the methodology for the late follow-up of angioplasty patients has been satisfactorily established. Are angioplasty patients to be subject to periodic reangiography indefinitely? Such answers need to be secured for single-vessel angioplasty first, then a controlled double-vessel angioplasty versus surgical revascularization study may be considered.

Left Main Artery Disease

It is the opinion of the Emory angioplasty physicians that abrupt closure of the left main artery cannot be tolerated by the patient nor can the patient be supported effectively to allow angioplasty surgical revascularization. Therefore, balloon dilatation of the left main artery involves risks of extensive myocardial injury that are not acceptable.

CASE EXAMPLE 3: A 56-year-old male auto factory worker with normal LV function and normal right coronary artery was referred from another state for angioplasty of a left main stenosis where he was felt not to be a surgical candidate due to "chronic pulmonary disease." Catheter manipulation at angioplasty dissected and closed the left main artery with prompt hemodynamic collapse. Continuous external and later internal cardiac massage was required to maintain life support. Transportation to surgery, anesthesia induction, surgical preparation, sternotomy, and cannulation for bypass were accomplished with the patient in refractory shock. Global LV discoloration and dysfunction persisted despite anterior descending and circumflex arteries being bypassed and blood flow reestablished within 69 minutes from onset of the dissection. Extensive infarction occurred. The patient was ultimately discharged in minimally compensated congestive heart failure and was totally disabled 172 days later, having required months of ventilator assist, renal dialysis, as well as cardiovascular and nutritional support.

Although this patient and one other are long-term survivors of acute total left main artery closure, both experienced extensive global infarction despite immediate support and the fastest surgical revascular-

ization effort possible. This experience in addition to the poor outcome with patients experiencing left main artery occlusion during coronary arteriography have been uniformly disappointing.

Procrastination after Angioplasty Complications

A disturbing trend in practice is the willingness of angioplasty physicians to accept complications or unsuccessful results occurring at angioplasty unless there is associated severe hemodynamic compromise.

CASE EXAMPLE 4: A 67-year-old lady underwent an angioplasty of the left anterior descending artery and encountered a dissection of the same artery. The patient experienced severe chest pain and ischemia but was hemodynamically stable. The pain and ischemia "settled down" on intensive therapy in the coronary care unit and surgery was "avoided." Recurrence of pain less than one month later prompted recatheterization which revealed total anterior descending artery and diagonal artery occlusion and extensive anterolateral and septal dysfunction. Surgical revascularization of LAD and the diagonal artery was uncomplicated with the epicardial myocardium found to be intact despite the prior infarct and dysfunction. The patient has had no further angina postoperatively.

Failure to send patients for emergency surgery after complicated angioplasty when large areas of myocardium are potentially jeopardized should be avoided. Whether motivated by a desire to hold down a team's emergency surgery rate or just to avoid interrupting their surgical colleagues' elective schedule, these efforts reflect a reluctance to acknowledge the impending injury and deny the patient an optimal revascularization result. Case 4 and six additional patients in the Emory series, who were referred for surgery from 2 or 24 hours after angioplasty when "intensive" medical therapy failed to control their symptoms, illustrate this point. All eventually did well although all had experienced myocardial infarction.

CONCLUSION

Despite careful patient selection and meticulous attention to technique, a small percentage of patients undergoing percutaneous angioplasty will develop acute myocardial ischemia distal to the instrumented artery. The infarction process starts when the angioplasty instrumentation produces an abrupt worsening of a coronary artery lesion on which myocardium is still dependent. This worsening whether by flap closure, dissection, elicitation of spasm, vessel disruption, immoblization, or thrombosis closes off flow that the myocardium needs and the infarction process begins. The coronary artery injury and acute myocardial ischemia resulting from balloon angioplasty may manifest itself immediately during the procedure or as a delayed onset and may have a variable degree of severity (1). Immediate diagnostic coronary arteriography allows for recognition of the cause of ischemic process and assessment of the mass of myocardium jeopardized by the obstructed artery. When myocardial infarction appears imminent or when the area of myocardium jeopardized is large, emergency surgical revascularization is the therapy of choice. Preoperative employment of intra-aortic balloon counterpulsation may serve to preserve ischemic myocardium and stabilize hemodynamic function. Prompt bypass revascularization and the use of preoperative intra-aortic balloon pumping has been associated with only 20 percent incidence of transmural infarction and no mortality in this difficult group of patients.

REFERENCES

1. Gruentzig A: Results from coronary angioplasty and implications for the future. Am Heart J 103:779–783, 1982.
2. Dorros G, Crowley MJ, Simpson J, et al: Percutaneous transluminal coronary angioplasty: Report of complications from the National Heart, Lung and Blood Institute PTCA Registry. Circulation 67:723–730, 1983.
3. Block PC, Myer RK, Stertzer S, Falbon JT: Morphology after transluminal angioplasty in human beings. N Engl J Med 305:382–385, 1981.
4. Murphy DA, Craver JM, Jones EL, Gruentzig AR, King SB III, Hatcher Cr Jr: Surgical revascularization following unsuccessful percutaneous transluminal coronary angioplasty. J Thorac Cardiovasc Surg 84:342–348, 1982.
5. Murphy DA, Craver JM, Jones EL, Curling PE, Guyton RA, King SB III, Gruentzig AR, Hatcher CR Jr: Surgical management of acute myocardial ischemia following PTCA. Role of the intra-aortic balloon pump. J Thorac Cardiovasc Surg (in press).
6. Murphy DA, Craver JM, King SB III: Distal coronary artery dissection following PTCA (in preparation).
7. Bentivoglio LG, Leo LR, Wolf NM, Meister SG: Frequency and importance of unprovoked coronary spasm in patients with angina pectoris undergoing percutaneous transluminal coronary angioplasty. Am J Cardiol 51:1067–1071, 1983.
8. Schofer J, Krebber HJ, Belifeld W, Mathey DG: Acute coronary artery occlusion during percutaneous transluminal coronary angioplasty: Re-opening by intracoronary streptokinase before coronary artery surgery to prevent myocardial infarction. Circulation 66:1325–1331, 1982.
9. Margolis JR: The role of the percutaneous intra-aortic balloon in emergency situations following percutaneous transluminal coronary angioplasty, in *Transluminal Coronary Angioplasty and Intracoronary Thrombolysis,* M Kaltenbach, A Gruentzig, K Rentrop, WD Bussman (eds). New York, Springer-Verlag, 1982, pp 144–150.
10. Harrison LH Jr, Gregg DL, Itscoitz SB, Redwood DR, Michaelis LL: Delayed coronary artery dissection after angiography. J Thorac Cardiovasc Surg 69:880–883, 1975.
11. Ford WB, Wholey MH, Zikria EA, Miller WH, Samadani SR, Koimattur AG, Sullivan ME: Percutaneous transluminal angioplasty in the management of occlusive disease involving coronary arteries and saphenous vein bypass grafts. J Thorac Cardiovasc Surg 79:1–11, 1980.
12. Curling PE, Waller JL, Murphy DA, Craver JM, Jones EL, Freniere S: Resuscitation, monitoring, and anesthesia for failed percutaneous transluminal coronary angioplasty (in preparation).
13. Reimer KA, Lower JE, Rasmussen MM, Jennings RB: The wavefront phenomenon of ischemic cell death. 1. Myocardial infarct size vs duration of coronary occlusion in dogs. Circulation 56:786–794, 1977.
14. Anderson JL, Marshall HW, Bray BE, Lutz JR, Frederick PR, Yarrowitz FG, Datz FL, Klausner SC, Hogan AD: A randomized trial of intracoronary streptokinase in the treatment of acute myocardial infarction. N Engl J Med 308:1312–1318, 1983.
15. Libow M, Cooke D, King SB III, Greuntzig A, Douglas J, Jones EL, Craver JM, McClees R: Left ventricular function after unsuccessful angioplasty and emergency coronary bypass surgery (in preparation).
16. Jones EL, Craver J, Gruentzig AR, King SB III, Douglas JS, Bone DK, Guyton RG, Hatcher CR Jr: Percutaneous transluminal coronary angioplasty: Role of the surgeon. Ann Thorac Surg 34:493–503, 1982.
17. Hollman J, Gruentzig A, Meier B, Bradford J, Galan K: Factors affecting recurrence after successful coronary angioplasty. Am J Cardiol (abstracts): March 1983.
18. Kutcher MA, Gruentzig AR, Turina M, Craver JM, Jones EL, Douglas JS, King SB III: Can emergency coronary bypass surgery following acute failure of coronary angioplasty prevent myocardial infarction? Am J Cardiology 49:956, 1982.

20

A Review of the NHLBI PTCA
Registry Data

MICHAEL J. COWLEY, M.D.
PETER C. BLOCK, M.D.

INTRODUCTION

Percutaneous transluminal coronary angioplasty (PTCA) was first performed in humans in September 1977 and the initial preliminary reports of its clinical application appeared shortly thereafter (1–3). The potential impact of this technique in the treatment of ischemic heart disease was recognized early on by members of the National Heart, Lung, and Blood Institute (NHLBI) and its advisory panels. In January 1979 a small group of investigators active in percutaneous coronary angioplasty met at the National Institutes of Health (NIH) and the concept of a PTCA registry was developed to provide a timely and efficient means of assessing the usefulness and limitations of this technique.

A position paper on coronary angioplasty by the NHLBI was published in March 1979 announcing the formation of a voluntary Interim Registry by the Cardiac Diseases Branch (4). At the first NHLBI workshop on coronary angioplasty held in June 1979, 12 centers had enrolled in the Registry and data had been submitted on 61 patients from five of these centers. Combined data were presented on results of coronary angioplasty in 205 patients from centers in Europe and the United States (5) and a summary of the workshop results and recommendations were

published in an editorial (6). From this, a permanent Registry evolved as a international, multiinstitutional collaboration involving centers from the United States, Europe, and other countries. Participation in the Registry was voluntary and all institutions performing coronary angioplasty were encouraged to join. Participation required submission of angioplasty protocol and the document of institutional review board approval. A steering committee composed of representatives from each clinical center, the data coordinating center, and the NHLBI was formed in July 1979.

As clinical application of coronary angioplasty spread, the number of participating centers increased rapidly. More than 35 centers were enrolled by January 1980. In February 1980 an executive committee was formed to guide the course of the Registry and to make future recommendations to the NHLBI. A permanent data coordinating center was established at the University of Pittsburgh. The organizational structure of the Registry is illustrated in Figure 1. The Registry continued to grow to 125 centers enrolled and 105 centers contributing data by September 1981, when the Registry was closed to new centers and discontinued enrollment of patients with single vessel disease. During the following year, enrollment was restricted to patients with

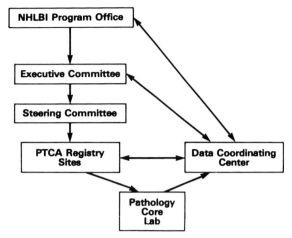

Figure 1. Organizational structure of the NHLBI PTCA Registry. The Steering Committee was composed of representatives from each participating clinical site. The Executive Committee included representatives from the Steering Committee, the NHLBI Program Office, and the Data Coordinating Center at the University of Pittsburgh. The Pathology Branch of the NHLBI served as the central Pathology Core Laboratory.

multivessel disease. In September 1982 all patient enrollment was closed; at that time, more than 3000 patients had been enrolled in the NHLBI PTCA Registry.

A second workshop on PTCA was held in June 1981 to review the status of coronary angioplasty. Data presented on 1500 patients were considered to establish the relative safety and efficacy of the procedure in the acute outcome of carefully selected patients. However, information on follow-up of these patients was insufficient to assess the longer term results, and the focus of the Registry shifted to concentrating on the follow-up outcome. By September 1982 virtually complete follow-up (98.5) for at least 1 year or longer on more than 2000 patients had been obtained from 65 centers. In June 1983 a third workshop on PTCA was held. This follow-up data as well as baseline data on 3079 patients were reviewed and future directions of coronary angioplasty beyond the current NHLBI Registry were discussed.

REGISTRY RESULTS

The total number of patients enrolled in the NHLBI PTCA Registry through the end of September 1982 reached 3079 patients. These results represent patients enrolled under general selection guidelines (7,8) and do not include patients enrolled under special protocols that are designed on such special category

of patients as patients with cardiogenic shock, intracoronary thrombolysis, or recent (within 3 weeks) myocardial infarction. From the outset, patients considered to be the most appropriate candidates for angioplasty were those with stable angina, single vessel disease, and proximally located discrete stenoses. Although there was variation in selection criteria at different centers, the majority of patients conformed to these guidelines.

These data were registered from 105 participating centers. The median case contribution was 11 patients (range 1 to 223) per center. The baseline clinical and angiographic characteristics of these patients are shown in Table 1. There were 2374 males (77%) and 705 females (23%); mean age was 53.5 years. Angina was present in 97% of these patients. Functional angina class of 3 or 4 according to the

TABLE 1 Baseline Characteristics of Registry Patients

Number of patients	3079
Mean age of patients (years)	53.5
Male gender	77.0%

Criteria	*%*
Angina	97.0%
Unstable angina	43.0%
Functional class (CHC)	
Class 1	4.0%
Class 2	29.3%
Class 3	39.3%
Class 4	27.3%
Prior MI	25.0%
Prior CABG	9.0%
History of hypertension	35.0%
Ejection fraction >50%	92.0%
Extent of disease	
Single vessel	72.5%
Multivessel	25.0%
Left main	2.5%
Dilations (no. attempts)	3341
Single attempt (no. patients)	2845 (92%)
Multiple attempts (no. patients)	234 (8%)
Vessels attempted	
LAD	62.0%
RCA	26.0%
Cx	7.0%
Graft	4.0%
Left main	1.0%

Key: CABG = Coronary artery bypass surgery; CHC = Canadian Heart Association Classification; Cx = circumflex artery; LAD = left anterior descending artery; RCA = right coronary artery.

Canadian Heart Association criteria was present in 67%. Unstable angina was present in 43%, prior myocardial infarction 25%, and prior bypass surgery 9%. The vast majority of patients had normal left ventricular function with 92% reported to have an ejection fraction greater than 50%. Single vessel coronary disease (greater than 50% narrowing of luminal diameter) was present in 72.5%, multivessel disease in 25%, and left main disease in 2.5%. A total of 3341 lesions were attempted in these 3079 patients, with single attempts in 2845 patients (92%) and multiple attempts in 234 patients (8%). The vessels attempted were the left anterior descending artery (62%), right coronary artery (26%), circumflex artery (7%), bypass vein graft (4%), and the left main coronary artery (1%).

ACUTE OUTCOME OF PTCA PROCEDURE

The clinical and angiographic results of coronary angioplasty are shown in Table 2. Angiographic success defined as 20% or greater increase in luminal diameter was achieved in 66.6% (2226/3341) of the lesions attempted. Primary clinical success, defined as angiographic criteria of success plus no myocardial infarction or bypass surgery during hospitalization, was achieved in 1878 patients (61%). The mean percent change in the luminal diameter in patients with successful angioplasty was 52%.

Angioplasty was unsuccessful in 1201 patients (39%) enrolled in the Registry. The most frequent reason for failure was inability to cross the stenosis, which occurred in 25% of patients, but unsuccessful attempts on these patients was 63%. The reasons for inability to cross the lesions in these patients were severity of stenosis (51%), sharp tortuosity (24%), inability of guiding catheter placement (12%), and other causes of failure to cross the lesion (14%). Inability to dilate the lesion once the lesion was crossed was a cause of failure in 8% of patients and 21% of failed attempts. Rigidity of the lesion was the most frequent reason (63%) for inability to dilate the lesion. In the remaining 6% of patients in whom angioplasty was unsuccessful, reasons for failure were multiple attempts (2.2%), abrupt reclosure after initially successful dilation (2.8%) and necessity for bypass surgery during the same hospitalization due to inadequate clinical results after angioplasty (1%).

FACTORS INFLUENCING SUCCESS RATE

Success rate of coronary angioplasty was influenced by a number of baseline clinical and angiographic factors (Table 3A). Baseline clinical factors associated with a significantly lower success rate by univariate analysis were: multivessel disease ($p < 0.001$), Class 3 or 4 angina ($p < 0.001$), unstable angina ($p < 0.05$), angina of longer than 6 months duration ($p < 0.001$), patients older than 60 years of age ($p < 0.002$), and female gender ($p < 0.002$). Angiographic factors that influence the success rate were severity of the lesion 90% or more in luminal diameter ($p < 0.0001$), angioplasty in coronary arteries other than the left anterior descending artery ($p < 0.001$), lesion calcification ($p < 0.001$), distally located lesions ($p < 0.001$), eccentricity of lesion ($p < 0.05$), and non-discrete lesions ($p < 0.01$). Procedural factors such as number of inflations and balloon pressure also significantly influenced success rate, although these variables themselves are partly dependent on outcome (9). An additional important factor influencing success rate was angioplasty experience of the investigator (10). Figure 2 shows the success rate as a function of experience ("learning curve"). Table 4 shows the differences in success rate at centers with experience of less than 50 cases and greater than 50 cases. The clinical success rate at the 87 sites with less than 50 cases was 55% whereas the success rate at the 18 sites with greater than 50 cases was 66% ($p < 0.0001$). At sites with more than 50 cases, the success rate for all cases above the first 50 cases was 70%, for all cases above the first 100 cases was 72%, and for cases above the first 150 cases was 77%. At these 18 centers where more than

TABLE 2 Results of Coronary Angioplasty

Criteria	%
Angiographic success (no. lesions)	67
Clinical success (no. patients)	61
Mean change of stenosis	52
Success rate by vessels	
LAD	65
RCA	56
Cx	46
Graft	60
Left main	68
Unsuccessful PTCA (no. patients)	39
Inability to cross	25
Inability to dilate	8
Other	6
Reclosure	3
Not all lesions successful	2
MI or continued symptoms	1

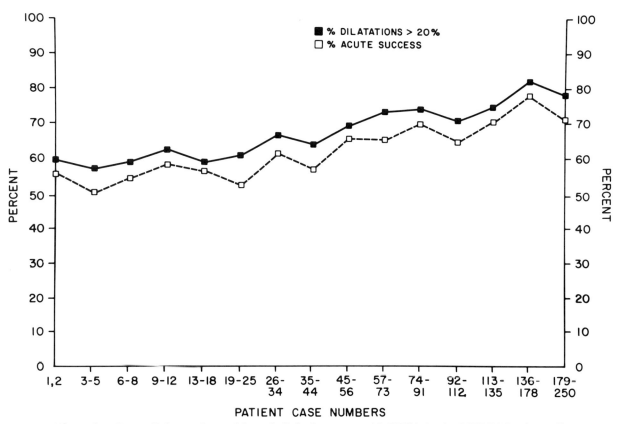

Figure 2. "Learning Curves" for angiographic and clinical success with PTCA in the NHLBI Registry. Success rate is plotted as a function of patient case number. Case sequence numbers are grouped to achieve approximately equal numbers of patients in each group. There is a significant progressive improvement in success rate with increasing experience with PTCA.

TABLE 3A **Factors for Reduced Success Rate (Univariate Analysis)**

Factors	Success Rate, %*	p Value
Lesion ≥90%	53 vs 71	<.0001
Experience <50 cases	55 vs 66	<.0001
Multivessel disease	55 vs 63	<.001
Class 3 or 4 angina	59 vs 66	<.001
Angina >6 months	57 vs 63	<.001
Nonproximal lesion	56 vs 65	<.001
Non-LAD attempt	55 vs 64	<.001
Lesion calcification	49 vs 71	<.001
Age >60 years	56 vs 63	<.002
Female gender	57 vs 62	<.002
Nondiscrete lesion	58 vs 67	<.01
Unstable angina	59 vs 63	<.05
Eccentric lesion	65 vs 70	<.05

*Success rate comparison for *Presence vs Absence* for each baseline factors.

50 cases were performed, the success rate for the first 49 cases was 59%, the success rate for the first 50 to 99 cases was 68%, for the first 100 through 149 cases was 69%, and all cases above the first 150 cases was 77%.

Multivariate analysis (Table 3) of risk factors as-

TABLE 3B **Factors for Reduced Success Rate (Multivariate Analysis)**

Factors	p Value
Lesion ≥90%	<.0001
Lesion calcification	<.0001
Nonproximal lesion	<.001
Experience with 50 cases or less	<.001
Multivessel disease	<.01
Circumflex attempt	<.01
Female gender	<.05
Angina >6 months	<.05

TABLE 4   Effects of Experience on Success Rate

Case Load	Success Rate, %
87 sites with <50 cases	55
18 sites with >50 cases*	66
1–49	59
50–99	68
100–149	69
150 and above	77
All cases above 50	70
All cases above 100	72
All cases above 150	77

*Breakdown of learning experience of the 18 sites with case experience of more than 50 cases each.

sociated with lower success rate was carried out on 2234 patients with complete data for variables identified as significant by univariate analysis. Ranked in order of importance, these factors were: severity of narrowing 90% or greater in lumen diameter (p < 0.0001), lesion calcification (p < 0.0001), nonproximal lesion (p < 0.001), experience of less than 50 cases (p < 0.001), multivessel disease (p < 0.01), circumflex lesion (p < 0.01), female gender (p < 0.05), and duration of angina greater than 6 months (p < 0.05).

COMPLICATIONS

One or more complications were reported in 652 (21.2%) patients out of 3079 patients enrolled in the Registry. A total of 1180 complications were reported. The type and frequency of complications are shown in Table 5. The most frequent compli-

TABLE 5   Complications Reported in the Registry

Category	Number	%*
Prolonged angina	236	6.8
Myocardial infarction	170	5.5
Coronary occlusion	151	4.9
Coronary dissection	135	4.4
Coronary spasm	130	4.2
Ventricular fibrillation	56	1.8
Hypotension	47	1.5
Bradycardia	43	1.4
Death	29	0.9
Peripheral arterial	22	0.7
Miscellaneous	110	3.6

*Percent of the total patients (n = 3079) enrolled. They are not mutually exclusive.

TABLE 6   Predictors of Complication

Univariate		Multivariate	
Category	p Value	Category	p Value
Female sex	<.001	Female sex	<.001
Lesion >90%	<.001	Unstable angina	<.001
RCA attempt	<.01	Age >60 years	<.05
Unstable angina	<.05	Lesion >90%	<.05
Age >60 years	<.05		

cation was prolonged angina, with 236 episodes reported in 211 patients (6.8%). Myocardial infarction occurred in 170 patients (5.5%), coronary occlusion in 151 patients (4.9%), coronary dissection in 135 patients (4.4%), and coronary spasm in 130 patients (4.2%). Ventricular fibrillation occurred in 56 (1.8%), hypotension in 47 (1.5%), bradycardia in 43 (1.4%), death (in hospital) in 29 patients (0.9%), peripheral artery complications requiring surgical repair in 22 patients (0.7%), and other miscellaneous complications in 110 patients (3.6%). Major complications occurred in 289 patients (9.4%), with death in 29 patients (0.9%), nonfatal myocardial infarction in 154 patients (5.0%), and emergency bypass surgery in the remaining 106 patients (3.4%).

Factors influencing the frequency of total complications are shown in Table 6. Factors associated with a significantly higher incidence of complications by univariate analysis include: female gender (p < 0.001), lesion severity ≥90% (p < 0.001), right coronary artery lesion (p < 0.01), unstable angina (p < 0.01), and age group 60 years and older (p < 0.05). Patients with unsuccessful angioplasty had a significantly higher incidence of complications than patients with successful dilation (32% vs 16%, p <

TABLE 7   Predictors for Type of Complications

Complication	Predictor	p Value
Emergency CABG	Eccentric lesion	<.001
Intimal tear or dissection	Female	<.001
	RCA attempt	<.001
	Multivessel disease	<.01
	Eccentric lesion	<.01
Occlusion	Eccentric lesion	<.01
	Tubular lesion	<.05
	Nondiscrete lesion	<.05
Myocardial infarction	Unstable angina	<.05
Prolonged angina	Unstable angina	<.05
	No prior MI	<.05
Coronary spasm	Younger age	<.05

0.01). The frequency of total complications and of major complications was also influenced by investigator experience (11). Predictors for the occurrence of specific type of complications are shown in Table 7.

ACUTE MORTALITY

Death in hospital occurred in 29 (0.9%) patients undergoing transluminal angioplasty. Analysis of each of the in-hospital deaths by an independent external review committee resulted in classification of 19 deaths as being directly related to angioplasty and the other 10 cases as probably being unrelated to angioplasty. Thus, angioplasty-related mortality was 0.6%. Mortality was 0.8% in patients with single vessel disease, 1.0% in patients with multivessel disease (excluding left main disease), and 3.8% in patients with left main disease (12). Baseline clinical and angiographic factors associated with increased risk of mortality by univariate and multivariate analysis are shown in Table 8. Factors identified by univariate analysis include: female gender ($p < 0.01$), prior bypass surgery ($p < 0.01$), left main coronary disease ($p < 0.01$), dilatation attempt of a bypass graft ($p < 0.05$), angina duration greater than 12 months ($p < 0.05$), and patient's age older than 60 years ($p < 0.05$). Significant independent predictors for increased hospital mortality identified by multivariate analysis were prior bypass surgery ($p < 0.01$), female gender ($p < 0.05$), and patient's age

TABLE 8A Factors for In-Hospital Mortality (Univariate Factors)

Category	Mortality*, %	p Value
Left main	3.8 vs 0.9	<.01
Female	1.8 vs 0.7	<.01
Prior CABG	2.9 vs 0.8	<.01
Graft attempt	2.8 vs 0.9	<.05
Angina >12 mo	1.7 vs 0.7	<.05
Age >60 years	1.7 vs 0.7	<.05

*Mortality rate comparison for *Presence* vs Absence of each factor.

TABLE 8B Factors for In-Hospital Mortality (Multivariate Factors)

Category	p Value
Prior CABG	<.01
Female sex	<.05
Age >60 years	<.05

older than 60 years ($p < 0.05$). For angioplasty-related deaths, female gender was the only predictor for increased hospital mortality (1.7% vs 0.3%, $p < 0.001$).

MYOCARDIAL INFARCTION

Acute myocardial infarction was reported in 170 patients (5.5%), with nonfatal infarction in 154 (5.0%). Myocardial infarction occurred most often in patients (53%) with complications requiring emergency bypass surgery (90/170), such as coronary occlusion or vessel dissection leading to deterioration of flow and progression to occlusion, but also occurred in 2.7% of patients having elective surgery and in 2.3% of patients who did not undergo bypass surgery after coronary angioplasty. Mortality rate in patients with myocardial infarction was 9.4% (16/170). Univariate factors associated with an increased frequency of myocardial infarction were unstable angina ($p < 0.05$), Class 4 angina ($p < 0.05$), eccentric lesions ($p < 0.05$), and nondiscrete lesions ($p < 0.01$). However, using multivariate analysis, the only independent predictor for myocardial infarction was unstable angina ($p < 0.05$).

EMERGENCY SURGERY

Emergency bypass surgery was performed in 202 patients (6.6%). The most frequent indications for emergency operation were coronary dissection in 46%, coronary occlusion in 20%, prolonged angina in 14%, and coronary spasm in 11%. Emergency surgery was most often necessary in patients in whom lesions could not be reached or traversed, but more than 25% of patients requiring emergency bypass surgery had initially successful dilatation followed by abrupt reclosure of the vessel. The mortality rate with emergency surgery was 6.4% and non-fatal myocardial infarction occurred in 41% of patients who underwent emergency bypass surgery, with Q waves developing in approximately 60% of patients with myocardial infarction. However, more than 50% of patients who underwent emergency surgery for severe ischemic events secondary to angioplasty did not develop evidence of infarction and had an uncomplicated postoperative course. Lesions eccentricity was the only baseline factor which was associated with a significant increase in frequency of emergency operation. The incidence of emergency surgery did decline with increasing experience of the angioplasty operator. However, although the trend was significant, the magnitude of decline was small (13).

INTIMAL TEAR VS DISSECTION

At the beginning of the Registry, plaque compression was considered the primary mechanism of successful angioplasty and angiographically evident intimal disruption or tear was classified as coronary dissection and considered to represent a complication of transluminal dilatation. It subsequently became clear that many patients with recognizable plaque disruption had a successful result without associated adverse effects and that a segment of intimal tear did not necessarily represent a complication. The term "coronary dissection" implies a complication, and intimal tear was therefore classified as a dissection in those patients in whom intimal tear was associated with major luminal obstruction, coronary occlusion, myocardial infarction, or deterioration of flow resulting in the need for urgent bypass surgery. Intimal tear or dissection was reported in 397 patients (13%) enrolled in the Registry. Dissection occurred in 135 of these patients (34%). Major complications (myocardial infarction, death, need for emergency support) occurred in 120 patients with intimal tear or dissection (30%). However, no major adverse effects were encountered in 277 patients (70%) with intimal tear or dissection and angioplasty was clinically successful in 218 of these 277 patients, representing 55% of patients with tear or dissection and 79% of those who did not have associated adverse effects. These results indicate that intimal tear is not uncommon with balloon angioplasty and does not represent a complication in the majority of patients unless adverse clinical effects occur. A distinction between intimal tear and pathological dissection is therefore appropriate and the term dissection should be reserved for situations in which intimal disruption leads to a complication (14).

Intimal tear or dissection occurred significantly more often in women ($p < 0.001$), with angioplasty of the right coronary artery ($p < 0.001$), in patients with multivessel disease ($p < 0.01$), with eccentric lesions ($p < 0.01$), and with nondiscrete lesions ($p < 0.05$). Tear or dissection was also more frequent in patients without prior bypass surgery ($p < 0.01$) and with tubular lesions ($p < 0.05$).

ELECTIVE SURGERY

Elective bypass surgery during the same hospitalization was performed in 583 patients (19%) following unsuccessful coronary angioplasty. Elective surgery was performed in 363 patients within 24 hours and in 220 patients longer than 24 hours after an attempted but unsuccessful angioplasty. Major complications (myocardial infarction or death) occurred in 4.6% of patients with elective surgery; nonfatal myocardial infarction in 2.7% and operative mortality in 1.9%.

LONG-TERM FOLLOW-UP

Virtually complete follow-up information was obtained in a subset of 2272 patients from 65 clinical sites; each of these centers achieved greater than 90% follow-up (mean = 98.5%) of at least 1 year (mean = 18 months) (15). Follow-up information was collected on the anniversary date of the original procedure performed and was tabulated on vital status, presence and frequency of angina, hospitalizations for chest pain or myocardial infarction, and need for medication, repeat angioplasty, or coronary bypass surgery. Follow-up was obtained in 2272 patients and the mean follow-up time for the group was 18 months.

Initial angioplasty was successful in 1397 patients in the follow-up cohort (61.5%); the duration of follow-up periods in the successful group were 906 patients for 1 year, 349 patients for 2 years, 108 patients for 3 years, and 34 patients for 4 years. The follow-up information was available in 1354 (97%) of 1397 patients. The clinical results during the follow-up period in patients with initially successful angioplasty are listed in Table 9.

During the latest year of follow-up, 74% of patients did not require additional revascularization

TABLE 9   **Follow-up Results**

Criteria	%
No additional revascularization	73.9
Symptom-free	51.6
Repeat PTCA, no CABG	11.4
Symptom-free	7.4
Total no CABG	85.3
Symptom-free	59.0
CABG	12.9
Symptom-free	7.9
Angina status at last follow-up	
No angina	71.0
Improved	20.0
Unchanged	6.0
Worse	3.0
Myocardial infarction	3.6 (2.0%/yr)
Death	1.8 (1.0%/yr)
Myocardial infarction or death	5.4 (3.0%/yr)

and 52% were symptom-free. Repeat transluminal angioplasty was performed in 11.4% and 65% of these patients were symptom-free at latest follow-up. Eighty-five percent of patients who underwent initially successful percutaneous angioplasty did not require bypass surgery during follow-up, 81% were event-free (no myocardial infarction or death), and 59% were event-free without angina. Coronary bypass surgery was performed in 12.9% who had initially successful angioplasty and 61% of these were symptom-free at subsequent follow-up.

Myocardial infarction during follow-up occurred in 3.6% of patients who had initially successful angioplasty and death occurred in 1.8% of patients (*n* = 25). The rate for combined events (myocardial infarction or death) was 5.4%.

During the final year of follow-up in patients with initially successful angioplasty, angina was improved or eliminated in 91% (angina-free in 72%, improved in 20%), unchanged in 6%, and worse in 3% of patients when compared with preangioplasty status. Differences in follow-up results by extent of coronary artery disease are shown in Table 10. In the 1397 patients who had initial success of angioplasty, 1115 had single vessel disease and 282 had multivessel disease.

Patients with single vessel disease had a higher proportion (75%) not requiring additional revascularization during the follow-up period than patients with multivessel disease (69%, p < 0.05), and a larger percentage of single vessel disease patients were symptom-free without additional revascularization (54% vs 41%, p < 0.001).

The frequency of repeat angioplasty was not sig-nificantly different (p = NS) between single vessel (12%) and multivessel (9%) disease. Patients with multivessel disease had a higher incidence of bypass surgery during follow-up (17% vs 12%, p < 0.05). Myocardial infarction occurred slightly more often in multivessel disease (4.95% vs 3.3%), but the difference was not statistically significant during this relatively short period of follow-up. However, patients with multivessel disease had a significantly higher mortality rate during follow-up (4.95% vs 1.0%, p < 0.001) and a higher combined event rate (9.9% vs 4.3%, p < 0.001) than patients with single vessel disease.

PREDICTORS OF LATE MORTALITY

Factors influencing late survival after successful angioplasty were similar for the total follow-up cohort and for the group with initially successful angioplasty (Table 11). Factors identified by univariate analysis that were associated with increased late mortality were: Male gender (p < 0.01), history of myocardial infarction prior to angioplasty (p < 0.01), presence of left main disease (p < 0.01) or multivessel disease (p < 0.01), left main attempt (p < 0.01), Class 3 or 4 angina (p < 0.05), ejection fraction less than 50% (p < 0.05), history of hypertension (p

TABLE 10 Follow-up Results: Single Vessel Vs Multivessel Disease

Category	Single Vessel	Multivessel	p Value
Number of patients	1115	282	
No revascularization	75%	69%	<.05
Symptom-free	54%	41%	<.05
Repeat PTCA	12%	9%	
Symptom-free	8%	6%	
Total no CABG	87%	78%	<.05
Symptom-free	62%	48%	<.05
CABG	12%	17%	<.05
Myocardial infarction	3.3%	4.95%	
Death	1.0%	4.95%	<.001
Myocardial infarction or death	4.3%	9.9%	<.001

TABLE 11A Factors Associated with Late Mortality (Univariate Factors)

Factors	Mortality	p Value
Left main disease	8.5% vs 1.8%	<.01
Multivessel	3.6% vs 1.5%	<.01
Single vessel	1.3% vs 4.0%	<.01
Male	2.4% vs 0.6%	<.01
Prior MI	3.7% vs 1.4%	<.01
Left main attempt	13.3% vs 1.4%	<.01
Class 3 or 4	2.2% vs 1.1%	<.05
Smoker	2.2% vs 1.4%	<.05
Hypertension	2.9% vs 1.4%	<.05
LVEF <50%	4.5% vs 1.5%	<.05

TABLE 11B Factors Associated with Late Mortality (Multivariate Factors)

Factors	p Value
Left main disease	<.01
Male gender	<.05
Class 3 or 4	<.05
Hypertension	<.05

< 0.05), and history of smoking (p < 0.05). Using multivariate analysis for all late deaths, left main disease (p < 0.01), male gender (p < 0.05), functional Class 3 or 4 angina (p < 0.05), and history of hypertension (p < 0.05) were identified as significant independent predictors of late mortality after angioplasty.

REPEAT PTCA

A second transluminal angioplasty was performed in 241 patients during follow-up after initially successful angioplasty. A second or third lesion dilatation was performed in 38 of these 241 patients (13%) and clinical success was achieved in 61% of them. In 203 patients, repeat angioplasty of the same lesion was performed for restenosis (16). Comparison of results with repeat angioplasty and results with initial angioplasty in the Registry are shown in Table 12. Repeat angioplasty was successful in 173/203 patients (85.2%). The repeat angioplasty was unsuccessful due to inability to pass the lesion in 13 patients (6.4%), and due to inability to dilate the lesion in 8 patients (3.9%). Major complications occurred in six patients (3.0%), with nonfatal myocardial infarction in three (1.5%), and emergency bypass surgery in four patients (2.0%). No deaths occurred in patients having repeat angioplasty. During follow-up after successful repeat angioplasty, 73% were event-free and 42% were asymptomatic. Restenosis by angiographic criteria occurred in 33% which did not differ from the recurrence rate after a successful first procedure. Bypass surgery was subsequently performed in 22%, and a third angioplasty in 3% (16). These data indicate that repeat angioplasty is associated with a high success rate and a low complication rate and that the majority of patients will have sustained clinical improvement after repeat angioplasty.

RESTENOSIS

The incidence of stenosis or recurrence following angioplasty is dependent upon the definition used. Clinical recurrence requiring additional revascularization (repeat angioplasty or coronary bypass surgery) occurred in 24.3% of patients who had initially successful angioplasty. Clinical recurrence appeared most often within the first 3–6 months after the procedure. Restenosis by angiographic criteria occurred more frequently (17) than clinical evidence of recurrence. Angiographic restenosis was assessed in a subset of 557 patients from 27 centers at which follow-up angiography was performed in 75% or more patients who had successful angioplasty.

Restenosis defined by 30% or more decrease of luminal diameter of the immediate postdilatation lumen size occurred in 30% of these follow-up patients and restenosis defined by 50% or more decrease in luminal diameter of the initial angiographic improvement occurred in 33% of these patients. Baseline clinical factors associated with an increased restenosis rate are listed in Table 13. The factors associated with restenosis as defined by univariate analysis included: male sex (36% vs 22%, p < 0.01), angina duration of less than 6 months (38% vs 29%, p < 0.05), history of diabetes (47% vs 32%, p < 0.05), and lesion severity of 90% or greater (39% vs 32%, p < 0.05). Technical factors associated with restenosis included initial pressure gradient of 40 mmHg or greater (37% vs 24%, p < 0.01) and postdilatation pressure gradient of 20 mmHg or greater (44% vs 31%, p < 0.01). Predictors for restenosis by multivariate analysis in 439 patients with complete data were: male sex (p < 0.05), bypass graft stenosis (p < 0.05), initial Class 3 or 4 angina (p < 0.05), and no prior myocardial infarction (p < 0.05). No effect of different postangioplasty drug regimens (antiplatelet therapy, warfarin, calcium antagonists) on restenosis rate was found in these analyses.

TABLE 12 **Comparison of Results with Initial PTCA and Repeat PTCA**

Category	PTCA 1	PTCA 2	p Value
Number of patients	3079	203	
Clinical success	61.0%	85.2%	<.01
Major complications	9.3%	3.0%	<.01
Nonfatal MI	5.0%	1.5%	<.05
Emergency surgery	6.6%	2.0%	<.10
Death	0.9%	.0%	<.01
Restenosis	33.0%	33.0%	(ns)

TABLE 13 **Factors for Restenosis**

Factors	Univariate, p	Multivariate, p
Male sex	<.01	<.05
Angina <6 months	<.01	<.15
No prior MI	<.05	<.05
History of diabetes	<.05	<.15
Class 3 or 4 angina	<.15	<.05
Bypass graft PTCA	<.15	<.05

SUMMARY AND PERSPECTIVES

From its inception, the NHLBI PTCA Registry represented a unique approach to the collection of scientific information. As a voluntary effort, the Registry was dependent entirely on the sustained interest and enthusiasm of the participating investigators who contributed their time and effort. This format provided an extremely cost-effective method for an early, broad-based clinical evaluation of a new and promising technique and should serve as a valuable model for future investigations of other new methods of therapy. The NHLBI PTCA Registry provided a timely and efficient means of assessing the short-term safety and efficacy of coronary angioplasty during its early days of clinical application. Long-term follow-up information confirms that a salutary result is maintained in the majority of patients following successful transluminal angioplasty. The Registry had an important role in establishing coronary angioplasty as a satisfactory alternative to coronary bypass surgery in selected patients with symptomatic coronary artery disease, particularly in patients with single vessel disease. Factors influencing early and late outcome after coronary angioplasty have also been identified. However, the NHLBI PTCA Registry does not provide sufficient information to assess the efficacy of transluminal angioplasty in patients with more extensive coronary artery disease or with more complex coronary anatomy.

The Registry data represent experiences with percutaneous coronary angioplasty during its first 4 years of clinical application. Since its closure of new patient enrollment in 1981, important technical improvements in catheter systems design (smaller profile catheters, high-pressure balloons, steerable dilatation systems) have been introduced and have undoubtedly contributed to the current success rates in excess of 80–90% at many centers (18).

The NHLBI PTCA Registry represents an important source of information on coronary angioplasty and should serve as a useful reference for additional studies. Better understanding of the mechanisms of successful dilatation, restenosis, and identification of methods to prevent recurrence are important areas for basic investigation. Careful studies of the use of transluminal angioplasty in patients with multivessel disease, complex coronary lesions, acute ischemic syndromes, and in patients with poor left ventricular function are also needed to determine if there are subsets of these patients in whom coronary angioplasty will prove a safe and effective therapy.

REFERENCES

1. Gruentzig AR, Myler RK, Hanna ES, Turina MI: Transluminal angioplasty of coronary artery stenosis (abstract). Circulation 56:84, 1977.
2. Gruentzig AR: Transluminal dilatation of coronary artery stenosis. Lancet 1:263, 1978.
3. Gruentzig AR, Senning JA, Siegenthaler WE: Nonoperative dilatation of coronary artery stenosis: Percutaneous transluminal coronary angioplasty. N Engl J Med 301:61–68, 1979.
4. Levy RI, Jesse MJ, Mock MB: Position on percutaneous transluminal coronary angioplasty (PTCA). Circulation 59:613, 1979.
5. Proceedings of the workshop on percutaneous transluminal coronary angioplasty. US Dept HEW NIH Publication 80–2030, 1980.
6. Levy RI, Mock MB, Willman VL, Frommer PL: Percutaneous transluminal coronary angioplasty. N Engl J Med 301:101–103, 1979.
7. Levy RI, Mock MB, Willman VL, Passamani ER, Frommer PL: Percutaneous transluminal coronary angioplasty—A status report. N Engl J Med 305:399–400, 1981.
8. Kent KM, Bentivoglio LG, Block PC, Cowley MJ, Dorros G, Gosselin AJ, Gruentzig AR, Myler RK, Simpson J, Stertzer SH, Williams DO, Fisher L, Gillespie MJ, Mullin SM, Mock MB: Percutaneous transluminal coronary angioplasty: Report from the NHLBI Registry. Am J Cardiol 49:2011–2020, 1983.
9. Faxon DP, Ryan TJ, McCabe CH, Kelsey SF, Detre K: Determinants of a successful percutaneous transluminal coronary angioplasty (NHLBI–PTCA Registry). Abstract. Am J Cardiol 49:905, 1982.
10. Kelsey SF, Mullin SM, Detre KM, Cowley MJ, Gruentzig AR, Kent KM: The effect of investigator experience on percutaneous transluminal coronary angioplasty. Am J Cardiol (in press).
11. Dorros F, Cowley MJ, Simpson J, Bentivoglio LG, Block PC, Bourassa M, Detre K, Gosselin AJ, Gruentzig AR, Kelsey SF, Kent KM, Mock MB, Mullin SM, Myler RK, Passamani ER, Stertzer SH, Williams DO: Percutaneous transluminal coronary angioplasty: Report of complications from the NHLBI PTCA Registry. Circulation 67:723–730, 1983.
12. Dorros G, Cowley MJ, Janke L, Kelsey SF, Mullin SM, VanRaden MJ: The in-hospital mortality occurring in the National Heart, Lung and Blood Institute Percutaneous Transluminal Coronary Angioplasty (PTCA) Registry. Am J Cardiol (in press).
13. Cowley MJ, Dorros G, Kelsey SF, VanRaden MJ, Detre KM: Emergency coronary bypass surgery following coronary angioplasty: NHLBI PTCA Registry experience. Am J Cardiol (in press).

14. Cowley MJ, Dorros G, Kelsey SF, Van Raden MJ, Detre KM: Acute coronary events associated with percutaneous transluminal coronary angioplasty. Am J Cardiol (in press).

15. Ken KM, Bentivoglio, LG, Block PC, Bourassa MG, Cowley MJ, Dorror G, Detre K, Gosselin AJ, Gruentzig A, Kelsey SF, Mock M, Mullin SM, Passamani E, Myler RK, Simpson J, Stertzer SH, VanRaden M, Williams DO: Long-term efficacy of percutaneous transluminal coronary angioplasty (PTCA): Report from NHLBI–PTCA Registry. Am J Cardiol (in press).

16. Williams DO, Gruentzig AR, Kent K, Detre K, Kelsey S, Shalone: Efficacy of repeat percutaneous transluminal coronary angioplasty for coronary restenosis. Am J Cardiol (in press).

17. Holmes DR Jr, Vlietstra RE, Smith HC, Vetrovec GW, Kent KM, Cowley MJ, Faxon DP, Gruentzig AR, Kelsey SF, Detre KM, VanRaden MJ, Mock MB: Restenosis after percutaneous transluminal coronary angioplasty: A report from the NHLBI Registry. Am J Cardiol (in press).

18. Meier B, Gruentzig AR: PTCA learning curve: Skill, technology or patient selection. Am J Cardiol (in press).

21

Transluminal Angioplasty in Patients Undergoing Streptokinase Infusion

GEORGE W. VETROVEC, M.D.
MICHAEL J. COWLEY, M.D.
HERMAN K. GOLD, M.D.

INTRODUCTION

Myocardial infarction is a major cause of morbidity and mortality associated with atherosclerotic cardiovascular disease. Numerous strategies designed to salvage ischemic muscle during acute infarction have been investigated (1). Intracoronary thrombolysis is a promising new technique to provide reperfusion to the jeopardized heart muscle during the acute phase of evolving infarction (2–5).

Significance of Intracoronary Thrombosis

Thrombus is frequently present at or near atherosclerotic lesions during the acute phase of transmural infarction. Documentation of the presence of intracoronary thrombus during acute myocardial infarction includes the following: (1) intra-arterial clot removal (6) during coronary bypass surgery for acute myocardial revascularization; (2) frequent observation of postmortem intracoronary thrombus (7,8) in patients dying soon after clinical infarction; and (3) thrombolytic recanalization of totally occluded ar-

teries during acute myocardial infarction (2–5). Because the long-term prognosis for patients sustaining a myocardial infarction is directly related to the extent of residual, functioning myocardium (9,10), numerous techniques to limit the extent of muscle loss during the acute phase of transmural infarction have been evaluated. Drug therapy designed to limit infarct size includes beta blocker and nitrates (11,12), which reduce oxygen demand and thereby spare the jeopardized myocardium; methods to achieve acute reperfusion include emergency coronary bypass surgery (13), mechanical recanalization (14), and thrombolysis (2–5). These latter methods are aimed at reestablishing coronary blood flow prior to complete, irreparable transmural myocardial damage.

Acute thrombolytic reperfusion can be achieved by both intracoronary and intravenous (15,16) administration of streptokinase (17) or urokinase (18,19) during the acute phase of myocardial infarction, but experience with urokinase is quite limited. Intracoronary administration of streptokinase is the most common method of coronary thrombolysis and is the most likely circumstance in which transluminal balloon angioplasty could be performed as an adjunct to thrombolysis treatment in patients with acute coronary occlusion.

Combined Thrombolysis and Angioplasty

Indications for balloon angioplasty in patients undergoing streptokinase infusion include failure to reperfuse with streptokinase alone and failure to maintain persistent reperfusion because of immediate reclosure. Furthermore, because individuals having a residual critical stenosis after thrombolysis have an extremely high incidence of early reclosure, such patients may benefit from transluminal dilation of the stenotic segment to maintain continued coronary perfusion and thus prevent myocardial damage secondary to reclosure after streptokinase infusion.

Indications for emergency coronary bypass surgery of patients undergoing streptokinase infusion are similar to those for emergency bypass surgery during coronary angioplasty but involve patients who are technically unacceptable for coronary angioplasty, secondary to either vessel or lesion morphology or extensiveness of overall coronary disease. In patients who have a less critical residual stenosis after streptokinase, transluminal dilation or bypass surgery may be delayed until a more elective setting in which the indication for a poststreptokinase balloon angioplasty will be based on the individual's symptoms or the extent of myocardium in jeopardy.

This chapter will discuss the technical aspects of reperfusion, as well as the effect of angioplasty in conjunction with streptokinase thrombolysis.

CHARACTERISTICS OF PATIENTS UNDERGOING STREPTOKINASE INFUSION

The effectiveness of coronary reperfusion during evolving stages of acute infarction is time-dependent: the earlier the reperfusion, the greater the likelihood of myocardial salvage. However, this time course may be variable. Factors potentially affecting the rate of myocardial loss during the acute phases of infarction include the presence and extent of collateralization and the length of time over which the coronary stenosis progresses to total occlusion. For example, if a coronary occlusion goes through a fluctuating phase between total and subtotal occlusion, such a process will create a "stuttering" infarction. In addition, the size of the overall infarct zone, as well as the presence of critical lesions in the coronary arteries that perfuse the distribution of

the border zone myocardium around the infarct zone, may also influence the extent and rapidity in which the myocardium is lost. Although some investigators (20) have suggested a time limit for reperfusion, in our initial streptokinase experience (21), despite a mean time to perfusion greater than 5 hours, significant improvement in left ventricular function was achieved in the successfully reperfused patients. Although it is true that the sooner reperfusion can be established, the greater the potential for myocardial salvage, the variable factors listed previously preclude an absolute cutoff time.

On the basis of the time constraints for effective reperfusion, it is extremely important that patients with evolving transmural infarction be identified quickly and transported to an appropriate medical facility where reperfusion can be attempted as early as possible. To accomplish this requires community as well as hospital and professional support. Public awareness of the urgency of early treatment of myocardial infarction must be enhanced, and effective emergency transport teams must be available. Finally, the emergency room or primary care physician must quickly identify patients who are candidates for this technique so that the cardiologist can be alerted. Even after the cardiologist is involved, early mobilization of the catheterization laboratory personnel is very important.

Initial Patient Evaluation and Management

Initial evaluation in the emergency room is important and should be thorough enough to identify the patient who is sustaining an acute transmural myocardial infarction and who is a potential candidate for thrombolysis. While this initial evaluation is proceeding, it is important to institute appropriate medical treatment for myocardial infarction. This includes oxygen, narcotics for chest pain, and anti-ischemic drugs, such as nitrates, beta blockers, or calcium blockers, depending on the clinical presentation and the hemodynamic condition of the patient. During this phase, coronary spasm should be excluded, and significant hypertension or tachycardia should be treated. Rapid heart rate, elevated systolic blood pressure, and increased myocardial contractility, those factors that increase myocardial oxygen demand, should be minimized with pharmacological therapy. In addition, nitrates and calcium blockers are useful in the presence of ST segment elevation and chest pain to exclude the possibility of prolonged coronary spasm (22,23).

While stabilizing the patient's condition, it is useful to begin discussing the possibilities of intracoronary streptokinase administration with the patient and family. If this discussion occurs immediately, one is able to obtain informed consent for the procedure prior to administering narcotics to the patient.

Indications for Thrombolysis

Before institution of streptokinase therapy, it should be determined that the patient is having a transmural infarction. Although preliminary data (24,25) have suggested possible potential efficacy for intracoronary streptokinase infusion in persons with unstable angina but without total occlusion of a coronary artery, the effectiveness of this treatment is still unclear. Thus, it is important to establish that the patient is having an evolving transmural infarction, i.e., symptoms of ischemic chest pain despite antianginal medications and electrocardiographic (ECG) changes of evolving transmural infarction. Potentially confusing clinical presentations include patients, frequently young, with early repolarization, whose chest pain is atypical and whose ECG shows no evolution over time. In addition, patients with pericarditis may have ST segment elevation, but this is usually diffuse and nonprogressive; chest pain with pericarditis is usually pleuritic or positional rather than ischemic, and a pericardial rub may be present. Conversely, in individuals who have evolving transluminal infarction, we generally proceed with the streptokinase procedure even though early Q waves may be developing. Partial ''R wave'' regeneration may occur late after successful reperfusion despite early Q wave development (26). Therefore, appearance of Q waves in the early stage of evolving infarction is not an absolute contraindication to reperfusion. Furthermore, coronary angiography prior to intracoronary streptokinase administration identifies total or subtotal occlusion of the infarct-related vessel, as well as the size of the vessel, and, thus, the extent of jeopardized myocardium. If this information confirms total or subtotal coronary occlusion and allows the clinician an opportunity to be certain about the extent of myocardium in jeopardy, the risk of streptokinase infusion is justified. There are occasional instances in which the infarct-related vessel may be quite small, and, therefore, one may elect not to proceed with reperfusion.

In addition to the indications for acute reperfusion during transmural myocardial infarction, there are important contraindications. Although symptom presentation of each patient has to be individually evaluated, Table 1 lists the relative contraindications to thrombolysis. These contraindications predominantly relate to the potential risk of major hemorrhage.

TABLE 1 Thrombolytic Reperfusion: Relative Contraindications

1. Completed event, i.e., pain-free with resolution of acute ECG changes
2. Surgery within 21 days
3. Recent trauma or major cardiopulmonary resuscitation
4. Recent streptokinase administration
5. Recent cerebral vascular accident
6. Active or recent gastrointestinal bleeding
7. Age >75 years (secondary to increased probability of complications)
8. Significant bleeding tendencies, e.g., congenital clotting abnormalities, tumors, chronic ulcer disease

STREPTOKINASE INFUSION TECHNIQUE

The following is a guide to the administration of intracoronary streptokinase infusion, including patient preparation, infusion techniques, and post-streptokinase management.

Preliminary Angiography

Cardiac catheterization should be carried out in the usual manner, using either Sones (27) or Judkins (28) technique, chosen according to available access to the vascular chamber, as well as experience of the operator and the laboratory personnel. In the case of Sones technique, careful local dissection of the antecubital area to minimize tissue trauma and to reduce the potential of bleeding complications should be performed. In the case of Judkins technique, it is important to avoid multiple vessel punctures before final vessel entry. Again, extra care in this regard will reduce the likelihood of local bleeding complications following the streptokinase procedure. An arterial sheath should be utilized for a transfemoral approach to minimize repeated vessel trauma with catheter exchanges, thereby decreasing the likelihood of local bleeding. Once access to an artery is gained, 5000 U of heparin is given intravenously for femoral studies and intra-arterially for brachial procedures. Diphenhydramine hydrochloride, 25 mg, and hydrocortisone, 100 mg, should also be given intravenously to lessen the risk of allergic reactions to streptokinase. Furthermore, sedatives may be administered to minimize ischemic pain and to reduce anxiety. Careful explanations about the procedure to the patient are also useful in allaying anxiety.

The catheterization procedure normally includes right heart pressure recordings, as well as coronary angiography. A balloon-directed right heart catheter is placed in the pulmonary artery for continued hemodynamic assessment throughout the procedure. Venous access also provides a mechanism for urgent cardiac pacing, if necessary. Once the right heart pressures are recorded, we have generally performed left ventriculography, provided the patient is hemodynamically and electrically stable. It has been our experience that limited left ventricular angiography may be performed safely. Utilizing a high contrast injection rate with low-volume contrast media provides a brief but reasonable left ventriculogram. A limited study is performed by using 25–30 mL of 76 percent renografin injected at 18–20 mL per second. Because cardiac output is reduced in most patients in this category, the rapid but low volume contrast injection fills out the ventricle rapidly but washes out slowly, providing adequate cardiac beats for evaluation of the left ventricular function.

Some institutions have available immediate echocardiography or nuclear scans, which permit noninvasive assessment of left ventricular function. These latter methods allow repeated left ventricular assessment with the same modality throughout the hospitalization. However, these studies must be done quickly to avoid time loss. Although it is possible to perform intracoronary reperfusion without establishing left ventricular function, there are patients with prior myocardial infarction or complex coronary anatomy in whom knowledge of left ventricular function is useful when coronary angioplasty or coronary bypass surgery is being considered. However, severe hypokinesia or even akinesia in the infarct area should not preclude attempted reperfusion, since the potential for reversibility is unknown at this time.

Initial coronary angiography is performed first in the coronary system least likely to be involved in the acute infarction; for example, in inferior wall infarctions, the left coronary artery is initially studied, although the right coronary artery is visualized first in an anterior myocardial infarction. Utilizing this technique, one is able to assess overall coronary anatomy, as well as collateral flow, thus completing the preliminary angiographic study with the catheter in the vessel system in which streptokinase is to be infused.

The coronary injections of contrast material should be limited but aimed at visualizing the greatest extent of anatomy with the least number of injections. For example, right coronary angiography may include a 30° cranial angulation in a left anterior oblique projection (to assess the proximal posterior descending) and a shallow right anterior oblique projection. These two projections will usually provide sufficient information regarding the right coronary artery, as well as the potential collateral flow to the left coronary system. Again, a minimal number of left coronary injections that provides the most useful information is ideal. Generally, a shallow right anterior oblique projection provides important information regarding the left main coronary artery and the left anterior descending and circumflex vessels. In addition, a left anterior oblique view with cranial angulation provides complementary data. Such screening views may be sufficient to decide about reperfusion and the status of other major branches; if not, subsequent projections should be selected according to the specific need of an area in question.

Most operators have routinely administered 100–200 μg of intracoronary nitroglycerin into the infarct-related coronary artery with a single follow-up injection to exclude coronary artery spasm. However, intracoronary injection of nitroglycerin seldom changes flow significantly in an artery that is completely occluded.

Intracoronary Thrombolysis

Methods of intracoronary thrombolysis vary greatly, particularly with the catheter delivery system. If possible, infusion of streptokinase directly to the location of thrombus appears most effective. Experimental studies have shown a reduced thrombolytic rate or a larger streptokinase dose requirement if infusion is remote from the site of thrombosis (29). However, long delays before infusion should be avoided if difficulties are encountered in selectively placing the tip of the infusion catheter near the occlusion point of the involved coronary artery.

Most infusion systems utilize the diagnostic coronary catheter, either for direct infusion through the diagnostic catheter or to deliver a smaller caliber cannula through its lumen and subselectively infuse streptokinase near the site of thrombus. Infusing through the diagnostic catheter may be associated with predominant infusion into the nonoccluded coronary artery, particularly in the left coronary system as flow is shunted away from the occluded vessel. To overcome such runoff, several available (number 3 French caliber) infusion catheters can be passed through the standard diagnostic catheter, utilizing an 0.018 inch or smaller guidewire to provide selective cannulation of the involved coronary artery. This method frequently allows the infusion catheter to be placed at the point of total vessel occlusion. Subselective catheter placement may be facilitated by use of angioplasty guidewires. These angioplasty wires are steerable and have relatively soft ends, reducing the chances of vessel dissection, particularly if the guidewire is advanced several centimeters beyond the infusion catheter tip, taking full advantage of the flexible portion of the wire.

An alternative option is to cannulate the affected coronary system with an angioplasty guiding catheter and to use a steerable balloon and guidewire system to cannulate the thrombosed vessel. Streptokinase can then be infused through the balloon catheter. The advantages of this technique are that the equipment is familiar to physicians with extensive angioplasty experience, it allows easy infusion of streptokinase and it facilitates balloon angioplasty, should a need arise during the acute phase of thrombolysis, without requiring catheter exchange.

Once the catheter has been adequately positioned and appropriate premedications administered (as described earlier), streptokinase is administered with an initial dose of 20,000–40,000 U in a bolus, followed by a continuous 2000–4000 U per minute infusion into the infarct-related vessel (2). Again, it should be emphasized that whatever variation of infusion technique is used, excessive time should not be spent attempting subselective catheter placement; early infusion at the most distal point possible should be the major goal.

Consideration has been given to recanalization of a thrombosed segment of an occluded artery with guidewire technique (2). As a general principle, it is useful to utilize a guidewire for advancing any subselective cannula at or near the stenosis. Should the guidewire reach and easily cross the stenosis, it may help facilitate advancing the subselective catheter and disrupting the thrombus. However, mechanical crossing of the obstruction should not be a major goal unless streptokinase infusion has failed to reopen the occluded artery.

Once the initial loading dose of streptokinase has been administered, infusion is continued by infusion pump or by hand. In our laboratory, we have tended to hand inject streptokinase slowly, frequently adding small amounts of contrast to allow visualization of the streptokinase infusion. Frequently, acute recanalization is manifested by cessation of chest pain or the sudden appearance of left ventricular rhythm disturbances. In addition, an angiographic test injection is performed every 30,000–50,000 U of streptokinase to determine the time and streptokinase dosage for reperfusion of the occluded artery. Once the vessel appears open, a single coronary angiogram is performed in a view that provides good delineation of the stenotic lesion and any local thrombus. Streptokinase infusion is continued, with another angiogram following an additional 30,000–50,000 U to see whether further "clearing up" of the stenosis has occurred. Once additional streptokinase does

not produce further improvement of the caliber at the stenosis, streptokinase infusion is discontinued. However, after the vessel becomes patent, if a large clot remains in the artery, further streptokinase infusion is warranted to promote further lysis of the clot. Most patients will have successful reperfusion between 200,000 and 400,000 U of streptokinase. Higher dosages may be necessary if urokinase is used. If reperfusion is not achieved by 400,000 U of streptokinase, thrombolysis is considered unsuccessful.

An additional benefit of follow-up angiograms throughout the streptokinase infusion is the ability to identify any evidence of early vessel reclosure. In those individuals in whom reperfusion leaves a very critical residual channel, particularly with compromised flow, reclosure may occur early. Careful follow-up angiograms over the first few minutes after opening the occluded artery allow one to identify reclosure and to prepare immediately for angioplasty. If the vessel recloses acutely after or during the streptokinase administration, further streptokinase may be tried. However, once reclosure occurs, additional streptokinase may not be successful in maintaining the patency of the stenotic segment that was temporarily opened up by streptokinase thrombolysis.

Follow-up Angiography

Following recanalization, additional selective coronary arteriograms may be performed to define overall anatomy better. At this time, such additional angiograms do not delay early reperfusion, and the patients may be more hemodynamically stable. These additional views may be helpful in considering follow-up therapy, including bypass surgery.

In our laboratory, because of research protocols, we have routinely repeated a left ventriculogram at the end of the reperfusion. This repeat study documents any change in overall wall motion and, again, assesses the probability of continued residual myocardium in jeopardy. However, the left ventricle immediately after reperfusion may show minimal or no improvement of left ventricular function compared to the preinfusion status. Return of myocardial function is often delayed up to several weeks (30). Thus, the usefulness of follow-up left ventriculography must be determined individually.

Patient Monitoring

Throughout the procedure, it is important to maintain close patient monitoring. Prior to reperfusion, most patients are hemodynamically unstable and thus need close observation to ensure that ventilation and peripheral perfusion are adequately maintained. If the patient is extremely unstable or in cardiogenic shock, one may elect to access both femoral arteries in a Judkins procedure or one femoral artery and the brachial artery in the Sones technique at the beginning of the procedure. Access to the additional femoral artery allows the passage of the intra-aortic balloon pump after streptokinase has been administered. Therefore, in selected individuals preliminary dual artery puncture may provide additional safety while avoiding the need for arterial access after systemic fibrinolysis occurs. However, if the patient is extremely unstable initially, immediate balloon pump insertion may be preferable.

Reperfusion arrhythmias occur in approximately 50–80 percent of patients (9,31). These may be simple premature ventricular contractions, ventricular tachycardia, or fibrillation. These arrhythmias occur despite lidocaine infusion throughout the procedure as part of routine acute myocardial infarction therapy. Reperfusion of the right coronary artery may produce acute "vagal" type responses with hypotension, bradycardia, and occasionally complete heart block. Standard treatment of these arrhythmias includes antiarrhythmics, cardioversion, atropine, and temporary pacing, as needed.

After streptokinase therapy, the majority of patients develop systemic fibrinolytic effects (32) despite low-dose, selective coronary administration of streptokinase. Thus, most patients have to be observed carefully for potential bleeding complications for the first 24 hours or more following streptokinase infusion. This potential for bleeding is further enhanced by the continuous administration of heparin to maintain therapeutic (twice normal) values for the partial thromboplastin time. However, continued administration of heparin seems to be most important to reduce the likelihood of acute reclosure in patients who underwent streptokinase treatment. Risk of serious bleeding events in the appropriately selected patients appears low (33). Heparin has been suggested to increase bleeding risk by one observer (34), but the significance of this is unknown and does not appear to override the need for continued anticoagulation after streptokinase reperfusion.

Practically all needle punctures for either blood sampling or medication should be limited after streptokinase infusion. Furthermore, any vessel puncture site must be watched closely with local pressure application until hemostasis occurs. In the case of Sones catheterization, care is necessary to ensure excellent tamponade of the arteriotomy site before full closure because surgical wounds are particularly at risk for bleeding following streptokinase. In addition, in patients undergoing femoral procedures,

we have routinely left the sheaths in place until the following day at a time when the fibrinogen has begun to normalize, and the PTT and PT are reasonably controlled.

A number of parameters are monitored closely following reperfusion. Daily ECGs are performed for 2 or 3 days. In addition, over the first 24 hours, CK determinations are obtained every 6 or 8 hours. These demonstrate a high, early CK peak consistent with early reperfusion. The height of this level is not an indicator of extent of heart muscle damage. In addition, twice daily hemoglobin determinations are useful to identify any early evidence of bleeding. Finally, coagulation parameters, including Pro-Time, PTT, and fibrinogen, are useful every 4–6 hours to assure adequate but not excessive anticoagulation over the first 48 hours.

ANGIOPLASTY FOLLOWING STREPTOKINASE

Although coronary angioplasty is technically feasible in a majority of patients who undergo streptokinase reperfusion, the most important question is when balloon angioplasty should be performed acutely. Angioplasty should be considered in *streptokinase failures*, those patients in whom streptokinase infusions produce no reperfusion or transient reperfusion with immediate reclosure. In both instances angioplasty is an option to bring an immediate recanalization of the occluded artery when streptokinase failed to reperfuse the artery. In addition, angioplasty may be considered when subacute vessel reocclusion is anticipated after thrombolysis. Conversely, if good reflow is achieved with a <90 percent residual stenosis, it may be preferable to defer angioplasty, since such stenoses may show striking further reduction in the residual stenosis over the ensuing 10 days after streptokinase infusion.

The following study (35) illustrates these points. Of 103 patients who underwent attempted intracoronary streptokinase at the Massachusetts General Hospital and the Medical College of Virginia, 35 (34 percent) failed to maintain stable reflow. Approximately half ($n = 18$) obtained transient reflow but did not have persistent patency at the termination of the procedure. This rate of streptokinase success is within the range observed by other investigators, as shown in Table 2. Thus, balloon angioplasty is a reasonable consideration in individuals without stable reflow following streptokinase infusion alone, a group comprising approximately 20–30 percent of patients undergoing attempted thrombolysis.

In addition, in the study noted, we compared the

TABLE 2   Success Rate for Intracoronary Thrombolysis

Investigator	Reperfusion, Percentage
Rentrop et al. (Ref. 2)	72
Reduto et al. (Ref. 3)	67
Cowley et al. (Ref. 21)	82
Mathey et al. (Ref. 5)	79

frequency of in-hospital recurrent ischemic events to the extent of residual coronary narrowing in patients who underwent streptokinase thrombolysis. Patients who underwent early bypass surgery for cardiogenic shock or who died of noncardiac causes within 10 days were excluded. There were 49 remaining patients with initial successful streptokinase reperfusion, 21 (41 percent) of whom had residual stenoses of greater than 90 percent immediately following thrombolysis. Of those 21, 16 (76 percent) had in-hospital ischemic events, including myocardial infarction in 3, recurrent angina in 12, and reocclusion without infarction in 1. The reocclusion patient died of ventricular fibrillation. In contrast, those patients who had a residual postreperfusion stenosis of less than 90 percent had a much more favorable in-hospital prognosis. Of these latter 28 patients, only 14 percent had recurrent ischemic events, 1 with recurrent angina and 3 with evidence of reocclusion without infarction. Thus, the difference in the rate of clinical ischemic event after streptokinase was significantly less for patients with a residual stenosis of less than 90 percent ($P \leq .001$). Based on this latter observation, our approach has been to consider acute balloon dilation or surgery at the time of the streptokinase procedure if the

TABLE 3   Indications for Expanded Intervention after Intracoronary Thrombolysis

Acute (during initial procedure)

Failure of thrombolytic reperfusion with persistent vessel occlusion

Immediate reclosure after successful reperfusion

Residual, critical ($\geq$90%) stenosis in infarct-related vessel

Extensive viable myocardium at jeopardy (e.g., left main, large LAD)

Subacute (1–14 days)

Recurrent angina with residual significant coronary artery disease

Extensive myocardium at jeopardy (e.g., left main, severe three vessel disease)

TABLE 4   Selection Criteria for Acute or Subacute Intervention

PTCA

Single vessel disease and/or multivessel disease with individually suitable lesions for dilation

Multivessel disease but a technically feasible infarct-related vessel with one of three indications (Table 3) for acute intervention

Bypass surgery

Indications for acute or subacute intervention but not technically suitable for angioplasty of one or more major ischemic vessels

Extensively jeopardized myocardium, particularly with complex collateral supply or markedly depressed left ventricular contractility, significantly increasing the risk of angioplasty (frequently left main or severe three vessel disease)

residual stenosis is greater than 90 percent, in the hope of limiting in-hospital recurrent ischemic events (see Table 3).

In light of the additional complexity of bypass surgery, we have considered surgery only for patients with a critical left main stenosis or patients with complex coronary anatomy making angioplasty unfavorable (Table 4). Examples include patients with severe multivessel disease in which acute reocclusion of the dilated segment might produce major ventricular compromise because of complex collateral distribution and those in whom an important artery continues to reclose after repeated dilations. Although emergency bypass surgery may reduce morbidity and mortality of acute infarcts, the potential risk of interoperative ischemia associated with infarction and the risk of bleeding problems immediately after streptokinase would seem to favor balloon angioplasty. Furthermore, even in individuals with multivessel disease, angioplasty of the predominantly affected stenosis may be a reasonable temporizing procedure during the acute event to allow for more elective surgery or multivessel angioplasty of the remaining disease.

Surgical Backup

During routine elective coronary angioplasty, surgical backup is required. With the unpredictable nature of acute myocardial infarction, this represents an additional logistic problem. The decision relating to the availability of surgery backup at this time must be based on the individual patient's anatomy, as well as the facilities available in the institution. However, it has been our philosophy always to have a surgeon aware of a poststreptokinase angioplasty. The decision about acute surgery for failed angioplasty after streptokinase relates to the clinical stability of the patient. If, despite reocclusion, the person is hemodynamically stable, with no sudden recurrence of pain or ECG changes suggesting further evolution of the infarction, it may be justifiable to continue medical treatment and to accept the result of occlusion. Conversely, if there is recurrent or continued evidence of myocardial ischemia with evolving infarction, urgent bypass surgery is warranted, particularly if there is a substantial amount of potentially viable heart muscle in the distribution of the affected coronary artery.

ILLUSTRATIVE CASES

Following are five cases illustrating the use of further interventions following acute thrombolytic reperfusion.

Case 1

A 73-year-old male had experienced 2 days of chest pain typical of coronary insufficiency. There were minimal anterior ST segment elevations but no enzyme changes. During hospitalization, recurrent prolonged chest pains associated with anterior ST segment elevation prompted urgent catheterization.

Catheterization

On initial coronary angiography, the left main artery was normal. The left anterior descending artery had a proximal 80–90 percent stenosis with a 70 percent midlesion. The left circumflex had tandem 60 percent and 80 percent stenoses at the origin of the first obtuse marginal branch. The dominant right coronary artery had an 80 percent proximal stenosis, followed by subtotal occlusion in the midportion with a filling defect in this area consistent with thrombus. The posterior descending artery was filled from left circumflex collaterals. The left ventricle had a 48 percent ejection fraction with moderate inferior akinesia.

Intervention

The right coronary stenosis was unchanged after intracoronary nitroglycerin. After 120,000 U of intracoronary streptokinase, ventricular arrhythmias occurred, and repeat angiography demonstrated reperfusion with prompt antegrade flow. After a total of 180,000 U of streptokinase, there was a residual 80 percent proximal right coronary artery stenosis, followed by a tubular 90 percent stenosis in the midvessel. Because the residual right coronary stenosis was severe, the patient underwent balloon angioplasty in an attempt to prevent early reclosure. Angioplasty was successful, leaving a residual 30–40 percent proximal stenosis and a 50 percent midstenosis.

In-Hospital Follow-up

The patient remained asymptomatic, and at angiography 10 days later, there was no significant change in the left coronary anatomy, although the proximal and midright coronary artery areas of previous dilation were each restenosed to 70–80 percent. The left ventricle revealed hypokinesis in the inferior wall with otherwise normal function, suggesting improvement in the previously akinetic inferior zone.

Because of significant multivessel coronary artery disease with evidence of early progression of the previously dilated right coronary artery stenosis, the patient underwent elective multivessel coronary bypass surgery within the same hospitalization.

Case 2

A 62-year-old male with new onset angina was admitted with prolonged chest pain only partially relieved by nitroglycerin. The initial ECG showed nonspecific ST-T changes with small Q's in V_1 and V_2. While in the coronary care unit, the patient developed ventricular fibrillation and was successfully cardioverted but had continued chest discomfort with anterior ST segment elevation.

Catheterization

The right coronary artery was large, dominant, and diffusely diseased with a maximal stenosis of 40–50 percent. The left main was normal, although the left anterior descending artery was totally occluded proximally without collateral filling. The left circumflex was large, with a 40–50 percent proximal stenosis. The left ventricular ejection fraction was 39 percent with anterior-apical akinesia.

Intervention

Intracoronary nitroglycerin produced no reperfusion. After 90,000 U of intracoronary streptokinase, the left anterior descending was reperfused with good distal runoff. After a total of 190,000 U, there remained a residual 95 percent stenosis proximal to the first septal perforator. In addition, there was a secondary mid left anterior descending (mid-LAD) 80–90 percent stenosis. Because of the high grade residual stenoses, angioplasty was attempted. The proximal lesion was successfully dilated (30–40 percent residual stenosis) although the secondary lesion could not be reached because of vessel tortuosity.

In-Hospital Follow-up

The patient remained asymptomatic for 12 days. At repeat catheterization left ventricular end-diastolic pressure was 7 mmHg. The left ventriculogram revealed anterior hypokinesis with apical akinesia and an overall ejection fraction of 48 percent. There was no change in coronary anatomy from the postangioplasty study. The secondary residual 80–90 percent mid-LAD stenosis remained. Angioplasty was again attempted, with successful dilation of the mid-LAD lesion leaving a residual 30–40 percent stenosis.

Late Follow-up

Follow-up angiography at 6 months demonstrated a 30 percent proximal LAD stenosis (initial angioplasty) with an early 40–50 percent mid stenosis (second angioplasty). The remainder of the coronary anatomy showed no disease progression. The left ventricle was normal in size with anterior hypokinesis, apical akinesis, and an ejection fraction of 52 percent.

Case 3

A 51-year-old male with a history of a remote inferior myocardial infarction complained of prolonged acute intermittent chest pain. The ECG revealed a previous inferior wall myocardial infarction with 1-mm anterior ST segment elevation in leads V_1–V_3. The chest discomfort initially resolved with parenteral nitrates. However, 2 hours later chest pain recurred with hypotension and further ST segment elevations anteriorly. Right heart catheterization revealed significant elevation of the pulmonary artery wedge pressure to 30 mmHg.

Intravenous Streptokinase

Because of the long time delay in hospital arrival and the unavailability of immediate catheterization, the patient received 500,000 U of intravenous streptokinase. Chest discomfort resolved with an increase in systolic blood pressure to 120 mmHg, a decrease of the PCW to 10 mmHg, and resolution of his acute ECG changes.

Heparin administration was begun, but anticoagulation parameters were normal at 4 hours, when the patient redeveloped chest discomfort with recurrent hypotension and elevation of the pulmonary wedge pressure. Following additional heparin and nitrates, pressure stabilized and the chest discomfort again resolved.

Catheterization

Urgent angiographic studies revealed proximal total occlusion of the right coronary artery with a 95 percent stenosis in the proximal LAD artery. There was collateral filling of a large, distal right coronary from the anterior descending artery. The circumflex artery was without significant disease. The left ventricle revealed an inferior akinetic zone with marked anterior wall hypokinesis. Overall ejection fraction was 39 percent.

Intervention

Although angioplasty was technically feasible, controlled revascularization with bypass surgery was considered preferable because of the critical nature of the anterior descending artery stenosis, the large amount of viable myocardium at risk, and the previously documented adverse hemodynamic effects during presumed anterior descending artery occlusion. The patient had emergency bypass surgery immediately after catheterization with good flow to the anterior descending and right coronary arteries and no further recurrent angina or infarction in the hospital.

Case 4

A 55-year-old man with unstable angina of new onset underwent elective catheterization in a hospital without angioplasty availability. On arrival in the catheterization laboratory, the patient developed chest discomfort with anterior ST segment elevation unrelieved by nitrates and nifedipine. The physician proceeded with catheterization, which revealed total occlusion of the anterior descending artery, which failed to recanalize with intracoronary nitroglycerin. The same artery was reperfused with 200,000 U of intracoronary streptokinase, leaving a significant residual 95 percent proximal stenosis. There were no other significant lesions involving the right or circumflex arteries. The anterior wall was akinetic. The patient was observed in the coronary care unit with the evolution of Q waves in V_1–V_4, marked early peak CK release consistent with reperfusion, but some preservation of anterior R waves. The patient was maintained on heparin and was transferred for urgent balloon angioplasty.

Intervention

On repeat angiography at the time of angioplasty left ventricular end-diastolic pressure was 10 mmHg, and left ventriculography revealed only moderately severe anterior-apical hypokinesia with an ejection fraction of 45 percent, suggesting viable anterior wall myocardium. The remainder of the coronary anatomy was essentially normal, except for 95 percent proximal LAD artery stenosis. The anterior descending artery was successfully dilated, leaving a final residual angiographic stenosis of 40 percent (Fig. 1).

Case 5

A 56-year-old white male awakened from sleep with prolonged chest pain. In the emergency room, he had a cardiac arrest with asystole and ventricular fibrillation. Normal sinus rhythm was restored after resuscitative measures, including cardiopulmonary resuscitation. Because of continued chest pain and ECG changes of an evolving inferior wall myocardial infarction, he was taken to the catheterization laboratory for possible reperfusion.

Catheterization

There was no significant disease in the left coronary system. The right coronary artery was totally occluded proximally with retrograde distal collateral filling nearly to the point of total occlusion. There was a fusiform filling defect in the right coronary artery identified on the preliminary arteriogram consistent with thrombus at the point of total occlusion (Fig. 2A).

Intervention

After diagnostic catheterization, coronary angioplasty was considered as an alternative to thrombolytic therapy for reperfusion in this patient, who had already had CPR. The right coronary artery

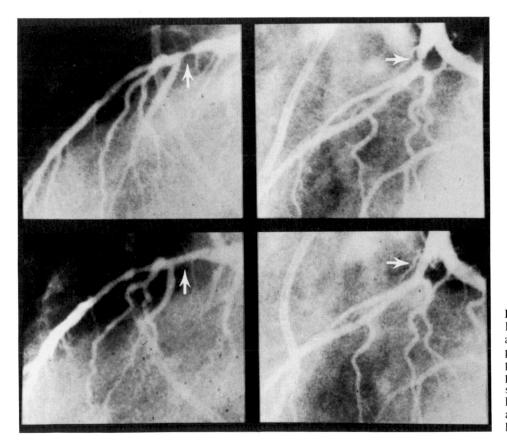

Figure 1. (Left to right) Lateral and LAO cranial angle views before (upper panels) and after (lower panel) the successful angioplasty of a significant residual LAD stenosis. The lesion was dilated 3 days after intracoronary streptokinase infusion.

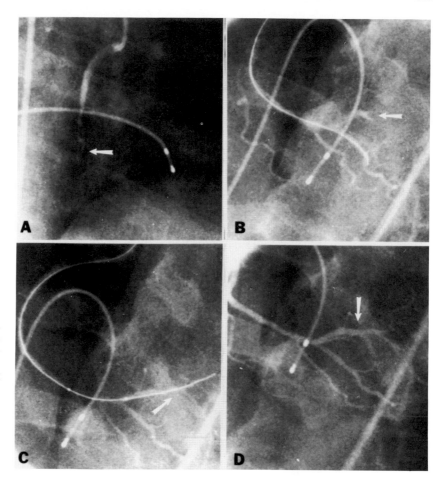

Figure 2. Recanalization by PTCA alone. **A.** Total RCA occlusion with intracoronary thrombus. **B.** RCA injection immediately after dilation with guidewire in the posterior descending branch. Note embolized thrombus (arrow) occluding posterolateral branch. **C.** Guidewire placement in the posterolateral branch to fragment thrombus. **D.** Arteriogram after PTCA, showing resolution of posterolateral branch occlusion.

stenosis was easily crossed with a 0.014 inch steerable guidewire and a 25–30 low profile steerable balloon catheter. After several inflations in the lesion, antegrade flow was reestablished to the posterior descending artery, but the posterolateral branch was now occluded by a dislodged embolic clot (Fig. 2B). The guidewire was then redirected into the posterolateral branch, and the balloon was advanced into this segment without inflations in an attempt to fragment the embolized thrombus mechanically (Fig. 2C). Follow-up angiography showed antegrade flow present in the posterolateral segment (Fig. 2D). The primary right coronary artery stenosis was patent but demonstrated a residual 60–70 percent stenosis with irregularity consistent with local thrombus or plaque disruption. Over the ensuing 10 minutes there was reocclusion of the dilated proximal right coronary artery stenosis, and redilation was instituted. After inflations to higher pressures, the primary right coronary lesion was reopened with a residual stenosis of 50 percent. There was no further evidence of acute reclosure in the laboratory. The distal posterior descending artery and posterolateral branches remained patent with good distal flow.

CLINICAL RESULTS OF TRANSLUMINAL ANGIOPLASTY IN CONJUNCTION WITH STREPTOKINASE INFUSION

Angioplasty Efficacy in Acute Occlusion

Ranges of reported success rates for intracoronary streptokinase infusion are illustrated in Table 2. In general, reflow with intracoronary streptokinase alone can be established in 70–80 percent of patients (2–5). As noted previously, the addition of acute angioplasty for patients not reperfused by streptokinase alone can significantly improve the primary success of reperfusion. In the collaborative study (35) of Massachusetts General Hospital (MGH) and Medical College of Virginia (MCV) discussed previously, angioplasty was performed on 16 or 35 patients without stable reflow after streptokinase. The low incidence of patients attempted represents early experience when streptokinase was considered the single therapy in most individuals. Furthermore, in this time frame of studies, steerable balloon catheters were not available, and, thus, many patients were less "approachable" with a balloon than is currently possible. Of the 16 patients with residual total occlusion attempted, 11 (69 percent) were successfully dilated. Even in this population in whom only half of the streptokinase failures were considered for angioplasty, the additional reperfusion success achieved with angioplasty increased the effective reperfusion rate from 66 percent (70 of 103 patients) with streptokinase alone to 77 percent with the addition of angioplasty. If current angioplasty techniques in patients who had thrombolytic failures are used, the combined reperfusion rate may approach 85–90 percent in laboratories that have experienced personnel and facilities for both recanalization techniques.

Acute reperfusion during transmural infarction has also been achieved by utilizing coronary angioplasty alone without preliminary streptokinase administration. Hartzler (36) has reported a success rate comparable to nonacute angioplasty for total and subtotal occlusion at the time of acute infarction. Furthermore, the risk of distal vessel (clot) embolization was said to be small. Case 5, discussed previously, illustrates the potential for clot embolization with angioplasty alone, although the balloon and guidewire system can be utilized at times to fragment such emboli. For persons with a large evolving acute infarction and a strong contraindication to thrombolytic reperfusion, balloon angioplasty alone may be considered in an attempt to provide reperfusion without the attendant risk of thrombolytic therapy. Further investigation is needed to determine the best method in selected patients for the earliest and most efficacious reperfusion.

Angioplasty for Severe Residual Stenosis

Several investigators (18,25–37) have demonstrated the effectiveness of angioplasty in patients with successful thrombolytic reperfusion but critical residual stenoses. In the MGH-MCV collaborative study (35), angioplasty was attempted in nine patients with a residual stenosis of greater than 90 percent and in an additional three patients with 60–80 percent re-

sidual stenoses following streptokinase. The latter three were included because of large residual intraarterial clot or severe plaque ulceration raising the question of early reocclusion. Successful angioplasty (≥20 percent decrease in stenosis) was achieved in nine (75 percent) of these patients. Of the three failures, only one resulted in vessel occlusion. This success rate for angioplasty was consistent with the reported success rate for nonsteerable balloon angioplasty being performed at that time. Likewise, today with steerable balloon systems, the initial success rate appears to be similar to that for elective angioplasty cases. However, acute on-site reocclusion requiring repeated dilations or higher inflation pressures appears more common in patients who are undergoing streptokinase infusion.

Frequency of Combined Angioplasty and Thrombolysis

The potential for utilizing transluminal angioplasty in combination with thrombolysis appears sizable. Considering a primary thrombolytic failure rate of 20–25 percent and a severe (≥90 percent) residual stenosis rate of at least 20 percent (approximately 30 percent in the MGH-MCV study) (35), as many as one-half of the patients undergoing catheterization for intracoronary thrombolysis are potential candidates for adjunct balloon angioplasty. It should be emphasized that patients with less than 90 percent residual stenoses warrant clinical follow-up because in certain instances marked resolution of residual clot occurs over days, leaving a nonsignificant stenosis that may require no further therapy.

Thus, the combined technique should be limited to specific indications, as outlined in previous discussions, but even with specific indications combined reperfusion techniques may be considered frequently.

Follow-up of Combined Angioplasty and Thrombolysis

Patient follow-up for acute angioplasty has suggested a recurrence rate similar to or higher than that for patients undergoing elective angioplasty. In the MGH-MCV collaborative trial (35), 11 patients were restudied at approximately an average of 5 months after the initial procedure. Significant restenosis or reocclusion was seen in 5 of 11 patients, or in 45 percent of the successful recanalizations. However, there were no deaths of the acute in-hospital angioplasty treatment group, nor were there deaths or recurrent infarcts in the patients who had successful angioplasty and late follow-up. Others (36) have also noted a significant clinical recurrence rate requiring repeat dilation in this category of patients.

It would seem from the foregoing observations that acute angioplasty at the time of thrombolysis can be performed with success and complication rates similar to those of elective angioplasty. Recurrence rate may be somewhat increased, but the increased overall primary reperfusion rate and the reduction in in-hospital recurrence of ischemic events justifies attempted angioplasty in selected patients undergoing streptokinase infusion during acute phases of myocardial infarction.

Significance of Reperfusion

Effective early reperfusion is important to salvage myocardium and, thus, ideally improve prognosis. Controversy regards the impact of reperfusion on changes of left ventricular function and prognosis. Predominantly unrandomized studies (2, 3, 5, 21) have shown improved late, if not acute, left ventricular function (Table 5). Many factors undoubtedly play a role in determining which patients are most likely to demonstrate the greatest improvement in left ventricular function. As would be expected, those patients with normal left ventricular ejection fractions prior to reperfusion are less likely to show a significant improvement following reperfusion (38). Increases in normal areas of myocardium may compensate for reduced contractile function of the jeopardized segment during the acute event. Thus, if a previously depressed segment improves its function, the global function is unchanged because of the subsequent reduction in the noninvolved segment. Conversely, patients with initially depressed left ventricular function are most likely to demonstrate improved global and regional wall motion following reperfusion. Furthermore, time of reperfusion and extent of collateral or antegrade flow to the infarct-related zone at the time of recanalization may be predictive factors in determining late salvage (39).

The Western Washington Randomized Trial (40) demonstrated a significant reduction in early mortality (3.7 percent) for the patients successfully reperfused compared to the mortality rate (11.2 percent) in the placebo group. However, significant changes in left ventricular ejection fraction were not demonstrated despite the improved prognosis. Furthermore, in most studies, the significance of reperfusion appears greatest in patients with anterior myocardial infarction or, perhaps more importantly, in those infarctions with the greatest amount of heart muscle in jeopardy. In studies that have looked at the prognosis on the basis of reperfusion established by streptokinase alone, prognostic and functional differences might be enhanced by the addition of acute angioplasty, which might significantly improve early reperfusion rates. Additional studies will be needed to determine what effect the combination of angioplasty with thrombolysis has on improving overall left ventricular function and prognosis.

Late Follow-up of Reperfused Patients

Only limited data (41,42) are available on the follow-up of patients undergoing acute reperfusion. Left ventricular function changes are at times delayed, as the stunned myocardium only slowly recovers over the initial days to weeks following reperfusion (30).

Late follow-up on prognosis has suggested that patients with good initial reflow may continue to do well. However, preliminary data from the Western Washington Trial (42) suggest that at 1 year the survival benefits of early reperfusion, though still present, are no longer statistically significant. Factors relating to this observation may well be multiple, but it would seem that these results should further emphasize the importance of maintaining improvements in left ventricular function and prognosis achieved with acute reperfusion by angioplasty or bypass surgery to maintain adequate myocardial flow. Studies (43,45) of the prognosis in patients with coronary artery disease generally indicate to various degrees that the greater the amount of viable myocardium at jeopardy, the more likely bypass surgery is to prolong life. Similar consideration will undoubtedly have to be applied to the patients who had undergone reperfusion as well. Further studies will be necessary to determine the late significance of early reperfusion. At the present time, it would seem prudent to emphasize the maintenance of adequate perfusion that is established early during the acute phases of coronary artery occlusions.

TABLE 5 Change in Global Ejection Fraction Early and Late for Successful Reperfusion

Investigator	Initial Ejection Fx, Percentage	Late Ejection Fx, Percentage	P
Rentrop et al. (Ref. 2)	39	45	≤.001
Reduto et al. (Ref. 3)	46	55	≤.002
Cowley et al. (Ref. 21)	41	45	≤.001
Khaja et al. (Ref. 4)	48	45	NS

Intravenous Reperfusion

The application of intracoronary thrombolysis and angioplasty during acute infarction is limited by available resources. The expense and the limited availability of facilities and personnel to perform acute revascularization of large segments of the population are significant concerns. Thus, the possibility of peripheral administration of thrombolytic agents to achieve early reperfusion without the need for emergency catheterization is attractive.

Historically, intravenous streptokinase was utilized in the European Cooperative Trial (46) for myocardial infarction with a significant 6-month reduction in mortality but a high complication rate at least partially related to prolonged high dose streptokinase administration. In the trial, the investigators did not attempt to determine the effect of streptokinase on vessel patency or reperfusion rates and, thus, the pathophysiologic significance of the treatment regimen was unknown. However, since the success of intracoronary streptokinase and particularly the realization that even limited, local intracoronary infusions of streptokinase produce systemic fibrinolytic effects, the possible use of intravenous streptokinase has been investigated. Results have been quite variable. Most studies (47–49) have suggested a slightly lower reperfusion rate when streptokinase was given intravenously, i.e., 50–60 percent of patients, as compared to 70–80 percent of those having intracoronary thrombolysis. However, a number of factors may affect these results. Time to attempted intravenous reperfusion may be most important. Ganz (50) demonstrated that patients receiving intravenous (IV) streptokinase within 3 hours of onset of symptoms had a 96 percent clinical reperfusion rate. Early reperfusion was diagnosed by sudden resolution of chest discomfort, reversal of ECG changes, and the frequent appearance of reperfusion arrhythmias. Eliminating the time required to perform a catheterization and begin intracoronary reperfusion may reduce the interval between the onset of coronary occlusion and establishment of reflow in those individuals in whom recanalization is achieved by IV administration of streptokinase. Thus, although IV treatment may induce reperfusion more slowly than intracoronary infusion, reperfusion can be started early; thus, for those patients in whom reperfusion is established, the amount of myocardial salvage may be enhanced. Current regimens for IV treatment use from 500,000 to 1,500,000 U of streptokinase over several hours. This lower dosage, although producing systemic fibrinolytic effects, appears to produce fewer complications than the regimens with higher doses and longer IV infusion times.

Thus, the IV streptokinase treatment potentially increases the target population. However, it must be recognized that complications similar to those produced by intracoronary reperfusion, including bleeding, reperfusion arrhythmias, and the potential for early reclosure, do occur in such patients. Thus, these cases entail similar considerations to those involved in intracoronary thrombolytic therapy, including potential contraindications and careful monitoring immediately after the procedure. Furthermore, because the same potential for continued ischemia and possible reclosure caused by persistent severely stenotic lesions exists, continued anticoagulation with heparin and early angiography, with the possibility of expanded intervention, including angioplasty or bypass surgery are probably important to maintain benefits established by IV reperfusion. Furthermore, some investigators (48) have considered combined IV and intracoronary evaluation for those individuals who show no evidence of acute reperfusion 1 hour after intravenous therapy. This, again, requires the availability of the catheterization laboratory team. For communities that have no catheterization laboratory, IV thrombolytic therapy may be used in the future, with early transfer to facilities with a catheterization laboratory. Intravenous streptokinase for acute myocardial infarction remains investigative, awaiting further investigations of its efficacy and possible future Food and Drug Administration (FDA) approval.

One additional exciting new possibility for IV thrombolysis involves tissue plasminogen activator (t-PA). Plasminogen activator is undergoing pilot evaluations (51–53) to determine its potential efficacy in both intracoronary and IV thrombolysis. The attractive feature of this agent is that it is more clot-specific and produces thrombolytic effects with probably lower potential for systemic fibrinolytic effects. Pilot data have been most encouraging, and future studies must determine whether this is a feasible alternative for early reperfusion in patients with evolving acute myocardial infarction.

FUTURE DIRECTIONS

The introduction and early evaluation of intracoronary and now systemic thrombolytic therapy appear to have a significant impact on patients sustaining myocardial infarction. However, further studies to determine the most important subgroups who benefit from this therapy are necessary. Furthermore, initial investigations have suggested that the gains achieved by early reperfusion must be maintained by more definitive therapeutic interventions, including angioplasty and bypass surgery. Late follow-up of these patients will be important to determine the most appropriate management techniques. However, at the present time, individuals undergoing reperfusion need careful early and late observation to assess prognostic direction of the reperfused area. Since the potential systemic risks of thrombolytic coronary reperfusion have been accepted, it seems quite warranted to proceed with appropriate additional studies and interventions in an attempt to maintain the improvement brought about by the thrombolytic therapy alone. The impact of agents such as t-PA may significantly improve the availability and applicability of the coronary reperfusion therapy. In many patients sustaining myocardial infarction, early expanded interventions will continue to be necessary to maintain the early effects of reperfusion.

REFERENCES

1. Rackley CE, Russell RO Jr, Mantel JA, Rogers WJ: Modern approach to the patient with acute myocardial infarction. Curr Probl Cardiol I:49, 1977.
2. Rentrop P, Blanke H, Karsch KR, et al: Selective intracoronary thrombolysis in acute myocardial infaraction and unstable angina pectoris. Circulation 63:307, 1981.
3. Reduto LA, Freund GC, Gaeta JM, Smalling RW, Lewis B, Gould KL: Coronary artery reperfusion in acute myocardial infarction: Beneficial effects of intracoronary streptokinase in left ventricular salvage and performance. Am Heart J 102(6):1168, 1981.
4. Khaja F, Walton JA Jr, Brymer FJ, et al: Intracoronary fibrinolytic therapy in acute myocardial infarction: Report of a prospective randomized trial. N Engl J Med 308(22):1305, 1983.
5. Mathey DG, Kuck K-H, Tilsner V, Krebber H, et al: Nonsurgical coronary artery recanalization in acute transmural myocardial infarction. Circulation 63:489, 1981.
6. DeWood MA, Spors J, Notske R, Mouser LT, Burroughs R, Golden MS, Lang HT. Prevalence of total coronary occlusion during the early hours of transmural myocardial infarction. N Engl J Med 303:897, 1980.
7. Horie T, Sekiguchi M, Kirosawa K: Coronary thrombosis in pathogenesis of acute myocardial infarction. Br Heart J 40:150, 1978.
8. Davies MJ, Woolf N, Robertson WB: Pathology of acute myocardial infarction with particular reference to occlusive coronary thrombi. Br Heart J 38:659, 1976.
9. Page DL, Caulfield JB, Kastor JA, DeSanctis RW, Sanders CA: Myocardial changes associated with cardiogenic shock. Br Heart J 32:728, 1970.
10. Harnarayan C, Bennett MA, Pentecost BL, Brewer DB: Quantitative study of infarcted myocardium in cardiogenic shock. Br Heart J 32:728, 1970.
11. Rude RE, Muller JE, Braunwald E: Efforts to limit the size of myocardial infarcts. Ann Intern Med 95:736, 1981.
12. International Collaborative Study Group: Reduction of infarct size with the early use of timolol in acute myocardial infarction. N Engl J Med 310:9, 1984.
13. Phillips S, Kongtahworn C, Zeff RH, Vinson M, et al: Emergency coronary artery revascularization: A possible therapy for acute myocardial infarction. Circulation 60:241, 1979.
14. Feit F, Rentrop KP: Thrombolytic therapy in acute myocardial infarction. Cardiovasc Rev Rep 4:426, 1983.
15. European Cooperative Study Group for Streptokinase Treatment in Acute Myocardial Infarction: Streptokinase in acute myocardial infarction. N Engl J Med 301:797, 1979.
16. Schroder R, Biamino G, Enz-Rudiger VL, et al: Intravenous short-term infusion of streptokinase in acute myocardial infarction. Circulation 57:536, 1983.
17. Ganz W, Gelf I, Shah PK, et al: Intravenous streptokinase in evolving acute myocardial infarction. Am J Cardiol 53:1209, 1984.
18. Yasuno M, Saito Y, Ishida M, Suzuki K, Endo S, Takahaski M: Effects of percutaneous transluminal coronary angioplasty: Intracoronary thrombolysis with urokinase in acute myocardial infarction. Am J Cardiol 53:1217, 1984.
19. Simon TL, Ware JH, Stengle JM: Clinical trials of thrombolytic agents in myocardial infarction. Ann Intern Med 79:712, 1973.
20. Ganz W, Buchbinder N, Marcus H, et al: Intracoronary thrombolysis in evolving myocardial infarction. Am Heart J 101:4, 1981.
21. Cowley MJ, Hastillo A, Vetrovec GW, Hess ML: Effects of intracoronary streptokinase in acute myocardial infarction. Am Heart J 102:1149, 1981.
22. Maseri A, L'Abbate A, Varoldi F, et al: Coronary vasospasm as a possible cause of myocaradial infarction: A conclusion derived from a study of "preinfarction" angina. N Engl J Med 299:1271, 1978.
23. Oliva PB, Brechenridge JC: Arteriographic evidence of coronary arterial spasm in acute myocardial infarction. Circulation 56:366, 1977.
24. Vetrovec GW, Leinbach RC, Gold HK, Cowley MJ: Intracoronary thrombolysis in syndromes of unstable ischemia: Angiographic and clinical results. Am Heart J 104:946, 1982.
25. Mandelkorn JM, Wold NM, Singh S, Schechter JA, Kersh RI, Rodgers DM, Workman MB, Bentivoglio LG, LaPorte SM, Meister SG: Intracoronary thrombus in nine transluminal myocardial infarctions and in unstable angina pectoris. Am J Cardiol 52:1, 1983.
26. Blanke H, Scherff R, Karsch KR, Levine RA, Smith H, Rentrop P: Electrocardiographic changes after

streptokinase-induced recanalization in patients with acute left anterior descending artery obstruction. Circulation 68:406, 1983.

27. Sones FJ, Shirey EK: Cine coronary arteriography. Mod Concepts Cardiovasc Dis 31:735, 1962.

28. Judkins MP: Percutaneous transfemoral selective coronary arteriography. Radiol Clin North Am 6:467, 1968.

29. Kammatsuse K, Lando U, Mercier JC, Fishbein MC, Swan HJC, Ganz W: Rapid lysis of coronary thrombi by local application of fibrinolysin, abstracted. Circulation 59–60(suppl II):II–216, 1979.

30. Braunwald E, Kloner RA: The stunned myocardium: Prolonged, postischemic ventricular dysfunction. Circulation 66:1146, 1966.

31. Goldberg S, Greenspon AJ, Urban PL, et al: Reperfusion arrhythmia: A marker of restoration of antegrade flow during intracoronary thrombolysis for acute myocardial infarction. Am Heart J 105:26, 1983.

32. Cowley MJ, Hastillo A, Vetrovec GW, Fisher LM, et al: Fibrinolytic effects of intracoronary streptokinase administration in patients with acute myocardial infarction and coronary insufficiency. Circulation 67:1031, 1983.

33. Merx W, Dorr R, Rentrop P, Blanke H, Karsch KR, Mathey DG, Kremar P, Rutsch W, Schmutzler H: Evaluation of the effectiveness of intracoronary streptokinase infusion in acute myocardial infarction: Post procedure management and hospital course of 204 patients. Am Heart J 102:1181, 1981.

34. Timmis GC, Gangadharan V, Ramos RG, Hauser AM, Westveer DC, Stewart J, Goodfliesch R, Gordon S: Hemorrhage and the products of fibrinogen digestion after intracoronary administration of streptokinase. Circulation 69(6):1146, 1984.

35. Gold HK, Cowley MJ, Palacios IF, Vetrovec GW, Akins CW, Block PC, Leinbach RC: Combined intracoronary streptokinase infusion and coronary angioplasty during acute myocardial infarction. Am J Cardiol 53:122C, 1984.

36. Hartzler GO, Rutherford BD, McConahay DR: Percutaneous transluminal coronary angioplasty: Application for acute myocardial infarction. Am J Cardiol 53:117C, 1984.

37. Meyer J, Merx W, Schmitz H, Erbel R, Kiesslich T, Dorr R, Lambertz H, Bethge C, Krebs W, Bardos P, Minale C, Messmer BJ, Effert S: Percutaneous transluminal coronary angioplasty immediately after intracoronary streptolysis of transmural myocardial infarction. Circulation 66(5):905, 1982.

38. Ferguson DW, White CW, Schwartz JL, Brayden GP, Kelly KJ, Kioschos JM, Kirchner PT, Marcus ML: Influence of baseline ejection fraction and success of thrombolysis on mortality and ventricular function after acute myocardial infarction. Am J Cardiol 54:702, 1984.

39. Rogers WJ, Hood WP, Mantle JA, Baxley WA, Kirklin JK, Zorn GL, Nath HP: Return of left ventricular function after reperfusion in patients with myocardial infarction: Importance of subtotal stenoses or intact collaterals. Circulation 69(2):338, 1984.

40. Kennedy JW, Ritchie JL, Davis KB, Fritz JK: Western Washington randomized trial of intracoronary

streptokinase in acute myocardial infarction. N Engl J Med 309:1477, 1983.

41. Urban PL, Cowley MJ, Goldberg S, Vetrovec G, Hastillo A, Greenspon AJ, Walinski P, Maroko P: Intracoronary thrombolysis in acute myocardial infarction: Clinical course following successful myocardial reperfusion. Am Heart J 108:873, 1984.

42. Kennedy JW, Ritchie JL, Davis K, et al: Western Washington intracoronary SK trial of acute myocardial infarction: One year survival (abstract). Circulation 70(suppl II):324, 1984.

43. European Coronary Surgery Study Group: Long-term results of prospective randomized study of coronary artery bypass surgery in stable angina pectoris. Lancet 2:1173, 1982.

44. Takaro T, Hultgren HN, Detre KM, Peduzzi P: The Veterans Administration Cooperative Study of Stable Angina: Current status. Circulation 65(suppl II):60, 1982.

45. Chaitman BR, Fisher LD, Courassa MG, et al: Effect of coronary bypass surgery on survival patterns in subsets of patients with left main coronary artery disease: Report of the Collaborative Study in Coronary Artery Surgery (CASS) Am J Cardiol 48:765, 1981.

46. European Cooperative Study Group for Streptokinase Treatment in Acute Myocardial Infarction: Streptokinase in acute myocardial infarction. N Engl J Med 301:797, 1979.

47. Schroder R, Biamino G, Leitner ERV, Linderer T, Bruggemann T, Heitz J, Vohringer HR, Wegscheider K: Intravenous short-term infusion of streptokinase in acute myocardial infarction. Circulation 67(3):536, 1983.

48. Rogers WJ, Mangle JA, Hood WP Jr, Baxley WA, Whitlow PL, Reeves RC, Soto B: Prospective randomized trial of intravenous and intracoronary streptokinase in acute myocardial infarction. Circulation 68(5):1051, 1983.

49. Blunda M, Meister SG, Schechter JA, Pickering NJ, Wolf NM: Intravenous versus intracoronary streptokinase for acute transmural myocardial infarction. Cathet Cardiovasc Diagn 10:319, 1984.

50. Ganz W, Geft I, Shah PK, Lew AL, Rodriguez L, Weiss T, Maddahi J, Berman DS, Charuzi Y, Swan HJC: Intravenous streptokinase in evolving acute myocardial infarction. Am J Cardiol 53(9):1209, 1984.

51. Van de Werf F, Bergmann SR, Fox KAA, de Geest H, Hoyng CF, Sobel BE, Collen D: Coronary thrombolysis with intravenously administered human tissue-type plasminogen activator produced by recombinant DNA technology. Circulation 69(3):605, 1984.

52. Van de Werf F, Ludbrook PA, Bergmann SR, Tiefenbrunn AJ, Fox KAA, de Geest H, Verstraete M, Collen D, Sobel BE: Coronary trombolysis with tissue-type plasminogen activator in patients with evolving myocardial infarction. N Engl J Med 310(10):609, 1984.

53. Collen D, Topol EJ, Tiefenbrunn AJ, Gold HK, et al: Coronary thrombolysis with recombinant human tissue-type plasminogen activator: A prospective, randomized, placebo-controlled trial. Circulation 70(6):1012, 1984.

22

Antiplatelet and Anticoagulant Drugs in Percutaneous Transluminal Angioplasty

LOUIS ROY, M.D.
HOWARD R. KNAPP, M.D.
GARRET A. FITZGERALD, M.D.

INTRODUCTION

Percutaneous transluminal angioplasty (PTA) is a technique finding increased application in the relief of arterial stenosis, whether atherosclerotic (1), dysplastic (2), or congenital (3). It is an attractive alternative to surgery in a number of circumstances, and occasionally an attempt can be made via percutaneous angioplasty to improve blood flow through a vessel which cannot be approached surgically. At this writing, transluminal angioplasty has been utilized on nearly every major artery, including coronary, renal, mesenteric, iliac, basilar, subclavian, and so forth. Atherosclerotic narrowing of arteries supplying renal transplants as well as coronary artery bypass grafts has been successfully managed using this technique.

Although the potential of transluminal angioplasty as a major therapeutic modality is substantial, several significant problems are dampening enthusiasm for its widespread application. Immediate complications of angioplasty are not uncommon (7 percent) and include thrombosis, arterial spasm, intimal dissection, and distal embolization. Equally serious from the viewpoint of the eventual usefulness of angioplasty is a high restenosis rate (25–30 percent within 6 months in the coronary artery) and the acceleration of the atherosclerotic process seen both at the angioplasty site (4) and distal to it (5).

Pharmacologic interventions to reduce the incidence of such complications have been adopted empirically, with some apparent success. During the last few years there has been an increased appreciation of the role of platelets in the genesis of atherosclerosis and subsequent thrombotic complications thereof (6–8). It seems likely that once we have a better understanding of platelet–vessel wall interactions, we can devise a rational basis for pharmacologic intervention and overcome the major obstacles limiting an expanded role for angioplasty. In this chapter, we will briefly present what is currently known about platelets, blood vessels, and transluminal angioplasty, as well as the pharmacology of currently used antiplatelet agents. We shall also consider the role antiplatelet therapy might play in transluminal angioplasty both today and in the future.

PLATELET STRUCTURE

An overview of platelet function will be helpful in order to understand the pharmacology of antiplatelet agents, as well as their application in angioplasty. The platelet population in circulating blood is heterogeneous with regard to size and density. They are disk-shaped anuclear cell fragments averaging 3.6 μm in diameter and 0.9 μm in thickness, when measured freely rotating in suspension. Smaller dimensions are found on stained blood smears where platelets of 2.5 μm diameter make up less than 10 percent of the population in normal subjects. Increased numbers of larger platelets (megathrombocytes) are found in a variety of conditions associated with accelerated platelet destruction, and less dense degranulated platelets are also seen in such settings, suggestive of enhanced in vivo platelet activation.

Platelets contain a complex canalicular system, which is a network of tubules and vesicles running throughout the cytoplasm and connecting to the cell surface. In close association with this tubular system are the three types of platelet granules. The alpha granules contain mitogenic factors, fibrinogen, and platelet factors 4 and 5. Dense granules store serotonin, nucleotides, calcium, phosphate, and catecholamines. Finally, the platelets have true lysosomes containing a variety of enzymes.

The release of platelet granules is an energy-requiring process and many similarities have been found between platelet energy metabolism and that of smooth muscle. Platelets also contain a complex contractile apparatus which is involved in platelet shape change and possibly aggregation, and which is based on an actin-myosin system similar to that of muscle.

PLATELETS, ENDOTHELIUM, AND HEMOSTASIS

Circulating platelets do not normally adhere to each other or to intact vessel wall endothelium. Vascular injury disrupts the endothelium, exposing collagen, microfibrils, and other components of the subendothelial tissue to flowing blood. The immediate platelet response to this is adhesion to the subendothelium by a process dependent upon the presence of a particular plasma protein, the von Wille-

brand factor. This is quickly followed by a change in the shape of the platelet from discoid to spherical with cytoplasmic protrusions extending along the exposed collagen fibrils of the subendothelium.

Following adherence to the injured vessel wall, platelets proceed to attach to one another (aggregation) and to release their granule contents. The shape change and initial phase of aggregation precedes the release reaction under controlled conditions in vitro but the sequence of events in vivo is uncertain. At any rate, release and aggregation result in the formation of the initial platelet plug at the site of endothelial disruption. In parallel with the above processes, prostaglandin endoperoxides and thromboxane A_2 are formed from their precursor, arachidonic acid. This is liberated from platelet membrane phospholipids via activation by specific phospholipases. These compounds can stimulate the release reaction as well as aggregate platelets by a mechanism independent of ADP release.

While platelet plug formation is occurring, there is a simultaneous generation of the enzyme thrombin, which can activate platelets at lower concentrations than are required to convert fibrinogen to fibrin. Thrombin stimulates the platelet release reaction as well as thromboxane formation but can still cause platelet aggregation via specific membrane receptors if both ADP release and thromboxane synthesis are blocked.

Two interrelated mechanisms can result in thrombin generation at the site of vascular injury. Tissue factor release from damaged blood vessels can result in factor X activation via the extrinsic coagulation system, and the surface-mediated activation of factor XII and prekallikrein result in the cascade of the intrinsic coagulation system. Eventually the platelet plug becomes thoroughly infiltrated with the insoluble protein fibrin to complete the hemostatic process (Figure 1).

Platelet plug formation and blood coagulation are inseparable processes in vivo and the interactions between platelets and the proteins of the coagulation cascade has been the focus of much recent work. A number of coagulation proteins are stored and released in platelet granules, including factor V, fibrinogen, and platelet factor 4. Platelets may participate in the surface-associated proteolytic activation of several coagulation enzymes and the membrane surface of activated platelets serves as a protected environment which prevents inactivation of factor X while greatly stimulating its activity. Platelet factor 4 (antiheparin factor) binds endogenous heparin and may prevent inactivation of factor X_a by heparin-activated antithrombin III. Platelet factor 3 is a lipoprotein moiety of the platelet membrane which is required for at least two steps in

VASCULAR AND INTRAVASCULAR CONSEQUENCES OF PLATELET ACTIVATION

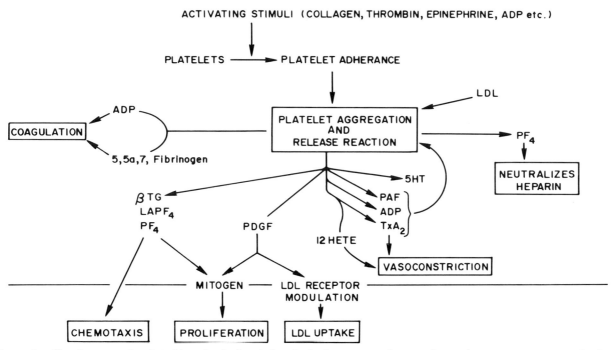

Figure 1. Platelets are stimulated by physical and chemical triggers to undergo a shape change, aggregate, and release a variety of biologically active compounds. These include thromboxane A_2 ($T_x A_2$), adenosine diphosphate (ADP), and serotonin (5HT), which may amplify the aggregation process in addition to acting as antagonists on vascular smooth muscle. Other mediators such as platelet-derived growth factor (PDGF) may influence cellular proliferation and lipid metabolism on the endothelial surface. Platelet aggregation is also closely related to activation of the coagulation cascade. Additional abbreviations: LDL, low density lipoprotein; PF_4, platelet factor 4; $LAPF_4$, long acting platelet factor; βTG, β-thromboglobulin; 12 HETE, 12-hydroxy-eicosatetraenoic acid (the major lipoxygenase producer of arachidonic acid in platelets).

blood coagulation; the association of factor IX_a and VIII to result in factor X activation and the interaction between factors X_a and V to yield the pro-thrombinase complex.

The normal endothelium does not serve as a site for platelet attachment and recently has been found to exert several controlling influences on platelets and platelet-mediated coagulation reactions. A vitamin K–dependent anticoagulant, protein C, is activated by thrombin in concert with an endothelial cell cofactor, thrombomodulin. Once activated, protein C is able to inactivate platelet-bound factor V_a, which may lessen factor X_a-mediated thrombin generation. Like platelets, endothelial cells release derivatives of arachidonic acid, their main product being prostacyclin (PGI_2). This substance is a very potent inhibitor of platelet function, and it is likely that the production of proaggregatory thromboxane by platelets and antiaggregatory PGI_2 by endothelium

may modulate the formation and size of the platelet plug. The possibility for pharmacologic manipulation of this "balance" is currently stimulating a great deal of research effort and will be discussed further in the section on antiplatelet drugs.

PLATELETS, THROMBOSIS, AND ATHEROSCLEROSIS

A number of lines of evidence have led to an appreciation of the essential role platelets play in the development of atherosclerosis. Platelets do not adhere to normal endothelium but do adhere to exposed subendothelial structures through a process requiring von Willebrand factor. Platelets which accumulate at sites of endothelial injury undergo the release reaction and among the many products secreted are cationic protein growth factors. These substances stimulate the mitosis of cultured smooth muscle cells and fibroblasts but not of endothelial cells. Growth factors are stored in the platelet granules and their synthesis is believed to occur in megakaryocytes. In addition to inducing mitosis in selected cell types, they also appear to have chemotactic activity for fibroblasts and smooth muscle cells and to stimulate pinocytosis. It has been suggested that they serve the function of "wound hormones" due to their overall effect on stimulating fibroblast mitosis and collagen production as well as smooth muscle cell proliferation.

Another putative stimulator of fibroblast mitosis which has been isolated from human platelets is low-affinity platelet factor 4 (LA-PF$_4$). Whether LA-PF$_4$ or platelet-derived growth factor(s) actually function as mitogens in vivo is uncertain, but both do exert metabolic effects in vitro on specific cell types believed to be important in atheroma formation.

In addition to mitogenic factors, platelets release several granule proteins which bind to endothelial cells and may alter their permeability and anticlotting properties. These include LA-PF$_4$, platelet factor 4 (PF$_4$) and β-thromboglobulin. Platelets also release their lysosomal heparinase, which degrades heparin on the endothelial cell surface (Figure 2).

Several studies in experimental animals support a role for platelets in the atherosclerotic process. Lowering the platelet count below 7000 platelets per mm^3 inhibits the development of intimal thickening following active endothelial injury in rabbits. Interestingly, re-endothelialization after injury is not altered by thrombocytopenia. In monkeys with hemocystinemia, dipyridamole appears to slow the accelerated rate of platelet destruction and intimal proliferation. This same antiplatelet agent enhances formation of atherosclerotic plaque in rabbits fed a cholesterol-rich diet, while aspirin seems to have a protective effect on the development of coronary thrombosis in monkeys fed an atherogenic diet (9–11).

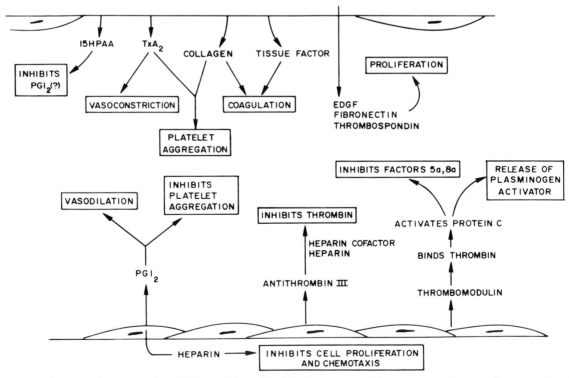

VASCULAR FACTORS WHICH MODIFY THROMBOSIS AND ATHEROGENESIS

Figure 2. Production of prostacyclin (PGI$_2$) antithrombin III, heparin. and thrombomodulin contributes to the thromboresistance of intact endothelium. These mechanisms are deficient following endothelial disruption when release of collagen, tissue factors. and perhaps thromboxane A$_2$ (T$_x$A$_2$) may predispose to activation of platelets and the clotting system. Release of growth factors may stimulate cellular proliferation at the origin of injury and lipid peroxides such as 15-hydroxyperoxy fatty acids in atheromatous plaques may limit the capacity for prostacyclin biosynthesis.

Another condition which provides insight into mechanisms of experimental atherosclerosis is that of von Willebrand's disease. In patients with this condition, a deficiency of the von Willebrand factor (necessary for platelet adhesion to subendothelium) is expressed as a prolonged bleeding time and a predisposition to a bleeding diathesis. An apparently identical disease has been found in pigs, and it has been shown that such pigs are resistant to cholesterol-induced atherosclerosis but not to the endothelial damage elicited by cholesterol feeding (12). When the coronary arteries of pigs with von Willebrand's disease receive balloon injury while the pigs are on an atherogenic diet, atherosclerosis developed at the same rate as in normal pigs on the same diet. The rate of atherogenesis in the aortas of the von Willebrand pigs was delayed compared to normals, however (12). There is much less information on whether such a relationship exists in humans, but several reports on autopsy findings in patients with severe von Willebrand's disease have documented extensive atherosclerosis (13). The near absence of the von Willebrand factor in humans, therefore, does not by itself appear to protect them against the development of atherosclerosis.

One situation in man where platelet-inhibitory drugs seem to have a beneficial effect is in the markedly accelerated atherosclerosis which develops in the coronary circulation of cardiac transplant patients. This syndrome appears to be related to chronic immune injury of the endothelium with subsequent platelet activation. Addition of dipyridamole (100 mg qid) to the usual immunosuppressive treatment in such patients significantly prolonged their survival (14), compared to a prior group of patients who did not receive antiplatelet therapy. However these studies were neither randomized nor double-blind.

There are, then, a number of situations in both animals and man where antiplatelet therapy appears to favorably influence the rate of atherosclerotic vascular occlusion. This is not true in all circumstances but the evidence is suggestive enough to stimulate further work along these lines.

ers of the arterial wall rather than the previously postulated compression of atheromatous material (15). In experimental animals, it has been shown that the balloon inflation causes endothelial denudation followed rapidly by a dense deposition of platelets and fibrin over the denuded area (16). Aspirin pretreatment has only a minimal effect on the extent of platelet adhesion while heparin has no effect, although it does prevent subsequent fibrin deposition. Rheological factors are quite important in determining the extent of platelet-fibrin adhesion to damaged endothelium so every attempt is made to maximize blood flow past the angioplasty site. The procedure itself, of course, exposes large amounts of very thrombogenic material to the blood stream and embolization of platelet-fibrin microthrombi distal to the angioplasty may result in small punctate myocardial infarctions. Coronary spasm both at the angioplasty site and distal to it can develop into a clinically significant problem. The role of platelet derived thromboxane A_2 in this type of arterial spasm is uncertain.

As re-endothelialization and repair processes proceed at the angioplasty site there is often an angiographically demonstrable "fibrous contraction" whereby fibrosis of the reparative tissue around the split atheroma causes a further increase in the luminal diameter of the treated vessel. The complex relationships between platelet adhesion and release, endothelial repair, and atherosclerosis are poorly understood at the present time and it may be that certain platelet-related processes are necessary or beneficial in the overall picture of angioplasty. Once we understand the individual processes being stimulated by transluminal angioplasty, then rational pharmacological intervention can be attempted. Until then, our best information indicates that decreasing platelet activation and limiting the extent of platelet-fibrin thrombus formation are desirable goals. The following part of this chapter describes the pharmacology of currently available antiplatelet and anticoagulant agents, and the final section attempts to formulate a rational approach to the use of such agents in PTA.

ANGIOPLASTY AND PLATELET ACTIVATION

In the preceding discussion it was pointed out that the coagulation cascade and platelet activation were essentially inseparable processes in vivo. Both are triggered in the course of angioplasty. The mechanism of arterial dilatation induced by transluminal angioplasty has recently been demonstrated to involve splitting of the atherosclerotic plaque and underlying media with stretching of the remaining lay-

PHARMACOLOGY OF ANTIPLATELET AND ANTICOAGULANT DRUGS

Aspirin

The association between aspirinlike drugs and prostaglandins became clear in 1971. Aspirin and indomethacin were shown to inhibit prostaglandin release from human platelets and the perfused dog spleen and to prevent prostaglandin synthesis in cell-free preparations of guinea pig lung. These observations had been preceded by the demonstration by Piper and Vane (17) that aspirin inhibited the release of an unknown compound, "rabbit aorta constricting substance" (later identified as thromboxane A_2), from guinea pig lungs during anaphylaxis. Inhibition of prostaglandin synthesis by "aspirinlike" compounds has since been demonstrated in more than 30 different systems.

Acidic nonsteroidal anti-inflammatory drugs, particularly aryl acids, form the predominant group of prostaglandin synthetase inhibitors in man. Most inhibit the cyclooxygenase enzyme, which catalyzes prostaglandin formation, by competing with the substrate arachidonic acid at the active site. Aspirin inhibits prostaglandin synthesis by quantitatively and selectively acetylating cyclooxygenase at its functional site. Following the observation that ^{14}C in the acetyl but not the carboxyl position of aspirin irreversibly binds to platelets in platelet-rich plasma (18), Roth and Majerus (19), using acetyl-^{3}H-aspirin of high specific activity, demonstrated permanent acetylation of a particulate platelet protein with a molecular weight in the range of 85,000 daltons. The acetylation reaction reached saturation at a low aspirin concentration (30 μM) which correlates with its effect on human platelets in vivo and in vitro. Two other platelet protein fractions were also acetylated, but the reaction was not saturated. When acetyl-^{3}H-aspirin was incubated with sheep and bovine seminal vesicles, a single protein with a molecular weight of 85,000 daltons was acetylated (20). Since acetylation of the protein was associated with loss of cyclooxygenase activity and since arachidonic acid, a natural substrate of cyclooxygenase, inhibited acetylation by aspirin, it was assumed that aspirin inhibited cyclooxygenase by acetylating its active site. Subsequently, Roth and Soik (21) showed that aspirin caused formation of N-acetyl serine and suggested that aspirin acetylated the NH_2 group of the N-terminal serine of the oxygenase portion of the enzyme. These studies also demonstrated that cyclooxygenase is a dimer consisting of two subunits of 63,000 daltons each. It was postulated that one subunit contains the oxygenase activity, while the other contains the peroxidase activity and is not acetylated by aspirin. The acetylation also is blocked partially by preincubation with indomethacin, which presumably alters the active site (22).

A consideration of the human pharmacology of the antiplatelet and anticoagulant drugs used in conjunction with angioplasty procedures is necessary prior to evaluating their efficacy in this clinical setting. Antiplatelet drugs have been defined as those which interfere with a measurable index of platelet function (22). However, the available indices are either very indirect or liable to be confounded by artifact. The most commonly studied parameters have been platelet aggregation, platelet adherence to foreign surfaces (e.g., glass beads), and the platelet survival time. Effects on platelet aggregation and adherence may be measured in vitro or ex vivo (on blood drawn from a volunteer who has taken the drug of interest). Platelet survival time is a measure of platelet turnover in vivo. Following injection of radiolabeled platelets, the decrease of the tracer in plasma is measured as a function of time. However, although derived from conditions in vivo, such calculations are highly dependent on mathematical modeling and do not necessarily reflect alterations in the degree of platelet activation in vivo. More recently, assays have been developed which measure platelet granule constituents such as β-thromboglobulin and platelet factor 4.

These compounds are released from platelets following activation in vivo. Although the assays for these compounds are readily available and easy to perform, their usefulness as indices of drug action in vivo is somewhat limited. In particular, blood withdrawal, particularly via long catheters, is likely to be associated with a substantial artifact due to platelet activation ex vivo. This problem is likely to be even more pronounced where samples are obtained from chronically indwelling lines. The major cyclooxygenase product of the platelet, thromboxane A_2, is too short lived (t 1/2 30 seconds) to be measured directly. Although assays directed toward its stable hydration product thromboxane B_2, have been developed they are of little use as markers of in vivo platelet function. When blood is allowed to clot, 200–300 ng/ml of this compound are typically found in serum. However, the majority of studies report circulating concentrations of 100–200 pg/ml in plasma. It can be readily appreciated that only 1–2 percent platelet activation ex vivo could account for the bulk of this so-called endogenous material. An alternative to measuring thromboxane B_2 in plasma is the measurement of enzymatic metabolites of this compound in either plasma or urine. Plasma measurements of such metabolites circumvent the problem of nonenzymatic degradation during blood withdrawal, whereas urinary metabolites offer a noninvasive approach to measuring prostanoid production (23). The efficacy of antiplatelet drugs may be assessed using both in vitro and in vivo models of thrombosis.

The dose of aspirin required to inhibit platelet cyclooxygenase in man is of obvious clinical interest. Burch et al.(24) studied the ability of ^{3}H acetyl aspirin to acetylate washed platelets periodically obtained from volunteers. These radioligand binding studies suggested that more than 50 percent inhibition was obtained by daily doses of 20 mg aspirin. An effect on megakaryocytes seemed to occur as no unacetylated platelet enzyme was evident in the circulation for two days following aspirin ingestion.

The appreciation that prostacyclin, a potent inhibitor of platelet aggregation (25) was the major arachidonic acid metabolite formed by vascular endothelium led to speculation that a dose of aspirin that inhibited cyclooxygenase in vascular tissues would be potentially undesirable. Two pieces of evidence suggested that a critical dose of aspirin might selectively inhibit platelet thromboxane formation. Burch et al.(24) found that platelet cyclooxygenase was very sensitive to aspirin inhibition. Subsequently, the same workers found that aortic and coronary artery microsomes were 1/250th as sensitive to aspirin as the enzyme in intact platelets and 1/60th as sensitive as a platelet microsomal preparation (26). Baenziger et al. (27) also found that cyclooxygenase in vascular tissues in culture was 20- to 40-fold less sensitive to aspirin inactivation than is the cyclooxygenase of platelets. The cyclooxygenase of human cultured fibroblasts and arterial smooth muscle cells was inactivated by 9 to 12 percent; whereas platelet cyclooxygenase was inhibited by 92 percent at corresponding doses of aspirin. However, others (28) found endothelial cell cyclooxygenase to be as sensitive to aspirin as the enzyme in platelets. Whereas platelets were unable to synthesize a new cyclooxygenase, the enzyme turnover in endothelial cell culture was very high and therefore its acetylated form was rapidly replaced. These observations suggest that the rapid resynthesis of vascular cyclooxygenase might render a thrombogenic effect of aspirin unlikely in clinical practice.

Does high-dose aspirin constitute a potential risk of thrombosis? Aspirin treatment of endothelial cells cultured from the human umbilical vein increases the adherence of ^{51}Cr platelets in the presence of thrombin and thrombin-stimulated prostacyclin release, i.e., inversely related to thrombin-stimulated platelet adherence to the endothelial monolayer (29). In vivo, Reyers et al.(30) failed to identify a beneficial or harmful effect of high-dose (50–200 mg/kg) or low-dose (2.5–10 mg/kg) aspirin on arterial and venous models of experimental thrombosis. However, high-dose aspirin therapy (200 mg/kg) significantly augmented experimentally induced thrombus size in rabbits compared to animals treated with low-dose (10 mg/kg) aspirin, sodium salicylate (200 mg/kg), and untreated controls (31). The thrombogenic effect of high-dose aspirin was lost if the interval between thrombus induction and drug administration was greater than 2.5 hours, perhaps reflecting the

high turnover of vascular cyclooxygenase. Aspirin (100 mg/kg ip) and indomethacin (5 mg/kg ip) enhanced platelet aggregation in mouse mesenteric vessels, supporting observations that the same drugs increased platelet aggregation in transected rabbit mesenteric vessels.

What relevance have these studies for the inhibition of platelet-mediated vascular occlusion by aspirinlike drugs in humans? Despite the marked potency of prostacyclin in vitro, recent studies have shown that very little is actually generated under physiological circumstances in humans (32). Such observations seemed to suggest that the quest for a dose of aspirin which might spare prostacyclin biosynthesis while inhibiting thromboxane formation might be of little therapeutic relevance. However, such a low rate of prostacyclin secretion in healthy individuals might merely reflect a low frequency and/or intensity of stimuli to its production under physiological circumstances. Platelet granule constituents such as β-thromboglobulin and platelet-derived growth factor and thrombogenic stimuli such as thrombin and epinephrine, potently stimulate prostacyclin release from endothelial cell cultures in vitro. In support of this hypothesis, it has recently been demonstrated that prostacyclin biosynthesis is actually increased in patients with severe atherosclerosis and evidence of platelet activation in vivo (33).

Thus far, aspirin has been shown to provide statistically significant protection in five human syndromes of vascular occlusion: prevention of stroke in patients with transient ischemic attacks (34), prevention of shunt occlusion in patients undergoing hemodialysis (35), prevention of death and myocardial infarction in patients with unstable angina (36), prevention of graft occlusion in patients undergoing coronary artery bypass (37), and prevention of accelerated atherosclerosis in cardiac graft recipients (14). In each of these studies the daily dose of aspirin has exceeded 320 mg. Studies of the human pharmacology of aspirin have shown that dose-related biochemical selectivity is relative rather than absolute. When administered chronically, doses less than 100 mg cause considerably less depression of prostacyclin biosynthesis coincident with maximal inhibition of thromboxane formation than do higher doses (38–40). However, some depression does occur, even with doses as low as 80 mg (39).

Such results were obtained using a variety of approaches to the measurement of prostacyclin and thromboxane A_2 biosynthesis (23). These include so-called tissue-specific, capacity-related indices and tissue nonspecific measurements of actual in vivo biosynthesis. Arachidonic acid metabolites are formed phasically in response to diverse chemical and traumatic stimuli. They are not stored within tissues following synthesis and although biologically potent, are extremely evanescent. In general, the capacity of tissues to form prostanoids greatly exceeds their actual biosynthetic rate in vivo. Because they are short lived, the primary compounds themselves are difficult to measure so assays generally have been directed towards more stable, inactive hydration products or enzymatically formed metabolites. Capacity-related indices employed in studies of aspirin effects have included the formation of thromboxane B_2 in serum, and release of 6-keto-PGF_{12} by vascular biopsies ex vivo (38,40). Studies on the effects of aspirin on in vivo prostanoid biosynthesis have relied on measurements of the urinary metabolites of these compounds (39).

Other aspects of the pharmacology of aspirin have been addressed elsewhere (41). However, with regard to patients undergoing angioplasty, it is important to bear in mind potential interactions with other antithrombotic medications. For example, aspirin may enhance the action of oral anticoagulants.

In summary, aspirin is a potent platelet inhibitory drug in vitro which acts by irreversibly inhibiting platelet cyclooxygenase and has been shown to be of benefit in certain platelet-mediated vascular occlusive episodes in man. A theoretical limitation of aspirin in doses in excess of 100 mg has been coincident depression of prostacyclin biosynthesis. Some evidence suggests that prostacyclin may be an important local regulator of platelet-vascular interactions in man. However, the comparative efficacies of ''low-'' and ''high-''dose aspirin have yet to be tested, so the only available evidence from clinical trials pertains to daily doses of 324 mg and above.

Dipyridamole

Dipyridamole (2,6-bis-diethylamino-4-8-dipiperidinopyrimido-5,4-d-pyrimidine) is a pyrimidopyrimidine derivative that has both vasodilating and antiplatelet properties. Although the increase in blood flow resulting from vasodilation may facilitate its antithrombotic effect in vivo, the antithrombotic effect of dipyridamole is independent of its hemodynamic action as it is demonstrable in rigid synthetic tubing models of thrombosis. Shortly after it was first administered to animals and man, it was recognized that dipyridamole retarded the breakdown of adenosine, a potent coronary vasodilator. More recently, the inhibitory effects of dipyridamole on platelet aggregation have prompted research into its involvement in the formation of prostacyclin and thromboxane. Dipyridamole inhibits platelet phosphodiesterase, the enzyme responsible for conversion of cyclic AMP to 5'-AMP. The platelet release reaction is inhibited by compounds which elevate platelet cAMP. The antiaggregatory effects of prostacyclin are also closely related to the stimulation of platelet adenylate cyclase and are enhanced by phosphodiesterase inhibitors. Theophylline, a weak inhibitor of phosphodiesterase, has been shown to enhance the disaggregatory action of prostacyclin. Moncada and Korbut (42) have suggested that dipyridamole and other phosphodiesterase inhibitors might act as antithrombotic agents by potentiating the antiplatelet effect of endogenous prostacyclin. However, no clear cut evidence that this actually occurs in vivo has yet been provided. Attempts to address the hypothesis have generally relied upon superfusion-bioassay techniques which are insufficiently precise to quantify alterations in prostacyclin biosynthesis. An additional confounding factor may have been the wide interindividual differences in systemic bioavailability of the Persantine brand of dipyridamole in humans (43).

Although dipyridamole inhibits ADP-induced aggregation and thromboxane synthesis at 50 μM or more, blood levels in vivo rarely exceed 10 μM. An additional mechanism of action has been suggested by recent studies of the ability of cultured endothelial and smooth muscle cells from bovine pulmonary artery to metabolize 8-^{14}C ADP (44). Cellular uptake of radioactivity was almost completely inhibited by 10 μM dipyridamole, and inosine (an adenosine metabolite) was largely replaced in the medium by adenosine. This was accompanied by an increased antiaggregatory activity of the conditioned medium that could be matched by authentic adenosine at the same concentrations. Thus, dipyridamole may inhibit platelet aggregation by increasing local concentrations of adenosine via inhibition of erythrocyte and platelet uptake mechanisms.

Some controversy also pertains to the antithrombotic efficacy of dipyridamole in vivo. Although dipyridamole has been shown to be effective in some animal models of thrombosis, the evidence for its efficacy in man is less clear cut. Oral dipyridamole prolongs platelet survival in patients with increased platelet consumption in association with arterial thromboembolism, arteriovenous prosthetic cannulas, vasculitis, and prosthetic aortic grafts (45). However, dipyridamole does not significantly alter the combined consumption of platelets and fibrinogen in patients with venous thrombosis. Dipyridamole may, in high doses, inhibit thrombus formation in extracorporeal shunts (46). In situations where the benefit from dipyridamole therapy seems less equivocal, as in the preservation of coronary bypass graft patency, it has been combined, empirically, with aspirin. Little evidence is available which favorably compares dipyridamole plus aspirin with aspirin alone and studies which relate plasma concentration to drug effect in thrombotic syndromes are lacking. In volunteer studies, the antiplatelet effects of dipyridamole are often difficult to demonstrate. Since the steady state levels of dipyridamole vary widely between individuals after oral dosing, empiric combination with aspirin, recently within the same capsule, may be somewhat premature.

Dipyridamole is usually unassociated with adverse effects at normal therapeutic doses. Gastrointestinal upset accompanied by nausea, vomiting, and diarrhea occurs occasionally as do rashes, headache, and vertigo. Excessive doses produce hypotension. A "coronary steal" phenomenon was suggested as the mechanism by which acute myocardial ischemia was provoked by intravenous dipyridamole in three patients with coronary artery disease (47). Acute psychiatric disturbance has been reported in two patients with acute myocardial infarction who received dipyridamole. A recent preliminary report suggests that dipyridamole, despite an antiplatelet effect, may enhance atherosclerotic plaque formation in rabbits fed an atherogenic (48) diet whereas treatment with aspirin plus dipyridamole reduced lipid accumulation in vein bypass graft segments in hyperlipidemic monkeys.

Sulfinpyrazone

Sulfinpyrazone [1,2,diphenyl-4-(2-phenylsulfinylethyl)-3,5-pyrazolidinedione] is a phenylbutazone analog. Like aspirin, it inhibits the cyclooxygenase enzyme and alters prostanoid formation. However, sulfinpyrazone differs from aspirin in several important features. Whereas aspirin irreversibly inhibits cyclooxygenase for the lifetime of the platelet, sulfinpyrazone does so reversibly. Thus, compared to aspirin, platelet thromboxane formation rapidly recovers following oral dosing. Second, sulfinpyrazone is a much less potent inhibitor of platelet function, at least ex vivo, than aspirin. Third, in the doses employed in clinical studies, sulfinpyrazone is much less selective, with regard to prostacyclin biosynthesis, than is aspirin. For example, chronic administration of sulfinpyrazone 800 mg/day for 7 days leads to a mean 63 percent reduction of serum thromboxane B_2 and a mean 45 percent reduction of prostacyclin metabolite (PGI-M) excretion. Aspirin 20 mg/day for 7 days reduces serum thromboxane B_2 by 95 percent and PGI-M excretion by only 15 percent (35,49). An interesting feature of the pharmacology of sulfinpyrazone is that its platelet inhibitory and uricosuric effects may be largely mediated in vivo by the sulfide and sulfone metabolites rather than by the parent compound in vivo.

In the experimental homocystinemic baboon, daily sulfinpyrazone administration effectively prevents the development of atherosclerotic lesions. Also, hyperlipidemic monkeys treated with sulfinpyrazone have shown markedly reduced endothelial lesion formation compared with hyperlipidemic, untreated controls. Clopath et al. (49) tested the possible effect of sulfinpyrazone on atherogenesis in the rat and in miniature swine. The rats were treated for nearly 8 weeks with sulfinpyrazone at a dose of 160 mg/kg/day PO. The atherosclerotic lesions were produced in the aorta by immunological injury and cholesterol feeding. None of the treated animals showed any aortic lesions, although five of nine control rats showed early intimal changes. The drug did not alter the serum cholesterol levels. In the miniature swine, the abdominal aorta was repeatedly injured with a balloon catheter during oral treatment with 30 mg/kg sulfinpyrazone twice daily. The drug exerted no protective effect against the intimal lesions that had been produced, nor did it have any effect on the vascular concentrations of collagen, elastin, or cholesterol. The effect of sulfinpyrazone on intimal lesions in rabbits more closely resembled that observed in rats (50). After treatment with either sulfinpyrazone, aspirin, or dipyridamole, the iliac artery of the rabbits was injured using a balloon catheter. Sulfinpyrazone was very effective in reducing the volume of the neointima, but neither aspirin nor dipyridamole showed any such effect.

Some evidence has been provided to suggest that sulfinpyrazone possesses a variety of additional pharmacological actions. These include antiarrhythmic, fibrinolytic, and free radical scavenging properties. It is also said to protect endothelium from the effects of toxic insults and to reduce the humoral immune response to antigen challenge. The evidence in support of these properties is, however, tenuous and their relevance to the antithrombotic action of the drug in vivo has not been established. Sulfinpyrazone has been studied in the secondary prevention of myocardial infarction in large scale clinical trials (51,52). Although some benefit was apparent, the interpretation of these data is seriously constrained by problems of trial design. Sulfinpyrazone may potentiate the action of coumarin-type anticoagulants, probably due to its high degree of protein binding.

Like aspirin, sulfinpyrazone is a gastric irritant and therefore should be avoided in patients with peptic ulcer. There is some indication that the ulcerative effect of sulfinpyrazone is weaker than that of aspirin and the nonsteroidal anti-inflammatory drugs. This may be attributable to the fact that sulfinpyrazone itself is also a weaker inhibitor of prostaglandin synthetase in the stomach than aspirin.

Generally, sulfinpyrazone is well tolerated in the therapeutic doses, even over protracted periods. Allergic reactions, episodes of transient renal failure and blood dyscrasias have been observed clinically, but the incidence of such complications is low.

In summary, sulfinpyrazone is a poorly selective, reversible cyclooxygenase inhibitor. Although it may possess other properties of relevance to the prevention of platelet mediated syndromes in vivo, this remains to be conclusively established. There seems little reason to choose sulfinpyrazone over aspirin in angioplasty patients.

Thromboxane Synthase Inhibitors and Thromboxane A_2 Endoperoxide Receptor Antagonists

Thromboxane synthesis inhibitors offer two theoretical advantages over cyclooxygenase inhibitors such as aspirin and sulfinpyrazone. First, they do not inhibit prostacyclin biosynthesis coincident with that of thromboxane. Second, prostacyclin biosynthesis may actually be increased by these compounds. Inhibition of platelet thromboxane formation may lead to accumulation of the prostaglandin endoperoxide intermediates, PGG_2 and PGH_2. Experiments in vitro have provided evidence that with optimal proportions of platelets and endothelial cells, the platelets can actually donate the endoperoxide precursor to endothelial prostacyclin synthetase and thereby increase generation of prostacyclin (53). Although such a pharmacological profile would seem advantageous, the results available from clinical trials of the first of these compounds studied in man have not been particularly encouraging. Several factors may have contributed to these results. First, excessive generation of thromboxane A_2 may not have been of major importance in the syndromes and/or in the particular patients under study. In most of the investigations the patients were gathered loosely under a diagnostic tag without an attempt to obtain evidence of platelet activation in vivo prior to randomization to drug therapy. Second, there is the theoretical possibility that the accumulated prostaglandin endoperoxides substituted for the proaggregatory actions of thromboxane A_2. In some volunteers administered thromboxane synthase inhibitors, platelet aggregation can still be readily induced ex vivo despite complete inhibition of thromboxane formation. Inhibitors of the shared thromboxane A_2/ endoperoxide receptor inhibit aggregation in these individuals, suggesting a role for the endoperoxides (55). Finally, these drugs may be insufficiently potent or action insufficiently prolonged. Certainly, in the doses currently employed, their effects on platelet function in "responders" are less marked than aspirin (54,55).

The importance of an endoperoxide shunt in the action of these drugs remains unestablished. Urinary PGI-M excretion is increased by some of these compounds, but to a minor extent (54–56). Whether this reflects an increase in vascular synthesis or endoperoxide rediversion within some organ capable of making both prostacyclin and thromboxane A_2 is unclear. Certainly, the increment in prostacyclin generation is insufficient for it to have a systemic rather than a local effect on platelet-vascular function (32). These drugs have not yet been evaluated either postangioplasty or in the preservation of bypass graft patency. It seems likely that demonstration of efficacy in these settings will await the development of a subsequent, more potent, generation of thromboxane synthase inhibitors.

A potentially more promising therapeutic strategy is offered by the development of thromboxane A_2/ endoperoxide receptor antagonists (57,58). These drugs could potentially inhibit thromboxane-mediated vasospasm and platelet aggregation. Potent agents have been tested in vitro and early studies of their efficacy in animal models in vivo look promising. A potential therapeutic combination of the future is that of a potent synthesis inhibitor with a receptor antagonist.

Heparin

Heparin is composed of a heterogenous group of straight chain acidic mucopolysaccharides with average molecular weights of about 15,000. It is contained in the mast cell granules of many mammalian tissues and is released during mast cell degranulation, but its physiologic function is unknown.

Intravenous injection of heparin has a number of pharmacological effects, the most prominent being the stimulation of a rapid clearance of triglycerides from the plasma and an impairment of blood coagulation. The disappearance of heparin anticoagulant activity from the blood follows apparent first-order kinetics, but in a dose-dependent manner with t 1/2 ranging from 1 to 3 hours. Heparin is largely metabolized in the liver but either hepatic or renal dysfunction will significantly prolong its half-life. Because of its large molecular weight and strongly ionic character, it crosses membranes poorly and cannot be absorbed from the gastrointestinal tract, transferred across the placenta, or secreted into maternal milk.

The anticoagulant activity of heparin is mediated via its interaction with a particular plasma α_2 globulin, antithrombin III. This protein is one of several proteinase inhibitors in the plasma, and it acts to inhibit the activated forms of clotting factors II, IX, X, XI, XII, and XIII. When antithrombin III interacts with thrombin it forms an irreversible complex which results in the inactivation of both proteins. Heparin markedly increases the rapidity of this reaction by participating in the formation of a ternary complex between itself, antithrombin III, and thrombin. The presence of low concentrations of heparin seems to stimulate the activity of antithrombin III against thrombin and factor X_a, allowing the therapeutic use of "low-dose" heparin given subcutaneously.

Since heparin stimulates the formation of thrombin–antithrombin III complexes, its use results in a decrease in circulating antithrombin III activity. Such a reduction in antithrombin III may result in a paradoxical heparin-induced increase in the tendency to thrombosis in vivo. Thrombosis resulting from heparin therapy can also be found in patients who become thrombocytopenic during its administration. An appreciable percentage of patients treated with intravenous heparin for over a week will develop significant thrombocytopenia with evidence of increased peripheral platelet consumption. While the role of antiheparin antibodies in this phenomenon is controversial, platelet aggregation and the release reaction are apparently stimulated and can enhance the possibilities for thrombosis.

Most patients undergoing coronary angioplasty will be treated with heparin for only a short time, but one must keep the possibility of heparin-induced platelet alterations in mind, in addition to the more common hemorrhagic complications of its use. Coagulation abnormalities resulting from an excess effect of heparin are rapidly reversed by the intravenous administration of protamine sulfate, but this is not true for the effects of heparin on platelet aggregation.

Oral Anticoagulants

Oral anticoagulants are antagonists of vitamin K and inhibit blood clotting by reducing the hepatic synthesis of the vitamin K–dependent coagulation factors II, VII, IX, and X. The most widely used of these agents is sodium warfarin, which serves as the prototype of this drug class. The therapeutic effect of warfarin is delayed 8–12 hours after oral administration because of the time lag to a significant decrease in the circulating levels of the active forms of the clotting factors. Larger initial doses of oral anticoagulants only hasten the onset of hypoprothrombinemia to a limited extent, since the rate of degradation of active circulating prothrombin is not altered. There is a 1–3-day delay between a peak plasma concentration of warfarin and its maximum affect on circulating prothrombin levels.

A number of factors are able to decrease or potentiate the activity of oral vitamin K antagonists.

Interactions with other drugs are frequent, with altered plasma protein binding as well as altered metabolism having been described. Increased metabolism of warfarin with decreased anticoagulant effect can be caused by barbiturates, rifampin, or other inducers of drug-metabolizing enzymes. Drugs which inhibit warfarin metabolism and thus prolong its half-life and effect include metronidazole, disulfiram, and cimetidine.

Since vitamin K is obtained via both dietary intake and production by intestinal bacteria, factors decreasing availability from either of these sources can result in a lower requirement for warfarin needed to achieve a desired level of anticoagulation. Clinically, it usually requires a combination of lowered vitamin K intake plus antibiotic suppression of intestinal flora to significantly increase the sensitivity to warfarin anticoagulation.

Altered production or metabolism of clotting factors can also alter the required dose of warfarin. Increased catabolism of clotting factors is seen in hypermetabolic states such as hyperthyroidism. Hepatic disease, including passive congestion due to cardiac failure, results in decreased synthesis of clotting factors and increased sensitivity to warfarin.

During pregnancy, the increased activity of maternal clotting factors results in resistance to the actions of oral anticoagulants. The fetus, however, is very sensitive to these drugs because of its limited capacity to produce clotting factors and the fact that warfarin readily crosses the placenta. It is recommended, therefore, that heparin (which does not cross the placenta) be used rather than warfarin during pregnancy.

The concomitant use of antiplatelet agents and warfarin-type drugs on a chronic basis is somewhat hazardous and usually cannot be recommended. Hemostasis appears to be adequately maintained when either platelet function or the coagulation cascade is defective, but impairment of both functions has often resulted in catastrophic hemorrhage. Many antiplatelet agents also predispose to gastrointestinal bleeding, and some such as phenylbutazone, also potentiate the action of warfarin by displacing it from plasma proteins and thus increasing the concentration of free drug in the plasma.

ANTITHROMBOTIC DRUGS AND ANGIOPLASTY

The use of anticoagulant and antiplatelet drugs in association with angioplasty has evolved empirically. A variety of regimens have been employed, generally involving heparin administration at the time of the procedure followed by oral anticoagulants and/or aspirin (59). However, only sparse data on the natural history of vascular patency post angioplasty are available as no placebo controlled trials of any antithrombotic interventions in this setting have been performed. In part, this has reflected the rapid evolvement of angioplasty techniques which would have constrained the performance of any prolonged prospective trial. However it has now created a situation where "ethical" objections to placebo-controlled trials may be based on widespread empirical practice. Very few blinded comparisons of antithrombotic strategies have been performed. Zeitler (60) reported a double-blind comparison of patients who receive aspirin 500 mg tid prior to and after coronary angioplasty either with (n=90) or without (n = 87) 5000 IU heparin at the time of the procedure. Both the acute thrombosis rate (7.0 vs 4.6 percent) and the short-term postprocedure occlusion rates (12.5 vs 7 percent) were slightly lower in the group which did not receive heparin. More recently, Thornton et al. (61) studied 24 patients who received heparin during angioplasty and were then randomized to either 325 mg aspirin or to coumadin sufficient to maintain a prothrombin time 2–2.5 × control. Both groups received 650 mg aspirin the day before angioplasty and the "coumadin" group continued to receive 325 mg aspirin daily until their prothrombin time had been satisfactorily prolonged. All patients were followed for at least 9 months. Restenosis occurred in 36 percent of the coumadin-treated patients and in 27 percent of the aspirin-treated group, a difference which did not attain statistical significance at the 5 percent level. There was a slightly lower restenosis rate with aspirin in patients who had at least a 6 month history of angina.

These data suggest that long-term coumadin treatment is no better than aspirin in this setting and raise two important questions. First, is either treatment better than placebo? From an ethical standpoint we feel that this is a reasonable question to address. Both anticoagulant and antiplatelet therapy are associated with documented morbidity, so it is difficult to justify exposure to them without establishing a resulting benefit. Additionally, our lack of

understanding of the atherosclerotic and the reparative processes which occur postangioplasty does not allow us to predict the eventual effects of such intervention. The reluctance to perform a placebo-controlled study at this stage in evolution of transluminal angioplasty might be addressed by a sequential trial design (62) which would permit early detection of an increased occlusion rate in either group. However, in practice, it is unlikely that a placebo controlled study will ever be performed in angioplasty. Should aspirin therapy be shown to significantly reduce occlusion rates a second issue is raised; what dosage or formulation optimizes the antithrombotic efficacy of this drug. Doses of aspirin less than 100 mg/day are relatively selective with respect to preservation of prostacyclin biosynthesis while inhibiting thromboxane formation. If inhibition of thromboxane formation is the primary mechanism by which aspirin is of benefit to these patients, then lower doses would be equally effective but less likely to cause side effects which might adversely influence compliance. In this regard the early reports of a controlled study in which aspirin 100 mg/day significantly reduced coronary graft occlusion are relevant. Alternatively, specific antagonists of the shared thromboxane A₂/ endoperoxide receptor alone or perhaps in combination with thromboxane synthase inhibitors deserve evaluation as adjuvant therapy postangioplasty.

For the present, we would recommend continuation of careful heparin use during the angioplasty procedure and the addition of aspirin 325 mg/day based on the controlled data of Thornton et al. (61), as it seemed possibly superior to coumadin in patients with a prolonged history of angina. Side effects tend to be more serious during anticoagulant therapy and more patients are noncompliant in taking coumadin than antiplatelet drugs. We would also recommend the performance of a randomized, double-blind evaluation of both high- and low-dose aspirin therapy. This would provide information on a dose of aspirin which yields the best long-term angioplasty results as well as a base for subsequent evaluation of more antithrombotic strategies (e.g., thrombin receptor and thromboxane receptor antagonists) as they become available.

REFERENCES

1. Dotter CT, Judkins MP: Transluminal treatment of arteriosclerotic obstruction. Circulation 30:654–670, 1964.
2. Millan VG, Mast WE, Madias NE: Non-surgical treatment of severe hypertension due to renal-artery internal fibroplasibay percutaneous transluminal angioplasty. NEJM 300:1371–3, 1979.
3. Sperling DR, Dorsey TJ, Rowen M, Gazzaniga AB: Percutaneous transluminal angioplasty of congenital coarctation of the aorta. Am J Cardiol. 51:562–564, 1983.
4. Waller BF, McManus BM, Gardinkle HJ, et al: Status of the major epicardial coronary arteries 80 to 150

days after percutaneous transluminal coronary angioplasty. Am J Cardiol 51:81–84, 1983.

5. Hollman J, Austin GE, Gruentzig AR, et al: Coronary artery spasm at the site of angioplasty in the first 2 months after successful percutaneous transluminal coronary angioplasty. J Am Coll Cardiol 6:1039–1045, 1983.

6. Saunders RN: Evaluation of platelet-inhibitory drugs in models of atherosclerosis. Ann Rev Pharmacol Toxicol 22:279–295, 1982.

7. Thomas WA, Kim DN: Atherosclerosis as a hyperplastic and/or neoplastic process. Lab Invest 48:245–255, 1983.

8. Niewiarowski S, Rao AK: Contribution of thrombogenic factors to the pathogenesis of atherosclerosis. Prog Cardiovasc Dis 26:197–22, 1983.

9. Moore A, Friedman RJ, Singal DP, et al: Inhibition of injury-induced thromboatherosclerotic lesions by anti-platelet serum in rabbbits. Thromb Diath Hemorrh 35:70–81, 1976.

10. Harker LA, Ross R, Slichter SJ, et al: Homocystine-induced atherosclerosis: The role of endothelial cell injury and platelet response in its genesis. J Clin Invest 58:731–741, 1976.

11. Pirk R, Chediak J, Glick G: Aspirin inhibits development of coronary atherosclerosis in cynomolgus monkeys (macacca fasicularis) fed on atherogenic diet. J Clin Invest 63:158–162, 1979.

12. Fuster V, Bowie EJ, Lewis JC, et al: Resistance to arteriosclerosis in pigs with von Willebrand's disease: spontaneous and high cholesterol diet induced arteriosclerosis. J Clin Invest 61:722–730, 1978.

13. Griggs TR, Reddick RL, Sultzer D, et al: Susceptibility to atherosclerosis in aortas and coronary arteries of swine with von Willebrand's disease. Am J Pathol 102:137–145, 1981.

14. Griepp RB, Stinson EB, Shumway NE, et al: Control of graft atherosclerosis in human heart transplant recipients. Surgery 81:262–269, 1977.

15. Block PC, Fallon JT, Elmer D: Experimental angioplasty: Lessons from the laboratory. Am J Radiol 135:907–912, 1980.

16. Pasternak RC, Baughman KL, Fallon JT, Block PC: Scanning electron microscopy after coronary transluminal angioplasty of normal canine coronary arteries. Am J Cardiol 45:591–598, 1980.

17. Piper PJ, Vane JR: Release of additional factors in anaphylaxis and its antagonism by antiinflammatory drugs. Nature (London) 233:35–39, 1969.

18. Al-Mondhiry H, Marcus AJ, Spaet TH: On the mechanism of platelet function inhibition by acetysalicylic acid. Proc Soc Exp Biol Med 133:632–636, 1970.

19. Roth GJ, Majerus PW: The mechanism of the effect of aspirin on human platelets. I. Acetylation of a particulate fraction protein. J Clin Invest 56:624–632, 1975.

20. Roth GJ, Stanford N, Majerus PW: Acetylation of prostaglandin synthetase by aspirin. Proc Natl Acad Sci (USA) 72:3073–3076, 1978.

21. Roth GJ, Soik CJ: Acetylation of the NH_2-terminal serine of prostaglandin synthase by aspirin. J Biol Chem 253:3782–3784, 1978.

22. Weiss HJ: Antiplatelet therapy. NEJM 298:1344–1347, 1978.

23. FitzGerald GA, Pedersen AK, Patrono C: Analysis of prostacyclin and thromboxane biosynthesis in cardiovascular disease. Circulation 67:1174–1177, 1983.

24. Burch JW, Stanford N, Majerus PW: Inhibition of platelet prostaglandin synthetase by oral aspirin. J Clin Invest 61:314–319, 1978.

25. Bunting SR, Gryglewski R, Moncada S, Vane JR: Arterial walls generate from prostaglandin endoperoxides a substance (prostaglandin x) which relaxes strips of mesenteric and coeliac arteries and inhibits plate aggregation. Prostaglandins 12:897–915, 1976.

26. Burch JW, Baenziger NL, Stanford N, Majerus PW: Sensitivity of fatty acid cyclooxygenase from human aorta to acetylation by aspirin. Proc Natl Acad Sci (USA) 75:5181–5184, 1978.

27. Baenziger NI, Dillender MJ, Majerus PW: Cultured human skin fibroblasts and arterial cells produce a labile platelet inhibitory prostaglandin. Biochem Biophys Res Commun 78:294–301, 1977.

28. Jaffe EA, Weksler BB: Recovery of endothelial cell prostacyclin production after inhibition by low doses of aspirin. J Clin Invest 63:532–535, 1979.

29. Czervionke RL, Smith JB, Fry GL, Hoak JC, Haycroft DL: Inhibition of prostacyclin by treatment of endothelium with aspirin. J Clin Invest 63:1089–1092, 1979.

30. Reyers I, Mussoni L, Donati MB, de Gaetano G: Failure of aspirin at different doses to modify experimental thrombosis in rats. Thromb Res 18:669–674, 1980.

31. Kelton JG, Hirsh J, Carter GJ, Buchanan MR: Thrombogenic effect of high-dose aspirin in rabbits. J Clin Invest 62:892–895, 1978.

32. FitzGerald GA, Brash AR, Falardeau P, Oates JA: Estimated rate of prostacyclin secretion into the circulation of normal man. J Clin Invest. 68:1272–1275, 1981.

33. FitzGerald GA, Smith B, Pedersen AK, Brash AR: Prostacyclin biosynthesis is increased in patients with severe atherosclerosis and platelet activation. NEJM 310:1065–1068, 1984.

34. The American-Canadian Cooperative Study Group. Persantine aspirin trial in cerebral ischemia. Stroke 14:99–103, 1983.

35. Jarter JR, Burch JW, Majerus PW, Stanford N, Delmez JA, Anderson CB, Weerts CA: Prevention of thrombosis in patients on hemodialysis by low-dose aspirin. NEJM 301:577–579, 1979.

36. Lewis HD Jr, Davis JW, Archibald DG, et al: Protective effects of aspirin against acute myocardial infarction and death in men with unstable angina: results of a Veterans Administration Cooperative Study. NEJM 309:396–403, 1983.

37. Chesebro JH, Clements IP, Fuster V, et al: A platelet-inhibitor-drug trial in coronary-artery bypass operations: benefit of perioperative dipyridamole and aspirin therapy on early postoperative vein-graft patency. NEJM 307:73–78, 1982.

38. Patrignani P, Filabozzi P, Patrono C: Selective cumulative inhibition of platelet thromboxane produc-

tion by low-dose aspirin in healthy subjects. J Clin Invest 63:532–535, 1979.

39. FitzGerald GA, Oates JA, Hawiger J, Maas RL, Roberts LJ II, Lawson J, Brash AR: Endogenous biosynthesis of prostacyclin and thromboxane and platelet function during chronic administration and aspirin in man. J Clin Invest 71:676–688, 1983.

40. Weksler BB, Pett SB, Alonso D, Richter RC, Telzer P, Subramranian V, Tack-Goldman K, Gay WA: Differential inhibition by aspirin of vascular and platelet prostaglandin synthesis in atherosclerotic patients. NEJM 308:800–805, 1983.

41. FitzGerald GA, Sherry S: The pharmacology and pharmacokinetics of platelet active drugs under current clinical investigation. Adv Prostaglandin Thromboxane Leukotriene Res 10:107–172, 1982.

42. Moncada S, Korbut R: Dipyridamole and other phosphodiesterase inhibitors act as antithrombotic agents by potentiating endogenous prostacyclin. Lancet 1:1286–1289, 1978.

43. Mahoney C, Wolfman KM, Cocchetto DM, Djornsson TD: Dipyridamole kinetics. Clin Pharmacol Ther 31:330–338, 1982.

44. Crutchly DJ, Ryan US, Ryan JW: Effects of aspirin and dipyridamole on the degranulation of adenosine diphosphate by cultured cells derived from bovine pulmonary artery. J Clin Invest 66:29–35, 1980.

45. Hamberg M, Samuelsson B: On the metabolism of prostaglandins E_1 and E_2 in the guinea pig. J Biol Chem 247:3495–3502, 1972.

46. Cucuianu Mp, Nishizawa EE, Mustard JF: Effect of pyrimidopyrimidine compounds on platelet function. J Lab Clin Med 77:958–974, 1971.

47. McGregor M, Fam WM: Regulation of coronary blood flow. Bull NY Acad Med 42:940–950, 1966.

48. Dembinska-Kiec A, Rucker W, Schonhofer PA: Effects of dipyridamole in experimental atherosclerosis. Action on PGI_2, platelet aggregation and atherosclerotic plaque formation. Stroke II:117, 1980.

49. Clopath P, Horsch AK, Dieterie W. Effect of sulfinpyrazone on development of an atherosclerosis in various animal models, in *Cardiovascular Actions of Sulfinpyrozone—Basic and Clinical Research*, M McGregor, JF Mustard, M Oliver, S Sherry (eds), Miami, Symposium Specialist Inc, 1979, pp 121–138.

50. Baumgartner HR, Studer A: Platelet factors and the proliferation of vascular smooth muscle cells, in *Atherosclerosis, Proceedings of the Fourth International*

Symposium Atherosclerosis, Tokyo, 1976, G Gots, Y Hata, Y Klose (eds), 1977, Berlin, Springer-Verlag, pp 605–609.

51. The Anturane Reinfarction Trial Research Group: Sulfinpyrazone in the prevention of sudden death after myocardial infarction. NEJM 302:250–256, 1980.

52. Aspirin Myocardial Infarction Study Research Group: A randomized controlled trial of aspirin in persons recovered from myocardial infarction. JAMA 243:661–669, 1980.

53. Marcus AJ, Weksler RB, Jaffe EA, Broekman MJ: Synthesis of prostacyclin from platelet-derived endoperoxides by cultured human endothelial cells. J Clin Invest 66:979–986, 1980.

54. FitzGerald GA, Brash AR, Oates JA, Pedersen AK: Endogenous prostacyclin biosynthesis during selective inhibition of thromboxane synthase in man. J Clin Invest 72:676–688, 1983.

55. FitzGerald GA, Oates JA: Selective and nonselective inhibition of thromboxane formation. Clin Pharmacol Ther 35:633–640, 1984.

56. Reilly IAG, FitzGerald GA: Platelet function, prostanoids and combined inhibition of thromboxane synthase and cyclooxygenase: A double blind controlled study. Proc IXth Int Cong Pharmacol (London) 1984.

57. Jones RL, Wilson NH: Ligand binding to thromboxane receptors on human platelets: correlation with biological activity. Br J Pharmacol 79:953, 1983.

58. LeBreton GC, Venton DL, Enke SE, Halushka PV: 13-Azoprostanoic acid: a specific antagonist of the human blood platelet/endoperoxide receptors. Proc Natl Acad Sci (USA) 76:4697, 1979.

59. Murray PD, Garnic JD, Bettmann MA: Pharmacology of angioplasty and intravascular thrombolysis. AJR 139:795–803, 1982.

60. Zeitler E: Drug treatment before and after percutaneous transluminal recanalization, in *Percutaneous Vascular Recanalization. Technique Application Clinical Results*, E Zeitler, A Gruentzig, W Schoop (eds), Berlin-Heidelberg-New York, Springer-Verlag, 1978.

61. Thornton MA, Gruentzig AR, Hollman J, King SB, Douglas JS: Coumadin and aspirin in prevention of recurrence after transluminal coronary angioplasty: A randomized study. Circulation 69:721–727, 1984.

62. Armitage P: *Sequential Medical Trials*, Oxford, Biochemical Scientific Publishers, 1975.

23

Pathophysiologic Considerations of Transluminal Coronary Angioplasty

DAVID P. FAXON, M.D.

INTRODUCTION

In 1964, Charles Dotter and Melvin Judkins described the nonoperative technique of transluminal angioplasty to treat patients with severe femoral and popliteal atherosclerotic obstructions (1). Since this initial report, major advances in the development of new catheters have greatly increased its application. Perhaps the most notable is the development of a coaxial balloon catheter by Andreas Gruentzig in 1974 (2), allowing the technique to be greatly expanded to the renal and coronary arteries (3). Despite these developments, our understanding of the mechanisms by which angioplasty restores flow has remained rudimentary until recently. This chapter will explore what is known about the acute mechanisms as well as the chronic changes following angioplasty. Finally, directions for future study will be outlined.

ACUTE MECHANISMS OF DILATION

Initially, numerous mechanisms were suggested to explain how angioplasty restores flow. It was once accepted that remodeling of the atherosclerotic plaque is achieved by compressing the plaque mass into a smaller volume with release of its fluid contents (4).

Others raised concerns about possible embolization of the atherosclerotic debris (5). Recently, investigators have suggested that angioplasty may work by stretching the vessel wall or by splitting and fracturing the neointimal plaque (6). Examination of these potential mechanisms and clarification of the pathophysiology of angioplasty have come primarily from postmortem and animal studies (7).

Postmortem Studies

Remodeling and compression of the atheroma were initially suggested by the examination of the histopathology of iliac and femoral vessels of patients who underwent dilation. These studies revealed little intimal destruction, no evidence of dissection, and a larger internal lumen (4). However, these limited observations were made in patients dying weeks to months after angioplasty at a time when the healing process might have obscured any acute changes. Other studies have also supported the compression theory (5,7). Lee et al. reported minimal destruction of the arterial wall with compression of the neointimal plaque in a study of postmortem coronary vessels. Marked intimal tears were not noted in this study despite successful angioplasty (7). Kaltenbach et al. examined the effect of 5 kg/cm^2 pressure on fresh human postmortem aortic and iliac vessels (8). This weight was chosen since it closely simulates the pressure exerted by an angioplasty balloon. The authors concluded that the compressed vessel became thinner and contained less fluid within the vessel wall.

Additional postmortem studies of human atherosclerotic arteries by Castaneda-Zuniga et al. argued against the concept of compression (9). In their study angioplasty caused vessel stretching, cracking of the intima, and separation between intima and media. In other studies in normal dogs, these same researchers demonstrated no evidence of compression. Instead, they found angiographic and histologic evidence of stretching of the vessel wall. They hypothesized that the balloon dilation stretches and distends the intima, media, and, to some degree, the atheromatous material (9). Additional studies by Freudenberg, Baughman, and others have verified these findings by revealing various degrees of stretching, plaque injury, and dissection (10–14).

Animal Studies

The results of postmortem studies have not been uniformly accepted because of the variability in findings and the inherent differences between the intact atherosclerotic plaques in the viable tissue and the postmortem pathologic specimen. Because of these limitations, several investigators have examined the effects of balloon angioplasty on in vivo animal models. Studies in normal canine coronaries have shown that the balloon dilation of an artery causes minimal intimal damage but extensive endothelial desquamation (14–15). Using scanning electron microscopy, Pasternak et al. demonstrated a dense platelet adhesion in the areas of endothelial denudation following angioplasty (15). Although the deposition of platelets was not inhibited by heparin or aspirin, it was by dextran.

Again extrapolating these findings in normal arteries to changes occurring in atherosclerotic vessels is difficult and has prompted a number of investigators to adapt the rabbit model of atherosclerosis to study the mechanisms of angioplasty (16–18). Block et al. demonstrated that the rabbit model has the advantages of low cost and rapid development of stenotic lesions in the iliac vessels, which are comparable in size to human coronary arteries. He demonstrated localized desquamations and intimal plaque disruptions in the vessels dilated with balloon (16). One of the major limitations of the rabbit model is that the lesions are highly cellular and not typical of human atherosclerotic lesions.

In order to circumvent these objections, we developed three different atherosclerotic lesions in the rabbit model that were suitable for the study of angioplasty mechanism (17). Using the techniques of balloon de-endothelialization and an indwelling catheter, we developed three different histologic patterns of atherosclerosis. The first type of atherosclerotic lesion was highly cellular and lipid-laden. The second was fibrous in *histology* and eccentric in lesion configuration, and the third had a mixture of fibrous and cellular material and was eccentric in configuration. In each model, significant iliac narrowing could be demonstrated by angiography, and successful angioplasty was accomplished in each (Fig. 1). Histopathologic evaluation of these arterial segments disclosed a variety of vascular injury that correlated with the initial histopathologic types. In foam cell lesions, we observed stretching of the vessel wall with prominent neointimal fracture extending into the media (Fig. 2). Occasionally, dissections could also be seen. In eccentric fibrous or mixed lesions, stretching of the nondiseased portion of the vessel wall could be demonstrated with thinning and necrosis of the normal media (Fig. 3). Further studies from our institution have verified these observations and suggest that the increase in luminal diameter following angioplasty is primarily due to vessel wall stretching (19). Morphometric analysis of the dilated and nondilated segments has shown no evidence of compression of the atheroma (Fig. 4). When fracture of the neointima occurs, it appears to be related to the compliance of the atheroma, as well as the vessel wall, which is determined by the underlying histopathologic characteristics.

Studies from this rabbit model have also relieved

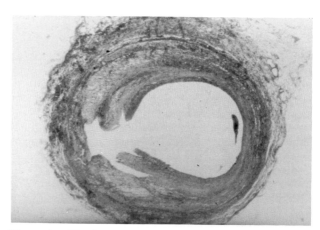

Figure 1. The iliac artery of a rabbit 24 hours after transluminal angioplasty. Gross neointimal fracture into the media is evident with a secondary inflammatory response in the media.

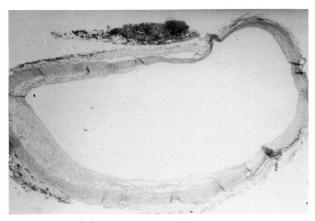

Figure 2. The effect of transluminal angioplasty on an eccentric atherosclerotic lesion of a rabbit. Stretching of the vessel wall, particularly the least diseased portion is evident.

fears that angioplasty may dislodge atherosclerotic material. Sanborn et al. (20) and Block et al. (21) separately reported that there is no evidence of embolization of atherosclerotic material occurring during or after the balloon angioplasty. Using the highly sensitive liquid chromatography, Sanborn also demonstrated that an evidence of cholesterol embolization could not be detected from the fluids collected downstream from the angioplasty site.

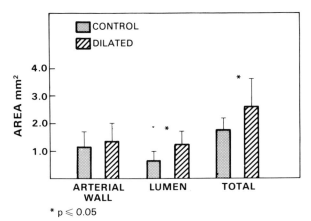

Figure 3. The results of morphometric analysis of histologic sections from dilated and control (nondilated) atherosclerotic iliac arteries of rabbits. The arterial wall area (intima and media) did not differ significantly between groups. However, the intimal lumen and the total vessel area were larger, implying that the vessel was stretched and that compression of the atherosclerotic arterial wall did not occur.

Human Studies

Extrapolations of these animal and postmortem studies to angioplasty in humans can be made only with caution since postmortem tissues and athero-

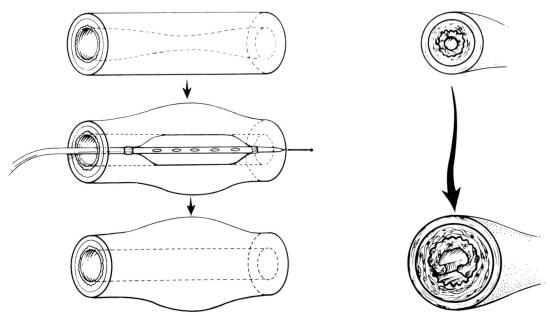

Figure 4. The mechanism of angioplasty appears to be vessel wall stretching without compression or embolization of the atheroma. Intimal fracture is a secondary process that occurs when the neointima is noncompliant.

sclerotic lesions found in animal models are unlike human atheromatous plaques and may not respond to angioplasty as in vivo human tissue does. Block addressed this problem by examining the histopathology of three patients who died 2 hours to 9 days after angioplasty (22). Histologic sections revealed intimal splits with a dissecting hematoma in one patient. Isner and Fortin have raised concerns about the validity of these observations (23). They examined coronary arteries from 70 patients in whom angioplasty was not performed and found intimal fracture that resembled the previously described pathologic findings.

Emerging Current Concept

Despite continued controversy, it seems that a reasonable concept of how angioplasty works can be formulated. On the basis of morphometric studies from our laboratory, the findings in the experimental animal models and autopsy studies can be reconciled. As previously reported (19) and shown in Fig. 5, angioplasty appears to work by stretching the vessel wall with a fusiform dilatation or localized aneurysm formation. Plaque fracture appears to depend upon the histopathology of the lesion, with the most compliant portion of the vessel wall stretching. If the lesion is eccentric, then the most nondiseased portion of the vessel wall will stretch. If the lesion is concentric, then the nonelastic atheroma may fracture and split, creating a variety of histopathologic changes. However, two features remain constant after balloon angioplasty: (1) desquamation of the endothelium and (2) stretching of the vessel wall.

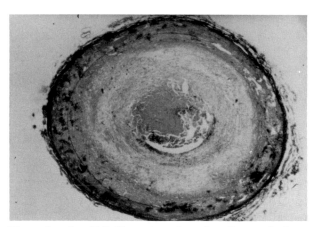

Figure 5. A rabbit iliac artery 4 weeks after angioplasty shows marked neointimal thickening with a thrombosis nearly totally occluding the vessel lumen.

LATE CHANGES FOLLOWING ANGIOPLASTY

Advancements in catheter technology and improving skills of the experienced angiographers have steadily raised the primary success rate of percutaneous coronary angioplasty. The initial group success rate of 65 percent has improved to greater than 90 percent in experienced hands (24) over the last couple of years. In contrast to improvements in the immediate results, the long-term success of coronary angioplasty has not been altered significantly. Although the vast majority of patients is clinically improved immediately after angioplasty, restenosis is now recognized as an important factor in limiting the long-term clinical benefit of the procedure. Follow up studies of coronary angioplasty have reported that restenosis occurs in as many as 25–35 percent of patients who have had initially successful procedures (24–25). The factors responsible for restenosis and methods of reducing this problem remain largely undefined.

Postmortem Studies

The information available from pathohistologic studies of the postmortem specimens is limited. In an autopsy study of three patients who died days to months after successful angioplasty, Waller et al. did not find any specific abnormalities of the arterial wall and could not readily demonstrate the site of angioplasty (26). However, despite these limited findings, histopathologic study on patients who have had documented restenosis has not been reported.

Animal Studies

On the basis of the available rabbit model, our group has devised a chronic model of atherosclerosis that has helped to provide insights into the mechanism of restenosis (27). We first developed atherosclerosis in both iliac vessels of 16 New Zealand white rabbits by means of balloon de-endothelialization and a 2 percent cholesterol diet for 6 weeks. Under an experimental protocol, 9 iliac arteries were balloon-dilated, and another 10 iliac arteries were chosen to serve as the undilated controls. Angioplasty resulted in a significant increase in luminal diameter in all dilated vessels. However, 4 weeks after successful angioplasty and continued atherogenic diet, restenosis was demonstrated in all of the dilated vessels. In contrast, vessels that served as controls showed no change in luminal diameter despite a con-

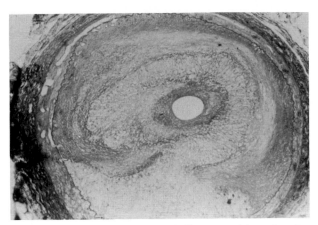

Figure 6. Restenosis of a rabbit iliac vessel 4 weeks after angioplasty. A pattern of fibrocellular tissue resembling the original neointimal tear is seen, with spaces filled in by abundant loose connective tissue severely narrowing the lumen.

tinued atherogenic diet. Histopathologic correlates showed several patterns.

Most prominent was presence of intraluminal clot in four animals, as well as multilaminated neointimal thickening, which occasionally filled the area of initial intimal damage (Figs. 6 and 7). These experimental findings support the hypothesis that both clot

formation and new atherosclerosis are mechanisms in the pathophysiology of restenosis.

Recently we have examined the role of antiplatelet therapy in the prevention of restenosis in this animal model (28–29). Twenty-five New Zealand white rabbits had iliac artery stenosis created by balloon de-endothelialization and high- (2 percent) cholesterol diet. All animals had a successful angioplasty of the stenosis in the iliac artery, resulting in an average increase in luminal diameter of 81 percent (Fig. 8). Nine animals served as controls, nine received sulfinpyrazone (100 mg per day), and seven received combination of aspirin (32 mg per day) and dipyridamole (25 mg per day). Four weeks later, *repeated* angiography revealed significant restenosis in all control animals, with an average decrease in luminal diameter of 64 percent. The animals on antiplatelet therapy, however, exhibited significantly less restenosis: only an 18 percent decrease in luminal diameter for the sulfinpyrazone group and a 32 percent decrease for the aspirin/dipyridamole group. Importantly, both drug regimens appeared to prevent clot formation, as well as reduce intimal proliferation (Figs. 9 and 10) under microscopic examination of the necropsy tissue. These results support the hypothesis that the interaction between platelet and vessel wall following intimal damage is an important factor in promoting restenosis. The mechanism by which these drugs prevent restenosis is uncertain. Neither regimen inhibits platelet adhesion to the neointima. However, both inhibit cyclooxygenase and may prevent aggregation and release of vasoactive substances and platelet growth

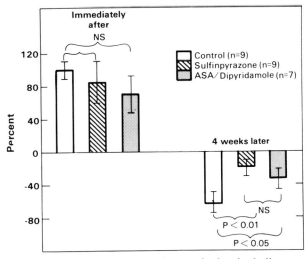

Figure 7. The percentage change in luminal diameter calculated from serial angiograms. Immediately after angioplasty, the luminal diameter increased and did not differ in dilated vessels of control and drug-treated animals. Four weeks later the luminal diameter was significantly narrower in all control animals; the drug-treated animals showed significantly less restenosis.

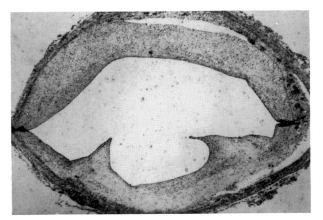

Figure 8. The effect of sulfinpyrazone 4 weeks after angioplasty. The neointimal flaps created by the angioplasty mechanism are still visible. No evidence of further neointimal thickening is evident, and intraluminal clot is absent.

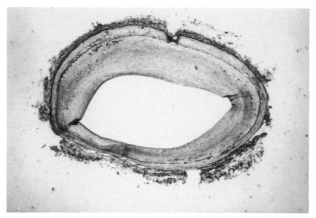

Figure 9. The effect of aspirin plus dipyridamole on restenosis. No intraluminal clot is seen; the neointima is less thick, with a widely patent lumen.

factor (30). The findings of our study closely parallel results of experimental and clinical trials of antiplatelet drugs in the prevention of intimal hyperplasia in coronary vein grafts (31–32). This is not surprising since both procedures result in significant denudation of vascular endothelium.

FUTURE DIRECTIONS

We have learned much about the possible mechanisms of angioplasty during the past few years; however, many questions remain unanswered. Is fracture or splitting of the neointima harmful, and could it lead to early or late vessel occlusion? In addition, can the vessel be stretched without causing vascular damage, and might this be a preferable consequence? Also, a great deal of uncertainty exists concerning the mechanical aspects of angioplasty. As described elsewhere, the balloon size, degree of stenosis, length of the lesion, and inherent tissue properties of the vessel wall may play important roles in a successful dilation (33). Which factors are most important, and how can we modify them in an individual case to reduce the restenosis rate? Although current practice is to use a variety of antiplatelet agents, anticoagulants, and calcium blocking agents before and after balloon dilation, the value of these individual agents is uncertain. Future studies are needed to define the optimal procedural factors and the best drug regimen to maximize the success of this procedure.

Of the many questions that remain unanswered, the mechanism of restenosis seems most important. Restenosis following angioplasty is not an unexpected event, given the response-to-injury hypothesis of atherosclerosis first introduced by Virchow and later modified by Ross and Glomset (34). Since angioplasty results in desquamation of the endothelial barrier, exposure of the subendothelium to circulating lipids might predispose vessels to further progression of atherosclerosis (35). The role of platelets in promoting further atherosclerosis and the mechanism by which smooth muscle cells proliferate to create further vascular occlusion must be studied (30). In addition, clot formation might result in acute or chronic occlusions, as well as promoting further atherosclerotic development. Finally, the

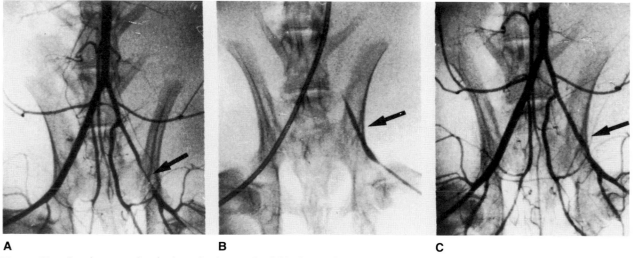

A   **B**   **C**

Figure 10. In vivo transluminal angioplasty of rabbit femoral artery stenosis. The artery stenosis is created experimentally (see text). **A.** Before. **B.** During angioplasty. **C.** Immediately after the procedure.

importance of both acute and latent vascular spasm should be examined, since clinical studies to date have suggested that patients with variant angina have a higher likelihood of restenosis following successful angioplasty (36).

Further study is necessary to define the individual importance of these factors and to understand their interrelations. Although we have learned a great deal about the acute mechanisms of angioplasty over the last few years, the long-term clinical benefits of this technique can be expanded only through careful and controlled experimental studies.

REFERENCES

1. Dotter CT, Judkins MP: Transluminal treatment of atherosclerotic obstruction: Description of a new technique and a preliminary report of its application. Circulation 30:654–670, 1964.
2. Gruentzig A: Die perkutane Rekanalisation chronischer arterieller Verschulesse mit einem neuen doppellumigen Dilatationskatheter (Dotter Prinzip). Fortschr Geb Roentgenstr Nuklearmed 124:80–86, 1976.
3. Kaltenbach M, Gruentzig A, Rentrop K, Bussman WD: Transluminal coronary angioplasty and intracoronary thrombolysis in *Coronary Heart Disease,* vol. 4. Berlin, Springer-Verlag, 1982.
4. Dotter CT, Rosch J, Judkins MP: Transluminal dilation of atherosclerotic stenosis, abstracted. Surg Gynecol 127:784–790, 1968.
5. Leu HJ, Gruentzig A: Histopathologic aspects of transluminal recanalization, in Zeitler E, Gruentzig A, Schoop W (eds.) *Percutaneous Vascular Recanalization.* Berlin, Springer-Verlag, 1978; pp 39–50.
6. U.S. Department of Health, Education, and Welfare: *Proceedings of the Workshop on Percutaneous Transluminal Coronary Angioplasty,* NIH Publication 80:2030, 1980.
7. Lee G, Ikeda RM, Joye JA, Bogren HG, DeMaria AN, Mason DT: Evaluation of transluminal angioplasty of chronic coronary artery stenosis. Value and limitations assessed in fresh human cadaver hearts. Circulation 61:77–83, 1980.
8. Kaltenbach M, Beyer J, Klepzig H, Schmidts L, Hubert: Effect of 5 kg/cm² pressure on atherosclerotic vessel wall segments, Kaltenbach M, Gruentzig A, Rentrop K, Bussman WD (eds): in *Transluminal Coronary Angioplasty and Coronary Thrombolysis in Coronary Heart Disease,* vol. 4. Berlin, Springer-Verlag, 1982, pp 189–193.
9. Casteneda-Zuniga WR, Formanek A, Tadavarthy M, Vlodavar Z, Edwards JE, Zollikofer C, Amplatz K: The mechanism of balloon angioplasty. Radiology 135:565–571, 1980.
10. Freudenberg H, Lichtlen PR, Engel HJ: Transluminal angioplasty of coronary arteries: An analysis of the most important complications by a post-mortal study in the human heart. Z Kardiol 70(1):39–44, 1981.
11. Baughman KL, Pasternak RC, Fallon JT, Block PC: Transluminal coronary angioplasty of postmortem human hearts. Am J Cardiol 48:1044–1047, 1981.
12. Simpson JB, Robert EW, Billingham ME, Myler R, Harrison DC: Coronary transluminal angioplasty in human cadaver hearts, abstracted. Circulation 58:II–80, 1978.
13. Zarins CK, Lu CT, Gewertz BL, Lyons RT, Rush BS, Glagove S: Arterial disruption and remodeling following balloon dilation. Surgery 92(6):1086–1095,1982.
14. Saffitz JE, Totty WG, McClennan BL, Gilula LA: Percutaneous transluminal angioplasty: Radiologic-pathological correlation. Radiology 141(3):651–654, 1981.
15. Pasternak RC, Baughman KL, Fallon JT, Block PC: Scanning electron microscopy after coronary transluminal angioplasty of normal canine coronary arteries. Am J Cardiol 45:591–597, 1980.
16. Block PC, Baughman KL, Pasternak RC, Fallon JT: Transluminal angioplasty: Correlation of morphologic and angiographic findings in an experimental model. Circulation 61:778–785, 1980.
17. Faxon DP, Weber VJ, Haudenschild CC, Gottsman SB, McGovern WA, Ryan TJ: Acute effects of transluminal angioplasty in three experimental models of atherosclerosis. Atherosclerosis 2:125–133, 1982.
18. Leveen RF, Wolf GL, Villanaeva TG: New rabbit model for the investigation of transluminal angioplasty. Invest Radiol 17(5):470–475, 1981.
19. Sanborn TA, Faxon DP, Haudenschild CC, Gottsman SB, Ryan TJ: The mechanism of transluminal angioplasty: Evidence for aneurysm formation in experimental atherosclerosis. Circulation 68:1136–1140, 1983.
20. Sanborn TA, Faxon DP, Waugh D, Small DM, Haudenschild CC, Gottsman SB, Ryan TJ: Transluminal angioplasty in experimental atherosclerosis: Analysis for embolization using an in vivo perfusion system. Circulation 66:917–922, 1982.
21. Block PC, Elmer D, Fallon JT: Release of atherosclerotic debris after transluminal angioplasty. Circulation 65:950–953, 1982.
22. Block PC, Myler RK, Stertzer S, Fallon JT: Morphology after transluminal angioplasty in human beings. N Engl J Med 382–386, 1981.
23. Isner JM, Fortin RV: Frequency in nonangioplasty patients of morphologic findings reported in coronary arteries treated with transluminal angioplasty. Am J Cardiol 51:683–703, 1983.
24. Gruentzig AR, Meier B: Percutaneous transluminal coronary angioplasty: The first five years and the future. Int J Cardiol 2:319–321, 1983.
25. Jutzy KR, Berte LE, Alderman EL, Ratts J, Simpson JB: Coronary restenosis rates in a consecutive patient series one year post successful angioplasty, abstracted. Circulation 66:II–331, 1982.
26. Waller BF, McManus BM, Garfinkel JH, Kishel JC, Scheidt EC, Kent KM, Roberts WC: Status of the

major epicardial coronary arteries 80 to 150 days after percutaneous transluminal coronary angioplasty: Analysis of three necropsy patients. Am J Cardiol 51(1):81–84, 1983.

27. Faxon DP, Sanborn TA, Weber VJ, Haudenschild CC, Gottsman SB, McGovern WA, Ryan TJ: Restenosis following transluminal angioplasty in experimental atherosclerosis. Arteriosclerosis: 4:190–195, 1984.

28. Sanborn TA, Faxon DP, Haudenschild CC, Gottsman SB, Ryan TJ: Sulfinpyrazone inhibition after experimental angioplasty, abstracted. J Am Coll Cardiol 1(2):644, 1983.

29. Faxon DP, Sanborn TA, Haudenschild CC, Ryan TJ: Effect of antiplatelet drugs on restenosis following experimental angioplasty. Am J Cardiol 53:72C–76C, 1984.

30. Packham MA, Mustard JF: Pharmacology of platelet-affecting drugs. Circulation 62(5):26–41, 1980.

31. Metke MP, Lie JT, Fuster V, Josa M, Kaye MP: Reduction of intimal thickening in canine coronary bypass grafts with dipyridamole and aspirin. Am J Cardiol 43:1143–1148, 1979.

32. Chesebro JH, Clements IP, Fuster V, et al: A platelet-inhibitor drug trial in coronary-artery bypass operations: Benefit of perioperative dipyridamole and aspirin therapy on early postoperative vein-graft patency. N Engl J Med 307:74–77, 1982.

33. Abele JE: Balloon catheters and transluminal dilation: Technical considerations. AJR 135:901–906, 1980.

34. Ross R, Glomset JA: The pathogenesis of atherosclerosis. N Engl J Med 295:369–377, 420–425, 1976.

35. Guyton JR: Smooth muscle growth and cholesterol accumulation in atherogenesis, in Guyton AC, Hall JE (eds): *Cardiovascular Physiology*, vol. 4. Baltimore, University Park Press, 1982, pp 3–49.

36. David PR, Waters PD, Scholl JM, Crepeau J, Lesperance J, Thudson G, Bourassa MG: Percutaneous transluminal coronary angioplasty in patients with variant angina: Circulation 66:695–701, 1982.

24

Effects of Coronary Angioplasty on Coronary Circulatory Dynamics and Myocardial Perfusion

DAVID O. WILLIAMS, M.D.
HENRY GEWIRTZ, M.D.
ALBERT S. MOST, M.D.

INTRODUCTION

Coronary angioplasty is a mechanical solution for a mechanical problem. Carefully performed intracoronary balloon inflation disrupts obstructive atheroma and increases the cross-sectional area of the coronary artery lumen, thus improving blood transport through the vessel.

The degree to which coronary artery diameter, determined angiographically, is reduced has been the most commonly utilized index for quantitating the extent of atheromatous coronary obstruction and has served as the clinical "gold standard" in defining angioplasty success (1,2). Despite its universal acceptance, this potentially quantifiable index has certain limitations. First, the extent of coronary narrowing cannot be easily measured "on-line" during the balloon angioplasty procedure. Thus, by this technique, it is difficult to determine whether the procedure is successful prior to its completion. Second, the histological appearance of atheroma following balloon angioplasty indicates that the involved vessel segments are substantially traumatized (3,4). The thickened intima is split, and localized crescentic disruption of the media is commonly observed. Angiographically, areas of coronary stenosis that were sharply definable before angioplasty may become hazy and demonstrate nonopaque linearities, extension of contrast into the vessel wall, and widened segments with reduced opacity. Even if precise measuring devices are utilized, an accurate assessment of the extent of residual coronary narrowing is virtually impossible because of the ill-defined configuration of the coronary artery lumen that results from balloon angioplasty. Finally, a comparison between the extent of coronary stenosis, determined by quantitative angiography, to coronary blood flow has been performed. This investigation indicated that the severity of angiographic stenosis does not accurately reflect actual flow impairment (5).

The trans-stenotic pressure gradient is an alternative objective parameter that reflects lesion severity. Constant monitoring of the pressure gradient during the angioplasty procedure provides on-line quantitative data to assist in obtaining and defining success. Since the intracoronary pressure distal to the stenosis is measured by means of the dilation catheter itself, the results obtained from this method undoubtedly do not reflect the true trans-stenotic pressure gradient. As a consequence, the value of this method has been questioned.

Although measurement of intracoronary pressure gradients and the extent of angiographically derived coronary obstruction are important in performing the angioplasty procedure and assessing its outcome, neither is fully capable of assuring that the angioplasty procedure has achieved its ultimate goal, i.e., relief of myocardial ischemia. Thus, the objective of this report is to review other methods that have been used to determine whether coronary angioplasty can achieve its goal. A secondary objective is to investigate in greater detail the predictive value of the trans-stenotic pressure gradient as an indicator of successful outcome.

EFFECTS OF CORONARY OBSTRUCTION ON CIRCULATORY DYNAMICS

Can coronary angioplasty restore normal coronary circulatory dynamics and myocardial perfusion? To answer this fundamental and important question, the physiological effects of coronary stenosis on myocardial perfusion must be reviewed. Significant coronary artery atherosclerosis impairs myocardial perfusion by adding a resistance, usually fixed but with a potentially dynamic component. Physiological resistance of the coronary artery is responsible for coronary flow reserve and autoregulation of myocardial blood flow. The effects of a high grade, fixed coronary narrowing on subendocardial, subepicardial, and transmural myocardial blood flow at rest and during stress have been described (6,7). In our laboratory, we have determined the values of regional transmural blood flow in a group of awake pigs with a fixed 82 percent diameter stenosis located in the left anterior descending coronary artery. As illustrated in Figure 1, values of blood flow in myocardium served by a stenotic coronary artery at rest may be only minimally depressed. In this particular animal, basal values of myocardial blood flow in the stenotic zone were similar to values in a normal zone. Preservation of resting flow resulted from compensatory vasodilatation of coronary arterioles. Normal values of myocardial blood flow may be observed even in subendocardial layers where flow reserve is already physiologically reduced in

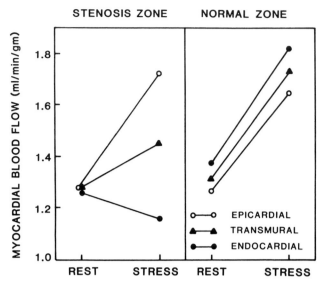

Figure 1. Response of myocardial blood flow in the awake pig with 82 percent artificial stenosis of the left anterior descending coronary artery. Values of blood flow at rest in the stenosis zone are similar to those within the normal zone. With pacing tachycardia stress, uniform increases in epicardial, transmural, and endocardial flow are observed within the normal zone. Disparate responses are observed in the stenosis zone. Epicardial flow increases to a similar extent as the normal zone. Endocardial flow, on the other hand, declines. The modest increase in transmural flow represents the net effect of the endocardial and epicardial responses.

order to compensate for the ventricular systolic compressive forces that limit blood flow in this region to only the diastolic phase of the cardiac cycle. These experimental results in the animal are concordant with the clinical observation that some patients with significant coronary artery narrowing often do not experience angina pectoris at rest or even with low level activity.

In this same animal model, values of myocardial blood flow were also determined after 10 minutes of pacing tachycardia stress. Significant differences are observed in the response of myocardial blood flow in the stenotic zone when compared with the normal zone. In the *normally* perfused zone, both the subendocardial and subepicardial flows were increased substantially during tachycardia stress. Since the extent of flow increase was similar in both subendocardial and subepicardial layers, there was no change in the endocardial-to-epicardial flow ratio, as demonstrated in Figure 2. In myocardium served by a *stenotic* coronary artery, however, total transmural flow demonstrated a small increase with pacing. The lack of significant change in total trans-

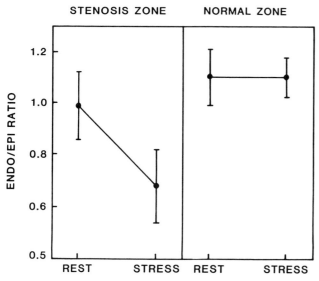

Figure 2. Change in endocardial-epicardial flow ratio (endo/epi) as a result of tachycardia stress in the presence of coronary stenosis. In the normal zone, endo/epi is slightly greater than 1.0 and remains unchanged during stress. In the stenosis zone, the endo/epi value is reduced at rest and declines further with stress (see text).

mural flow was due to the discordant changes in flow that occurred between subendocardial and subepicardial layers. Subepicardial flow increased significantly while subendocardial flow declined. These contrasting changes in regional transmural flow accounted for the very pronounced decline in endocardial-to-epicardial flow ratio.

The discordant change in regional transmural blood flow may be explained in terms of the relative amount of coronary vasodilatory reserve that remains in subepicardial and subendocardial layers. As noted previously, subendocardial flow reserve is partially reduced physiologically since all the subendocardial blood flow must occur during the diastolic phase of the cardiac cycle. No such physiological erosion of flow reserve is present in subepicardial layers since ventricular systolic compressive forces are minimal in that region.

Epicardial coronary narrowing results in an additional utilization of flow reserve, in order to maintain resting flow. This burden must be borne equally by both the subepicardial and subendocardial layers. This additional demand may fully exhaust the already diminished subendocardial flow reserve capacity while only partially depleting subepicardial reserve. Thus, pacing tachycardia stress, by further increasing myocardial oxygen requirement, imposes an additional demand upon the already depleted flow

reserve. Compensatory vasodilatation in this setting can only occur in subepicardial layer. The net decrease in vascular resistance in subepicardial layers, compared to subendocardial layers, results in shunting or stealing of blood from the inner to the outer layers of the myocardial wall. Hence, blood flow to subepicardial layers increases at the expense of the subendocardium, which becomes ischemic. These changes in transmural flow in the ischemic region are in accordance with the clinical observation that patients with chronic coronary artery disease may experience angina pectoris with stress exercise, and the electrocardiographic manifestation of ST segment depression indeed is an indicator of subendocardial ischemia.

EFFECTS OF PTCA ON CORONARY CIRCULATORY DYNAMICS

With these fundamental principles in mind, it is appropriate to review the various approaches and techniques to determine whether coronary angioplasty restores normal coronary circulatory dynamics and myocardial perfusion. Methods to determine the beneficial effects of coronary angioplasty have included direct and indirect means of assessing myocardial perfusion, both at exercise and during stress. Indirect methods have included exercise electrocardiography, thallium scintigraphy, and radionuclide assessment of global left ventricular function. Direct measurements include translesional pressure gradients, coronary blood flow, myocardial energetics and metabolism, and, more recently, coronary blood flow velocity. Despite the differences inherent in various methods to assess the efficacy of coronary angioplasty, the results uniformly indicate that coronary angioplasty has the potential of fully restoring normal coronary circulatory dynamics.

Normalization of the electrocardiographic response to exercise following successful coronary angioplasty was initially reported by Gruentzig in his original description of the technique (1). Subsequently, Scholl et al. described normalization of the electrocardiogram during exercise stress in 13 of 20 patients who had a "positive" stress test prior to the angioplasty procedure (8). Furthermore, the duration of exercise and the rate-pressure double product achieved were significantly increased. Such salutary results have also been observed during late follow-up. Williams et al. compared peak heart rate achieved and exercise duration among patients in whom angioplasty was successful with individuals in whom the procedure was unsuccessful who were treated with either coronary bypass surgery or continued medical therapy (9). At 1 year of follow-up, exercise duration and peak heart rate achieved were increased only in patients with successful angioplasty and coronary bypass surgery. In addition, Gruentzig described an increase in work capacity in patients who had successful dilation at a mean follow-up of 13 months, as determined by bicycle ergometry (10). Thus, coronary angioplasty can relieve electrocardiographic evidence of myocardial ischemia and improve exercise tolerance. Furthermore, these salutary effects may be sustained during late follow-up.

Thallium 201 scintigraphy, performed both during exercise and at rest, has also been employed to assess further the efficacy of transluminal angioplasty. Hirzel et al. performed thallium imaging before and following coronary angioplasty in 33 patients in whom the procedure had been considered angiographically successful (11). Before angioplasty, all but three patients demonstrated an exercise-induced scintigraphic defect. The location of the defects correlated with the distribution of the diseased coronary artery targeted for angioplasty. Following angioplasty thallium 201 imaging revealed "no sustained" defects during exercise in those who had abnormal zones before successful angioplasties. Furthermore, quantitative assessment of the scintigraphic activity revealed a significant increase in counts in all but two patients. Subsequently, Scholl et al. confirmed these observations, reporting normalization of previously abnormal exercise thallium 201 scintigrams in 15 of 21 patients after successful angioplasty (8). Thus, data obtained from thallium scintigraphy, a noninvasive, indirect method of assessing regional myocardial perfusion, indicate that balloon angioplasty can correct the impaired myocardial perfusion resulting from coronary artery stenosis.

Kent et al. employed the radionuclide gated blood pool scan to assess left ventricular function in patients before and after percutaneous coronary angioplasty (12). Regional wall motion abnormalities of the left ventricle were observed during exercise in 94 percent of patients prior to coronary angioplasty. Global left ventricular ejection fraction did not increase in these patients during exercise. After successful angioplasty, however, regional left ventricular dysfunction during exercise was observed in only 8 percent of patients. Furthermore, global left ventricular ejection fraction increased significantly.

Direct assessments of the effects of successful balloon dilation upon coronary circulatory dynamics have been performed by Williams et al. (13,14). Coronary blood flow, myocardial lactate extraction, and oxygen consumption were determined at rest

and during tachycardia stress on the patients before and after the successful angioplasty.

Before dilation, rapid pacing resulted in the development of angina pectoris in six of seven patients who participated in the investigation. Lactate extraction, which was normal at rest (Fig. 3), declined in each patient, with three patients demonstrating lactate production. For the group, the mean value of lactate extraction fell from 28 ± 25 percent (mean ± 1 SD) to 3 ± 18 percent. Following an-

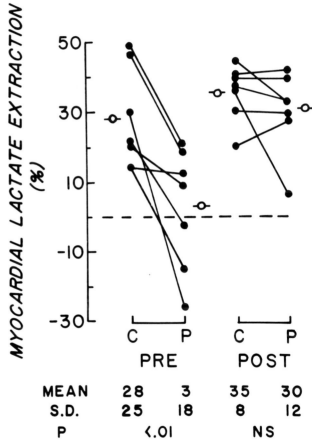

	PRE		POST	
	C	P	C	P
MEAN	28	3	35	30
S.D.	25	18	8	12
P		<.01		NS

Figure 3. Changes in myocardial lactate metabolism in patients before (PRE) and following (POST) successful PTCA. Values above the dashed line indicate myocardial lactate extraction; values below the line indicate lactate production. C = control resting heart rate; P = pacing tachycardia. Before PTCA, patients demonstrate normal values of myocardial lactate extraction. With pacing, each patient demonstrates a decrease in myocardial lactate extraction, with three patients exhibiting lactate production. Following PTCA, resting values of lactate extraction are again normal. Rapid pacing no longer results in lactate production or a significant decline in lactate extraction. (Reprinted from Ref. 14.)

gioplasty, resting values of lactate extraction were again in a normal range and did not differ from those prior to percutaneous angioplasty. Rapid atrial pacing in this same group of patients resulted in a decline of lactate extraction in only one patient; no change was observed in the mean value of lactate extraction for the group. Thus, these data indicated that before balloon angioplasty, tachycardia stress resulted in the development of anaerobic metabolism. After successful angioplasty, aerobic metabolism was preserved, even during tachycardia stress.

The changes in regional coronary blood flow are shown in Figure 4. Before angioplasty, pacing resulted in an increase in coronary blood flow in five patients and a decrease in two. After coronary dilation, no patient's blood flow declined during pacing, and each demonstrated a value of flow greater than that prior to successful dilation. For those patients in whom flow decreased before angioplasty, it is possible that vasodilatative reserve was fully exhausted to such a degree that stress resulted in a

"steal" of blood flow to noncompromised myocardial zones. Partial vasodilatory reserve was preserved in those patients capable of an attenuated flow increase. After angioplasty, reserve was restored, so that each patient now demonstrated a net increase in flow with stress. These observations indicate that successful angioplasty is capable of restoring coronary vasodilatory reserve, indicating that the functionally significant coronary stenosis has been relieved.

The effect of angioplasty on the time course of coronary blood flow response to abrupt tachycardia was also assessed in patients with and without coronary artery disease, before and after alpha blockade, and in patients undergoing successful angioplasty (15). This investigation was based on the observation that coronary blood flow increases more rapidly in response to abrupt pacing tachycardia in normal subjects than in patients with denervated hearts and normal subjects after alpha blockade (16). Fully innervated hearts maintain a basal level of alpha-adrenergic–mediated coronary vasoconstrictor tone. This tone may be promptly withdrawn in the presence of tachycardia stress to enhance coronary blood flow. In patients with coronary artery disease, it is quite likely that basal alpha-adrenergic constrictor tone or the peripheral vascular resistance would be chronically withdrawn in an attempt to compensate for the flow-limiting coronary stenosis. Alpha blockade in such individuals should have no effect on the time course of coronary blood flow response. Furthermore, if coronary angioplasty fully relieves significant coronary obstruction, then alpha tone might be restored in such a way that coronary blood flow would increase more rapidly after successful angioplasty.

Figure 5 demonstrates the changes in coronary blood flow over a 30-s time interval after the abrupt onset of pacing tachycardia in normal subjects before and after alpha blockade. At 5 s of rapid pacing, the value of blood flow was lower after alpha blockade than in the control state. There were no differences in flow at 10, 20, or 30 s. These observations are in agreement with the concept that basal alpha tone is present in patients with unobstructed coronary arteries and that this tone can be withdrawn to augment flow response to stress.

Figure 6 demonstrated the effect of abrupt pacing on the time course of coronary blood flow in patients with coronary artery disease before and after alpha blockade. Unlike that of the normal subjects, no difference in flow was observed 5 s into pacing after alpha-adrenergic blockade. This response indicates that basal alpha-adrenergic tone had been withdrawn, most likely as an attempt to compensate for flow-limiting stenosis. In Figure 7, changes in coro-

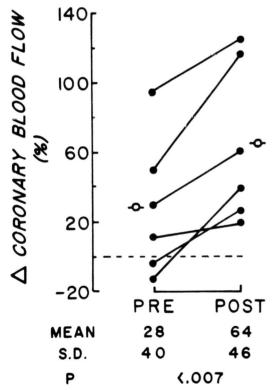

Figure 4. Changes in regional myocardial blood flow before (PRE) and following (POST) successful coronary angioplasty during atrial pacing tachycardia. Each patient demonstrates an increase in coronary blood flow response to tachycardia stress following successful PTCA. (Reprinted from Ref. 14.)

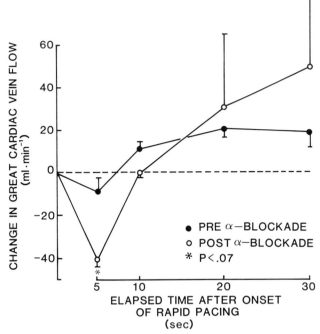

Figure 5. Changes in cardiac vein flow in patients with coronary artery disease following the abrupt onset of pacing tachycardia before (closed circles) and after (open circles) alpha blockade. Values above the dashed line indicate a net increase in flow; values below the dashed line indicate a net decrease when compared to control. Blood flow is significantly greater at 5 s of rapid pacing after alpha blockade. (Reprinted from Ref. 15.)

nary flow following abrupt tachycardia are observed in patients before and after coronary angioplasty. Peak flow following angioplasty occurred sooner. The response before angioplasty resembles that observed in normal subjects after alpha-adrenergic blockade and in patients with coronary artery disease not undergoing coronary angioplasty. The response after angioplasty resembles the flow pattern before alpha-adrenergic blockade in individuals with normal coronary arteries. These observations suggest a restoration of basal alpha-adrenergic tone after coronary angioplasty, another indication that flow-limiting effects of an epicardial coronary stenosis have been relieved.

VALUE OF THE TRANS-STENOTIC PRESSURE GRADIENT

Since angioplasty technology permits measurement of intracoronary pressures through a fluid-filled catheter lumen, it is feasible to utilize the difference

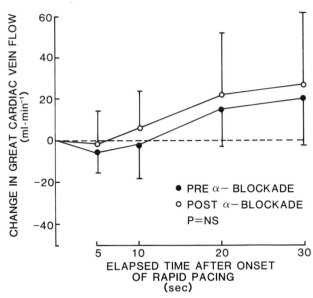

Figure 6. Changes in great cardiac vein flow following pacing tachycardia in patients with coronary artery disease before and after alpha-adrenergic blockade. No significant difference in flow is observed at any time interval, indicating that basal alpha-adrenergic tone has been withdrawn. (Reprinted from Ref. 15.)

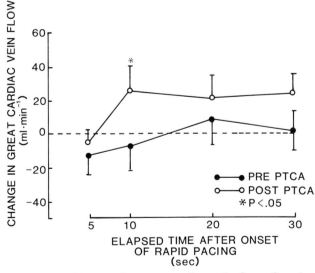

Figure 7. Changes in great cardiac vein flow after the abrupt onset of pacing tachycardia in patients before and after PTCA. After PTCA, greater values of flow are observed 10 s after the onset of rapid pacing. The more rapid attainment of peak flow implies a restoration of basal alpha-adrenergic tone following successful PTCA. (Reprinted from Ref. 15.)

between the intracoronary pressure proximal and distal to an area of narrowing as a method of determining the efficacy of coronary angioplasty. Indeed, a correlation does exist between the translesional pressure gradient, as measured with a dilating catheter across the narrowed segment, and the degree of coronary stenosis, as assessed angiographically. Of 234 dilated lesions, the mean arterial diameter reduction was 79 ± 12 percent prior to angioplasty, with a translesion pressure gradient of 54 ± 14 mmHg (Fig. 8). After angioplasty, angiographical luminal narrowing decreased to 27 ± 13 percent, and the gradient declined to 15 ± 0 mmHg. Linear regression analysis of these patients revealed a correlation ($r = .72$) between percentage of stenosis and translesional gradients ($p < .001$).

Measurement of the translesional pressure gradient requires the presence of the dilation catheter within the lumen of the stenosis. The presence of the catheter reduces the luminal area available for blood flow, and thus the catheter itself contributes to the pressure gradient. Furthermore, the relation between pressure gradient and the degree of stenosis is nonlinear. Thus, the effect of the dilation catheter on the pressure gradient is not uniform and varies according to the degree of coronary stenosis. For example, in the setting of an 80 percent stenosis, the presence of the dilation catheter may result in an effective 95 or 100 percent stenosis with a high

gradient. If, on the other hand, the stenosis is 15 percent, then the additional reduction in coronary cross-sectional area resulting from the dilation catheter will not create a significant obstruction and thus will not in itself result in a pressure gradient. With this consideration in mind, the translesional pressure gradient may more accurately reflect the extent of stenosis following angioplasty than that preceding it.

Since quantitative angiography (i.e., assessing diameter of a lesion) cannot be conveniently performed during the angioplasty procedure, the trans-stenotic pressure gradient can potentially provide on-line information to permit a judgment as to the relief of obstruction. In practical terms such data may serve as a guide for the need to perform additional balloon inflations. We have analyzed the value of the pressure gradient in predicting the relief of a significant stenosis. A successful angioplasty was defined angiographically as coronary narrowing of less than 50 percent diameter reduction. As shown in Table 1, a final gradient of less than 40 mmHg identified 98 percent of patients demonstrating angiographic success. The vast majority (92 percent) of successfully dilated patients had a final gradient of less than 30 mmHg. Angiographic success was not achieved in four patients, each of whom demonstrated a residual gradient in excess of 20 mmHg. Thus, the trans-stenotic pressure gradient can be helpful in mediating successful angioplasty. A gradient of less than 20 mmHg indicates an angiographically successful outcome. That some patients demonstrate success with higher gradients may reflect the potentially spurious nature of gradient determination or, on occasion, the misleading nature of the angiogram.

Recently the use of digital subtraction radiographic techniques has corroborated the ability of on-line trans-stenotic pressure gradient determina-

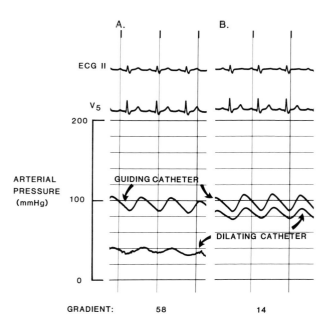

Figure 8. The trans-stenotic pressure gradient, as measured by the guiding catheter and dilating catheter before (panel A) and after (panel B) successful PTCA. The pressure gradient declines from 58 to 14 mmHg.

TABLE 1   **Pressure Gradient and Successful PTCA**

Pressure Gradient after PTCA, mmHg	Patients with Final Diameter Stenosis $\leq 50\%$		Patients with Final Diameter Stenosis $\geq 50\%$	
	Number	(Percentage)	n	(%)
<40	226	(98)	4	(100)
<30	212	(90)	4	(100)
<20	177	(77)	0	(0)
<10	58	(25)	0	(0)
Total	226	(100)	4	(100)

tion to predict success. O'Neil et al. compared coronary flow reserve, an index of the functional significance of coronary stenosis, to the pressure gradient following angioplasty (17). Successful angioplasty, in this investigation defined as normalization of coronary flow reserve, was closely correlated with a final pressure gradient of 25 mmHg or less.

CONCLUSIONS

This report has demonstrated that percutaneous coronary angioplasty can achieve its primary goal, namely relieving myocardial ischemia that results from atherosclerotic narrowing of epicardial coronary arteries. Should the rigorous and often elaborate methods reviewed here be utilized routinely to determine the outcome of an angioplasty procedure? Obviously not. A reasonable and pragmatic approach, however, can be developed; when possible, objective, noninvasive confirmation of myocardial ischemia should be obtained before the angioplasty procedure and a basis for comparison for a similar assessment following the procedure. Furthermore, the trans-stenotic pressure gradient can be measured readily at the onset of and during the angioplasty procedure. A realistic goal is to achieve a gradient of less than 20 mmHg.

REFERENCES

1. Gruentzig AR, Senning A, Sigenthaller WE: Nonoperative dilation of coronary artery stenosis: Percutaneous transluminal coronary angioplasty. N Engl J Med 301:61, 1979.
2. Kent KM, Bentivoglio LG, Block PC, Cowley MJ, Dorros G, Gosselin AJ, Gruentzig AR, Myler RK, Simpson J, Stertzer SH, Williams DO, Bourassa MG, Kelsey SF, Detre KM, Mullin S, Passamani E: Percutaneous transluminal coronary angioplasty: Report from the Registry of the National Heart, Lung and Blood Institute. Am J Cardiol 49:2011, 1982.
3. Lee G, Ikeda KM, Joye JA, Bogen HC, DeMaria AN, Mason DT: Evaluation of transluminal coronary angioplasty of chronic coronary artery stenosis: Value and limitation assessed in fresh human cadaver hearts. Circulation 61:77, 1980.
4. Block PC, Myler RD, Stertzer S, Fallon JT: Morphology after transluminal angioplasty in human beings. N Engl J Med 305:382, 1981.
5. White CW, Wright CB, Doty DB, Hiratza LF, Eastham CL, Harrison DG, Marcus ML: Does visual interpretation of the coronary arteriogram predict the physiologic importance of coronary stenosis? N Engl J Med 310:819, 1984.
6. Gewirtz H, Most AS: Production of a critical coronary arterial stenosis in closed chest laboratory animals. Am J Cardiol 47:589, 1981.
7. Gewirtz H, Williams DO, Most AS: Influence of coronary vasodilation on the transmural distribution of myocardial blood flow distal to a severe fixed coronary artery stenosis. Am Heart J 106:674, 1983.
8. Scholl J, Chaitman BR, David PR, Dupras G, Brevers G, Val PG, Crepeau J, Lesperance J, Bourassa MG: Exercise electrocardiology and myocardial scintigraphy in the serial evaluation of the results of percutaneous transluminal coronary angioplasty. Circulation 66:380, 1982.
9. Williams DO, Singh A, Most AS: Sustained efficacy of coronary angioplasty documented by stress testing at one year. J Am Coll of Cardiol 1(2):725, 1983.
10. Gruentzig AR, Schlumpf, Siegenthaler W: Late functional results after success or failure of coronary angioplasty (PTCA). Circulation 64(suppl IV):IV-193, 1981.
11. Hirzel HO, Neusch K, Gruentzig AR, Leutolf UM: Short and long-term changes in myocardial perfusion after percutaneous transluminal coronary angioplasty assessed by Thallium-201 exercise scintigraphy. Circulation 63:1001, 1981.
12. Kent KM, Bonow RO, Rosing DR, Ewels CJ, Lipson LC, McIntosh CL, Bacharach S, Green M, Epstein SE: Improved myocardial function during exercise after successful percutaneous transluminal coronary angioplasty. N Engl J Med 306:441, 1982.
13. Williams DO, Riley RS, Singh AK, Most AS: Restoration of normal coronary hemodynamics and myocardial metabolism following percutaneous transluminal coronary angioplasty. Circulation 62:653, 1980.
14. Williams DO, Riley RS, Singh AK, Gewirtz H, Most AS: Evaluation of the role of coronary angioplasty in patients with unstable angina pectoris. Am Heart J 102:1, 1981.
15. Williams DO, Riley RS, Singh AK, Most AS: Coronary circulatory dynamics before and following successful coronary angioplasty. J Am Coll Cardiol 1(5):1273, 1983.
16. Orlick AE, Ricci DR, Alderman EL, Stinson E, Harrison DG: Effects of alpha-adrenergic blockade upon coronary hemodynamics. J Clin Invest 62:459, 1978.
17. O'Neill WW, Walton JA, Bates ER, Colfer HT, Aueron FM, LeFree MT, Pitt B, Vogel RA: Criteria for successful coronary angioplasty as assessed by alterations in coronary vasodilatory reserve. J Am Coll Cardiol 3:1382, 1984.

25

Coronary Distribution Fraction: An Anatomic Model and Quantification Algorithms

G. DAVID JANG, M.D.
S. MATTS E. SJOLANDER, PH.D.

INTRODUCTION

Coronary revascularization can be achieved either by surgical vein bypass (1,2) or percutaneous transluminal angioplasty (3). The primary purpose of coronary revascularization is to preserve the viable myocardium, which is jeopardized by the diseased coronary artery.

Severity of coronary artery disease is often classified according to the number of vessels diseased (4–18) because the extent of coronary artery disease influences the patient's prognosis. The severity of myocardial pump failure and hemodynamic consequences is proportional to infarct size, and the larger an infarct segment, the greater is mortality (19–20).

Although coronary bypass surgery can easily revascularize multivessel disease, percutaneous coronary angioplasty does have a relative limitation in multivessel disease. The classic indications for percutaneous coronary angioplasty have been in single vessel disease, although multivessel disease angioplasty is expanding rapidly.

Despite the well-documented facts that the mul-tivessel disease has a larger myocardium at risk and thus has a poorer prognosis than single vessel disease in general, it would not be correct to assume that all single vessel disease has the same degree of risk or the same prognosis. A single coronary artery may perfuse a small or a very large segment of myocardium. Thus, when assessing the arteriogram prior to revascularization, there is a need for a fine-tuned method that is superior to simple counting of vessels. This would be especially true in the case of percutaneous coronary angioplasty, for which single vessel disease will remain the best indication. Furthermore, multivessel disease for balloon angioplasty is only an accumulation of single vessel dilations.

A coronary artery and its branches are distributed over a specific region or segment of left ventricle. The distribution fraction of a coronary artery over the epicardial geometry of left ventricle should reflect the myocardial mass within the geographic distribution of the target coronary artery. If the global

425

and segmental surface areas of left ventricle can be computed, the distribution fraction of a myocardial segment perfused by a coronary artery can be obtained. The epicardial coronary artery and its branches, which are irreversibly fixed to a certain regional muscle segment of left ventricle, can be opacified into an arteriogram, which in turn can provide a basis for mapping the coronary distribution territory for computation of coronary distribution fraction. Coronary distribution fraction (CDF) is defined as a surface fraction derived by dividing a regional surface of a coronary artery with the global surface of the left ventricle under study.

This chapter introduces an anatomic model and computer algorithms for quantification of coronary distribution fraction. The original noncomputer method (21) and computer-assisted methods (22–24) of quantification have been reported. The algorithm presented here has been developed to calculate the regional and global surface geometry of left ventricle on the basis of anatomic data from the routine cine coronary arteriogram. Coronary distribution fraction is especially useful in assessment and quantification of the myocardial perfusion segment occupied by an individual coronary artery but also can be utilized in multivessel disease.

Coronary distribution fraction may be an ideal index of the relation between a segmental myocardium in jeopardy and its potential infarct size. This technique can be applied to coronary revascularization in general and transluminal coronary angioplasty in particular. We hope that this technique will be found useful by the readers in their clinical and experimental applications.

MATHEMATICAL MODEL

In order to quantify a regional coronary artery distribution, it is necessary to convert a two-dimensional arteriographic image into a three-dimensional geometry that accurately represents the spatial relation of left ventricle and epicardial coronary arteries. To achieve this goal, we adopted a truncated prolate spheroid as a three-dimensional surface model of left ventricle. The general equation for the surface of a prolate spheroid is

$$\frac{x}{a^2} + \frac{y}{b^2} + \frac{z}{c^2} = 1$$

However, an intact prolate spheroid does not represent the left ventricular muscle geometry because the base of left ventricle is essentially composed of two valves, but not of myocardium.

For the prolate spheroid model of the left ventricular epicardial geometry, we adopted dimension ratios of 5:3:3 for the major ($2a$) and the minor ($2b$, $2c$) axes, where $c = b$ in the preceding equation. From the prolate spheroid with these dimensions, one of the nose cones with an equivalent length of one-fifth of the major axis was truncated and eliminated. Thus, the remaining four-fifths of the prolate spheroid has a truncated shape resembling a normal left ventricle in end-diastole, with the nose cone representing the apex and the truncated end representing the base of left ventricle (Fig. 1). Using this truncated prolate spheroid model, we developed a computer algorithm that can compute surface areas of the left ventricular epicardium.

MAPPING TECHNIQUE OF ARTERIOGRAM

As opacified by contrast media, an epicardial coronary artery has a direct contact and spatial fixation to the epicardial layer of left ventricle. For this reason, the outer dimension of left ventricle occupies an area within an imaginary shell delineated by various epicardial branches of the right and left coronary arteries, except in the septum, where septal arteries are embedded within the septal myocardium, rather than on the right ventricular side of the septal endocardium. The left ventricular silhouette, which is superimposed with the opacified coronary arteries, provides further clues for the shape and dimension of left ventricle for mapping purposes.

In order to convert arteriographic information to digitized data for computer calculation, the arteriographic anatomy should be mapped.

Following is the step-by-step approach to mapping the coronary arteriogram:

1. To map the outline of the left ventricular silhouette (Fig. 2) in maximum end-diastole as the left ventricular silhouette is superimposed with opacified coronary arteries in the RAO-30° projection, using the LAO-60° projection as a correction reference.
2. To map the borders around the perfusion territory of a target coronary artery distal to the obstructing lesion so that the borders of coronary distribution will superimpose within the outline of the left ventricular silhouette, which was obtained in step 1 (Fig. 3).
3. To compute the three-dimensional surface area of coronary distribution from the two-dimensional mapping data acquired in the two preceding steps, using the computer algorithm (Fig. 4).

TRUNCATED-PROLATE SPHEROID MODEL

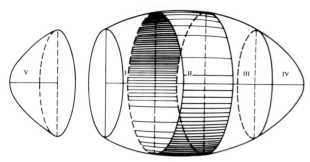

Figure 1. This truncated model is created from a prolate spheroid with $x{:}y{:}z$ axes in the ratio 5:3:3. When one of the nose cones is truncated at one-fifth the length of the x axis, the surface geometry of the remaining four-fifths of the model roughly resembles the epicardial surface geometry of a normal mammalian left ventricle in maximum diastole.

In the RAO-30° projection, the septum belongs to the anterior half (i.e., anterior hemisphere) of the left ventricular myocardium. In this projection, the border mapping of the anterior descending artery and the posterior descending artery occurs in the septum (Fig. 2). The border between the anterior descending artery and the circumflex artery in the RAO-30° projection is usually located on the pos-

PLANAR MAPPED GEOMETRY
(RAO-30° VIEW)

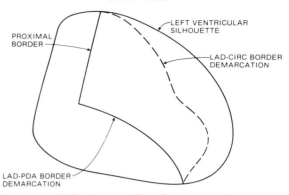

Figure 3. This planar outline of artery mapping is produced by the steps illustrated in Figure 2. The LAD-PDA border is in the anterior hemisphere of the left ventricular geometry, which consists mostly of the septum; and the LAD-CIRC border is in the posterior hemisphere, which consists of free wall.

terior half (i.e., posterior hemisphere) of the ventricle. This border usually divides the diagonal branches and the obtuse marginal branches.

The LAO-60° view, which is a perpendicular projection to the RAO-30° view, is carefully compared as a correction reference during the mapping process to ensure that the border mapping in the RAO-30° projection is indeed correct. This cross-checking of the border demarcation is especially critical when

CORONARY ARTERY MAPPING
(RAO-30° VIEW)

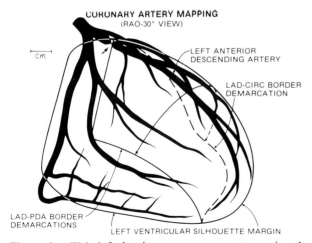

Figure 2. This left dominant coronary anatomy is schematically displayed as a coronary arteriogram in the RAO-30° projection in maximum diastole. Mapping of the left anterior descending artery (LAD) distal to the proximal lesion involves three manual steps: (1) defining the left ventricular silhouette, (2) defining the LAD-posterior descending artery (PDA) border, and (3) defining the LAD-circumflex (CIRC) border. All three steps are taken from the same arteriographic frame.

TRANSFORMATION OF MAPPING DATA
TO CROSS-SECTIONAL GEOMETRY

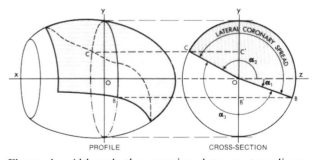

Figure 4. Although the mapping data are two-dimensional, the real coronary artery distribution occupies a surface territory over the spherical epicardium. This schematic illustration depicts how the cross-sectional dimensions are abstracted from the planar profile view of Figure 3 by our algorithm to calculate the surface geometry of a prolate spheroid. This is possible because the cross-sectional epicardial geometry of the ventricle is roughly circular in maximum diastole.

the border between the diagonal branches and the obtuse marginal branches is located very close to the silhouette crest of left ventricle in the RAO-30° projection.

These approaches to artery mapping can be used for the right coronary, circumflex, and left anterior descending arteries. However, coronary artery mapping of this kind requires a conceptual understanding of coronary anatomy in three dimensions as it relates to the surface geometry of left ventricle.

CONCEPT OF QUANTIFICATION

In this discussion one should assume that CDF has been defined before as

$$CDF = \frac{Ar}{At} \qquad (1)$$

where CDF = coronary distribution fraction
Ar = regional LV coronary artery distribution area
At = global LV coronary artery distribution area

Although coronary artery distribution pertains to an epicardial surface geometry of the left ventricle, the coronary distribution fraction CDF, as defined above, is also a measurement of the segmental myocardial mass distribution.

If we define the myocardial mass fraction by the equation:

$$MMF = \frac{Mr}{Mt} \qquad (2)$$

where MMF = myocardial mass fraction
Mr = regional LV myocardial mass
Mt = total LV myocardial mass

we can show that $CDF = MMF$ in the following way: Assuming that the LV muscle has a constant thickness in end-diastole and that the LV has a symmetrical shape we can state that muscle volume is proportional to epicardial surface area, that is:

$$V = k \cdot A \qquad (3)$$

where V = myocardial volume
k = proportionality constant
A = epicardial surface area

Introducing myocardial density p and using Eq. (3) we find the myocardial mass M by

$$M = p \cdot V = p \cdot k \cdot A \qquad (4)$$

Combining Eqs. (1), (2), and (4) we have

$$MMF = \frac{Mr}{Mt} = \frac{p \cdot k \cdot Ar}{p \cdot k \cdot At} = \frac{Ar}{At} = CDF$$

When coronary artery distribution is expressed in a fraction or a percentage, it implies a fraction of left ventricular muscle mass perfused by a coronary artery, as illustrated in Figures 5 and 6.

Since coronary distribution fraction is a numerical unit, a given distribution fraction of the right coronary artery may be directly compared with the same distribution fraction of the left coronary artery. Likewise, a distribution fraction of one species may be compared with that of another species.

From the case illustrated in Figure 2, the following data were obtained by the computer algorithm described in this paper:

$$RDA = 91 \text{ cm}^2$$
$$\text{Total LV surface area} = 215 \text{ cm}^2$$
$$CDF = 0.42$$
$$\left(\text{or } 42\% = \frac{91}{215} \right)$$

where RDA is the regional distribution area, and CDF is the coronary distribution fraction.

LAD SURFACE DISTRIBUTION

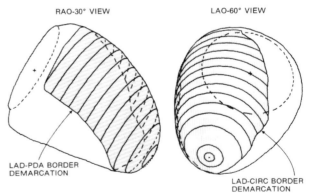

RAO-30° VIEW   LAO-60° VIEW

LAD-PDA BORDER DEMARCATION

LAD-CIRC BORDER DEMARCATION

Figure 5. These drawings illustrate the epicardial surface distribution of the anterior descending artery of Figure 2. The two perpendicular views, RAO-30° and LAO-60°, give an added sense of dimension that a single view cannot. The LAD-CIRC border is demonstrated better in the LAO-60° view, whereas the LAD-PDA border is clearer in the RAO-30° view. These graphics are computer-generated.

ANTERIOR DESCENDING ARTERY DISTRIBUTION
(LAO-60° VIEW)

SURFACE DISTRIBUTION

MASS DISTRIBUTION

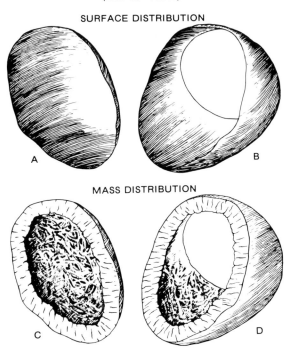

Figure 6. These illustrations are LAO-60° derivations of Figures 3 and 5. Top pictures illustrate the LAD-epicardial distribution (**A**) separated from the rest of the left ventricular epicardial surface (**B**). The LAD-myocardial mass distribution (**C**) is cut away from the rest of the left ventricular myocardium (**D**). The fractional ratio of **A** and **B** is the same as that of **C** and **D**.

COMPUTER ALGORITHM

We will present the quantification algorithm only in mathematical notations.

Digitization of Mapping Data

When the borders of a coronary artery distribution are mapped, digitizing the mapped geometry (Fig. 7) is the first step.

A mapped left ventricular silhouette is registered in two arrays and stored in the processor. To expedite processing and to save computer memory, left ventricle is rotated into a horizontal position.

This transformation of the ventricle (Fig. 7) can be expressed by the following equations (23):

$$x^1 = (x - x_0)\cos\phi - (y - y_0)\sin\phi$$
$$y^1 = (x - x_0)\sin\phi + (y - y_0)\cos\phi$$

DIGITAL ROTATION AND TRANSLATION

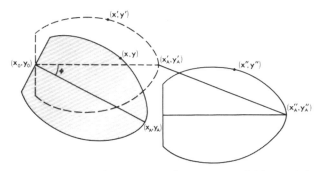

Figure 7. The planar mapped geometry of Figure 3 is rotated as an integral part of digitizing. The major axis of the ventricle is aligned parallel to the x coordinate. Translation is not necessary for the distribution computation, but it is useful for a graphics display mode.

The data points finally stored in the processor are 1.0 mm apart in the x coordinate, and missing gaps between points are filled in by a linear interpolation.

Computation of Coronary Artery Distribution

The computation of coronary artery distribution in the algorithm computes (1) the global left ventricular surface area, (2) the regional distribution area, and (3) the coronary distribution fraction.

Global Surface Area of Left Ventricle

The transformed ventricle is mathematically divided into 1.0-mm thick slices perpendicular to its major axis. Assuming that the left ventricle has a circular cross-sectional geometry in end-diastole, each muscle segment has the epicardial surface on the lateral side of the cross-sectional disc. This lateral surface of each disc can be approximated by the lateral surface of a frustum of a right circular cone (Fig. 8) that has a height of 1.0 mm (Fig. 8A). In our case example (Fig. 2), the major axis of the ventricle is approximately 10 cm in length; thus, 100 custom-fit frustums of right cones are required to approximate the surface area of this ventricle.

The lateral surface (Fig. 8B,C) of the frustum of a right cone (that fits each cross-sectional slice of left ventricle) can be expressed by the following equation (23):

$$A_i = \pi(R_1 + R_2)S$$
$$S = [(R_1 - R_2)^2 + h^2]^{1/2}$$

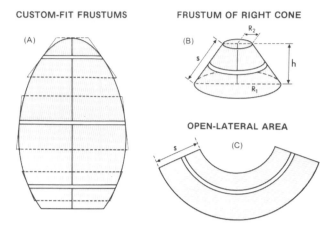

CUSTOM-FIT FRUSTUMS

(A)

FRUSTUM OF RIGHT CONE

(B)

OPEN-LATERAL AREA

(C)

Figure 8. These drawings illustrate how a frustum of a right cone approximates the lateral surface of a thin (i.e., 1.0 = mm width) cross-sectional slice of the left ventricle. **A.** The custom-fit frustum of a right cone for the lateral surface of individual cross-sectional slices. **B.** Dimensional designations for the equation of a frustum of right cone. **C.** Open lateral area of a frustum of right cone. A narrow frustum has its own respective dimensions, like a larger frustum.

where A_i = lateral surface of frustum
R_1, R_2 = radii of bottom and top cut surfaces of frustum, respectively
S = slant chord of frustum
h = height of frustum

The sum of lateral surface areas of all the individual slices will represent the total left ventricular surface:

Total LV surface = $(A_1 + A_2 + \cdots + A_n) \div M^2$

where A_1 is the lateral surface of the frustum, and M is the magnification factor.

Regional Distribution Area

The distribution area of a coronary artery occupies only a regional segment of the left ventricular surface. The size and location of a myocardial segment will be determined by the distribution zone and size of an artery. A regional distribution area has a vertical spread and a lateral spread.

The *vertical spread* pertains to an extent from the proximal (from base) to the distal (to apex) ends along the major axis of left ventricle. The vertical spread is an integral part of computing the regional distribution area in the algorithm, and it is accounted for by the number of ventricular muscle slices included in the target distribution zone.

The *lateral spread* is a circumferential extent of the epicardial surface of a cross-sectioned left ventricular disc, which is occupied by the distribution territory of a target coronary artery. This lateral spread occupies an arc of the cross-sectional circumference, defined by the border demarcation already determined by the mapping process (Fig. 4). Therefore, a lateral spread is the circumferential extent of the surface arc that is defined by the angle of the sector. The angle of a sector is determined by the vectors that connect the two lateral borders to the center of the cross-sectional circle.

The sector of a lateral spread can be defined by the following equations (Fig. 4):

$$\alpha_1 = \text{arcsin (OB}')$$
$$\alpha_2 = \text{arcsin (OC}')$$

where OB$'$ and OC$'$ are the sine values of two angles, α_1 and α_2, respectively. The angles are defined by two vectors, OB and OC (Fig. 4).

The lateral coronary spread and the lateral surface area of each slice are expressed in the following equations:

$$\text{LCS} = \frac{\alpha_2 - \alpha_1}{2\pi}$$
$$L_i = (\alpha^2 - \alpha^1) \times \frac{A_i}{2\pi}$$

where LCS = lateral coronary spread
A_i = total lateral surface area of each muscle slice
$\alpha_2 - \alpha_1$ = angle of sector
L_i = lateral surface area distributed by target artery in each muscle slice

Summation of coronary distribution areas of all the individual muscle slices will give a regional distribution area:

$$\text{RDA} = (L_1 + L_2 + \cdots + L_n) \div m^2$$

where RDA is the regional distribution area, and M is the magnification factor.

Coronary Distribution Fraction

If the regional distribution area is divided by the total left ventricular surface area, it gives the coro-

nary distribution fraction of the target artery that was mapped at the beginning of the computation process:

$$CDF = \frac{(L_1 + L_2 + \cdots + L_n)}{(A_1 + A_2 + \cdots + A_n)}$$

where CDF = coronary distribution fraction
L_i = lateral surface area distributed by target vessel in each muscle slice
A_i = total lateral surface area of each muscle slice

VERIFICATION OF THE COMPUTER METHOD

An algorithm that calculates a numeric value such as coronary distribution fraction should be accurate and reproducible in its intended operation. Unlike in verification of a chamber volume calculation (25–28) of left ventricle, a plastic case method is not suitable in verifying the surface area measurements of a prolate spheroid geometry. Therefore, we devised the following methods to verify our quantification algorithm, using Fortran V language in the following studies.

Accuracy of Computer Measurements

We subdivided the surface area of a truncated prolate spheroid, as illustrated in Figure 9. Although Figure 9 has a truly planar (two-dimensional) surface geometry of a truncated ellipse, the same picture also depicts a profile view of the surface area of a three-dimensional truncated prolate spheroid that fits into the truncated ellipse. Using Figure 9, we mathematically calculated the spheroid surface area of each segment projected in the planar view and expressed the results as a percentage of the global surface area of the model. The lower half of the truncated prolate spheroid is a mirror image of the upper half, and there are two hemispheres.

Next, we created 20 geometrically regular surface areas by various combinations of segments shown in Figure 9. Geometrically regular surface areas of these 20 test samples allowed for mathematical calculations. Then, using the computer method, we again calculated the same segmental areas of these 20 samples. Thus, each segmental surface area was calculated by two independent methods.

When the mathematically calculated surface fractions on these 20 samples were correlated with the computer-calculated surface fractions (using the al-

THEORETICAL CDF FRACTIONS

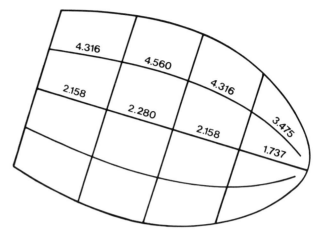

Figure 9. This planar view of a truncated ellipse can also represent a true profile view of a truncated prolate spheroid that fits into this truncated ellipse. Thus, each grid segment of this figure also represents the fitting surface segment of the three-dimensional prolate spheroid. The number of each box is the spherical surface in percentage of the total surface of the truncated prolate spheroid. These fractions are mathematically calculated.

gorithm described in this chapter) the correlation coefficient (r) was .99, with $p < .0001$ (Fig. 10). Thus, this computer method has very high accuracy.

Reproducibility of Computer Measurements

A computer program that has an ability to calculate various regular surface configurations accurately should have a high probability of accurately calculating various irregular surface configurations when the surface area is subdivided into 1.0-mm slices, as in the approaches of our algorithm.

If indeed the program has a high accuracy in calculating both regular and irregular configurations, it should also have a very high degree of reproducibility when two consecutive but independent calculations are made by the same method on the same set of samples.

In order to test the program's reproducibility in computing the irregular surface areas, we created 25 sample cases of various irregular surface configurations, using the same truncated prolate spheroid model (Fig. 11). We made two consecutive but independent measurements on these 25 samples with the computer program. As depicted in Figure 12,

tanin C, Dodge HT: Effects of increasing left ventricular filling pressure in patients with acute myocardial infarction. J Clin Invest 49:1539–1550, 1970.

11. Rackley CE, Russell RO Jr: Left ventricular function in acute myocardial infarction and its clinical significance. Circulation 45:231–244, 1972.

12. Parmley WW, Diamond G, Tomoda H, Forrester JS, Swan HJC: Clinical evaluation of left ventricular pressures in myocardial infarction. Circulation 45:358–366, 1972.

13. Bruschke AV, Proudfit EL, Sones FM Jr: Progress study of 590 consecutive non-surgical cases of coronary disease followed 5–9 years: II. Ventriculographic and other correlations. Circulation 47:1154–1163, 1973.

14. Nelson GR, Cohn OF, Gorlin R: Prognosis in medically-treated coronary artery disease: Influence of ejection fraction compared to other parameters. Circulation 52:408–412, 1975.

15. Weber KT, Janiki JS, Russell RO Jr, Rackley CE: Identification of high risk subsets of acute myocardial infarction. Am J Cardiol 41: 197–203, 1978.

16. Reeves TJ, Oberman A, Jones WB, Sheffield LT: Natural history of angina pectoris. Am J Cardiol 33:423–430, 1974.

17. Cohn LH, Boyden CM, Collins JJ Jr: Improved long-term survival after aorto-coronary bypass for advanced coronary artery disease. Am J Surg 129:380–385, 1975.

18. Bruschke AV, Proudfit WL, Sones FM Jr: Progress study of 590 consecutive non-surgical cases of coronary disease followed 5–9 years: I. Arteriographic correlations. Circulation 47:1147–1153, 1973.

19. Page DL, Caulfield JB, Kastor JA, DeSantis RW, Sanders CA: Myocardial changes associated with cardiogenic shock. N Engl J Med 285:113–137, 1971.

20. Caulfield JB, Leinback R, Gold H: The relationship of myocardial infarct size and prognosis. Circulation 53(suppl I):141–144, 1976.

21. Jang GC, Brody WR, Alderman EL: Computation of Coronary Distributional Area, abstracted. Presented to 64th Scientific Assembly and Annual Meeting of Radiological Society of North America, November 29–December 3, 1978.

22. Jang GC, Sjolander SME, Zimmerman CD, et al: A computer program for computation of regional coronary distributional area, abstracted. Proceedings of VIII European Congress of Cardiology, number 1427. Basel, Karger, 1980, p 119.

23. Sjolander SME: "The Coronary Distribution Zone—Mathematical Modeling of the Left Ventricular Epicardial Surface for Area Calculation, Three-Dimensional Rotation and Display." Ph.D. dissertation, Loma Linda University, Loma Linda, CA, 1980.

24. Sjolander SME, Jang GC: A computer model of the left ventricular epicardial surface for quantification of

jeopardized myocardium. Comput Programs Biomed 14:127–132, 1982.

25. Kennedy JW, Reichenbach DD, Baxley WA, Dodge HT: Left ventricular mass: A comparison of angiocardiographic measurements with autopsy weight. Am J Cardiol 19:221–223, 1976.

26. Rackley CE: Quantitative evaluation of left ventricular function by radiographic techniques. Circulation 54:862–879, 1976.

27. Dodge HT, Hay RE, Sandler H: An angiographic method for directly determining left ventricular stroke volume in man. Circ Res 11:739–745, 1962.

28. Sandler H, Hanley RR, Dodge HT, Baxley WA: Calculation of left ventricular volume from single plant (A-P) angiocardiograms. J Clin Invest 44:1094–1095, 1965.

29. Greene DG, Carlisle R, Grant C, Bunnell IL: Estimation of left ventricular volume by one-plane cine angiography. Circulation 35:61–69, 1967.

30. Gupta RL: "Modeling the Coronary Circulation." Ph.D. dissertation, State University of New York at Buffalo, Buffalo, NY, 1974.

31. Gould KL, Lipscomb K, Hamilton GW: Physiologic basis for assessing critical coronary stenosis: Instantaneous flow response and regional distribution during coronary hyperemia as measure of coronary flow reserve. Am J Cardiol 33:87–94, 1974.

32. Kloke FJ, Mates RE, Copley DP, Orlick AE: Physiology of coronary circulation in health and coronary artery disease, in Yu PN, Goodwin JF (eds): Progress in Cardiology, vol 5. Philadelphia, Lea & Febiger, 1976, pp 1–17.

33. McMahon MM, Brown BG, Cukingnan R, et al: Quantitative coronary angiography: Measurement of the "critical" stenosis in patients with unstable angina and single-vessel disease without collaterals. Circulation 60:106–113, July 1979.

34. Jang GC, Brody WR, Harrison DC, Alderman EL: Myocardial damage and size of coronary distribution, abstracted. Circulation 57–58(suppl II):122, October 1978.

35. Jang GC: The relation of inferior infarction to coronary distributional area and collateral vessels, abstracted. Circulation 59–60 (suppl II):160, October, 1979.

36. Jang GC, Brody WR, Mitchell WA, Alderman EL: The pathophysiologic correlation between coronary collaterals and myocardial mass fraction in patients with single-vessel occlusion of the coronary artery, abstracted. Invest Radiol 17(4):52, July–August, 1982.

37. Jang GC, Meltzer RS, Alderman EL, Harrison DC: Comparative analysis of coronary distributional area assessed by arteriogram, planimetry, myocardial weight in canine hearts, abstracted. Clin Res 27(1):6A, February 1979.

Index

Page numbers in *italic* indicate illustrations.